Society and exploitation through nature

Society and exploitation through nature

Martin Phillips and Tim Mighall

LONDON AND NEW YORK

First published 2000 by Pearson Education Limited

Published 2013 by Routledge
2 Park Square, Milton Park, Abingdon, Oxon OX14 4RN
711 Third Avenue, New York, NY 10017, USA

First issued in hardback 2017

Routledge is an imprint of the Taylor & Francis Group, an informa business

ISBN 13: 978-1-138-40851-7 (hbk)
ISBN 13: 978-0-582-27725-0 (pbk)

British Library Cataloguing in Publication Data
A catalogue record for this book can be obtained from the British Library.

Library of Congress Cataloging-in-Publication Data
Phillips, Martin.
Society and exploitation through nature / Martin Phillips and Tim Mighall.
p. cm.
Includes bibliographical references and index.
ISBN 0-582-27725-6
1. Nature -- Effect of human beings on. 2. Human ecology -- Political aspects. 3. Economic development -- Environmental aspects. 4. Environmental policy. 5. Environmental degradation -- Political aspects. 6. Conservation of natural resources. I. Mighall, Tim. II. Title.
GF75.P5 2000 99–25396
333.7'13--dc21 CIP

Typeset in Garamond by 30.

Contents

Acknowledgements

The origins of this book start back in 1991 when we were both appointed lecturers in the Department of Geography at the, then, Coventry Polytechnic and given/landed with the task of developing a course entitled 'Society and Nature', which had been suggested but never taught by one of our predecessors (thanks, Peter!). Neither of us had a particular research interest in society and nature, one of us being a physical geographer and the other a human geographer who, although married to a physical geographer, was rather willing to dismiss it as being positivistic science with no sense of social responsibility. After engaging in a crash course in reading we somehow managed to bring something together that at the very least made us, and as evidenced by feedback cards, our students, think. During the course of the next few years the course evolved into something that in its bare outline carries much of the argument of this book. It was at this point that Sally Wilkinson, then editor at Longman, approached us and asked whether we would be interested in writing a book based on the course. This we readily agreed to, little realising that some four years later we would finally be trying to finish the manuscript off. Sincere apologies are owed to Sally and to her successor Matthew Smith for all the delays to the book. We could point to such things as changes in jobs, teaching commitments, the priorities given to us by the RAE, and to society and nature just being such a big topic, but really all we want to do is thank you for sticking with us. Thanks are also due to Julie Knight for all her hard work in the production of this text.

Many thanks are also owed to those we live with and who have had to live with this book for so long. Tim would like to thank and dedicate this book to Susan and his parents for all their encouragement, love and support. Thanks also to Penny, Nick, Suzanne, Thomas, Sandra, Derek, Reilly, members of the Geography Subject Area at Coventry University, Tim Healy and MTM for their support and advice. Martin would simply like to acknowledge his debt to Kate for her love, support and advice, and to say to Rebecca and Ruth, 'Yes, Daddy has finished that book now and can come and play.'

Publisher's Acknowledgements

The publishers wish to thank the following for permission for use of the material detailed below:

Figures 2.1, 2.3, Table B2.5: Mannion, *Global Environmental Changes*, 2nd edn, 1998, Harlow: Pearson Education.

Figure 2.2: Zvelebil and Dulukhanov, The transition in farming in eastern and northern Europe. *Journal of World Prehistory* 5(3), 233–278.

Figure 2.5: *Climate Change and Human Impact*, p. 219, (1992) Chambers F.M., fig. 18.2c, © 1992 with kind permission from Kluwer Academic Publishers.

Figure 2.6: Reprinted with permission from Hong, Candelone, Patterson and Boutron, 1994, Greenland ice evidence of hemispheric lead pollution two millenia ago by Greek and Roman civilisations, *Science*, 265, 1841–1843. © 1994 American Association for the Advancement of Science.

Figure 2. 10: Simmons, L, in Silvertown and Sarre, *Environment and Society* (London: Hodder & Stoughton).

Figure 2.12: from *The Geography of the World Economy* by Knox and Agnew, Arnold 1998. Page 265. Source Duignam and Gann: The United States and Africa: A History, 1989.

Figure 3. 1: Meadows, Meadows, Randers and Behrens (1992) *Beyond the limits: global collapse or sustainable future?* (London: Earthscan); Figures 3.2a, b: Brown *et al.* (1997) *State of the World 1997* (London: Earthscan); Figure 3.7: Brown *et al.* (1998) *State of the World 1998* (London: Earthscan); Tables 3.3, 3.6: Brown (1994) Facing food insecurity. In Brown *et al. State of the World 1994* (London: Earthscan) 177–197; Table 3.4: Brown (1997) Facing the prospect of food scarcity. In Brown *et al. State of the World 1997* (London: Earthscan), 22–41; Table 4.4: Abramovitz (1997) Valuing nature's services. In Brown *et al. State of the World 1997* (London: Earthscan) 95–114.

Figure 3.3 © 1998 Laurie Grace

Figure 3.8a, b: J.L. Simon, *The State ofHumanity*, Blackwell Publishers: Oxford.

Figure 3.10, Tables 3.7, 3.9, B3.1, B3.2, B3.3, B3.6a, b, c: IPCC Secretariat.

Figures 3.11a, b: Woodward, F.J. (1992) A review of the effects of climate on vegetation. In Peters and Lovejoy (eds) *Global Warming and Biological Diversity* (New Haven: Yale University Press).

Figure 3.12: Reprinted with permission from *Nature*, Nriagu (1979) 279, p. 409. © 1979 Macmillan Magazines Limited.

Figure 3.15: Reprinted from *Energy Policy*, 24(7), ApSimon and Warren, Transboundary air pollution in Europe, 631–640, © 1996 with permission from Elsevier Science.

Figure 4.1: © *New Scientist* 1993.

Figure 4.2: Smith and Greene, in Smith and Warr, *Global Environmental Issues* (London: Hodder & Stoughton).

Figure 4.3: In Turner *et al.*, *The Earth as transformed by human action*, 1990, Cambridge University Press.
Figure 4.4: Vaccuming the Seas. *The Environmental Magazine*, July/August 1996, 28–35.
Figure 5.2: IBRD (1998) *World Bank Atlas*.
Figure 5.3: Figure on p. 10: structure of exports from *A geography of the Third World*, 2nd edn, by Dickenson, J. P. London: Routledge, 1996.
Figure 6.2: The Green Party.
Figures 6.4, 6.5, 6.7: Reprinted from *Journal of Rural Studies*, 7(3), Pattie, Russell and Johnston, Going Green in Britain? 285–299, © 1991, with permission from Elsevier Science.
Table 2.1, Figures B2.5a, b, c: from *The Emergence of Agriculture* by Smith © 1995, 1998 by Scientific American Library. Used with permission by W.H. Freeman and Company.
Table 3.1: World Development Report (1997) The state in a changing world, 0 1952-1114-6, Oxford University Press.
Table 3.5: Ellis and Mellor, *Soils and Environment*, London: Routledge (1995).
Table 3.11: Williams and Leibhold, *Journal of Biogeography*, 22, 665–671, Herbivorous insects and global change, © 1995 Blackwell Science Ltd.
Table 4.1: SORG (1996) *Stratospheric Ozone 1996* (London: HMSO). Crown copyright is reproduced with the permission of the Controller of Her Majesty's Stationery Office.
Figure B1.1: with kind permission of the British Antarctic Survey, Cambridge.
Figures B2.4a, b: Institute of Archaeology, University of Oxford.
Figure B2.6a: Bell and Walker, *Later Quaternary Environmental Change*, 1992, Harlow: Pearson Education.
Figure B2.6b: *Climate changes and human impact of the landscape*, 1993, p. 141, Chambers (ed.), Fig. 12.4, © Chapman and Hall 1993, with kind permission from Kkiwer Academic Publishers.
Figure B3.3: Reprinted with permission from *Nature*, Jones and Shanklin, Continued decline of total ozone over Halley, Antarctica since 1985, 376, © 1995 Macmillan Magazines Limited.
Figure B3.5: *The Independent on Sunday*/Michael Roscoe.
Figure B6.4: Figure 10.1, p. 212 from *Liberation ecologies: environment, development, social movements* edited by Peet, R. and Watts, M. London: Routledge, 1996.
Plate 2. 1: with kind permission of S&S Homes Ltd, Milton Keynes.
Plates 5. 1, B6.2, 6. 1 a, b, c: © Greenpeace.
Plate B6.3 photo by Graham Turner, © Guardian Newspaper; Figure 5.5: first published 14 February 1992, Environment section © Guardian Newspaper.
Table B3.7b: Reprinted from *Global and Planetary Change*, 14, Anisimov and Nelson, Permafrost distribution in the northern hemisphere under scenarios of climate change, 59–72, © 1996 with permission of Elsevier Science.
Table B4.1: *Environment* 38(1) 1–20, 36–45 (1996). Reprinted with permission of the Helen Dwight Reid Educational Foundation. Published by Heldref Publications, 1319 18th Street NW, Washington D.C., 20036–1802. © 1996.

Though every effort has been made to trace the owners of copyright material, in a few cases this has proved impossible and we take this opportunity to apologise to any copyright holders whose rights may have been unwittingly infringed.

List of Figures

List of Tables

List of Plates

List of Boxes

List of Box Figures

List of Box Tables and Box Plates

Tables

Plates

Chapter 1

Themes in the Study of Society and Nature

1.1 Introduction

Notions of environment and nature and their relationship to contemporary and future societies are subjects of considerable interest, concern and debate at the present moment. Unsurprisingly therefore a whole host of publications has emerged addressing these issues, including many student textbooks. So why did we feel the need to write another? Generally it stemmed from a feeling that two approaches dominate presentations of environmental issues. First, a large number of textbooks tackle the environment from broadly physical science and managerial perspectives: they adopt what Unwin (1992), drawing on the work of Habermas (1978), characterises as an 'empirical-analytical' approach to nature. Nature is seen as a distinct, material, entity which: i) is separate from, although affected by, society; ii) operates through sets of universal processes; and iii) can, at least potentially, be understood and manipulated by people. The textbooks adopting this perspective focus on describing the processes that create environmental resources or problems for society and outlining instances whereby these resources and/or problems are managed or mismanaged (e.g. see Mannion, 1991b, 1998; Pickering and Owen, 1994; Kemp, 1994 and O'Riordan, 1995a). The second approach may be seen as broadly social scientific and also more philosophical, in that it seeks to identify how people think about and interact with what they consider to be nature. Much of this work is quite historical in character, highlighting how concepts of 'what is nature?' vary over time (e.g see Pepper, 1984; and also from a less historical perspective Simmons, 1993a; 1997). Such writing is also, at times, couched in quite abstract levels of argument (e.g. see Pepper, 1993; Smith, 1984). Although we feel that both these sorts of books have their value, and indeed as will become apparent we have made extensive use of them in writing this one, we also feel that there is a need to link together the philosophical, the social science and physical science perspectives. In doing so we have tried to write in a manner that is accessible to first-year undergraduates studying geography, environmental studies and science, and others interested in the issues of society and nature and their interconnections and interactions.

In this book one particular linkage between the social and the natural will be explored, that of 'social exploitation'. We have consciously avoided structuring the book on an issue by issue basis but have instead sought to look for some commonalities and develop some critical lines that draw philosophical distinctions and concepts into the discussion of particular case studies. In this first chapter the philosophical argument surrounding the concepts of society, nature and exploitation will be discussed. Subsequent chapters will explore: i) how the exploitative relationship between society and nature may have changed over time; ii) whether this relationship is facing problems of sustainability in the contemporary world; iii) whether is it possible to manage this relationship to overcome or alleviate the problems it is creating; iv) why this relationship is established and who benefits from it; and v) whether it is possible to have alternative relationships between society and nature.

1.2 Society and nature: the meanings behind the terms

Terms like nature and society, although frequently used, are highly complex. Soper (1995, p. 1) begins her book, *What is nature?*, by observing that the term nature is 'at once both very familiar and extremely elusive'. She claims it is an idea that people employ with 'such ease and regularity' that at times it appears as if people have 'some "natural" access to its intelligibility', and yet it is also seen to be a concept 'so various and comprehensive in its use as to defy ... definition'. Some people have nevertheless tried, with Lovejoy (1935), for instance, isolating 66 different senses of the term 'nature' or 'natural'. Furthermore, as Neil Smith (1984) has remarked, the term 'nature' not only includes a whole variety of meanings, but many of these meanings are contradictory to others. Hence:

> Nature is material and it is spiritual, it is given and made, pure and undefiled; nature is order and it is disorder, sublime and secular, dominated and victorious; it is totality and a series of parts, women and object, organism and machine. Nature is the gift of God and it is the product of its own evolution; it is a universal outside history and is also the product of history, accidental and designed, wilderness and garden (Smith, 1984, p. 2).

Smith suggests, however, that it is possible to resolve all these definitions and contradictions into two basic viewpoints. According to the first viewpoint nature is '*external*' to human activity: it is the realm of objects that lie outside human activity. This concept, for instance, is embodied in the notion of 'natural landscapes' as being landscapes unaffected by human activity.

The second, and sometimes related, conception of nature is as an '*inherent state*'. Objects and processes are said to have an 'inherent and essential quality' (Williams, 1980, p. 68). This conception of nature as an inherent state is built upon two other notions. First, nature is seen as being universal: that is objects and processes are said to behave in a particular and *unchanging* way. For instance, if someone says that it is in the 'nature' of people to be selfish or greedy, or kind and considerate, it implies that this is a universal characteristic of people: all people everywhere and throughout human history have had this characteristic. Seen in this light the concept of nature is profoundly '*ahistorical*': it implies that, at least in essence, things do not change through history. A second assumption of the notion of nature as 'inherent state' is that of 'one dimensionality': objects and processes are seen to have a single basic character or 'essence'. Diversity is reduced to a single characteristic: that of nature. Indeed, the concept of nature is frequently used as a vehicle for generalisation: hence, for instance, people talk not only of 'human nature' and 'animal nature', but also frequently argue that the two can be *reduced* to each other: both people and animal in a sense share a common 'nature'.

One or other (or both) of the two basic notions of nature, as an *'external object'* and as an *'inherent state'*, can be said to underlie most of the individual definitions of nature you are likely to find. However, both these notions of nature have been heavily criticised. The notion of *'external nature'*, for instance, has been criticised from at least three directions. First, it has been argued that even the most extreme wilderness areas have been affected by human activity: Antarctica, for example, sometimes portrayed as the last wilderness, is being changed by direct and indirect human activity (see Box 1.1). The notion of external nature has also been challenged by 'evolutionary' and 'ecological' theories that see humankind as part of nature. Charles Darwin's *Origin of species*, for example, both implied that humans were descended from animals and argued that they were conditioned by similar circumstances of existence, such as the 'struggle for existence' and the 'survival of the fittest' (see Livingstone, 1992). Third, it has been argued that technological developments have meant that people have become in a sense less human, in that there are blurred boundaries between the human and non-human (see Box 1.2).

Box 1.1 Antarctica – a place of external nature?

As Dodds (1996, p. 63) has commented, by the middle of the twentieth century Antarctica had become one of 'the last "exciting blank spaces"' in the world. It was the only continent that, as Dodds puts it, 'remained largely beyond the measurements, classifications, and naming practices of European science' (p. 66). It was by the mid-twentieth century certainly a place of European and neo-European exploration, but it was then, and is still now, a place that had been neither totally walked over nor completely claimed by some sovereign power. Even today approximately a quarter of the continent has not been claimed by nation state or international group, largely as a consequence of the *Antarctic Treaty,* which came into force in 1961 and froze claims for sovereignty as they were at that point. The treaty also maintained that the continent should 'for ever be used exclusively for peaceful purposes only', and should not become either 'the scene or object of international discord' nor a place to conduct nuclear explosions or used for the disposal of nuclear wastes (Antarctic Treaty, 1959). In 1991 a *Protocol on Environmental Protection* was added to the treaty (Antarctic Treaty, 1991b). The protocol designated Antarctica as 'a natural reserve' and in the meeting at which the protocol was drawn up it was argued that 'Antarctica is the largest unspoiled continent on Earth' (Antarctic Treaty Consultative Committee, 1991). In these international agreements, and also in more everyday 'geographical imaginations', Antarctica is seen as a place of 'pristine nature', that is a place lying beyond human activity and influence.

The *Antarctic Treaty* and the *Protocol on Environmental Protection* have not only seen Antarctica as a place of pristine nature but also actively sought to limit human activity so as to maintain this status. It is, however, questionable as to whether this has been achieved. The *Antarctic Treaty*, for example, applied itself to the land area, 'including all ice shelves', that lay south of 60 degrees south latitude. However, it explicitly excluded itself from 'the exercise of rights... with regard to the high seas within that area' (Antarctic Treaty, 1959, Article VI). The seas of Antarctica are, however, highly rich in nutrients and support a high level of plant and invertebrate animal life, which in turn act as food for considerable numbers of fish, birds and mammals. Early Antarctic explorers often sought to exploit commercially elements of Antarctic nature, for example by hunting seals, whales and penguins for their skins, blubber and oil respectively (Walton and Morris, 1990). By the early twentieth century considerable numbers of British and Norwegian ships were in Antarctic waters hunting for whales, largely as a consequence of having exhausted whale stocks in the Atlantic Ocean. These whaling ships were then joined by Japanese, German and Russian ships. However, by the 1930s catches of the world's largest whale, the blue whale, had started to decline due to over-hunting, followed by catches of fin, sei and humpback whales (Smith and Greene, 1991). In the 1960s attention turned to fish and krill, although once again there has been the suggestion of over-exploitation leading to declining catches (see Walton and Morris, 1990). At least part of the nature of the Antarctic, its oceans, has hence clearly come under the influence of human activity.

It is not only the sealife of the Antarctic that has been seen as an exploitative resource. In the 1970s attention turned in a period of 'oil crisis' to consider whether there might be oil and natural gas reserves in Antarctica. A number of exploratory studies have been undertaken, albeit with rather differing interpretations about the extent of potential stocks (see Beck, 1990; Walton and Morris, 1990). A similar story can also be told with regard to mineral resources, with some studies arguing that there is clear potential for the commercial exploitation of high-value minerals such as plutonium. In 1980 negotiations started amongst members of the *Antarctic Treaty* to establish a system to regulate mineral activity in the Antarctic. There was, however, considerable disagreement about both how mineral activity should be regulated and, increasingly in the late 1980s and early 1990s, about whether any mineral exploitation should be allowed. In 1989 the governments of Argentina, New Zealand, the United States of America, the United Kingdom and Uruguay favoured 'environmentally regulated' commercial exploitation of Antarctic mineral resources, while the Australian and French governments favoured a complete ban on mineral exploitation (MacKenzie and Joyce, 1990). Although a *Convention on the Regulation of Antarctic Mineral Activities* (Antarctic Treaty, 1991a) was drawn up, the signatories of the *Antarctic Treaty* chose to adopt the *Protocol on Environmental Protection* which prohibited '[a]ny activity relating to mineral resources, other than scientific research' (Antarctic Treaty, 1991b). While this protocol may be seen to have put a considerable break on human activity, three caveats need to be made.

First, as quoted above, the *Protocol* explicitly excludes mineral activity related to scientific research from its regulation. There have been a steadily growing number of research bases established in Antarctica, many of them conducting geological drilling and excavation. It has come to be recognised that this activity is affecting the Antarctic environment, not only through transforming local topographies but also through the release of pollutants into wider environments. Scientific research activity often draws on a considerable amount of mechanised technology, ranging from drilling machines, through to snow-cats, skidoos, helicopters and aeroplanes. These all emit substances into the local environment, not least oxides of nitrogen, carbon and sulphur. Scientific research activity also involves people, who require the provision of accommodation, lighting, heating and food, and who also produce waste products. Human activity and the provision of the means to sustain it clearly transforms local areas from 'pristine nature' and the disposal of waste can have quite widespread environmental impacts. In research bases waste material is flown or shipped out for disposal beyond Antarctica.

A second important feature of the *Protocol* is that it explicitly allows another human activity to occur in Antarctica, namely tourism. Tourism has grown considerably in Antarctica over the 1980s and 1990s (see Wace, 1990), it being estimated that during the 1991–2 summer season some 6,200 tourists visited Antarctica, compared to some 2,000 a decade earlier (Carvallo, 1994). Tourist visits to Antarctica make much of it being a pristine, wilderness area, away from society. With the growth of tourism, however, come the similar environmental impacts of pollution and environmental transformation as outlined with respect to scientific research. In 1988, for instance, an Argentinian tour boat ran aground spilling some 250,000 gallons of fuel into the Antarctic Ocean. Tourism clearly has a greater potential to expand into a mass activity than does scientific research, and as a result the extent of its impact on the nature of the Antarctic could be much more marked.

A third point to make about human activity and the Antarctic is that so far we have largely been concentrating on human activity *within* Antartica. However, it has also become clear that Antarctica has been impacted considerably by human activities occurring elsewhere in the globe. Most notably, it was in the atmosphere above the Antarctic icesheets that so-called 'ozone holes' were first recognised. Ozone is a molecular form of oxygen (O_3) that is formed in the upper atmosphere but that can quickly be destroyed by small quantities of other chemicals such as oxides of nitrogen, hydrogen and chlorine (see Chapter 3 for more information on this). In the Antarctic the level of ozone would 'naturally' vary seasonally, related to the formation during the winter months of a vortex of cold air that both prevents ozone-laden air from moving in from lower latitudes and also leads to the formation of large clouds of ice particles (known as polar stratospheric clouds) that release ozone-destroying gases when they melt in the Antarctic spring (again see Chapter 3 for more detail). Each year the level of ozone would fall slightly from a summer peak in November to a spring low in

September. However, in the late 1970s a number of researchers began to report that ozone levels were falling to significantly lower levels and were remaining at low levels for much longer (see Figure B1.1). It was argued that this was in large part due to the human activity, and in particular the use of chlorofluorocarbons (or CFCs) within industrial products such as aerosols, plastics and coolants in refrigerators and air-conditioning systems. Attention was also drawn to the release of methane and nitrous oxides from an expanding world agriculture. Although produced beyond the borders of Antarctica, the ozone-destroying gases appear to have been spread by the 'natural' processes of atmospheric circulation into this area and, as a consequence, led to a significant lowering in the level of Antarctic ozone.

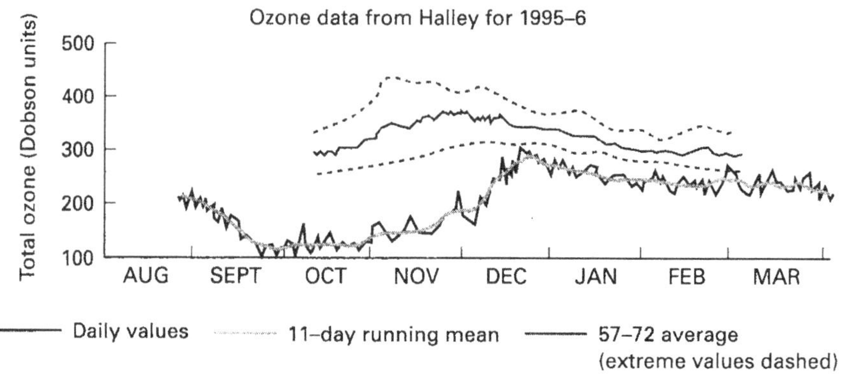

Figure B1.1 Atmospheric ozone levels in the atmosphere at Halley Base, Antarctica (Source: British Antarctic Survey (http://www.nbs.ac.uk/public/icd/jds/ozone/graphs. html))

Overall, therefore, it can be argued that the notion of the Antarctic as being a place of 'pristine nature' outside human influence is severely questionable. Furthermore, Antarctica's image as being a place of nature is clearly influencing its use by society, albeit in potentially quite contradictory ways. In informing international treaties and protocols the image of Antarctica as a place of pristine nature may well be limiting human activity and its transformative effects. On the other hand, this image also attracts scientists – Antarctica has, for example, been described as 'a pristine laboratory, of world-wide significance' (Antarctic Treaty Consultative Committee, 1991) – and tourists, whose activity is transforming Antarctica.

Box 1.2 The cyborg and the destruction of the human/ nature divide

Cyborgs have been defined as entities in which distinctions between humans, the natural and technology or machines is blurred. Cyborgs are seen as 'hybrid' entities. As Haraway (1990, p. 149) notes, contemporary science-fiction is full of cyborgs; amongst the most noted are arguably the 'Borg' in *Star Trek* and the 'replicants' portrayed in the film *Blade Runner*. The latter, for example, who are robots who not only appear to be virtually indistinguishable from people but who come to exhibit increasingly human characteristics, such as the emotions of anger and love. The film, produced in 1982, was set in the Los Angeles of 2010 and has been seen as a powerful vision of a now not so distant future. Shurmer-Smith and Hannam (1994), for example, argue that the film is a powerful 'dystopia' in that it gives an image of the future whereby current values and ideals held by people are seen to have been undermined by the course of history. In particular, they argue that *Blade Runner* 'has its roots in the great ecological questions which seek to understand the status of humans in nature' (Shurmer-Smith and Hannam, 1994, p. 70). For many people human life is seen as distinctively different and to be given greater value than either other living creatures or machines. *Blade Runner*, however, suggests that in the future a clear distinction between technology and humanity is far from clear cut.

Although robotic engineering has not advanced as far as portrayed in *Blade Runner* much of the power of the film lies in the way it resonates with many emergent features of contemporary life. Modern medicine, for example, makes increasing use of technology and can involve the implant of technology into the human body: witness, for example, the use of pacemakers, hip-joint replacements and heart valves. Technology is used in such medical techniques as *in vitro* fertilisation to influence the onset of human life, and is increasingly being used to modify biological material quite directly, as in genetic engineering. One of the most publicised examples of biotechnology in recent years is 'Dolly', a sheep which was 'cloned' from a single cell of an adult ewe (Campbell *et al.*, 1996). Technologies such as genetic engineering can also be seen to blur the human/animal divide, in that the organs of animals, such as pigs, have been proposed as potential replacements for many human organs.

Technologies such as cloning and genetic engineering can be seen as quite stark challenges to dualistic conventions of society and nature. It is, however, possible to argue that social life, forms of nature and experiences of nature have been continually reworked by technology. Thrift (1994) and MacNaghten and Urry (1998) have, for example, highlighted how new technologies of movement have created new 'structures of feeling' about space and nature:

> The very boundary of the railway flattened and subdued the existing countryside in a unique fashion. Rail travellers were propelled through space as mere parcels. The train was a projectile slicing through the landscape on level straight tracks, over bridges and embankments and through cuttings and tunnels. The landscape came to be viewed as a swiftly passing series of framed panoramas, rather than something which was to be lingered over, sketched or painted (MacNaghten and Urry 1998, p. 207).

Overall, it can be suggested that human life has always been dependent on non-human entities, has become increasingly dependent on technology and now has technologies that can significantly alter both non-human and human nature.

The notion of nature as an 'inherent state' has also been challenged from two directions. First, the notion of nature as unchanging has been criticised, particularly when being applied to people. Human behaviour, many people argue, has changed, and indeed some environmentalists would say must change if human society is to avoid environmental disaster. Second, the notion of 'one-dimensional' nature has been challenged by people such as Raymond Williams (1980). Williams argues that the precise meaning attached to the term nature frequently depends upon the objects to which it is applied. For instance, he notes that people talk of nature as being 'a cruel struggle for existence' when looking at predators in the food chain, or see nature as an 'interlocking system of mutual advantage' when looking at the interaction of bees and flowers. Williams, however, is extremely sceptical about whether there is a description of behaviour that could be applied to all objects. He states (1980, p.70): 'when I hear that nature is a ruthless competitive struggle I remember the butterfly, and when I hear that it is a system of ultimate mutual advantage I remember the cyclone.'

1.2.1 Viewpoints on society and nature

Williams and others reject the reductionism and ahistorism of the view of nature as an 'inherent state' largely because it creates what Pepper (1984) has described as a '*deterministic*' philosophy or view of the world. That is, the actions of all things, including people, are seen to be determined by their essences, their 'nature'. Pepper (1984) contrasted 'determinist' philosophies with what he terms '*free will*' philosophies. The former place stress on forces that lie beyond the individual: for instance, in this viewpoint whatever actions a person, animal or plant does are determined by their nature and not the result of any rational calculation by the agent itself. This viewpoint runs counter to the arguments of free will, or 'agency' focused, philosophies, which stress the element of choice or capability that individual agents have.

View of interrelation of society and nature	Society equals nature	Society and nature are distinct but interlinked objects		Nature equals society
	Holistic	Dualistic		Holistic
View of causal linkages	Nature creates society	Nature dominates society	Society dominates nature	Society creates nature
View of 'nature as external object'	Rejects	Accepts	Accepts	Rejects
View of 'nature as universal attribute'	Accepts	Accepts	Rejects	Rejects
Philosophical/ theoretical positions	Environmental constructionism	Environmental determinism	Social determinism	Social constructionism
Illustrative variants	Naturalism	Environmental determinism	Cultural/society determinism	Idealism Historical materialism

Figure 1.1 A framework for considering attitudes on nature and society–nature relations

Pepper suggests that the determinist/free will distinction can be used to understand some of the major ways in which people have sought to understand society and nature and interrelation. Figure 1.1 outlines how this argument may be advanced, and in particular stresses how the two views of nature identified in the first section, that is nature as 'external object' and nature as 'inherent state', can be linked to arguments about determinism and free will or human agency. At least four positions within the framework are worthy of comment. First there are the two positions on the extremes of the deterministic–free will relation. At the left-hand end, the determinist end, one has a position described as 'society equals nature'. This position implies that there are no distinctions between society and nature and that, effectively, society is determined by 'nature'. This at first may appear to be a contradictory statement: how can society be determined by nature if society and nature are one and the same thing? It is, however, possible to reconcile these statements if you recognise the distinction between nature as 'external object' and nature as 'inherent state'. The 'society equals nature position' basically rejects the notion of nature as external object, seeing everything in existence as being part of nature, but accepts the notion of nature as inherent state. Society, and indeed everything else in nature, is seen to be determined by inherent nature. This can be seen as a 'naturalist' perspective (Eder, 1996).

Rejection of the concept of nature as external object is also a feature of the extreme 'free will/human agency position', that is the position at the extreme right-hand end of Figure 1.1. However, in the viewpoint characterised as 'nature equals society' the concept of inherent nature is also rejected. Indeed, nature within this perspective is rather a meaningless term – or at least a term with no fixed meaning – in that according to this philosophical position everything is seen to be the outcome of human agency: nature is in a sense replaced by society. One version of this view is that there are no objects, no 'reality', external to people's consciousness, and hence if people are not conscious of an object it does not exist. This is often termed an *'idealist'* position. Other people see human activity as constructing the structure of the world, these are sometimes called *'materialists'*. Broadly, however, both idealists and materialists can be described as *'social constructionist'* and some may be seen to be what Best (1989) defines as *'strict social constructionists'* or what Demeritt (1998) defines as *'neo-Kantian constructionists'* in that they see that there is nothing in nature that is not socially created. Other people, such as Castree and Brown (Castree, 1995; Castree and Brown, 1998) argue that while what is taken to be nature and the form this nature takes in the world are both socially created it is important to recognise the limitation of both social representation and transformation, and that things of nature do have an agency – an effect – that is often neither socially recognised or amenable to social organisation or control.

Although there are advocates of both 'society equals nature' and 'nature equals society' viewpoints, the bulk of perspectives on society and nature probably lie somewhere in between these two positions. This is certainly the case in geography, where a lengthy debate raged in the late nineteenth and early twentieth centuries between so-called 'environmental determinists' and those called 'cultural' or 'social determinists'. Both these groups utilised the concept of nature as 'separate from society' but place differing emphasis on nature as a determinant and society as an agent. Environmental determinists stressed the role of 'nature', that is non-human environment, as a determinant of human behaviour. Cultural determinists, on the other hand, emphasised the role of human actions in creating 'nature', both materially, through for instance the ploughing of fields and the construction of buildings and towns, and also in the mind, through the creation of cultural interpretations of landscapes and regions. It is in this middle area of the interaction of 'nature' and 'society' that the present book will concentrate, although neither environmental or cultural determinism will be the adopted perspective.

One reason for our not adopting either position is that we would not wish to embrace the categorical separation of society and nature that such positions imply. We would argue that what is 'natural' and what is 'social' has been, and still is, highly variable, and hence we will talk of 'societies' and 'natures' rather than society and nature. As Whatmore and Thornes (1998) have argued, notions of nature act to describe, delimit and lump together, or 'incarcerate', a whole range of 'earthly inhabitants', including often both people and the earth itself. These various entities of nature may have quite disparate 'natures' in terms of their ways of acting and they interact in highly complex and often poorly understood and unexpected ways. In this book we will consider a whole range of 'natures', including the gases of the atmosphere, soils, plants, animals, ecosystems, landscapes and people.

This might seem to imply that the term nature is effectively redundant, but we want to suggest that these natures are interconnected, in the sense that they are widely conceived of as being 'parts of nature' and as such are thereby acted on by people in particular ways, ways that we will wrap together using the notions of 'exploitative' and 'non-exploitative' relations. In this book we will therefore be looking at the 'exploitative' and 'non-exploitative' relations between 'societies' and 'natures'. Before doing this, however, it is important to discuss briefly what is meant by the terms society and exploitation.

1.2.2 The concept of society

The term society is as equally contested as the term nature. So far we have been using the term society in a general sense as something involving people. While this may be a generally acceptable view, much argument has been made as to how this involvement should be conceptualised. Many people have adopted what may be described as an '*atomistic*' or '*individualist*' perspective on society, arguing that society is simply the outcome of the actions of individual people. A classic instance of this was Margaret Thatcher, Conservative Prime Minister of the United Kingdom between 1979 and 1990, who argued that there was 'no such thing as society', by which she meant there was nothing beyond individual activity. Other people argue however that society involves something more than the individual, in other words it involves some collective notions or forces. This is often described as a '*structuralist*' perspective of society, in that it sees human activity as conditioned in some way by forces that are greater than the individual. There are a range of different versions of structuralist thought, ranging from those that see structures as quite distinct, immutable and unobservable entities, through to more current, '*post-structuralist*' conceptions of structures as precarious, fluid and enacted and interpretable relations (see Phillips, forthcoming-c).

Applying these different conceptions of society to the issue of society–nature relations produces radically different interpretations. For instance, people with an individualistic interpretation of society might stress people's beliefs about nature and the *consequences* of each individual's actions, while a structuralist interpretation might place greater emphasis on the *conditions* under which people act. A third conception of society, one that may be termed '*dialectical*', sees society and individuals acting upon each other: society is not reducible to a collection of individuals, but nor is it isolated from them. Such a concept of society when applied to society–nature relations would stress both individual consequences and social consequence (see Figure 1.2).

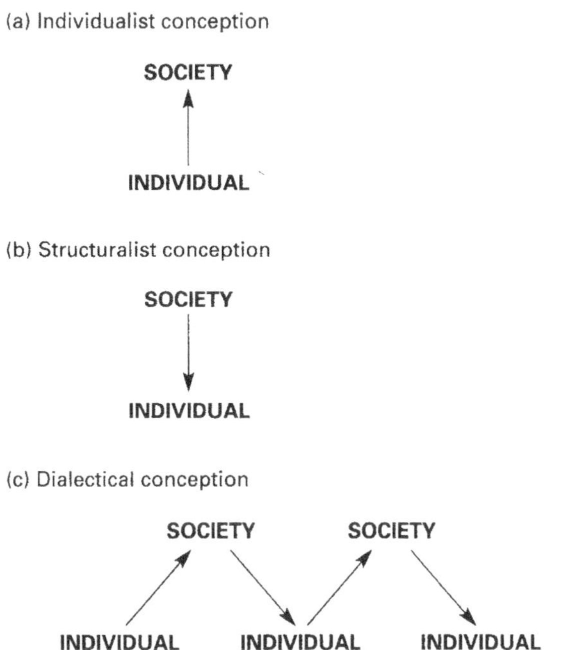

Figure 1.2 Individualist, structuralist and dialectical views on the construction of society (Source: Gregory, 1981)

These distinctions about how 'society' might be conceptualised can be integrated with the framework of society–nature relations discussed earlier. In Figure 1.1 the notion of society was used in a very general way. However, incorporating the individualistic/structuralist/dualistic concept allows further differences to emerge (see Figure 1.3). For an individualist, for instance, the 'human agency' end of the spectrum, described as 'society equals nature', will be interpreted as 'personal agency' or 'individual free choice': that is people will be seen as performing actions, including constructing nature, as a result of their own choice or desires. In a structuralist perspective the same position of 'society equals nature' will be interpreted very differently: people may not be able individually to create nature but they will be seen as part of a collective agency creating nature. Conversely, at the 'nature equals society' end of the framework, an individualist will see people as having 'natural behaviour' – a clear illustration of this is Dawkins' (1976) notion of people having a 'selfish gene' – while a structuralist would see societies as a whole having some natural logic or following some natural law. Figure 1.3 briefly outlines some of the characteristics of both the dialectic position on society and the dualistic concepts of society and nature interrelations.

View of interrelation of society and nature	Society is part of nature	Society and nature are distinct but interlinked objects		Nature equals society
	Holistic	Dualistic		Holistic
View of causal linkages	Nature creates society	Nature dominates society	Society dominates nature	Society creates nature
View of 'nature as external object'	Rejects	Accepts	Accepts	Rejects
View of 'nature as universal attribute'	Accepts	Accepts	Rejects	Rejects
View of society	View of nature/society			
Individualistic view	Individuals have natural behaviours	Individuals have natural behaviours, but these may be modified by an individual	Individual perceptions and behaviour impact the natural environment	Individual perceptions and actions construct nature
Dialectical view	Individuals and societies adopt laws of nature through interaction with each other through processes of nature	Individuals/societies have natural behaviours, but these may be modified by social norms/individual agency	Individual perceptions and actions within social structures impact the natural environment	Individual perceptions and actions within social structures construct nature
Structuralist view	Societies adopt laws of nature	Societies are conditioned by nature and affected by social norms	Social structures impact the natural environment	Social structures construct nature

Figure 1.3 A framework for considering attitudes on nature, society and society–nature relations

1.3 Exploitation: the fundamental relation between society and nature?

As mentioned earlier, the majority of discussions of society–nature relations within geography have arguably adopted what we have termed here a 'dualistic conception of society and nature'. That is they have seen nature and society as, in some way or another, distinct but interacting entities. It was also suggested that this interaction is often considered in terms of exploitative and non-exploitative relations.

The term 'exploitation' is, like those of nature and society, a complex and contested term. There are at least two ways in which the term can be used. First, one can use it in the general sense of '*use*'. Hence, for instance, one hears of the exploitation of a new energy source or raw material. Use may well involve transformation into something new and a series of authors have argued for the study of the way societies 'produce' new forms of nature (see particularly Smith, 1984). A second meaning for the term exploitation is to imply '*injustice*': hence, for instance, one hears of the exploitation of the Third World or of developing countries, or the exploitation of the working class, or the exploitation of women, or the exploitation of black people, or the exploitation of children, or the exploitation of animals. This second meaning is frequently associated with the first, hence exploitation is injustice created through the use of particular objects. In this book we want to make use of both senses of the term exploitation: we will be looking at the 'social exploitation' of nature both in the sense of how people or societies make use of that which is classified as nature and how this use may create injustice, both between people and nature and between people.

The preceding section has outlined in a very abstract way different interpretations of nature and society. It is useful also to focus on some slightly less abstract understandings of nature. In particular we will outline *one* way in which ideas about natures and societies and their interrelations have been classified. This classification, first proposed by O'Riordan (1976) in his book *Environmentalism*, suggests that western ideas about the environment can be divided into one of two basic categories: '*technocentrism*' and '*ecocentrism*'. These broadly can be seen to be ideas relating to the distinction between 'exploitative' and 'non-exploitative' society and nature relations. Technocentrism encourages, and attempts to justify, social exploitation of nature. Ecocentrism, on the other hand, argues for the establishment of 'non-exploitative relations' with nature.

The advocacy of an 'exploitative' or 'non-exploitative' society–nature relationship can be seen as the key distinction separating a 'technocentric' and an 'ecocentric': that is separating someone who thinks in technocentric ways from someone who adopts a more ecocentric perspective. It is not however the only distinction one can make: both ecocentrics and technocentrics employ a series of further arguments to justify their positions. It is therefore necessary to consider each of these positions in more detail.

1.3.1 The nature of technocentrism and its attitude to nature

According to O'Riordan (1976) technocentrism is '*anthropocentric*', meaning that it is 'people centred'. The centrality given to people takes two forms. First, the well-being of people is placed at the forefront within technocentrism, often at the expense of anything else, including nature. As Dobson puts it:

> If there is one word that underpins the whole range of Green objections to current forms of human behaviour in the world, it is probably 'anthropocentrism'. Concern for ourselves at the expense of concern for the non-human world is held to be a basic cause of environmental degradation and potential disaster (Dobson 1990, p. 62).

The reverse side of the centrality of people is that it is seen to lead technocentrism into a lack of concern for anything non-human: 'nature' is seen as an 'external' environment with no worth or value, except its ability to be manipulated or exploited by society. This is often described as an 'instrumental' attitude or relationship to nature

This anthropocentrism has been seen to be particularly evident in Occidental, or western, thought, and a range of otherwise quite divergent theories about the emergence of modern societies have all argued that an instrumental relationship to nature has been a key element in the development paths taken by these societies. For example, the social theories of Max Weber and Karl Marx, although taking very different perspectives on the character of modern societies, adopted quite similar instrumental attitudes towards nature (cf. Eckersley, 1992; Murphy, 1994). The reasons given for the emergence of this modern relationship to nature vary, although drawing on the discussion of society made earlier it can be said that many people stress changes in personal beliefs about human–nature relations while others stress changes in the circumstances in which people are situated. A clear example of the former group is White (1967), who suggested that between the eighth and twelfth centuries 'animistic beliefs', which saw everything in nature as having human and spiritual-like qualities, came to be supplanted by a belief in the separation of humanity and nature and in the need for human domination of nature. White pointed as evidence to the Book of Genesis where it is written:

> and God said unto them, 'Be fruitful, and multiply, and replenish the earth, and subdue it: and have dominion over the fish of the sea, and over the birds of the air, and over every living thing that moves upon the earth' (Genesis: 1, 28)

and later (Genesis: 9, 3)

> Every living thing shall be food for you; and as I gave you the green plants I give you everything.

White's arguments have, however, come under fierce criticism. People such as Doughty (1981), Glacken (1967) and Short (1991) have addressed the realm of ideas and claimed that White presented a partial account of the content of Christian thought and ignored, for example, the extent to which the Bible can be seen as encouraging a non-exploitative relationship with nature. As Glacken (1967) has argued, much of the Bible stresses human stewardship of nature: even in Genesis it is stated that 'God took man and put him in the Garden of Eden to work it and *take care of it*' (Genesis: 2, 15, emphasis added).

Another line of criticism is that White overemphasises the importance of Christianity and neglects the emergence of other belief systems that may be seen to encourage an exploitative relationship with that which is seen as the natural world. One alternative source of beliefs is the so-called 'scientific revolution' of the sixteenth and seventeenth centuries, which involved not only the emergence of practices that were seen to be distinctively 'scientific' – such as the laboratory experiment – but also changed philosophical ideas about the constitution of nature and took an increasingly instrumental attitude towards those things taken to be of nature. People were increasingly seen as being separate from nature: the philosopher Déscartes, for example, established the *Cartesian dualism* whereby people were seen to be distinctive in that they were capable of being a 'rational', 'thinking', subject. People were also seen as required to control nature: the scientific philosopher Francis Bacon, for instance, argued that the aim of science was to 'command nature' so as to provide the maximum level of resources to support people. Nature was also seen increasingly to act in a manner amenable to scientific analysis: people such as Bacon and, somewhat later, Sir Isaac Newton, promoted the notion that nature operated through universal procedures or laws. Finally, the practices of science encouraged instrumentalism: for the experimental method to work the elements going up to create a particular situation must be known about and controlled.

A third line of criticism of the work of White has been advanced, one that also has significance to those who posit the scientific revolution as being the origin of technocentrism. People such as Glacken (1967), Thomas (1984) and Pepper (1984) have all argued that White failed to consider the role of 'non-', or perhaps rather more precisely 'extra-', cultural factors in cultural constitution and change. In part this is the issue of cultural dynamics: why do cultures change? Why did Christianity replace medieval animism, for instance? There is also the issue of what constitutes a culture: is it just the presence of particular texts with a particular message, or is there much more to culture than this? The answer is very much yes, not least because if it is accepted that the Book of Genesis and other parts of the Bible can be interpreted both as encouraging an exploitative and a caring attitude to nature then what needs to be addressed is why one interpretation is favoured over another. For Glacken, Thomas and Pepper the answers to such questions lie not in the realm of religious ideas, but much more in the political, economic, cultural and social forces operating between the fifteenth and eighteenth centuries.

A second aspect of technocentrism's 'people centrality' is its belief in the power of people to master or control the environment. People are seen to be not only more important than everything else, they are also seen to have the ability to solve any problems the environment may throw at them. The history of society is seen to involve a progressive taming of nature to such an extent that today no problem is beyond human ingenuity to solve. This belief in the power of modern society is linked to two other characteristics of technocentrism: a belief in the power of science and a belief in managerial power.

The belief in the power of science is linked to the stress placed on 'objectivity'. Technocentrics often argue that environmental problems are best resolved by viewing them objectively, either by the use of an 'impartial' observer, or more generally through the use of techniques of observation said to be objective. This latter 'objectivity' is commonly (although some would say erroneously) associated with science, and has led to the design of a series of 'scientific' techniques to apply to solve environmental problems, perhaps most notably 'cost–benefit analysis 'and 'environmental impact assessments' (see Box 1.3). For some, such techniques provide a means of resolving social and environmental conflicts because of their 'objectivity'. However, people such as Pepper (1984) have argued that technocratic science, far from being a method for impartial observation, is essentially '*ideological*', that is tied into serving particular vested interests. Pepper identifies three ways in which science acts ideologically. First, he suggests that twentieth-century science, in the form of technical expertise and an appeal to universal laws, is used increasingly as a means of legitimating existing social conditions. Science tends to involve viewing things in terms of universal, unchanging and unchangeable forces. While this may seem to be justifiable when looking at some seemingly 'inanimate' nature it is arguably less so when applied to social problems and processes. Societies have changed in the past and people do have the capacity to change it in the future. To deny such a possibility serves to perpetuate the existing structure of society and 'naturalise' the winners and losers in the existing state of things.

Box 1.3 Cost–benefit and environmental impact analysis

Cost–benefit analysis involves the placing of a common measure, usually money, on the benefits and costs foreseen with any development. This form of analysis has been applied to a wide variety of social situations but is seen by its exponents as particularly suited to situations where significant elements of a social situation are uncommodified, that is they are not bought and sold in markets. One aspect of a growing concern with nature and environmental degradation has been a recognition that many aspects of nature are not formally bought and sold but are instead used as a 'free commodity'. It is argued that because people do not have to pay for many of the natures they consume they fail to recognise both the benefits and costs that are generated by this use. As a result, it is suggested, people tend to over-consume environmental resources and abuse the environment because they can ignore the environmental costs. Cost–benefit analysis seeks to increase awareness of non-commodified resources by giving them some form of monetary value. For a clear discussion of the range of methods by which this is done see Bateman (1993, 1995) and Turner *et al.* (1993).

Environmental impact assessments are broadly similar to environmental focused cost–benefit analyses, except that they do not insist on the development of monetary equivalents for non-commodified goods. Instead they present more qualitative statements about the costs and benefits of proposed developments. These assessments are open to criticism for presenting less 'clear–cut' assessments of the desirability or not of a proposal. On the other hand, many people object to cost–benefit analysis because it tries to quantify the unquantifiable: for instance by putting numerical values against such features as scenic beauty and the value of human, and non-human, life.

The second way in which science is linked in with particular social interests is the way the selection of which scientific research is done, or not done, is undertaken, generally, in accordance with the wishes and purposes of powerful groups. In effect, science therefore attempts to solve the problems of the owners and controllers of industry rather than those of any other groups. This can be clearly illustrated by the case of research on 'acid rain', that is rain or other forms of precipitation (such as hail, snow or fog) that have high levels of sulphur and nitrogen oxides. Although these oxides occur 'naturally', that is through non-human agency – through, for instance, thunderstorms and volcanic eruptions – their significance has been increased substantially by human activities such as the burning of fossil fuels like coal and oil (see Chapter 3 for more on this).

In Britain research on acid rain was until 1991 largely conducted by the *Central Electricity Generating Board* (CEGB). The CEGB, as well as being one of the largest organisations in Britain undertaking research on acid rain, was also this country's largest emitter of sulphur dioxide, which has generally been taken as the major cause of acid rain. The CEGB therefore clearly had a vested interest in the results of the research, and Elsworth (1984) has argued that the research undertaken by the CEGB was structured so that it did not so much ask the question 'What causes acid rain? but rather 'What, apart from sulphur oxide emission, could cause acid damage to the environment?'. The situation took a new turn with the privatisation of electricity generation, with neither of the then two largest privatised companies, *National Power* and *PowerGen*, undertaking any research on air pollution and acid rain, apparently on the basis that such research lies 'far beyond the boundaries' of the companies' own interests (Milne and Brown, 1991, p. 15).

A third way in which science can be said to serve the dominant social interests rather than universal interests is through the tendency of scientific practices to increase the amount of social and economic control exerted by select groups over the lives of ordinary citizens. It has been suggested that there has been a '*scientisation*' of public opinion and politics (see Habermas, 1987). This argument suggests that the procedures and assumptions used by scientists to do their research are becoming part of the way ordinary people conduct their life. One example of this has been provided by Yearley (1991), who claims that most of the current debate over the environment is based largely on a scientific perception of the world. He illustrates this argument through the instance of current widespread concern over the ozone layer:

> The ozone layer is only available as an object of knowledge because of our scientific culture. At ground level, ozone is relatively uncommon and remote from experience. The stratosphere where it is prevalent is even more remote. Knowledge about the hole in the ozone layer is only available through high-technology ventures into the atmosphere high over the poles (Yearley, 1991, p. 116).

In other words, Yearley suggests that we only 'know' about the ozone layer, and damage being done to it, through the work of scientists. Much of what we currently accept as 'fact' is not based on our direct experience, but on the work of scientists (see also MacNaghten and Urry, 1998). This gives scientists considerable control over the lives of the rest of the population. For instance, as noted in Box

1.1, scientists have argued that chlorofluorocarbons (or CFCs) are a key cause of atmospheric ozone depletion. Their 'findings' have led to both individual consumers giving up buying aerosol sprays (see Chapter 5 for more detail on this), and, albeit somewhat later, to governmental regulations against their production (see Chapters 4 and 5 for more details).

The role of science as ideology is further heightened by a belief in managerial efficiency and in 'progress'. These two beliefs are closely related, with technocentrics seeing a progressive taming of nature as an increase in efficiency, and an emphasis on continued improvement in the level of human appropriation of nature leading to a failure to evaluate precisely what may have been achieved. O'Riordan (1976) further argues that exponents of technocentrism tend to be politically influential because its advocates tend to move in the same circles as the politically and economically powerful, and because the dominant classes are 'soothed by the confidence of technocentric ideology and impressed by its presumption of knowledge' (O'Riordan 1976, p. 1).

The degree of confidence over the power of science and managerial efficiency does however vary within advocates of technocentrism. O'Riordan (1976; 1989) distinguishes between two forms of technocentrism: '*cornucopians*' and '*accommodators*'.

1.3.2 Two forms of technocentrism

Cornucopians, or 'interventionists' as O'Riordan also calls then, are seen to believe that people can solve any difficulties they face and create economic growth. This is the most radical form of technocentrism but perhaps not the most prevalent form, although it is, so O'Riordan (1989) argues, the view of those who control industry and commerce. By way of illustration, O'Riordan quotes from Simon and Kahn's (1984) book, *The resourceful earth*, which was an edited collection of papers written by academic researchers and people in business. The overall argument of the book was that 'the nature of the physical world permits continued improvement in humankind's economic lot in the long run, indefinitely' (Simon and Kahn, 1984, p. 3). The book was written as a response to claims that human society was on the verge of an ecological disaster stemming from overpopulation (see Ehrlich and Ehrlich, 1968; 1990; and Chapter 3 of this book). Simon and Kahn argued, however, that while at times human societies might face environmental problems, these problems were localised, and that 'the resilience in a well functioning economic and social system enables us to overcome such problems, and the solutions usually leave us better off than if the problem had never arisen' (Simon and Kahn, 1984, p. 3). In other words, while there may be localised problems, these could be dealt with, and in the process of dealing with them new ways of acting would be created that would not only deal with the local problem but would also contribute to global improvement.

O'Riordan (1989) adds that the essays in *The resourceful earth* generally suggest that the principal local hindrances to economic growth and the solution of environmental problems are generally state interventions aimed at environmental protection. O'Riordan argues that interventionism is hence associated with right-wing political views. Interventionism can, however, also be seen in some left-wing approaches. For example, Communist governments in Eastern Europe and the former Soviet Union placed considerable emphasis on economic growth with little concern for the impact on the environment, a policy that is now posing major problems for the new governments (see Chapter 5 for more on this). This approach can be seen to stem, at least in part, from claims by Karl Marx and more particularly Joseph Stalin, that a mark of human progress was control over nature because through this people were free from unnecessary labour. Overall, one can suggest that, as O'Riordan (1989, p. 86) puts it, interventionists although small in number are extremely powerful, controlling 'the levers of political and economic power in all countries, totalitarian, socialist or capitalist'.

While interventionists may be extremely powerful they are, so O'Riordan claims, substantially outnumbered by the second group of technocentrics, the 'accommodators' or 'environmental managers'. This group is said to be composed of middle-class people, such as executives, environmental scientists and white-collar employees. It shares with the interventionists a belief that nature is an external phenomenon to be valued only to the extent to which it can be manipulated for human advantage. Accommodators in particular favour manipulation by the application of science and management techniques. Accommodators, unlike interventionists, recognise that there are serious environmental problems that threaten a society based on economic growth. One frequently acknowledged problem is that of population growth and its balance with environmental resources. Accommodation can thus be traced back to the eighteenth- and nineteenth-century concern of people such as Malthus (1798) who saw population outstripping resources unless people, particularly the 'working classes', acted 'responsibly', and in particular exercised 'moral restraint' in the number of children they had. Contemporary accommodators also place considerable emphasis on 'responsible' actions to prevent environmental problems undermining the basis of economic growth. Accommodators, however, feel that these problems can be solved if people act in a 'rational' and responsible way. Advocates of accommodation hence argue for the development of more comprehensive and sophisticated planning techniques capable of incorporating assessment of risks and considering both short-term and long-term implications of any technological and social change. O'Riordan claims that accommodation:

> provides succour for liberal environmental academics and consultants. It is the heartland of cost–benefit analysis... It nourishes the environmental impact community within and outside government and industry. It has stimulated a new breed of ecological planner, armed with an environmental science training, and with an eye for beauty and heritage value. More recently, it has assisted the establishment of a new cadre of environmental mediators who claim to be able to negotiate between warring groups and resolve disputes without recourse to the courts or the legislature (O'Riordan, 1989, p. 88).

Accommodators can be said to have developed a variety of concepts with which to justify and secure their position. As well as cost–benefit analysis mentioned earlier, accommodationism has also been associated with the concept of '*sustainable development*', which argues that environmental resources need to be used in such a manner that wealth creation and economic growth can be maintained into the future. More recent concepts include the notion of '*green consumerism*', whereby people use their purchasing power to support the environmentally sensitive production of goods and services, and '*green capitalism*' in which firms make goods that are environmentally acceptable or deal with environmental pollution. Both these recent concepts have been criticised by non-technocentrics, or the 'ecocentrics' as O'Riordan called them.

1.3.3 The nature of ecocentrism and its attitude to nature and society

As mentioned earlier, ecocentrism can be seen to be a perspective that seeks to justify the possibility and value of non-exploitative relationships between society and nature. O'Riordan (1976, pp. 10–11), however, distinguishes five further characteristics of ecocentrism.

First, ecocentrism sets up a '*natural morality*'. Nature is not a neutral object open to manipulation but rather a source of limits and obligations that provide a set of rules for human behaviour. The 'natural' world is taken as a model for the 'human' world and prescriptions for political and social organisation are derived from a particular, 'ecological', point of view. This clearly has some links to the Judaeo-Christian notion of stewardship, although one can see at least two distinct notions of natural morality in modern ecocentrism. The first, and perhaps most reminiscent of the notion of stewardship, is the idea that an appreciation of nature provides a necessary check on 'the headlong pursuit of 'progress' which by and large, is the objective of the technocratic mode' (O'Riordan, 1976, p. 11). In addition there is a more extreme version of ecocentrism that displaces any humanistic concern and instead sees nature as the primary object of concern. Hence, for James Lovelock,

> It is the health of the planet that matters, not that of some individual species of organism. This is where Gaia and the environmental movements which are concerned first with the health of people part company. The health of the Earth is most threatened by major changes in natural ecosystems. Agriculture, forestry, and to a lesser extent fishing are seen as the most serious sources of this kind of damage ... the depletion of the ozone layer ... or the problem of acid rain ... are seen as real and potentially serious hazards but mainly to the people and ecosystems of the First World – from the Gaian perspective, a region which is clearly expendable ... As for what seems to be the greatest concern, nuclear radiation, fearful though it is to individual humans it is to Gaia a minor affair (Lovelock, 1988, p. xvii).

Gaia here refers to the earth, conceived as both a biological and a geological entity. Lovelock is espousing a rejection of anthropocentrism, or human centredness, and its replacement by an ecocentrism in the fullest sense of that term, whereby the interests of the earth as a whole are given priority over all other interests, including those of people.

A second characteristic of ecocentrism is that it talks of '*natural limits*' to economic growth. As Dobson notes,

> It is important to stress the word 'natural' because Green ideologies argue that economic growth is prevented not for social reasons – such as restrictive relations of production – but because the Earth itself has a limited carrying capacity (for population), productive capacity (for resources of all types) and absorbent capacity (pollution) (Dobson, 1990, p. 15).

Dobson adds that this notion of natural limits is encapsulated in the picture of the earth taken from Apollo 8 in 1968. This image and other similar ones have appeared in many ecological publications (e.g. Ward and Dubos, 1972) and, as O'Riordan puts it, it portrays

> a life filled, green and blue earth wrapped in a spiralling white cloud, slowly revolving in a black and inert void. These photographs ... compelled people to realise the earth's utter finiteness and the crucial life-supporting role of our biosphere (O'Riordan, 1976, p. 101); see also Cosgrove, 1994).

The early 1970s saw the emergence of various slogans, such as 'Only one earth', 'Lifeboat Earth' and 'Spaceship Earth' which attempted to encapsulate such realisations.

The notions of natural morality and natural limits are arguably the basic concepts underlying ecocentrism. O'Riordan, however, suggests three further characteristics that are associated with ecocentrism, albeit in various degrees. First, ecocentrics frequently make of use of a '*language of ecology*', that is they frequently adopt the language of professional ecologists, for instance by using concepts such as 'permanence', 'diversity', 'equilibrium', 'stability' and 'succession', developed initially within ecological system analysis. The degree to which it is valid to transpose concepts developed to analyse the 'natural' and 'human' worlds is very much open to debate, and depends in part on how one draws the society–nature distinction. In many academic disciplines such as geography there is considerable scepticism about the value of such a transposition, but much of this scepticism is sometimes lost in the transfer of the language into popular discourse.

A further characteristic associated with ecocentrism is a focus on '*reconciling means and ends*'. This point has been clearly demonstrated at the annual conferences of the British *Green Party* in the mid-1990s, which witnessed conflict between those who wished to restructure the party along a centralised line, reminiscent of virtually all the other political parties in Britain, and those who preferred a more decentralised form of organisation. This latter group are frequently taken to be the more 'traditional' Greens, with, for example, both O'Riordan and Dobson arguing that notions of decentralised and participatory democracy have been a central element of ecocentrism. Others, however, such as Bramwell (1989) have identified an authoritarian strand in the environmentalist argument, not least because the belief in natural limits is frequently translated into a 'doomsday scenario' whereby there is seen to be an impending environmental catastrophe of such dramatic proportions as to make any concern over reconciling means with socio-political aims largely irrelevant. The relationship of ecocentrism to political structure and actions is, hence, a far from straightforward one.

The final characteristic of ecocentrism identified by O'Riordan (1976) is the advocacy of '*self-reliance and self-sufficiency*'. Ecocentrics frequently suggest that there is a need to escape from contemporary ways of living, such as in large cities, and establish small, self-sustaining communities. The argument is that living in cities is an unnatural, harmful way of living and that people should live in smaller scale units in which one can more clearly see the ways of nature.

1.3.4 Two forms of ecocentrism

The five features associated with ecocentrism are, as we have already noted, given different emphasis by particular ecocentrics. O'Riordan suggests that from the differing emphasis given to these five characteristics it is possible to distinguish two groups of ecocentrics: i) the *'Communalists'* or *'Self-reliant, soft-technologists'* who promote the concept of a human society based on self-reliancy and self-sufficiency; and ii) the *'Deep ecologists'* or *'Gaianists'* who promote the concept of a human society conforming to ecological laws.

The communalists are ecocentrics who place particular stress on changes in organisation of society. Contemporary societies are seen to be structured in such a way as to make the degradation of nature an inevitable outcome. Jonathan Porritt, a well-known British environmentalist and one-time leader of the *Friends of the Earth* in Britain, for example, complains about industrial societies geared around 'the belief that human needs can only be met through *permanent* expansion of the process of production and consumption' (Porritt, 1986, p. 345). Porritt suggests that this belief is a *'super ideology'*, 'embracing ... capitalism, communalism, most variations of socialism and practically every other -ism in every part of the world', and that it will 'at some stage in the future' simply 'fall apart' because the earth will not provide sufficient resources to sustain continued expansion. Porritt argues that although some of the problems of industrial society can be avoided through reforming it, this is insufficient because of its emphasis on growth at all costs, and therefore there needs to be a 'replacement' of industrial societies with a more sustainable form of society.

Not only are contemporary societal practices seen by communalists to endanger nature, but nature is seen by communalists as providing blueprints for new, more ecologically sustainable forms of social organisation, often based on federated political structures. These forms of organisation are frequently taken to combine two 'natural features': diversity and integration. A further idea often borrowed from ecology and applied by communalists to society is that of stability: communal living is frequently represented as a form of social organisation that creates stability, both environmentally and socially.

Communalist ideas have long been advocated in western anarchist and socialist thought, with notable exponents being Peter Kropotkin, Robert Owen and Ebenezer Howard. Modern western champions include E.F. Schumacher, author of *Small is beautiful* (1973) and the contributors to *The Ecologist* magazine who wrote a *Blueprint for survival* (Goldsmith *et al.*, 1972). Both these texts place considerable emphasis on small communities, 'appropriate technology' and personal happiness and fulfilment. However, while these themes are also highlighted in anarchist writings, there are some fundamental differences. For instance, as Pepper (1984) argues when comparing Kropotkin's *Fields, factories and workshops tomorrow* (1899) with *Blueprint for survival*, while the latter argued for small-scale communities because they conformed to ecological principles, such concerns were marginal to Kropotkin who:

> *mainly* wanted to create a more fulfilling society in which the individual was not frustrated and dehumanised by large-scale mass organisation, where relationships of domination and hierarchy were replaced by mutually supportive and willingly-accepted relationships between people (Pepper, 1984, p. 191).

Pepper suggests that within Kropotkin's anarchism ecological imperatives were marginal. Bookchin (1979, p. 24), another self-identified anarchist, appears to give ecology a more central place, suggesting that 'the critical edge of ecology is due not to the power of human reason ... but to a still higher power, the sovereignty of nature'. However, as Adams (1990, p. 85) notes, Bookchin's '*ecological anarchism*' still placed nature as only one object of concern. Equally important to Bookchin, and in his view connected to the production of environmental crises, were the relations between people and the organisation of society. For Bookchin, modern society has a crisis of '*social ecology*' whereby not only are people becoming separated from nature but they live alienated from each other. As Adams (1990, p. 85) notes, Bookchin is 'anti-industry, anti-bureaucracy and anti-state', and he therefore shares some common intellectual ground with Schumacher. However, as Adams also notes, Bookchin and Schumacher advocate very different solutions to the socio-ecological problems they both identify. For Schumacher incremental change is always possible; for Bookchin revolutionary change is called for.

While Schumacher, Kropotkin and Bookchin can all be seen to advocate a communalists' perspective on society–nature relations, in the sense that they all see intimate connections between the organisation of society and nature, it is also clear that there are important differences in emphasis that relate both to how much primacy is given to nature and how societies are seen to function. Very similar issues are raised by another form of environmentalism exhibiting characteristics of communalism, namely '*ecofeminism*'.

The notion of ecofeminism emerged in the 1970s at a time when there was both widespread environmental concern in many 'western' countries and a rising feminist movement. As Mellor (1997) notes, ecofeminism 'interweaves' strands of environmentalist and feminist thought together. In particular, Mellor suggests that ecofeminism:

> takes from the green movement a concern about the impact of human activities on the non-human world and from feminism the view that humanity is gendered in ways that subordinate, exploit and oppress women (Mellor, 1997, p. 1).

The precise way that these concerns are interwoven, however, varies quite considerably and there are important differences in emphasis and argument within ecofeminism. MacNaghten and Urry (1998), for example, discuss distinct ways in which feminism and environmentalism may interconnect. First, attention has been drawn to the way nature is often given a female identity: nature is seen as a goddess, as in Gaia, or the earth is seen as mother, or as mistress. Second, and often connected to the first point, it is argued that the interactions between society and nature and between men and women are analogous: ecofeminists have drawn attention to and criticised the way that women and nature are frequently both constructed as being in need of 'control', 'mastery' and 'domestication', and seen as the object of a visual gaze. Third, some ecofeminists have argued that women are more 'natural' and 'closer to nature' due to their experience of menstruation, childbirth and child-care. Fourth, and relatedly, it has also been claimed by some ecofeminists that women are 'naturally' more caring than men, and therefore more likely to adopt ecocentric views than men. Fifth, it has been argued that women are often, particularly outside the western world, more involved in direct contact with nature through agriculture work and hence have a more intimate, and respectful, attitude towards nature.

While these five strands of argument can all be considered ecofeminist, it is rare to find an individual espousing all these arguments, and indeed there are clear tensions between ecofeminists that, in a similar manner to communalists, relate both to how much primacy is given to nature and to how societies are seen to function. Some of the ecofeminist literature argues that societies and their interaction with non-human elements of nature are constructed 'around bodies and their biological capacities' (Nesmith and Radcliffe, 1997, p. 384). In other words, men and women are seen as having inherent differences related to biology which influence how they relate to nature. Other ecofeminists, such as Plumwood (1988), argue that genders, and gendered relationships with non-human environments, are social rather than natural (i.e. biological) constructions.

Gaianists

The term Gaianism is taken from the Greek goddess '*Gaia*', who was said to be the Goddess of the Earth, the nurturing mother from whom sustenance was derived (O'Riordan, 1989, p. 90). The use of the term in the context of modern environmentalism stems from the work of Lovelock who, as noted earlier, espouses an extreme ecocentric position.

Lovelock was a scientist who was for a time employed by the *US National Aeronautics and Space Administration (NASA)* to design experiments to see if there was life on Mars. From this work Lovelock developed a scientific approach, which he termed '*geophysiology*', that sought to integrate the workings of living nature with the geochemical processes of the atmosphere, hydrosphere and lithosphere. In particular, Lovelock suggested that the geochemical processes and living organisms on earth work together to create states of equilibrium that are 'comfortable for life' (Lovelock, 1988, p. 19). This condition of equilibrium, although frequently taken for granted, is, Lovelock argues, rather surprising given that over geological history the output from the sun has been steadily increasing and the atmosphere is composed of an unstable mixture of reactive gases. Lovelock argues that there must be some mechanism that establishes this unexpected equilibrium and he suggested that this is 'the process of life' itself: living organisms act, so Lovelock claims, to regulate their environments so that life itself can continue.

In 1979 Lovelock used the term 'Gaia' to describe his theory (Lovelock, 1979). Initially the recourse to the name of a Greek goddess was merely a rhetorical device: Lovelock wanted a name to summarise his arguments. As he admitted later (Lovelock, 1988, p. 200), when he first used the term Gaia it was as a 'mere' scientist and he paid no attention to its religious connotations. He goes on to note, however (1988, p. 200), that although he developed his ideas as a scientific theory:

> many of its readers found it otherwise. Two-thirds of the letters received, and still coming in, are about the meaning of Gaia in the context of religious faith.

As will be discussed further in Chapter 6, Gaianism has come to signify much more than just a scientific theory of evolution. For some, such as Bunyard and Morgan-Grenville (1987) and Porritt (1984), for example, it provided some scientific legitimation for ecocentric views, while for others, including increasingly Lovelock himself, Gaia provides a spiritual basis for these views.

A similar ambivalence can be seen to characterise *'deep ecology'*. The term was coined by a Norwegian philosopher, Arne Naess, who sought to create an *'ecosophy'* – or environmental wisdom – that brought together philosophical analysis; the practical, 'scientific', knowledge of ecologists; and the experiential respect of the ecological field-worker or others living close to nature (see Naess, 1973; 1989). As Rothenberg (1989) argues, Naess's ecosophy or deep ecology can lead in quite different ways. On the one hand, the philosophical and practical dimensions have been drawn together whereby the term 'deep ecology' is used to refer to a socio-political movement in which people are concerned with establishing new, less exploitation-centred, forms of society–nature interaction. On the other hand, the philosophical and experiential dimensions have led to 'deep ecology' being interpreted very much as a spiritual/religious movement concerned, for instance, with 'a recognition of the "oneness" of creation and a subsequent "reverence for one's own life, the life of others and the Earth itself" (Dobson, 1990, p. 21).

The implications of 'deep ecology' and 'Gaianism' are deeply contentious. For some, such as Dobson (1990, p. 21), they represent an important challenge to existing ideologies and the structure of contemporary societies. For others they represent an ill-focused analysis of the problems of contemporary societies and a rejection of the benefits created through the exploitation of the environment. One German philosopher and sociologist, Jürgen Habermas, for instance, has suggested that the German Green movement should be seen less in terms of presenting a radical alternative to contemporary society, and more as a 'pathological' manifestation of the failure to present any alternative to capitalist modernisation (Habermas, 1975; 1987). We will look in more detail at these arguments in Chapter 6.

1.4 Society and nature: the focus of this book

This chapter has sought to introduced some of the complexities surrounding the concepts of society and nature and their interrelationship. To recap, it has been suggested that the term nature can be looked at in two distinct, albeit generally interlinked ways. First, one can apply the term nature to *'external object'*, that is anything that lies outside 'humanity', or 'society'. Second, one can use the term nature to imply an *'inherent state'*. These two senses of nature are both highly contested and can lead to the definition of quite different natures. Rather than talk about a singular 'nature' it is, we argue, better to think in terms of multiple natures.

After outlining the two basic conceptions of nature their application within a framework of 'determinism' and 'free will' or 'social agency' was undertaken. It was suggested that there were at least four important positions along the determinism-free spectrum with respect to society–nature relations; namely those of nature as society, nature determining society, society determining nature, society as nature. It was suggested that the precise meaning of these positions will also depend on how one conceives of society: whether one sees it in terms of an aggregation of individuals, analysable solely in terms of patterns of individual beliefs and actions; or as some supra-individual entity setting conditions of how people act; or as some dialectical outcome of individual actions and collective conditionings.

Attention was then paid to the issue of how society and nature are seen to be interrelated by those people who accept some notion of society–nature differential. It was suggested that society and nature are often seen to interact in terms of an exploitative relationship, where this is understood both in terms of a practical relationship of use and in more social and moral senses as a relation of exploitation. Different attitudes to the exploitation of what is taken to be nature is seen as being a central aspect of the technocentric/ecocentric distinction made by people such as O'Riordan. Two variants of technocentrism, 'cornucopians' or 'interventionists' and 'accommodators', and two forms of ecocentrism, 'communalists' and 'Gaianists', were identified; although it was also suggested that important differences within each perspective related to detailed constructions of nature and society and their interaction. We will discuss these issues in more detail, and in particular in Chapter 6. Prior to this, however, we will focus more generally on the issues of exploitation and nature, issues that we have suggested here underlie the basic technocentric/ecocentric distinction.

In Chapter 2 we will examine the extent to which it is possible to discern a steadily increasing exploitation of nature over time. A major strand of many technocentric attitudes is a view that societies have developed an increasing capacity to exploit nature over time. Ecocentrics are, however, rather more sceptical about whether this has been achieved, pointing to a range of negative feedbacks whereby nature may be seen to exercise control over society, or at the very least impose limits on it. In this chapter we will also explore the way that notions of nature may be important in establishing relations of exploitation between people as well as between people and natures.

Chapters 3 and 4 will take a more contemporary focus on the issue of social exploitation of nature. In particular the chapters will explore whether or not societies may be reaching some limit to its ability to exploit natures. Chapter 3 will explore the arguments between the 'neo-Malthusians', who believe that societies are heading for a global environmental disaster as a result of over-population and over-exploitation of the earth's resources, and the 'cornucopians', who hold polar opposite views. In particular the ability of societies to continue to produce enough food and energy for a growing human population will be discussed, and the environmental costs of industrial and agricultural activities outlined. In Chapter 4 attention will focus on whether it is possible to manage social exploitation of natures so as to avoid, overcome or at least ameliorate environmental problems. The first part of the chapter reviews the major forms of managing environmental problems, and examples of the successful implementation of each managerial approach are provided. The second part of the chapter examines the problems of implementing these managerial approaches, and focuses upon two major themes: first, the problems caused by scientific uncertainty and the complexity of natural agencies and, second, the problems caused by human behaviour.

In Chapter 5, the focus moves on to forms of social organisation and their connection with the creation and management or non-management of environmental problems. It is argued that in much of the discussion about environmental problems there is a failure to distinguish whose actions can be seen as causing the problem and who suffers the consequence. Throughout this chapter we have talked quite generally about society, although we have identified some different ways in which it can be conceptualised. In Chapter 5 we will focus on this issue much more directly and will argue that the notion of society needs to be both given historical and spatial specificity – that is recognising that there are different societies in different places and at different times – and disaggregated into social groups, social relations and social practices.

In Chapter 6 we will return to many of the issues raised in this introductory chapter. In particular we will look in more detail at arguments for non-exploitative society–nature relations. Attention will be paid to the extent to which these arguments are a recent development or whether there has been a long-standing, if rather neglected, strand of ecocentric argument. The chapter will also explore the social context and consequences of ecocentric arguments.

In the final chapter of the book we outline our own views about the social exploitation of nature and provide some thoughts about the five questions we have posed in this chapter.

Chapter 2

Stories of Exploitation

2.1 The social exploitation of nature: a single historical narrative or multiple and contradictory stories?

The previous chapter concentrated on ideas about how society and nature may interrelate, with particular emphasis being given to the notion that exploitation may be a key, albeit contested, element of connection. In this chapter attention will be directed at outlining particular examples about how people at different points in history have used, or 'exploited', nature. The particular examples to be examined are the emergence of agriculture and industry, the expansion of European imperialism and the post-industrial society. Each of these examples will be examined separately, but it is important to recognise that it is possible to see them as providing a history of the growing social control of nature. Particularly clear instances of this can be found within the 'ecological' histories of Simmons (1991; 1989) and Dodgshon (1987) (see Box 2.1).

Box 2.1 Ecological histories of society–nature relations: some examples

According to Odum (1975, p.1), ecology is derived from the Greek terms *oikos* meaning house and *logos* meaning the 'the study of', and thereby literally means 'the study of the earth's "households" including plants, animals, micro-organisms, and people'. Odum adds that ecology should be concerned with 'ecosystems', that is not only with 'households' or 'organisms' but also with their interrelationships, which he argued are best seen as being essentially flows of energy and material. He claimed, for instance, that:

> The interaction of energy and materials in the ecosystem is of primary concern to ecologists. In fact it may be said that the one-way flow of energy and the circulation of materials are the two great principles or 'laws' of general ecology, since these principals apply equally to all environments and all organisms, including man (Odum, 1975, p. 61).

Although, as for example Adams (1997) has stressed, there have been some quite different interpretations of ecology, the notions of ecosystems and a focus on energy flows have exerted a considerable influence on histories of society and nature relations.

▶

Simmons, for example, employs an ecological perspective and its emphasis on flows of energy as the organising theme for his books, *Changing the face of the earth* (Simmons, 1989) and *Earth, air and water* (Simmons, 1991). Energy, he argued slightly earlier, forms a 'medium of connection' (Simmons, 1981, p. 38) between people and the 'natural' environment. He suggested that people should be seen as part of ecosystems and as such as being bounded by the laws of thermodynamics in which energy cannot be created or destroyed but can change its form. Hence, coal can be burnt to create heat and light, which are new forms of energy. The heat from coal can be used to create steam, which can move turbines and generate electricity. In each case, however, no more energy can be created than was contained in the original coal, and indeed much energy may be lost from the process. This energy is not, however, lost from the total ecosystem but becomes *'bound energy'* or *'entropy'*, that is energy that is distributed through the ecosystem. Over time there is seen to be an increase in bounded energy or entropy. However, at present there is a large amount of unused 'bound energy' in the form of 'natural resources'. Much of this energy is described as *'non-renewable'* or *'stock'* energy resources and is contrasted with *'renewable'* or flow 'energy resources', although the distinction is rather a misnomer given that, as Simmons, emphasises, so-called non-renewable energy resources such as coal, natural gas and petroleum are produced by the energy of the sun via photosynthesis and are thereby new resources actively being produced today. The key issue, in terms of human use of the energy, is the relationship between the time-scale at which energy is being fixed in nature and the rate and efficiency at which this energy is being converted into energy of value to people. Simmons argues that it is possible to discern increasing levels of energy exploitation or 'capture' by societies through time, which are related to the use of distinct sources of energy, which in turn define distinct forms of social organisation (see Table B2.1).

Table B2.1 Simmons on energy and social history

Socio-economic stage	Distinctive sources of power	Per capita level of energy use
Primitive	Fire	2
Advanced hunters	Fire	5
Agriculturalists	Wind, water and animal power	12–26
Industrialism	Hydrocarbons (or fossil fuels)	77
Nuclear age	Nuclear power	230

(Source: after Simmons, 1989)

A very similar approach is taken by Dodgshon (1987), who argues, using *'systems analysis'*, that social history is characterised by an increase in social complexity based on an increasing amount of energy under societal control. He distinguished five distinct stages of 'social evolution': 'bands', 'egalitarian tribes', 'chiefdoms', 'regulated' or 'feudal' societies and 'market' societies.

It is, however, important to note that there may be problems in such a construction of the history of society and nature relations. First, such an account may ignore other facets of society–nature relations, such as non-exploitative ones in which nature is seen, for instance, as a source of moral or spiritual value. Ecological perspectives that focus on energy capture can be seen as employing a singularly technocentric evaluation of nature and at the very least may, therefore, be seen as rather *reductionist* history. Second, and relatedly, such a history may be overly optimistic about the degree to which societies have increased their control over natures. Simmons (1989), for example, seems to place what to many people appears as a surprising level of faith in nuclear power. As mentioned in Chapter 1, people adopting an ecocentric perspective are extremely concerned about the consequences of human activity both on nature and on people themselves. Rather than talk of increasing control of nature, deep ecologists and Gaianists, for example, warn of the 'revenge of nature', a notion that has also started to appear in some recent environmental histories such as that of Murphy (1994). There is a clear danger of a 'whiggish' or 'teleological' history that views the past from a particular end-point – such as the ability of contemporary society to alter nature – and emphasises events, actions and situations that fall into line with this interpretation, passing quickly over or ignoring things that are out of line.

This chapter will therefore proceed by discussing society–nature relations at four points in history/prehistory. The chapter will note how these historic/geographical forms can be linked into a single narrative of exploitation, but will also explore the possibility of constructing more complex, multi-dimensional and less linear 'historical geographies' of society and nature relations.

2.2 The emergence of agriculture

2.2.1 Lessons from prehistory?

Examining how our prehistoric ancestors related to their environments may seem to be a rather esoteric pastime. However, according to Bell and Walker (1992) looking at the impact of humans over long time-scales and the changing interaction between humans and nature produces two important forms of information. First we can indirectly observe people's capacity to change and adapt to natural environmental and climatic changes. Second we can identify how nature responds to human exploitation over a time-scale of thousands of years. As Bell and Walker state,

> environmental changes such as deforestation, erosion and the extinction of biota, occurred as a result of human activities, in many cases with histories extending back many thousands of years ... Contemporary interpretation of those histories both affects, and is affected by, our prognosis for the future (Bell and Walker, 1992, p. 224).

By scrutinising the palaeoenvironmental record (see Box 2.2) societies may identify positive and negative human–environment relationships, learn from them and hopefully use them to help us solve current environmental problems.

Box 2.2 Reconstructing past environments: the palaeoenvironmental record

A palaeoenvironment is an environment that existed in the past. Although an environment could remain unaltered for thousands of years, agents such as climate change and human activity can alter it. To determine the extent to which environments change through time we can reconstruct them in two ways. First through archaeology, which informs us about the social structure of past peoples, and second by using the geological and fossil records to reconstruct past physical environments. Deposits, especially peat and lake sediments, contain a valuable store of fossil plant, animal and inorganic material, e.g. pollen, diatoms, fungi, macrofossils, particulate material, charcoal and geochemical aerosols. These products accumulate and become preserved in sediments through time, providing scientists with a chronological record of palaeoenvironmental data. Rocks and sediments can also provide scientists with evidence of environmental changes such as soil erosion. The collection, analysis and application of that data, combined with archaeology, provides us with a record of past human impact, climate, volcanic eruptions (by finding tephra in sediments) and vegetational changes. For example palaeoecologists have reconstructed past vegetational histories using pollen data collected from sites located all over the world (see Edwards and MacDonald, 1991). The age of these sediments can be established using a variety of dating techniques such as radiocarbon dating, lead-210 dating and thermoluminescence dating, enabling a chronological sequence of environmental and climatic changes to be established. Data such as this enables quaternary scientists to estimate the magnitude and frequency of natural environmental changes, to assess the degree of human impact and evaluate the rate and sustainability of these impacts. Such reviews have already revealed the dynamic nature of climate, environmental changes and human impact on long and short time-scales, emphasising that nature is not static and perhaps should not be managed in that way (Chambers, 1993). For more information on the techniques and methods used to reconstruct past environments consult the texts by Birks and Birks (1980), Lowe and Walker (1997) and Bell and Walker (1992).

This section will argue that human exploitation and alteration of nature is not a recent process. The extent of human exploitation and impact can be recorded over a time-scale that extends back ten thousand years to the end of the last Ice Age, an epoch known as the Holocene (Box 2.3). Several different forms of prehistoric societies have been identified during this epoch. Each form of society is characterised by its own form of exploitation and the resultant impacts. Hunter-gatherers are thought to be one of the earliest forms of society and they exploited by using and/or transforming the environment through hunter-gathering activities and the use of fire. They existed during the Palaeolithic and, in north-west Europe, the Mesolithic period (c.10,000 to 5500 years BP) while forms of society during Neolithic (5500 to 4000 years BP), Bronze Age and Iron Age used agriculture as a major form of exploitation. Finally, the emergence of metalworking technology during the Bronze Age (4000 to 2500 years BP) is another form of prehistoric exploitation.

Box 2.3 Archaeological periods

The main archaeological periods in prehistory and their chronology for the British Isles are shown in Figure B2.3a. However, it is important to remember that the occurrence of these periods world-wide is time transgressive, that is the same 'period' is seen to have occurred at different times in different places. This is illustrated in Figure B2.3b on page 34, which shows the chronology for each period outlined for the Near East and the British Isles.

Mesolithic. The Mesolithic period refers to those hunter-gatherer societies that existed in Europe during the early to mid-Holocene when tundra was largely replaced by forests. Mesolithic hunter-gatherers did not occur in unforested parts of the old world. In these areas (e.g. the Near East) pre-Holocene *Palaeolithic* societies were displaced by Neolithic ones.

Neolithic. The development of Neolithic societies is often synonymous with the emergence of agriculture. However this period is distinguished from the Palaeolithic and Mesolithic periods by the production of pottery (ceramics) rather than agriculture. As Mannion (1997) points out, the beginnings of agriculture do predate ceramic Neolithic societies, and it is a term that is applied to early farming societies in the Old World rather than the New World. Neolithic societies first emerged in the Near East around 10,000 years ago and spread from there metachronously. Because most of the discussion presented here looks at the impact of the first farming societies in the Near East and western Europe, the term Neolithic will be used.

Bronze Age. This period is characterised by the development of metalworking technologies. Again this is metachronous with various metalworking centres emerging across Europe between approximately 8000 and 4000 years BP.

Iron Age. This period is characterised by the use of iron, which is believed to have first been used around 4400 years ago in Turkey but reached Europe and Britain 2500 years ago.

Subsequent periods following the Iron Age are shown in Figure B2.3a.

Source: Mannion (1998).

▶

	Years before present	Climatic period	Archaeological period	Technology	Food procurement	Environmental impact: Pollution	Environmental impact: Vegetation	Environmental impact: Soils (Erosion / Acidification)	Climate
Holocene Flandrian	1000	Sub-Atlantic	Modern, Medieval, Dark age, Romano-British		Permanent agriculture: Mixed arable + pastoral farming	Increasing ↑	↑	More widespread ↑	Deterioration
	2000		Iron Age	Metals: Iron					
Post-glacial	3000	Sub-Boreal	Bronze Age	Bronze					Stable
	4000			Copper		Localised	Regionally permanent		
	5000		Neolithic	Stone/flint axes, cleaves			Variable woodland clearance	Localised ↑	
	6000	Atlantic	Mesolithic		Hunter–gathering; Use of fire	No mining or metal-working	Very localised woodland clearance		Climatic optimum
	7000							Very limited, small-scale	
	8000	Boreal							Rapid amelioration
	9000						↑		
	10000	Pre-Boreal					Woodland development	Soil development	
		Younger Dryas	Upper Palaeolithic				Open tundra-like vegetation	Raw mineral soils	Cold
Late-glacial	11000	Allerød							Cool

Figure B2.3a A summary of the main archaeological periods, climate and environmental changes during the Holocene

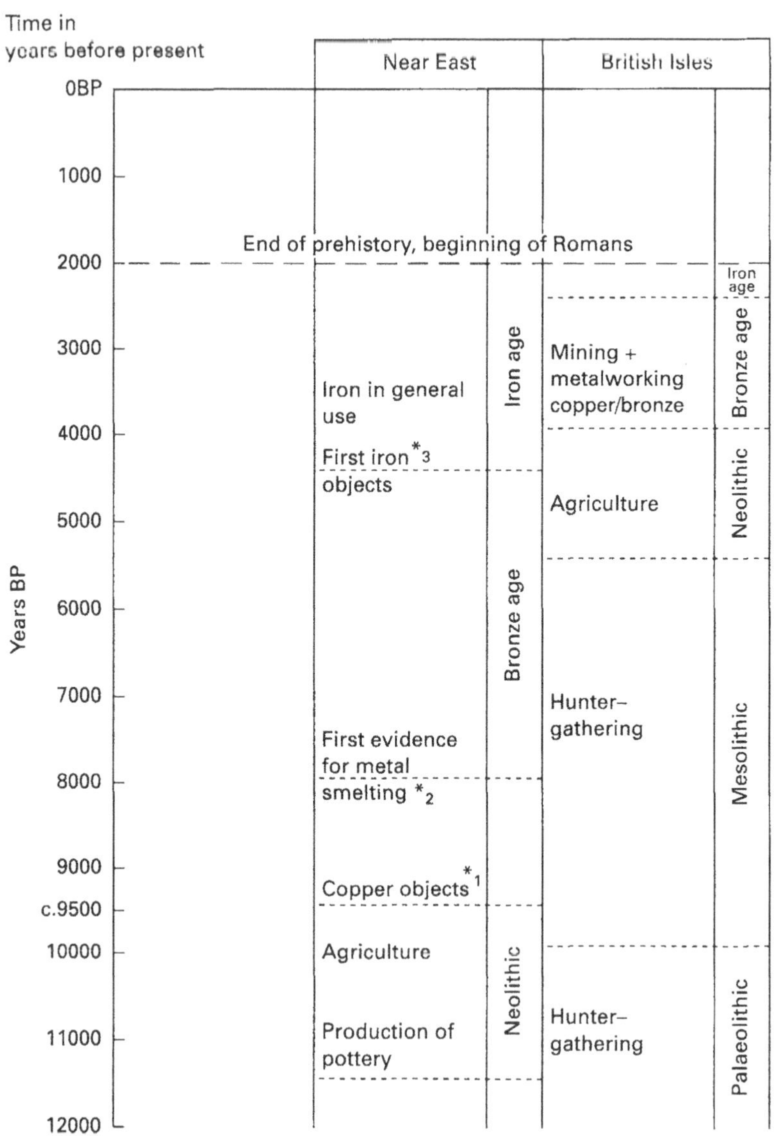

*1 De Laet (1996)

*2 Mellaart (1967)

*3 First authenticated iron objects c.4400 yrs BP (Collis, 1984)

Figure B2.3b A comparison of the advent of the main archaeological periods during the Holocene in the Near East and the British Isles

Each society is characterised by population growth and new forms of technology that increased the ability of humans to manipulate their environment. Hunter-gathering societies are believed to have used fire as an agent of environmental change while agriculturalists developed new techniques such as ploughing and by the Bronze Age metalworking technology was in use. Environmental impacts accompany these technological developments and they become more discernible as societies change through the Holocene. Delcourt (1987, p. 39) suggests that prehistoric human population affected biota in four ways: (i) changing the dominant structure within forest communities; (ii) extending or truncating the distributional ranges of both woody and herbaceous plant species; (iii) providing opportunities for the invasion of plants (ruderal) into disturbed areas; and (iv) changing the pattern of the landscape mosaic, especially the proportion of forested to non-forested land. Coeval with these four principal impacts were soil erosion and acidification, faunal changes and geomorphological landscape changes and evidence for pollution from the Bronze Age onwards.

The development of these forms of society and their impact on the physical environment is time transgressive and spatially variable but forms of hunter-gatherers and agriculturists have existed in most regions of the world at some point during the Holocene (see Delcourt (1987) and Edwards and MacDonald (1991)). The impacts of hunter-gatherer societies are often seen to be small scale and non-permanent, but with the emergence of agricultural and subsequently metalworking societies these impacts become larger in scale and permanent. This idea is argued by Simmons (1989; 1991), who suggests that hunter-gatherers had a minimal impact on their environments while agricultural societies are able to capture greater levels of energy from nature and so the environmental impact becomes more intense. However, it will be argued that this view is slowly being revised as research shows the hunter-gatherer societies were capable of creating localised permanent environmental changes and were beginning to separate themselves from nature.

However, the nature of human–environment relationships did fundamentally change with the emergence of Neolithic communities because agriculture represents a new form of exploitation. Neolithic and subsequent societies had the technology to inflict larger-scale changes to their physical environments. To demonstrate this, we will briefly consider the environmental impact of hunter-gatherer societies and then discuss the origins, emergence and environmental impacts of agriculture that shaped later prehistoric societies and continued into historical times.

2.2.2 Hunter-gatherers: did they permanently alter their physical environments?

Pre-Holocene human impact

Pre-Holocene human impacts are difficult to gauge, in part because the evidence has not been preserved and because low population densities and the forms of technology applied limited the ability of humans to significantly alter their environments. Climate and other natural agencies were the dominant forces shaping the physical world. One exception, however, might be the extinction of large mammals between 40,000 and 10,000 years ago when about 70 to 90 per cent of these species, including the mammoth, became extinct across Europe, Asia, America and Australia. One explanation for the Late Pleistocene disappearance of the mammoth between 12,000 and 10,000 years ago in Europe and America is hunting by humans. In North America the mammoths died out within 500 years of the arrival of 'Clovis hunters'. The discovery of stone spearpoints in mammoth burial sites provides circumstantial evidence for mammoth hunting. However, the ability of humans to hunt large mammal species to extinction over such a large area remains a contentious issue and a dramatic change in climate and vegetation may have been more instrumental in causing the demise of the mammoth at the end of the last Ice Age (Lister and Bahn, 1994).

The Holocene human impact: Mesolithic hunter-gatherers

In the introduction we presented the idea forwarded by Simmons (1989) that hunter-gatherers had a minimal impact on their environment. This view is becoming increasingly questioned. Edwards and Whittington (1997) for example suggest that the relationship between Mesolithic hunter-gatherers and the physical environment occurred on three possible scales: they were totally subservient to the nature of woodland they encountered; they had a minimal effect on distribution and composition of woodlands or they induced some irrevocable changes in woodland ecology and distribution. Evidence for all three changes can be observed in the palaeoecological record suggesting that relationship is variable.

These relationships can be seen in examples from the Mesolithic hunter-gatherer period in north-west Europe that occurred between 10,000 and c. 5500 years BP (Bell and Walker, 1992). Archaeological evidence from the British Isles and north-west Europe suggests that the Mesolithic economy involved hunting an array of animals including red and roe deer, badgers, foxes, aurochs, elk, wild pig and the collection of hazelnuts, acorns, water chestnuts (Zvelebil, 1994) and root and tuber tissue belonging to arrowhead and bindweeds and/or knotweeds (Kubiakmartens, 1996). At coastal locations shell middens have been found that contain remnants of limpets, seals, birds and fish (Mannion, 1997), and marine resources may have contributed up to 90 per cent of the Mesolithic diet in some north European groups (Price, 1987).

Birks (1988) argues that the evolution of the early Holocene landscape in north-west Europe was primarily shaped by climatic, topographic and edaphic factors. Cultural factors are thought to be less influential and many researchers suggest that hunter-gatherers created non-permanent, localised changes to vegetation while procuring food (e.g. Behre, 1988). To do so, Simmons (1996) argues that hunter-gatherers used fire as part of a highly organised and deliberate land management strategy. However, the evidence for the use of fire is equivocal. Chambers (1993) argues that the association of charcoal and artefact evidence for human presence for woodland recession does not always demand a casual connection. It is possible that charcoal in sediments of Mesolithic age represents natural fires, such as lightning strikes (especially when archaeological evidence is absent) or the charcoal could be derived from domestic sources such as cooking and heating (Simmons and Innes, 1996a, b). Fire is not always a precursor for Mesolithic woodland manipulation as charcoal is absent at the start some disturbance phases suggesting that if humans were responsible they must have used other methods, possibly girdling of trees and/or stripping them (Simmons and Innes, 1996a, b). Whether the use of fire was intentional or not, exploitation by Mesolithic societies appears to have played at least some part in altering vegetation, including changing the composition of woodland, providing space for the creation of ruderal plant communities and, to a lesser extent, clearing areas of woodland either to encourage browse to attract grazing animals or as a by-product of cropping woodland.

Non-permanent and localised phases of woodland disturbance during Mesolithic times are characterised by several features.

(1) Evidence from pollen and charcoal suggests that small-scale openings in woodland were caused by fire. However, firing of woodland may have helped alder establish itself in the British Isles during the early Holocene. Smith (1984) suggested that valley-bottom waterlogging was enhanced by increased runoff from areas cleared of woodland by fire and the association of a rise in alder pollen and charcoal has been noted in numerous pollen diagrams (Edwards and Whittington, 1997) but there is no consistent pattern.
(2) Woody shrubs increase, most notably hazel. Smith (1970) suggested that the rise of hazel around 9000 years BP was facilitated in part by coppicing and burning of pre-existing hazel stands by Mesolithic communities, thus promoting woody growth, profuse flowering and increased hazelnut yields. However, pollen and charcoal records do not always reveal a clear association between fire incidence and an increase in hazel (Edwards, 1990). Huntley (1993) argues that an absence of competition from other taxa and a favourable climate during the early Holocene was a more important factor in the establishment of hazel across the British Isles.
(3) Herbaceous plants increase in abundance. Pollen values of heather and grasses rise and the occurrence of herbaceous taxa nettles, tormentils, docks, sorrels and ribwort plantain are commonly associated with phases of Mesolithic woodland disturbance (Simmons and Innes, 1996a, b). These plants take advantage of increased light on the woodland floor and provide evidence for an increase in the floristic diversity during the early Holocene. Fire tolerant plants such as fireweed and bracken have also been recorded in pollen records.

Such phases may have been short lived, lasting only decades, or much longer, extending over hundreds of years (Simmons and Innes, 1996a, b) (see Box 2.4), but occurred on very small scales – tens of metres rather than hundreds of metres according to Turner *et al.* (1993). A series of studies have suggested that hunter-gatherer impact on woodland is characterised by a sequence of disturbance followed by recovery, which suggests that the impact of Mesolithic peoples was non-permanent.

Box 2.4 Examples of Mesolithic human impacts on vegetation

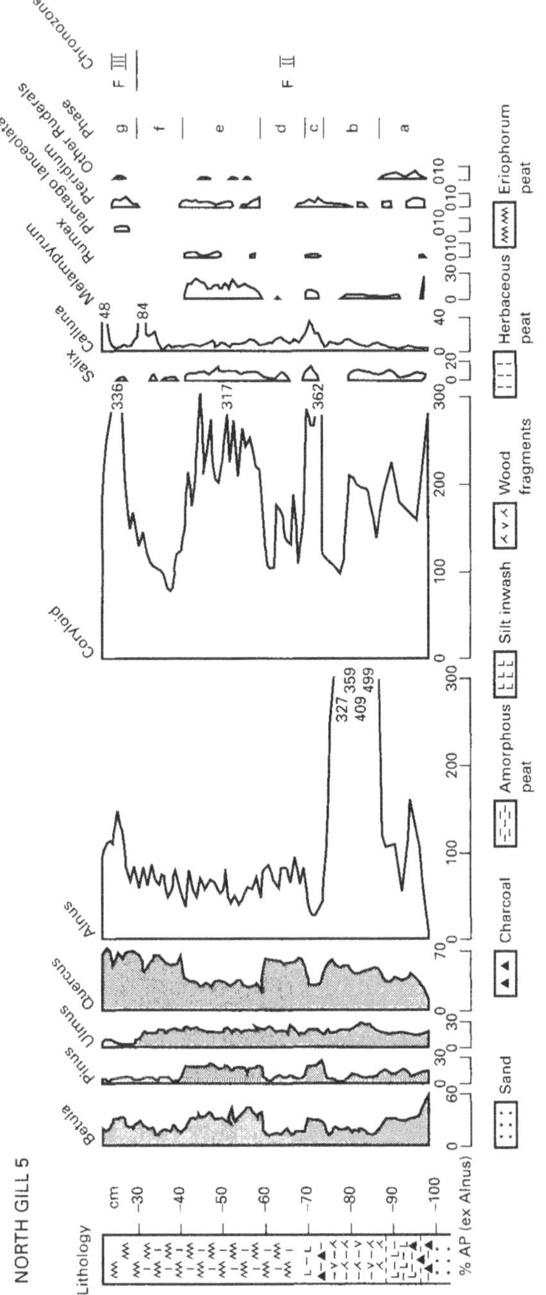

Figure B2.4a Pollen diagram for North Gill 5, North Yorkshire Moors, UK, showing selected taxa as percentages of total tree pollen. Taxa within the tree sum are shaded. (Source: Innes and Simmons, 1988)

Figure B2.4b A summary of the pollen, stratigraphic and radiocarbon evidence from Bonfield Gill Head, Bilsdale East Moor, UK (Source: Innes and Simmons, 1988)

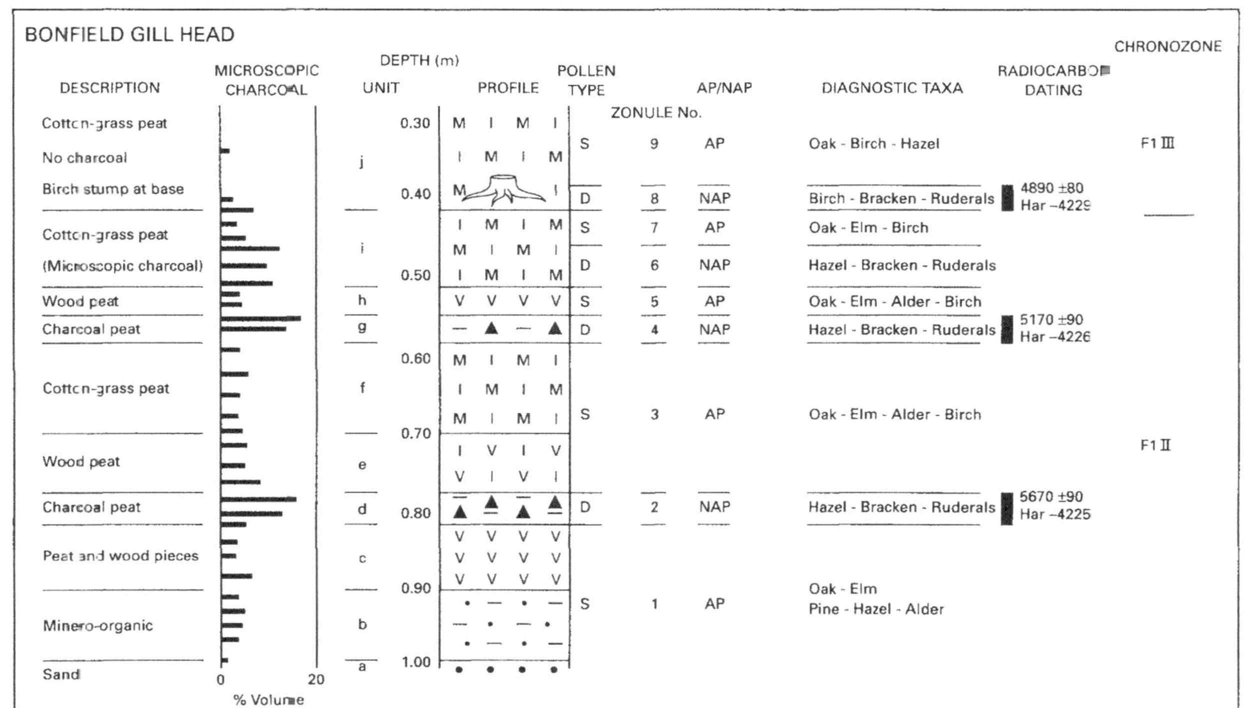

Notes: D = phases of human disturbance; S = phases of stability; AP = arboreal pollen; NAP = non-arboreal pollen. Stratigraphic symbols as in Figure B2.40

On the North Yorkshire Moors, where there is a high concentration of Mesolithic archaeological finds, there is abundant evidence of woodland disturbance (Innes and Simmons, 1988; Simmons, 1993b). Two examples are shown in Figures B2.4a and B2.4b. A series of disturbance phases occur during the Later Mesolithic at North Gill, where there is a clear association between lower tree values, higher charcoal values and the occurrence of light demanding herbs. Figure B2.4b shows a very similar pattern at Bonfield Gill Head. The frequency of fire at numerous profiles at North Gill and Bonfield Gill Head increases the possibility that humans were using fire to open up mesic and oak-dominated woodlands, especially as the return period for lightning fire in European deciduous forests is estimated to be every 6000 years (Simmons and Innes, 1996a, b).

At Farm Fields, Kinloch, Rhum provides evidence of the earliest known Mesolithic occupation site in Scotland. Here, carbonised hazelnut shells were radiocarbon-dated to 8590±95 years BP. Pollen data from a site located 300m away reveal sharp and sustained falls in alder and hazel pollen, an increase in grass pollen and a peak in charcoal (Hirons and Edwards, 1990). Once the disturbance ends woodland regenerates to its former level, indicating that the impact of Mesolithic people on woodland was minimal. This pattern of woodland manipulation is typical of upland sites of Mesolithic age on Dartmoor, the North Yorkshire Moors and lowland areas, most notably Starr Carr, Seamer Carr and Flixton in the Vale of Pickering, the Southern Pennines and Scotland (Caseldine and Hatton, 1993; Innes and Simmons, 1988; Simmons, 1996; Edwards and Whittington, 1997), but Mesolithic people were capable of permanently altering their physical environment (see text).

There is evidence to suggest that the Mesolithic hunter-gatherers could permanently alter their physical environment, albeit on a local scale. On Dartmoor at higher altitudes close to the treeline, Mesolithic use of fire was sufficiently severe to prevent the re-establishment of woodland. Soils in newly formed open areas were exposed to rainfall and higher rates of leaching promoted soil acidification, peat formation and, in more waterlogged areas, the formation of marsh and mire communities (Moore, 1988; 1993). Caseldine and Hatton (1993) suggest that Mesolithic societies used fire to sustain heathland for grazing on upland Dartmoor, leading to irreversible changes. Here woodland was transformed into blanket peat via a phase of acidic grassland over a period of 600 to 1000 years, suggesting that the successional pathways of vegetational regression were varied in response to human disturbance. Simmons (1990) argues that the action of hunter-gatherers shaped the contemporary upland Dartmoor landscape. Elsewhere in Britain vegetational communities appear to have persisted as a result of hunter-gatherer activities. Two examples are the maintainance of open birch woodland at Waun-Fignen-Felen in upland south Wales (Smith and Cloutman, 1988) and the preservation of a species-rich chalk grassland in the Wolds (Bush, 1988). This kind of evidence suggests that hunter-gatherers were responsible for localised, permanent transformations to the physical environment in marginal areas.

2.2.3 The advent of agriculture and Neolithic societies

The transition from hunter-gathering to agriculture is often seen as an event of major significance in the history of society-nature relations. This agricultural revolution made it possible for the feeding of larger populations, led to the development of sedentary, permanent settlements from which arose complex societies and the emergence of civilisation, and brought about changes in the way humans exploited their environment (Bell and Walker, 1992). This new form of exploitation is seen by Simmons (1989) to have allowed societies to capture more energy and also resulted in some different environmental impacts. The farming of domesticated plants and animals meant that humans deliberately began to select certain species as a food source, which unintentionally created genetically and morphologically new strains of plants and animals. In western Europe the conversion of forested areas created permanently open areas for agriculture, changed woodland structure, led to the expansion of ruderal plant communities and a change in soil properties. Before the environmental impacts of farming are discussed it is worth considering when, where and why agriculture began.

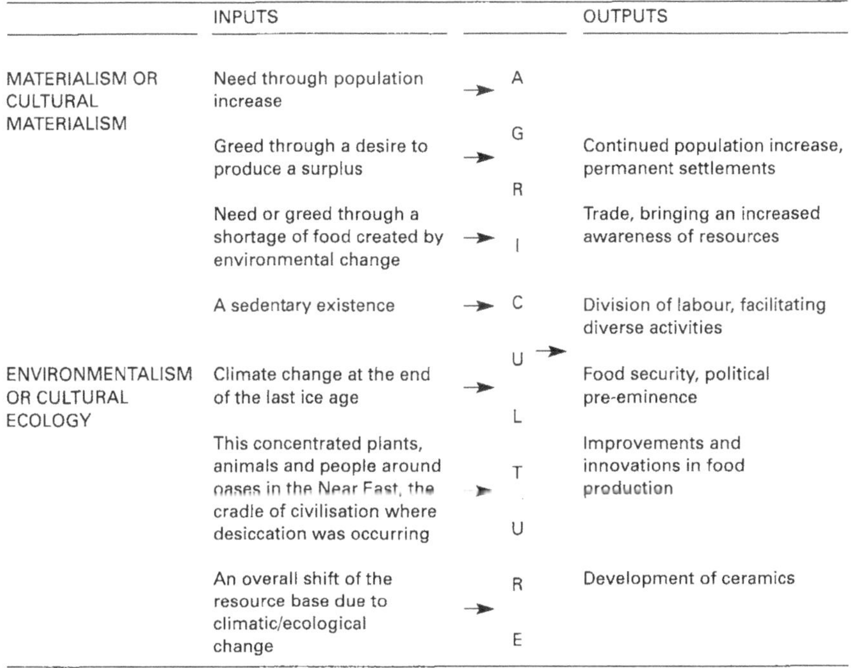

Figure 2.1 Two possible models for the emergence of agriculture (Source: Mannion, 1997)

The reasons why agriculture developed into the dominant method of food procurement is a subject of continuous and as yet unresolved debate (Mannion, 1999). Two possible models have been proposed to explain the emergence of agriculture (see Figure

2.1). One model proposes that environmental change provided a major stimulus for the inception of agriculture and this model can be considered a type of environmental determinism (see Chapter 1). This model suggests that climate change created uncertain ecological conditions that forced hunter-gatherer communities to domesticate plants and animals to increase their control on food supplies. McCorriston and Hole (1991), Moore and Hillman (1992) and Byrd (1994) provide evidence that climatic changes did occur just prior to the adoption of agriculture in the Near East around 10,000 years ago. The transition from the last glacial into the Holocene epoch is characterised by a period of very pronounced climatic instability and rapid oscillations in temperature. Between 11,000 and 10,000 years BP a cold, dry climate prevailed during a period known as the Younger Dryas (Bell and Walker, 1992) (Figure B2.3a) and it is argued that these conditions placed existing hunter-gathering methods under stress by reducing the availability of food and forcing people to turn to cultivation (Sherratt, 1997). Sherratt also suggests that climatic instability was a crucial factor in the beginning of agriculture in western Asia, and a decrease in monsoonal rainfall may have helped initiate rice farming in eastern Asia, but there is no evidence to suggest that a comparable climatic trigger initiated maize cultivation in America.

Cultural stimuli are considered to be more important by other archaeologists such as Vavilov (1992). Supporters of this materialism model argue that population growth and/or the desire to produce a surplus, reliable food supply, coupled with a shift to a sedentary existence, resulted in the adoption of farming. Smith (1997) argues that this gradual process of social and economic reordering meant that sedentary farming of domesticated plants and animals became a more conducive method of food procurement compared with hunting and gathering. Mannion (1995) notes that a 'Boserup-type', (see Box 3.2) population-led, materialist approach was possibly essential if human population became sedentary and caused a depletion in food resources. Hodder (1990) offers an alternative view, arguing that domestication was not driven solely by need but was a social and symbolic process by which the natural world was brought under the control of a social and cultural system. Farmers could also store surplus food obtained from harvests to minimise the risk of lean years and may have increased their social standing by making gifts of excess food. Mannion (1995) suggests, however, that there is little supporting evidence for hunter-gathering societies needing to convert to agriculture because of population growth. She cites the work of Roosevelt (1984), Wright (1994) and Molleson (1994), who looked at skeletal remains of Palaeolithic people that showed few signs of suffering from malnutrition or stress compared to Neolithic specimens, indicating that farming placed more strain on people. They all suggest that people would only adopt agriculture if wild sources became scarce. These ideas generally lend themselves to the cultural/social deterministic views outlined in Chapter 1 where society is beginning to separate itself from, and dominate, nature. Farming appears to represent a more structuralist, dualistic view as people had the capacity for free will thinking and the ability for planned, long-term action that frequently impacted on the environment. In fact it is probable that elements of both models contributed to the adoption of agriculture.

Compared with the question why, there is much greater certainity about when and where agriculture first commenced (see Box 2.5). Once farming was adopted in these centres it expanded into other areas. The proposed models of spread are discussed in the next section, with particular reference to north-west Europe.

Box 2.5 Where and when? Centres of domestication

The excavation of archaeological sites and the radiocarbon-dating of plant and animal remains has allowed archaeologists to reconstruct when and where agriculture first started.

Vavilov, Smith and Mannion suggest that each centre of domestication has some or all of the following characteristics:

(1) They were determined by the distribution of wild progenitors of plants and animals used in agropastoral farming. For example in the Near East, believed to be the earliest centre of domestication, there is a good correspondence between the location of wild and domesticated species. Between 10,000 and 8000 years BP sheep, goats, pigs and cattle and at least eight plant species (lentils, peas, chickpeas, bitter vetch, einkorn and emmer wheat, flax and barley) were domesticated in the Near East, an area where wild plants and animals were abundant.

(2) Centres of domestication were located in regions with high biological diversity.

(3) They had favourable environmental conditions, such as an abundance of dependable water sources and easily tilled, fertile soils.

(4) The location of cultivation may have been culturally determined as hunter-gatherers harvested these grasses prior to their domestication. These early forms of proto-domestication may have played an influential role in developing agricultural methods.

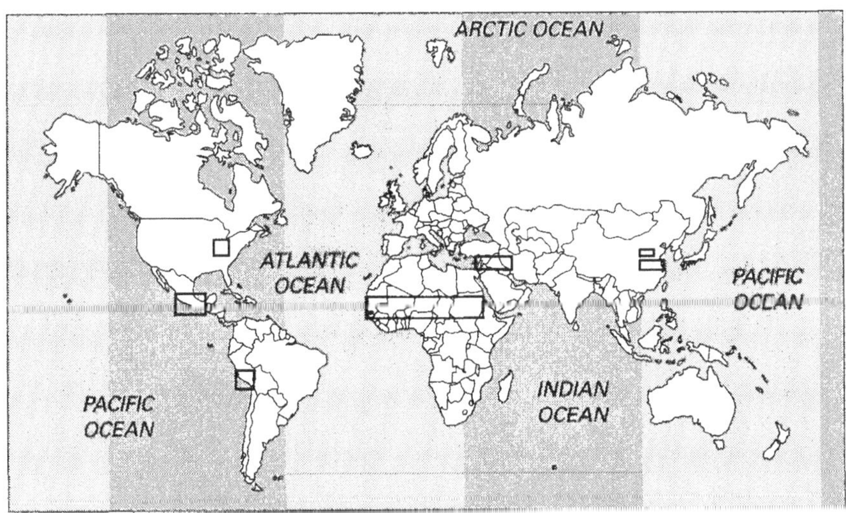

Figure B2.5a Centres of animal and plant domestication (Source: Smith, 1995)

Several people have proposed the location of centres of domestication (see

Mannion, 1998) and there is a good degree of correspondence between each proposal. Thus, one scheme by Smith (1995) is presented in Figure B2.5a which shows seven centres of domestication. Details of each centre can be obtained from various sources including Evans (1993), Harris (1996), Smith (1995) and Mannion (1997). While this scheme and the general details of the type of plants and animals that were domesticated at each centre are now known, it is pertinent to note that not all aspects of the pattern of domestication are fully understood. However a list of the main domesticates and when they are thought to have been first domesticated is provided in Table B2.5 and Figure B2.5b. It is noteworthy that the timing of domestication varies between each centre (Figure B2.5c) and at each centre societies followed different pathways through the transition from hunter-gatherer to farming, taking from around 6000 years in Mexico to 3000 years in the Near East (Smith, 1998). The time taken to domesticate plants and/or animals from their wild ancestors also varies. For example, recent AMS dating of macrofossils found in caves in Mexico suggest that squash was domesticated between 9000 and 7000 14_c years BP, predating the use of maize, beans and bottle gourd by 4000 years (Smith, 1997). However, archaeological evidence suggests that it took a further 6000 years before village-based farming economies began to make a substantial dietary contribution. At Cerro Juanaquena in Mexico, AMS dates on maize suggest that it arrived between 3500 and 3000 years ago, and domesticates were rapidly adopted across a broad area in the south-western United States. In eastern North America evidence for the initial use of domesticated squash around 5000 years ago is found at Phillips Spring site in Missouri, but the time span between this initial stage of domestication and the existence of farming is approximately 4000 years (Smith, 1998).

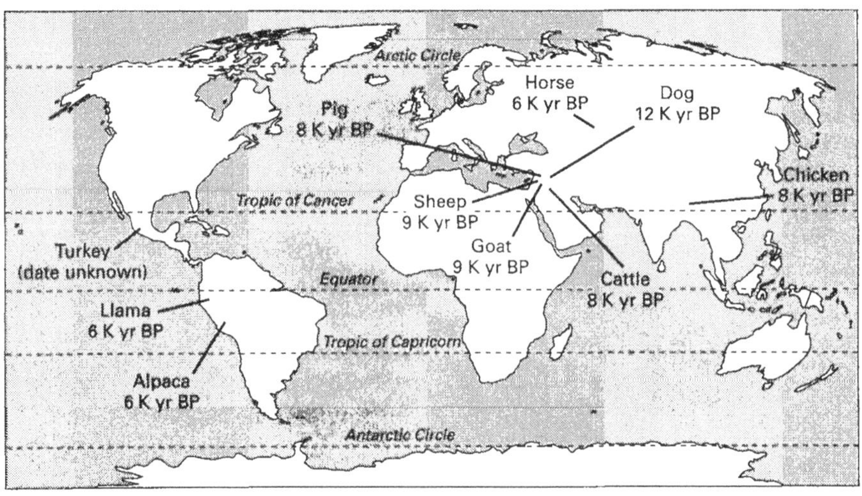

Figure B2.5b Places of origin and approximate dates of the domestication of common animals (Source: Smith, 1995)

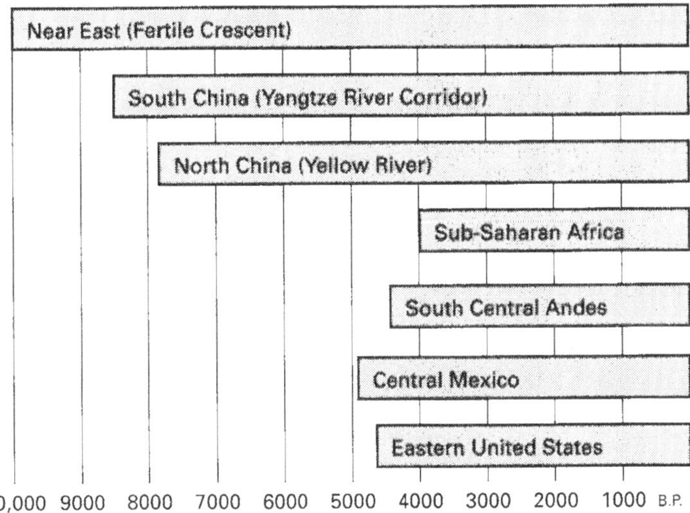

Figure B2.5c The approximate time period when plants and animals were first domesticated in the seven primary centres of agricultural development. (Source: Smith, 1995)

Several other issues remain unresolved. For example:

(1) The role of some animals is not fully known. The dog was the first animal to be domesticated but where, when and what from remains controversial. Dogs were probably used to hunt, guard and herd, as there is no evidence that dogs were butchered (Mannion, 1998).

Archaeologists are also uncertain as to whether certain animals were independently domesticated at different centres.

(2) Because of its extended range throughout Asia and Europe, the wild pig is believed to have been domesticated independently in the Near East and China.

(3) Not all the species hunted became domesticated, so the status of some animals is unclear. One example is the mountain gazelle. Zooarchaeological remains attest to the widespread hunting of gazelles for meat in the Near East. A close examination of bones suggests that diminution and increased variation occurred. Cope (1991) suggests that these morphological changes were the consequence of intensive sex-culling and suggests these cultural control practices were a form of 'proto-domestication'. A re-examination of the data by Dayan and Simberloff (1995) did not reveal any clear patterns and the issue of proto-domestication remains inconclusive.

(4) In other regions the timing of domestication is also unclear. In Sub Saharan Africa, African rice originated in the west savannah zone and is one of many important species including teff, finger and pearl millet, sorghum, cola nuts, coffee, yams and water-melon (Harlan, 1994), but the dating of the origins of these plants is uncertain (Mannion, 1997). Despite these uncertainties it is clear that the adoption of agriculture allowed society to manipulate their environment more intensively once it spread from each centre to peripheral areas (see text).

Table B2.5 Some of the world's most important crop plants and their approximate date of origin

Crop	Common name	Approximate date (K years BP)
A. Near East		
Avena sativa	oats	9.0
Hordeum vulgare	barley	10.2
Secale cereale	rye	9.0
Triticum aestivum	bread wheat	7.8
T. dicoccum	emmer wheat	9.5
T. monococcum	einkorn wheat	9.5
Lens esculenta	lentil	9.5
Vicia faba	broadbean	8.5
Olea europea	olive	7.0
Cannabis sativa	hemp	9.5
B. Africa		
Sorghum bicolor	sorghum	8.0
Eleusine coracana	finger millet	?
Oryza glaberrima	African rice	?
Vigna linguiculata	cowpea	3.4
Dioscorea cayenensis	yam	10.0
Coffea arabica	coffee	?
C. Far East		
Oryza sativa	rice	9.0
Glycine max	soybean	3.0
Juglans regia	walnut	?
Catanea henryi	Chinese chestnut	?
D. South-east Asia and Pacific Islands		
Panicum miliare	slender millet	?
Cajanus cajan	pigeonpea	?
Colocasia esculenta	taro	9.0
Cocos nucifera	coconut	5.0
Mangifera indica	mango	9.2
E. The Americas		
Zea mays	maize	5.0
Phaseolus lunatus	Lima bean	5.0?
Manihot esculenta	cassava	4.5
Ipomea batatus	sweet potato	4.5
Solanum tuberosum	potato	5.0
Capsicum annuum	pepper	8.5
Cucurbita spp.	various squashes	10.7?
Gossypium spp.	cotton	5.5

(Source: Mannion, 1997)

Core to periphery: the spread and transition to farming in north-west Europe

Two schools of thought pervade to explain the adoption of agriculture. The first, known as the 'diffusionist model', suggests that north-west Europe was colonised by farmers who displaced and absorbed hunter-gatherer communities. These early farmers are thought to have migrated from the 'core area of development' in the Near East. In this model local Mesolithic societies played little role in the adoption of agriculture and are viewed more as passive recipients of this innovation in response to social competition, a decline in wild resources or population growth (Zvelebil and Dulukhanov, 1991). Several variants of this model have been proposed to explain the adoption of agriculture in Europe, including the wave-of-advance model (see Ammerman and Cavalli-Sforza, 1984) and one of demic diffusion where farmers colonised areas that offered suitable resources to farm the land successfully (see Van Andel and Runnels, 1995).

Others propose an alternative model, where local Mesolithic societies play an active role in the transition to farming, and consider the possibility that early forms of domestication occurred within Europe (Dennell, 1983; Zvelebil and Rowley-Conwy, 1984; Barker, 1985). In this view, the apparent paucity of evidence for Mesolithic settlement in peripheral areas of Europe is considered to be misleading. Wild progenitors of plants and animals that were eventually domesticated were exploited by Mesolithic societies, and this has been construed as a type of husbandary. Agriculture therefore evolved in peripheral regions by combining the knowledge of local complex hunter-gatherer communities with Neolithic farming colonists, and this process varied from region to region according to farmer–forager interaction (Zvelebil and Dulukhanov, 1991) (Figure 2.2).

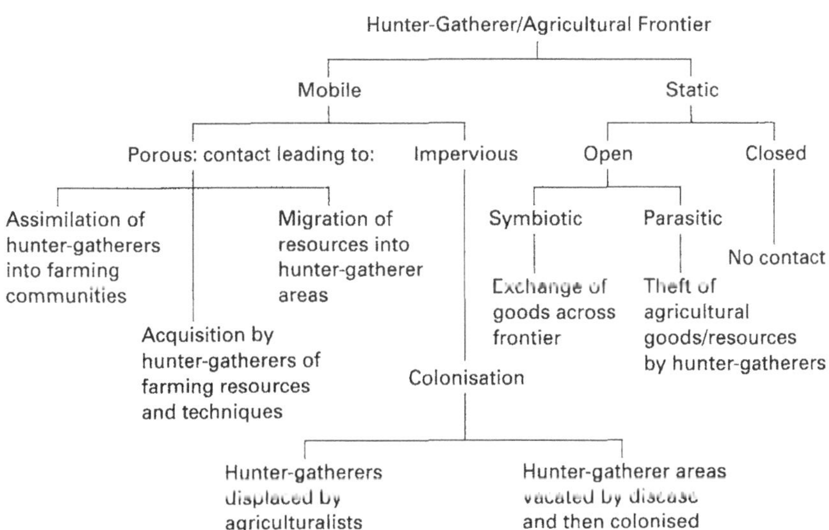

Figure 2.2 Possible types of farmer–forager interaction (Source: Zvelebil and Dulukhanov, 1991)

Zvelebil and Rowley-Conwy (1984) propose a three-stage model to explain how local Mesolithic societies gradually reverted to sedentary farming, which incorporates the possible variation in contact and exchange between farmers and hunter-gatherers:

(1) **The phase of availability**
Exchange of materials and knowledge occurs between farming and hunter-gatherer cultures but the latter do not utilise this form of food procurement. The phase ends when indigenous foragers begin to practise some elements of farming or farmers colonise an area. This phase could take several hundred years.

(2) **The phase of substitution**
This phase can occur in two forms. The arrival of sedentary farmers into an area who compete with foragers, or the foragers adopt farming and the two methods of food procurement operate simultaneously. Eventually foraging becomes a marginal or insignificant part of the economy.

(3) **The phase of consolidation**
expands into new areas and uses more intensive practices. The graduation to consolidation is characterised by the wealth of domesticated animal bones and cereals found at archaeological sites dated to the Neolithic.

Eventually an agricultural economy gradually developed between 9000 and 5000 years BP in north-west Europe (Figure 2.3).

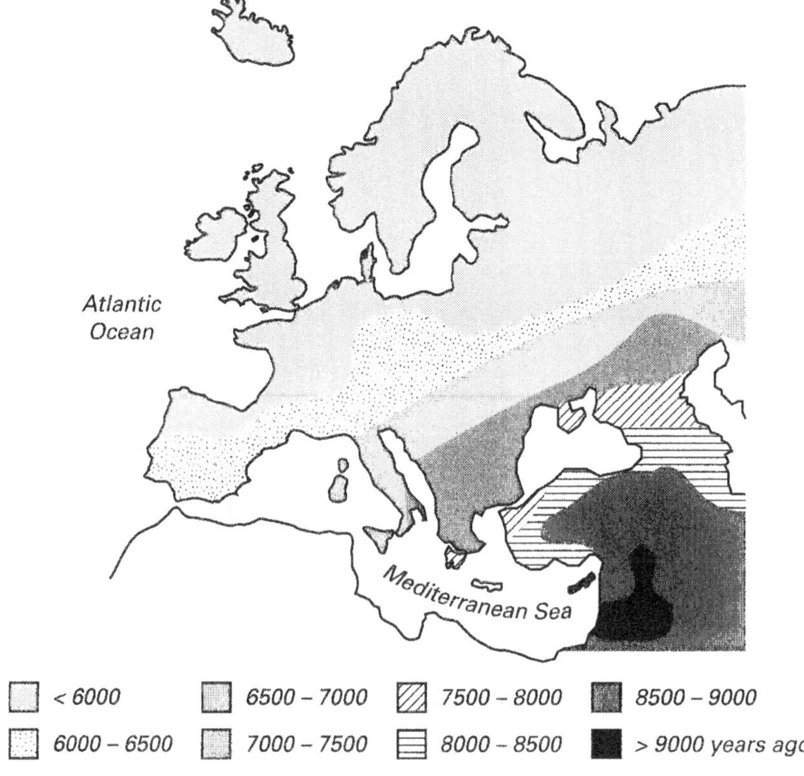

Figure 2.3 The spread of agriculture in Europe as indicated by radiocarbon dating of Neolithic agricultural sites (Source: Mannion, 1997)

Agriculture spread along a south-east–north-west axis starting in the southern fringes of Europe between 9000 and 7000 years BP. Agriculture was practised on the fertile soils of the plains of Thessaly by 9000 years BP (Van Andel and Runnels, 1995), under way in Macedonia, along the Dalmatian coast and on the Tavoliere plains of Italy by 7000 BP (Smith, 1995). Archaeological evidence from south-east and central Europe supports the diffusionist model for the adoption of farming. In Greece and areas of the south-east Balkans Neolithic settlements are concentrated in areas devoid of Mesolithic populations, suggesting that people moved into these uninhabited areas (Van Andel and Runnels, 1995). Local, indigenous Mesolithic societies played a more prominent role elsewhere in Europe (Zvelebil and Dulukhanov, 1991). Farming spread into central and western parts of Europe by 6500 to 6000 BP as far east as Russia, into the loess areas of northern Germany and further west into northern France (Smith, 1995). In northern and eastern Europe the transition to farming was a gradual process, possibly the result of deteriorating climatic conditions and progression into less favourable areas. It started between c. 6500 and 6000 BP and reached northern parts of Russia by 1500 BP, where the process was never completed. Moving north, shorter growing seasons, lower temperatures, higher precipitation and poor soils made farming more difficult. However, local hunter-gatherer communities, which may have begun to practise some form of animal husbandry, appear to have adopted exogenous plants to cultivate and domesticated animals to form a mixed hunting-farming economy that cannot be described as either Mesolithic or Neolithic in nature. This may explain why the transition was time transgressive and the longevity of each phase varies. Southern Scandinavia is characterised by a long availability phase and a short substitution phase, whereas north-east Europe is characterised by a long substitution phase. It is argued that the availability of abundant food sources delayed domestication but the loss of a single resource, if seasonally crucial, could place a foraging/hunting economy under stress. When under stress the foraging economy would be largely abandoned and replaced by farming – a more attractive alternative. Once in place the effects of farming would disrupt hunter-gathering economies sufficiently to persuade people to rely on agriculture to procure food. Greater social interaction between farming and hunter-gatherer communities may have led to a change in the social structure of hunter-gatherer cultures that was more compatible with agropastoral farming (Zvelebil and Rowley-Conwy, 1984; Zvelebil and Dulukhanov, 1991).

There is some debate as to when farming actually arrived in the British Isles. Traditionally, the mid-Holocene elm decline is dated to *c.* 5000 years BP (or *c.* 3200–3000 years BC) and was thought to mark the advent of agriculture. However, cereal pollen grains have been identified in pre-elm decline deposits, for example at Soyland Moor, central Pennines, dated to 5820±95 years BP (Williams, 1985) and at Cashelkeelty, Co. Cork, where a fall in tree pollen coincides with the occurrence of cereal pollen grains dated between 5845±100 years BP and *c.* 5370 years BP (Lynch, 1981), suggesting that agriculture may have been started much earlier by either indigenous Mesolithic societies or immigrant Neolithic agriculturists (Groenman-van Waateringe, 1983; Edwards and Hirons, 1984; Edwards, 1989). These occurrences may mark the beginning of the transition to agriculture or the 'availability phase', but problems with identifying cereal pollen grains accurately mean that these results and the interpretation made from them must be treated with caution (Edwards, 1988; Monk, 1993).

Once adopted Neolithic societies developed a number of different methods of farming and these resulted in several forms of environmental impact that intensified during the prehistoric period. The evidence to support this argument is the subject of the next section.

2.2.4 Impact of agriculture from Neolithic to the Iron Age in north-west Europe: an intensification of human impact?

Plant and animal domestication

Plant and animal domestication represents an important change in the relations between humans and nature because societies began to control parts of their environments, namely plants and animals. This control ultimately led to biological and genetic modification of the species that were exploited for food. Plants were controlled by planting them in specially prepared seedbeds and animals were controlled by herding and placing them in enclosures.

Smith (1995, p. 18) defines domestication as 'the creation of a new form of plant and animal – one that is identifiably different from its wild ancestors'. It is widely believed that human manipulation of these species resulted in morphological changes. Bones recovered from archaeological sites reveal that females dominated the domesticated population. Size diminution of all or part of an animal is also a common feature. For example the width of the metatarsal bone in goats and teeth size and snout length in pigs are smaller in domesticated animals. Archaeologists interpret the reduction in size as evidence of herding. The shape and size of grain (larger, plumper) and structure of the rachis are noticeable changes in wheat, while the domestication of squash in MesoAmerica is characterised by a change in fruit morphology, rind thickness, increased peduncle size and colour (Smith, 1997). These kind of morphological changes led to identifiable domesticated plants and Hillman and Davies (1990) suggest that these changes could occur within approximately 30 years in ideal conditions or up to several hundred years.

Smith (1995) suggests there are several core attributes of the animals selected for domestication, and they are outlined in Table 2.1. These attributes allowed humans to constrain the movement of animal populations, regulate their breeding and control their feeding, which in turn created the conditions for genetic and morphological changes. One example is sheep, which are submissive and behave in a herd structure combining both males and females. The isolation of selected plants from wild plants occurred with the deliberate planting of seeds in a prepared seedbed, creating distinctive subspecies with characteristics that are favourable to farming. These characteristics include the retention of seeds that are easy to harvest, rapid sprouting and thinner seed coats for quick germination and growth.

Table 2.1 Criteria for animal domestication

Criteria
Hardy: animals adjust to new conditions such as disease
In-born liking for humans
Comfort loving
Found useful to hunter-gatherers, eg already a good food source
Breed freely
Easy to feed

(Source: Smith, 1995)

Environmental impacts of prehistoric agriculture

Behre (1988) suggests that the arrival of agriculture marked a major shift in the way humans interacted with their environment, moving away from passive adaptation to active encroachment. Simmons (1991) argues that farming provided new food producing techniques to produce more surplus energy, so higher densities of humans could be supported on suitable land. Farming allowed human population to grow more rapidly when compared to earlier hunter-gatherer societies. This eventual transition from shifting forms of agriculture into permanent systems (i.e. Neolithic to Bronze Age in north-west Europe) also saw an intensification of the impact of agriculture on the environment. Technology such as the plough improved the ability of humans to exploit the land through agriculture. In north-west Europe forest had to be converted into open land and the plough disturbed the upper layers of the soil, which made them more susceptible to erosion. The scale of impacts appears to be regionally variable but by Bronze Age to Roman times there is evidence of these impacts becoming widespread. This is reflected in the increasing amount of woodland clearance and increasing rates of soil erosion that occurred from Neolithic through to Roman times.

The discovery and identification of pollen grains released from the anthers of cereals, grasses and weeds commonly associated with agricultural activity in the palaeoenvironmental record has shown that the introduction of arable and pastoral farming into the British Isles and Europe made a significant contribution to shaping the landscape by clearing large tracts of woodland (Behre, 1981). The need to create open space for arable agriculture made woodland clearance necessary. However, the conversion of forest to agricultural land was not a rapid process, as it is widely believed that the nature of Neolithic agriculture was largely akin to shifting cultivation ('landnam phases'), which allowed woodland to regenerate (Simmons, 1990a), although four models of early farming in Europe have been proposed (see Box 2.6).

Box 2.6 Models of prehistoric farming

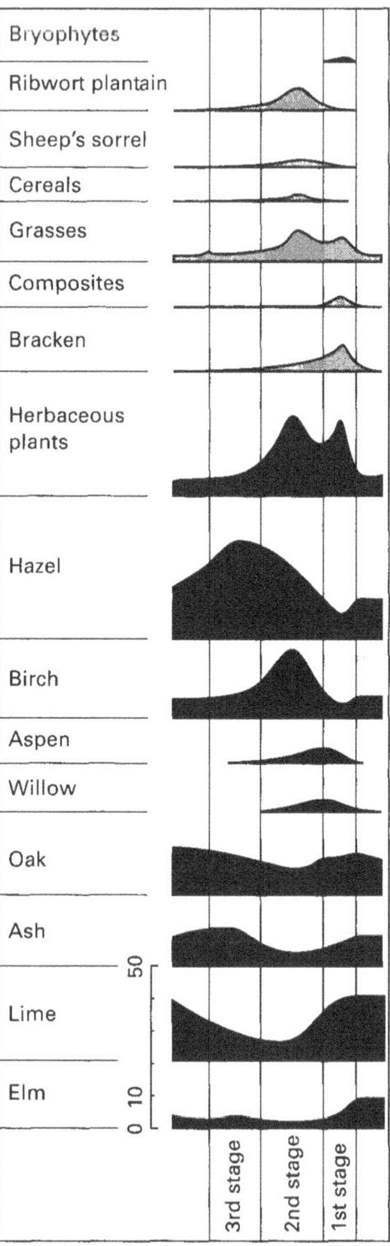

Figure B2.6a Generalised pollen diagram showing the three stages of a landnam phase (after Iversen, 1973) (Source: Bell and Walker, 1992)

Four models have been proposed to describe farming during the Neolithic period of north-west Europe.

(1) **The 'landnam' model**

First described by Iversen (1941), the landnam model presupposes that small areas of forest were cleared for agricultural purposes as part of a swidden or slash and burn system. Three stages can be identified (Figure B2.6a). First was the clearance of forest and a brief peak in herbaceous plants. Stage two is characterised by a decrease in lime, a peak in birch and the rise of grasses, ribwort plantain and cereals. Stage three sees the recovery of woodland and the expansion of hazel as the plot is abandoned. Thus a 'landnam' is typified by episodes of grazing and some cereal cultivation that would last for a few seasons, while three stages would cover approximately 50 years. However, radiocarbon-dating evidence of landnam episodes identified in pollen diagrams suggests they lasted much longer, possibly several centuries, and might be an aggregation of several clearances or the extenuation of clearings by post occupational grazing (Edwards, 1993). Slash and burn systems are thought to have been operational in parts of north-west Europe (British Isles, Scandinavia) and North America during prehistory (Bell and Walker, 1992).

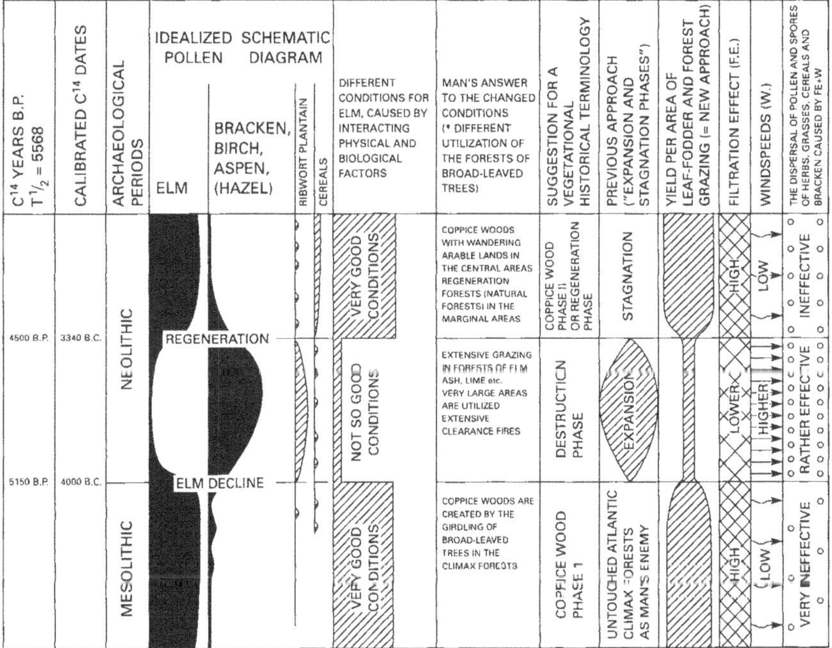

Figure B2.6b The forest utilisation model (after Göransson, 1983; 1986) (Source: Edwards, 1993)

(2) **The 'leaf foddering' model**
This model, proposed by Troels-Smith (1960), describes an economy based on stall feeding of cattle with leaves from pollarded trees. Elm and ash were the most valued leaves followed by hazel, lime and oak. Small plots were cultivated for barley and wheat. Troels-Smith (1960) suggests it preceded landnam episodes in Scandinavia and the excavation of a Neolithic settlement at Weier in Switzerland revealed evidence for leaf foddering (Robinson and Rasmussen, 1989).

(3) **The 'expansion–regression' model**
This model describes expansion phases as periods of forest disturbance and cultivation (characterised in pollen diagrams by suppressed tree pollen values and higher herbaceous pollen values). Berglund (1985; 1986) suggests that agriculture was largely restricted to expansion phases, whereas regression phases are characterised by stable forest cover and human impact is limited to grazing and coppicing. This model incorporates landnam episodes as expansion phases.

(4) **The 'forestutilisation' model**
A model proposed by Göransson (1986), which embraces all aspects of forest farming. Girdling (ring barking) of trees created patches of coppiced forest that provided browse for wild animals and winter fodder. Burning beneath girdled trees could form small plots for cereal cultivation. Thus forest clearance was not required to procure food. This form of food procurement and forest utilisation was interrupted by destruction phases, one of which coincided with the mid-Holocene elm decline (Figure B2.6b), opening up the landscape. When woodland eventually recovers, a coppice woodland could be exploited.

Neolithic farming activities resulted in the clearance of woodland, expansion of ruderal communities and blanket peat in Britain and Ireland. One example of Neolithic farming practice on the physical environment is the development of a field system covering around 1000 ha at Ceide Fields in Co. Mayo, western Ireland. Woodland appears to have been cleared primarily for pastoral agriculture with a farming economy based around cattle farming, but the occurrence of cereal pollen grains suggests that a limited amount of arable farming took place. Molloy and O'Connell (1995) suggest that landnam-type agriculture was practised between 4900 and 4500 years BP. This is characterised by a reduction of tree and shrub pollen and the increased representation of pollen and spores released from plants commonly associated with agriculture, including grasses, ribwort plantain, docks, hawkweeds and dandelions, buttercups, clovers and bracken. Once the site was abandoned there is evidence that woodland regenerated and therefore the impact of Neolithic farming was non-permanent. Successive woodland clearance and regeneration phases seem to have been common around the British Isles and Europe during the Neolithic but this is not always the case.

Neolithic farming activities could permanently alter their vegetational landscape. For example in the chalk area of southern Britain there is evidence for permanent woodland clearance at Winnal Moors, Winchester, as Neolithic peoples practised agriculture, and Bunting (1994) argues that woodland was cleared by Neolithic people on Orkney. Thus the impact of Neolithic peoples on their physical environment was variable and it is not until the Bronze Age or later that there is widespread evidence for permanent woodland clearance even in areas occupied by Neolithic farmers. For example pollen records suggest that the introduction of farming in Thessaly, Greece by Neolithic people did not cause widespread woodland clearance, which commenced c. 4000 years ago at the start of the Bronze Age (Willis and Bennett, 1994) when extensive exploitation of terrace and foothills occurred (Van Andel and Runnels, 1995). In other parts of the chalk area of southern Britain permanent woodland loss does not take effect until the Bronze Age or later (Thorley, 1981; Waton, 1982). Here, population pressure on the land's resources appears to have been an important factor contributing to deforestation and deterioration in soils across the area.

So there is good evidence to suggest that agriculture, as a dominant human form of exploitation, combined with a continued increase in human population, appears to have brought about permanent changes to woodland in the second half of the Holocene (Roberts, 1989). Eventually as areas were cleared of woodland Neolithic farming practices described in Box 2.6 must have been abandoned as people became permanently settled in an open landscape. One example of this is the development of field systems (known as the 'reaves' on Dartmoor) and small permanent communities during the Bronze Age on Dartmoor (Fleming, 1988).

Although permanent woodland clearance was widespread during the Bronze Age, Iron Age and Roman period in the British Isles, it was temporarily and spatially variable. Radiocarbon-dating suggests that differents areas became deforested at different times during these periods. For example extensive woodland clearance occurred in many areas of southern and central Scotland and areas of England during the Iron Age (Dumayne-Peaty, 1998). In northern England close to Hadrian's Wall and around the Antonine Wall in southern Scotland woodland removal commenced during the late Iron Age. Pollen diagrams for this region are characterised by small temporary clearances during the late Iron Age associated with an increase in indicators of agricultural activity, followed by extensive, permanent woodland removal during Roman times. Land was cleared of woodland by the Romans for agriculture, the construction of military garrisons and charcoal for metal smelting. Only a limited amount of woodland recovery took place after the abandonment of this frontier zone by the Romans, meaning that the landscape of northern England and southern Scotland remained open (Dumayne, 1993; Dumayne and Barber, 1994).

The examples presented above do support the idea that the emergence of new forms of prehistoric societies and increasing human population had an irreversible impact on woodlands. However, the scale of this impact must have been regional because Behre (1988) states that Europe was still covered by large areas of woodland by the onset of the Middle Ages (*c.* the eighth century).

So far the discussion has argued that the transition from hunter-gathering to farming enabled people to intensify their exploitation of the natural world by converting woodland into agricultural land. However, as noted in the introduction, the transformation of woodland into agricultural land also had other, less visual impacts. For example agriculture had dramatic effects on the distribution and presence of individual species of plants and animals. Between 5200 and 5000 years BP pollen diagrams across north-west Europe record a fall in elm pollen (Behre, 1988). Several hypotheses have been advanced to account for the loss of elm during the mid-Holocene with human activity, climate change and disease possibly being the most favoured (Ten Hove, 1968; Birks and Peglar, 1993; Birks, 1993). The discovery of *Scolytus scolytus*, remains of a beetle that carries a fungi *Ceratocytsis ulmi* that causes a disease similar to Dutch elm disease, in pre-elm decline deposits at Hampstead Heath in London provides evidence that a disease could have decimated elm populations (Girling, 1988). Woodland clearance by Neolithic people may have inadvertently created conditions conducive to the reproduction and spread of the beetle. Lime also declines with the arrival of agriculture. Turner (1962) suggests that lime occupied dry and fertile soils that were favoured by agriculturists and that its bark and leaves were exploited for leaf fodder (Behre, 1988). One tree that benefited from the clearance of woodland was beech, which migrated into the British Isles around 3000 years ago, colonising areas of abandoned farming land (Birks, 1989).

Animal, insect and molluscan faunas have also been altered with the introduction of agriculture. In particular those species with affinities for open space have benefited from the reduction in woodland cover, as have species associated with domesticated animals and plants. Beetles have been introduced by Norse communities in the North Atlantic Islands (Buckland *et al.*, 1991). Woodland clearance has restricted the distribution of many woodland species including the snail *Spermodea lamellata* (Evans, 1972; Bell and Walker, 1992). Large animals such as wild cattle, beaver, bear, wolf and boar all suffered habitat destruction, resulting in their eventual extinction during historical times (Bell and Walker, 1992).

The theme of an intensification of human exploitation and impact can also be seen when looking at soil erosion rates in prehistory. Since the Neolithic period humans and soil have had a close relationship as humans became dependent upon the soil to produce food (Groenman-van Waateringe and Robinson 1988). Woodland clearance and disturbance of the soil by ploughing made the soil more susceptible to erosion. This is reflected in increasing rates of soil erosion that occurred from Neolithic through to Roman times. Van Vliet-Lanoë *et al.* (1992) produced a graph showing the soil erosion efficiency of various factors in western Europe since 6000 years BP (Figure 2.4). The striking feature of this graph is that erosion efficiency first increases at 2000 BC and then dramatically at AD 800. The two major factors that led to higher rates of soil erosion, are deforestation and agricultural practices until about AD 1700. This graph visibly shows that soil erosion has steadily intensified since the advent of permanent agriculture in open field systems. Evans (1990) and Bell and Walker (1992) support a human-induced cause for increasing rates of soil erosion stating that colluviation in Britain was stimulated by human activity rather than climate change between 5000 and 1000 years BP. Evans (1990) has calculated that up to 2.5m of topsoil has been removed in England and Wales from the Bronze Age onwards, suggesting that prehistoric peoples had the capacity to instigate significant land use changes.

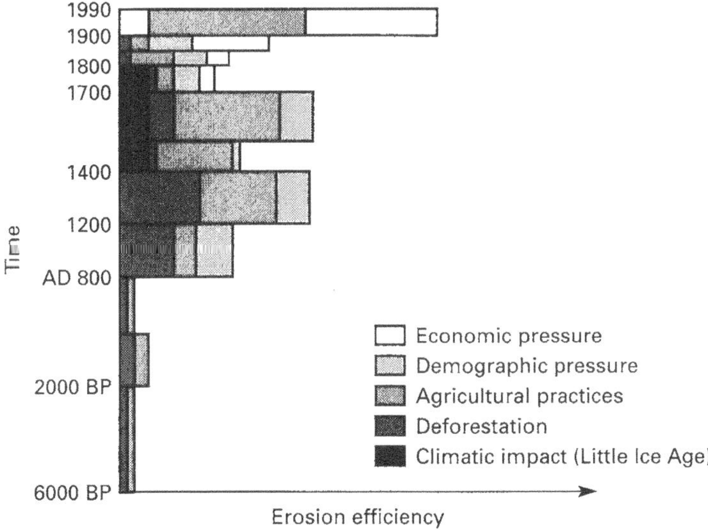

Figure 2.4 Soil erosion efficiency of various factors in western Europe (Source: Van Vliet-Lanoë *et al.*, 1992)

Further supporting evidence comes from the river valleys of the British Isles, northern Europe and the Mediterranean, all of which have yielded evidence of alluviation during prehistoric, Roman and Medieval times (Bell and Boardman, 1990; Brown, 1997), and increased levels of alluviation have occurred simultaneously with the development of extensive arable agricultural systems. In certain cases soil erosion levels in prehistory are similar in magnitude to the present day. For example, Brown and Barber (1985) have argued that the clearance of woodland to create agricultural land by late Bronze Age and early Iron age peoples (2900 to 2300 years BP) increased erosion rates from an estimated early Holocene rate of 20 to 120 tonnes per km^2/yr, a figure close to those of modern times.

However, it must be remembered that land use practices and climate operate together to erode the soil. Foster *et al.* (in press) have argued that agricultural activity creates conditions conducive to soil erosion, but ultimately extreme climatic events such as a storm transport topsoil and subsoil into the valley bottom and these events have occurred since woodland cover was removed (see Boardman, 1991, and Foster *et al.*, 1997 for modern examples).

It was noted earlier that in marginal environments Mesolithic hunter-gatherers could transform a forested environment into blanket peat. Soil acidification and blanket peat inception have often been linked with prehistoric human activity. While it must be remembered that soil acidification and blanket peat formation are natural processes (see Moore, 1988; 1993), both have often been linked with prehistoric human activity. Wiltshire and Moore (1983) suggest that woodland clearance or limited grazing could be sufficient to cross critical thresholds to initiate peat growth (Figure 2.5). There are now numerous examples that show evidence of human activity at the time of blanket peat inception. Radiocarbon-dating suggests that the time of peat initiation varies from site to site, but has also confirmed that prehistoric peoples were responsible for the initiation and spread of blanket peats in many parts of the British Isles (Moore, 1993). For example, blanket peat initiation has been dated to the Neolithic period on Exmoor and in Wales (Merryfield and Moore, 1974; Moore, 1975), and to the Bronze Age in Ireland and Norway (Smith, 1975; Solem, 1989), and human activity also has played a role in mire development in different parts of the world including Canada, Italy and southern Africa (Moore, 1987).

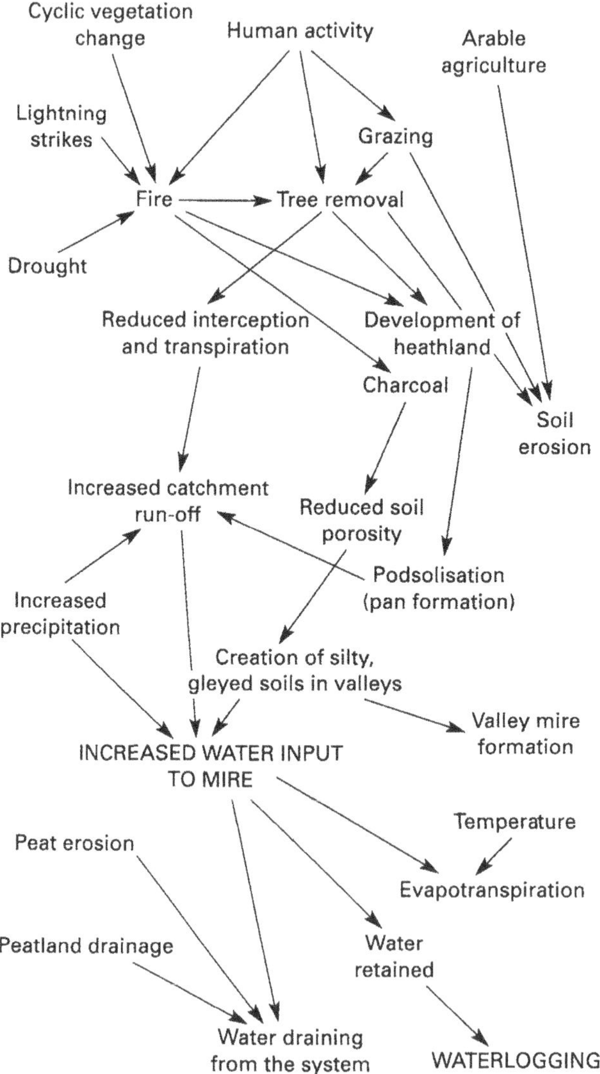

Figure 2.5 Hydrological processes leading to mire formation (Source: Moore, 1993)

Changes in soil properties have been linked with human activity in a similar way. French (1988), for example, argues that clearance of woodland and cultivation initiated a process of soil change including erosion and deposition of wind-blown and colluvial material while undisturbed argillic brown earth soils (characteristic of woodlands) buried underneath Neolithic monuments in the Lower Welland valley, East Anglia remained unchanged. The formation of heaths, dry grasslands and coastal heaths have also been linked with human activity (Kaland, 1986; Behre, 1988; Bell and Walker, 1992). Finally it is worth noting that humans also created soils for agricultural purposes. One example is plaggen soils. Plaggen soils are soils with unnaturally thick Al horizons that were created either to intensify production of winter rye or in periods of intense drought during the tenth century (Groenman-van Waateringe and Robinson, 1988).

As briefly mentioned earlier, the Bronze Age marks the use of metalworking technology and this created a new form of environmental impact, namely pollution. Environmental pollution is often regarded as a product of a modern industrial society but heavy metal pollution began with the use of fire and the first evidence for industrial pollution is simultaneous with the advent of mining and metalworking during the Bronze Age (Craddock, 1995). Mining and smelting played an important role in the economies of early societies (Nriagu, 1996). With the widespread use of metals in everyday life from approximately 4000 years ago has come the release of metals into the environment. Prior to the industrial revolution metal pollution occurred on a local scale (Macklin, 1992), and evidence for localised pollution during the early Bronze Age, *c.* 3500 years ago, has been forthcoming with small peaks in copper corresponding to known prehistoric mining in mid-Wales (Mighall, in press). The Bronze Age was followed by the use of iron during the Iron Age but it is not until Roman times that large amounts of metals were processed. Nriagu (1983; 1996) estimates that the Romans exploited 80,000 to 100,000 metric tonnes of lead, 15,000 tonnes of copper and 10,000 tonnes of zinc with minor use of mercury and tin. While metal mining areas appear to have suffered from local and regional pollution (see, for example, Lee and Tallis, 1973; Martin *et al.*, 1979; West *et al.*, 1997) evidence from the Greenland ice cores suggests that the cumulative effect of mining and smelting from Roman times created hemispheric-scale pollution. Lead and silver have been incorporated in the ice from Greek and Roman mining and smelting (Hong *et al.* 1994; 1996). Lead is present in the ice cores at concentrations four times greater than natural values from approximately 2500 to 1700 years BP (Figure 2.6) and silver records correspond with the discovery of silver mines in central Europe and Spain from around AD 1000. Metal use continued to increase throughout early historical times with the reopening of ancient mines in the eleventh century and development of large furnaces during the sixteenth century culminating with the industrial revolution.

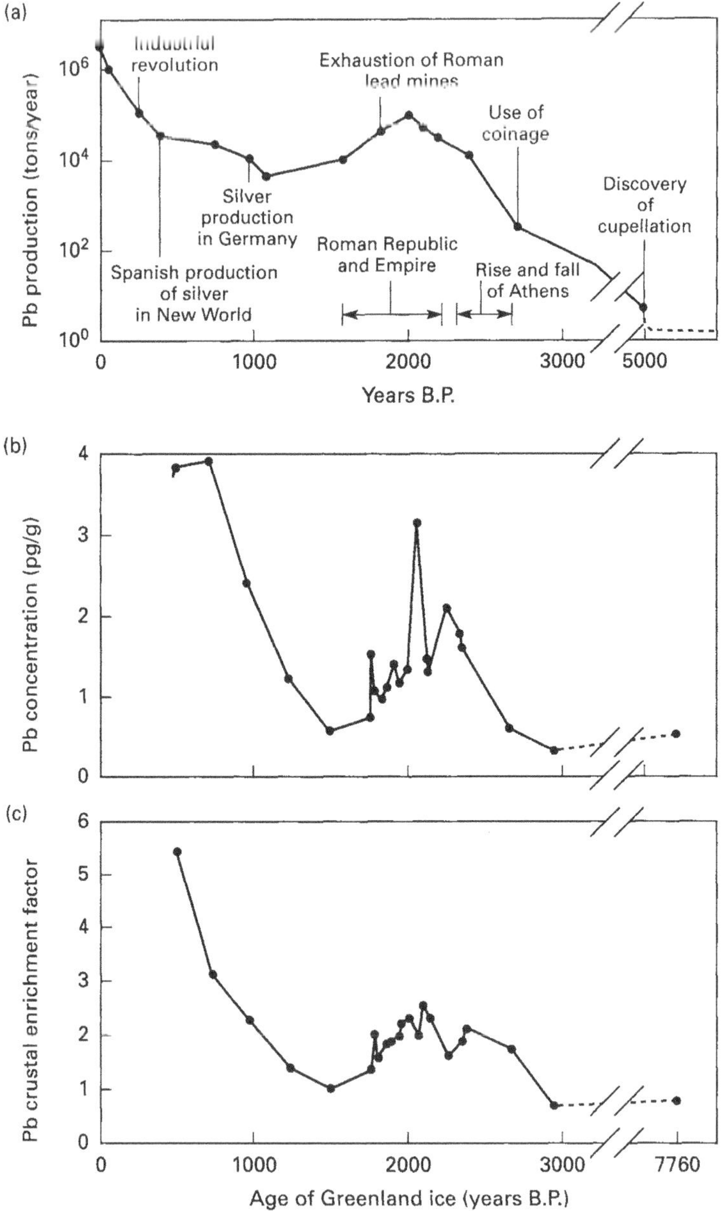

Figure 2.6 (a) Changes in world-wide lead production over the past 5500 years. (b and c) Changes in lead concentration and lead enrichment factor in central Greenland ice from 2960 to 470 years ago. Each data point was obtained from the analysis of a core length corresponding to exactly two years of ice accumulation, except in areas of poor ice quality (Source: Hong *et al.*, 1994)

As noted earlier, woodland was gradually cleared as increasing human population placed heavy demands on the forests for space for agriculture. However, mining and metalworking provided a new form of human exploitation (including smelting, glasswork, salt producing and tanning industries), all of which required large amounts of wood and charcoal (Behre, 1988). Simmons (1991) argues that although these industies were not on the scales of later times they did place pressure on resources, especially timber.

In response to this the level of destruction of woodland was partly offset by the introduction of woodland management schemes across Europe (Rackham, 1986) through coppicing. Coppicing can be traced back to Neolithic times and includes preserving timber supplies by coppicing (cutting close to the ground to produce a crop of long straight poles) and pollarding (cutting above the height of browsing cattle) (Bell and Walker, 1992, p. 168). Trackways uncovered in the Somerset Levels appeared to contain specifically grown timbers dating back to Neolithic times (Godwin, 1960). A form of advantageous woodland management helped maintain woodland cover surrounding the Bronze age mines at Mount Gabriel, Co. Cork, south-west Ireland (O'Brien, 1994). It is evident, therefore, that prehistoric peoples were capable of exploiting their natural environment without destroying it.

Human impact and prehistory: a global perspective

The general scheme of archaeological periods outlined in Figures B2.3a and B2.3b is not applicable to the New World, that is the Far East and the Americas. However, hunter-gatherer societies and farming as forms of human exploitation occurred in the New World, but temporal variations in human impact occur because the introduction of agriculture was also time transgressive (see Delcourt, 1987; Edwards and MacDonald, 1991).

However, the advent of agriculture did not trigger the widespread clearance of forest in all areas. Pollen analytical studies have shown that it was not until the arrival of European colonists into America in the eighteenth century that large areas of forest were cleared to make way for agricultural fields (Delcourt, 1987). Prior to this there is a limited amount of evidence for maize cultivation by Indians around 1200 and 1300 AD (Davis and Turner, 1986; McAndrews and Boyko-Diakonow, 1989). Prehistoric agricultural activity only registers in 7 per cent of pollen diagrams in North America (McAndrews, 1988). In North America native Indians did not keep domestic livestock and crop production supplemented hunting rather than replacing it (Bell and Walker, 1992). In South and Central America *Zea* cultivation and forest disturbances extend back *c.* 5300 years when a swidden type of agriculture was practised by late Archaic hunter-gatherers (Bush *et al.*, 1989).

Fire and agriculture appear to account for forest disturbances in Australasia. Early aborigines are thought to have used fire to manage their environment, although natural fires would contribute to changes in vegetation (Bell and Walker, 1992). In Papua New Guinea forest disturbance dated to *c.* 1200–1000 BP was caused by sylviculture and tree-fallowing associated with garden activities (Worsley and Oldfield, 1988). Woodland clearance is associated with an increase in grass and weed pollen stretching back around 5000 years in Africa (Edwards and MacDonald, 1991).

Such insights into the impact of early humans on the environment provide modern society with positive and negative lessons. Sustained and intensive exploitation of the environment can have devastating consequences. Evidence suggests that a combination of human-induced deforestation, soil erosion and depletion combined with climate change led to the eventual collapse of the Mayan civilisation in Central America (see Delcourt and Delcourt 1991; Healey *et al.*, 1983; Rue, 1987; Sabloff, 1995; Leyden *et al.*, 1998; Hodell *et al.*, 1995). It took five centuries for the forests to recover, indicating that the ecosystems were not degraded beyond their natural capacity (Abrams *et al.*, 1996). Agricultural practices causing widespread soil degradation have contributed to the decline of other ancient civilisations, including the Mesopotamians of the Tigris and Euphrates valleys in Iraq and the Harappans of the Indus valley in Pakistan (Olson, 1981).

2.2.5 Summary

This section has argued that prehistoric societies were capable of altering their environment. Simmons (1990b) asserts that the view that hunter-gatherer societies had no impact on their environment and that in Europe at least human-induced ecosystem change only began along with agriculture has now been largely abandoned. Mesolithic people irreversibly altered their environment by using fire and were not totally passive towards nature. These attempts left indelible marks on the landscape by changing the composition and structure of woodlands, acting as a trigger for soil acidification and peat growth, especially in marginal areas, and created openings in the forest that eventually provided a suitable environment for agriculture (Williams, 1989). Low population is possibly the main reason that the impact of Mesolithic societies was spatially restricted and many areas of forest remained in pristine condition, altered only by natural factors such as climate, natural fires, disease, soil development and topography.

Forms of human exploitation radically changed with the emergence of agriculture and later metalworking cultures, and the resultant impact was similar but more intense. In north-west Europe large tracts of woodland were cleared from arable and pastoral agriculture from Neolithic times onwards as complex agro-economies and today's cultural landscape began to evolve. Agriculture has become a dominant form of human exploitation. Today more than 70 per cent of all culitvated land is devoted to cereals and these provide the bulk of human nutrition (Reid, 1998). The strength of this exploitation began to mask natural factors, resulting in a significant alteration in the structure of vegetation and a deterioration of soils, which influenced the evolution of floodplains and river valley systems as well as having direct impacts on the plants and animals used to procure food.

The disciplines of Quaternary science, palaeoecology and archaeology provide society with information about long-term environmental change and human impact. This information is invaluable because it provides scientists with a temporal perspective for contemporary environmental problems and can be used to help model future environmental and climate change. In particular this data is currently contributing to two debates: 1) the direction, magnitude and rate of climate change; and 2) the sustainability of human impact (Chambers, 1993, p. 259), and both are relevant for nature conservation and the future management of all landscapes (cf. Mannion, 1997).

2.3 Industrialisation

2.3.1 The nature of industrialisation and the role of nature within industrialisation

Industrialisation, just like the emergence of agriculture, has to be seen as one of the most important cultural, economic and environmental changes to have occurred in human history. The term industrialisation has been defined as the process, or processes, by which industrial activity comes to play a dominant role in the economy of a society, be that society defined on a national, regional or local scale (see Smith, 1986, p. 227). The emergence of such a society has been the subject of considerable debate with the phrase 'industrial revolution' often being used, particularly with regard to industrialisation in Britain from the mid-eighteenth to nineteenth centuries. There are major problems with this phrase, implying as it does a quick, radical break. In the previous section, for example, we highlighted the presence of metalworking well into prehistory. However, the place and period of industrialisation covered by the term 'the industrial revolution' does provide a well-documented case with which to examine the connections between the growth of industrial activity and the use or exploitation of natural resources. In particular, there are a series of studies that raise the question of whether or not industrialisation involves societies finding a new way to exploit nature.

2.3.2 Industrialisation and the exploitation of nature

According to Gregory (1990), one of the dominant images of the period of industrialisation in Britain is of:

> blackened tubs of coal clanking from the pit-head and tipping into wagons and barges; of brooding factories shrouded in steam and smoke and echoing with the clang and clatter of machines; of bales of cotton piled high in warehouses and swung down into the holds of high-masted sailing ships (Gregory, 1990, p. 351).

Rather less prosaically, Simmons (1989, p. 196) has argued that industrialisation involved the manufacture of goods in factories, the manufacture of goods by machine, the manufacture of goods for sale outside the immediate neighbourhood, and finally, and most significantly for Simmons, a reliance on a specific form of energy, namely fossil fuels. For Simmons (1989, p. 196), industrialisation refers to the process whereby the 'stored photosynthetic energy of ... fossil hydrocarbons' is added to the solar energy of the sun and used by human societies to 'gain access to resources and to alter the structure and function of their surroundings'. In other words, for Simmons the basis of industrialisation is a change from human and animal energy sources to gaining energy from hydrocarbons, that is fossil fuels such as coal, oil and gas. The energy released from these sources provides the foundation for industrial production, permitting the powering of factory machines, the extraction of further 'natural' resources such as metal ores and chemicals, and providing the basis for the production of new 'resources' such as plastics and petrochemicals.

Simmons suggests that there are three stages in industrialisation: first, a period prior to the beginning of the nineteenth century, where there was what he terms 'a below the horizons use of hydrocarbons'; second, from about 1800 onwards, the emergence of industrial society; and finally, from '2.30 p.m. on 2nd December 1942, in Chicago, when Enrico Fermi achieved the first controlled fission of atomic nuclei in a chain reaction' (Simmons, 1989, p. 344) and thereby, for Simmons, marked the transition of human society from the 'industrial age' to the 'nuclear age'.

In the case of the first stage of 'industrialisation', the 'below the horizons use of hydrocarbons', most of society's energy resource was said to come from human and animal sources. This does not, however, mean that there were no other sources of energy. Water power, for instance, had long been used as a source of energy, with water wheels having been in use in Britain since at least the twelfth century (Donkin, 1976, p. 94). By the eighteenth century there were apparently some 143 water-driven cotton mills in Britain, together with numerous mills for other products (Darby, 1976, p. 61). There was also some use of hydrocarbons: according to Simmons (1989) coal was used by the Chinese at least 2000 years ago, while in the United Kingdom the Romans were certainly using it and there is some evidence to suggest that it was in use in the Bronze Age. By 1700 England and Wales had a coal output of three millions tons per year, which was about five times the total production from the rest of the world (Darby, 1976, p. 70). Coal was used in the manufacture of alum, salt-petre, gunpowder, soap, candles, brewing and salt extraction from sea water.

However, while there was use of non-animate energy sources, the predominant sources of energy in Britain prior to the end of the eighteenth century were essentially human labour power and the power of animals, or as Wrigley (1988) has termed it an '*organic economy*'. Wrigley suggests, however, that the eighteenth and nineteenth centuries saw a transition from this economy to a '*mineral-based, energy economy*'. In this economy the power of people and animals was supplemented by energy obtained from minerals, and in particular from hydrocarbons such as coal. Simmons (1989) remarks that from about 1800 there was a rapid expansion in Britain of coal production, although as early as 1709 Abraham Darby was replacing charcoal with coal at his Coalbrookdale iron foundry. It was, however, not until the second half of the eighteenth century that there was a more widespread change from charcoal to coal. Coal production in Britain increased from just under three million tons in 1700 to 30 million tons by 1830 and reached 50 million tons by 1850 (Gregory, 1990, p. 362), while per capita energy availability in Britain rose in the next half century from 1.7 tce (tonnes of coal equivalent) to over 4 tce by 1919 (Simmons, 1989, p. 196).

In addition to the intensification of coal production in Britain there was also an increase in the spatial spread of coal mining, and by implication, for Simmons at least, industrialisation. Industrialisation during the nineteenth century was concentrated in selected areas in Europe and North America, including Lancashire, Yorkshire, South Wales, Clydeside, Pennsylvania, Ruhr and Saar–Sambre–Meuse. In addition there was some further 'development' in Russia, Poland, the Ukraine and Japan, and a 'periphery of nations where industrialism was by no means absent but in which it never became a dominant way of life' (Simmons, 1989, p. 207). Today one of the world's largest coal producers and consumers is China, which is very much seeking to use this as a resource on which to build industrialisation.

The pattern of industrialisation is often seen to reflect the historical geography of coal extraction, with the degree of industrialisation reflecting the length of time an area has been able to access major coal reserves. However such an interpretation ignores at least two features about industrialisation and the exploitation of mineral energy.

First, mineral energy was not only derived from coal production but had a number of other sources including, most significantly, petroleum. Petroleum, just like coal, had a phase of 'below the horizons use' prior to its adoption by society. For instance in the eighteenth century 'coal oil' was used in Derbyshire for lighting. However such use was slight and the key event in the history of petroleum use was the drilling of the first modern oil well in Pennsylvania in 1859, together with the development of the internal combustion engine during the late nineteenth century.

It has been claimed by Porritt (1984) that oil effectively transformed the global economy through building up international transport and expanding production in all areas of the world. Certainly energy consumption has risen dramatically over the course of the twentieth century (see Figure 2.7). Furthermore, as Simmons (1991, p. 53) argues, petroleum is often found at places quite distant from previously established industrial centres. From its early focus in the United States oil production spread in the early twentieth century as a series of US and European companies gained concessions to extract oil from a range of former colonies of western countries, most notably in the Middle East. In the 1950s there was a further expansion in the areas of production as independent US firms or newly established European state-run companies entered the industry and sought new sources of oil outside the concessional areas (see Odell, 1963). North African states such as Algeria and Libya, for example, became sites of oil production and their importance increased during the closure of the Suez Canal. A futher expansion in areas of oil production occcurred following the 'oil crises' of 1973–4 and 1978–9 when oil-consuming industrialised countries such as the United States and the United Kingdom sought to lessen their dependence on overseas oil by expanding their own oil production.

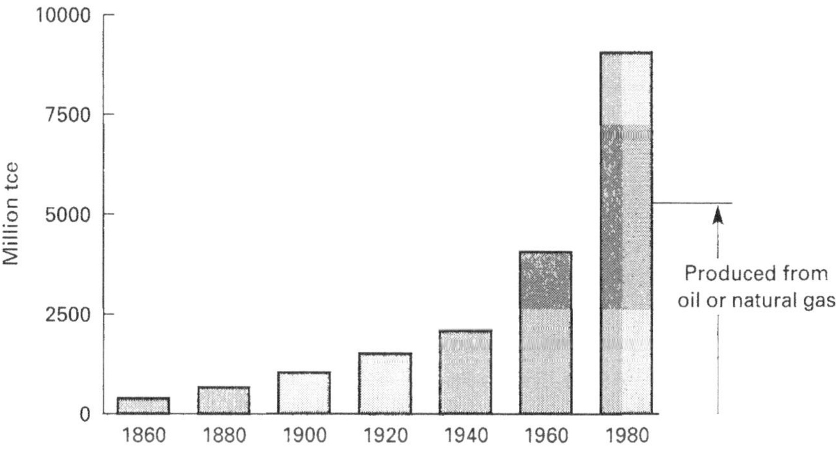

Figure 2.7 World energy consumption, 1860–1980 (after Mannion, 1991b)

The geographical expansion of oil production has, however, not been necessarily accompanied by an expansion in industry. This is well illustrated by the events of the 1970s, when members of the *Organisation of Petroleum Exporting Countries* (OPEC) decided to exercise some collective power by reducing their oil production and driving up prices. Many members of OPEC sought to use the large influx of income generated by the oil price rises to develop their own industry. There were, however, problems in doing this. First, as Halliday (1977) and Watts (1984) note, many members of OPEC were 'city states' or 'desert states' and thereby had limited natural and labour resources that could be put to use through financial investments. As a result much of the income of these states was actually invested in other counties, including those of the industrialised West. Rather than spreading industrialisation, there was a reconcentration of industry in existing centres. A third set of OPEC members, which Halliday and Watts describe as 'normal states', did have sufficient land and labour resources to absorb financial investment and many of these, such as Venezuela, Indonesia, Iraq, Iran and Nigeria, did invest in industrial development. However, in many cases this investment has not proved to be particularly successful. One reason for this was that oil prices soon came to fall, in part due to the development of new source areas of oil production such as Alaska and the North Sea, and more generally because there was a recession in developed industrial economies that reduced demand for oil. As a consequence those members of OPEC with substantial industrialisation programmes found that they faced falling incomes at just the time investments were needed. Many borrowed heavily and fell seriously into debt, while others curtained their plans for industrialisation.

While mineral raw materials, and in particular coal and petroleum, create enormous increases in energy and output and may induce localised industrialisation, it is also important to consider not only the production of energy but its transmission. Wrigley (1962; 1988) has suggested that the key feature of industrialisation was not so much the increase in energy made available by fossil fuels, but that this energy could be transformed into mechanical energy. This was achieved, so Wrigley argues (1962, p. 12), by the invention of the steam engine: 'More than any other single development, the steam engine made possible the vast increase in individual productivity which was so striking a feature of the industrial revolution.' The steam engine both permitted the extraction of coal from greater depths and allowed the power of machinery, such as textile looms. In these two ways steam engines created what Wrigley terms 'positive feedback systems' in which each action is said to make subsequent actions easier (Figure 2.8). Wrigley contrasts such 'positive feedbacks' with the situation in an 'organic economy', where the application of additional human energy can lead to diminishing rather than increasing returns (via the sub-division of land-holding) and where economic growth is made continually more difficult (see Figure 2.9).

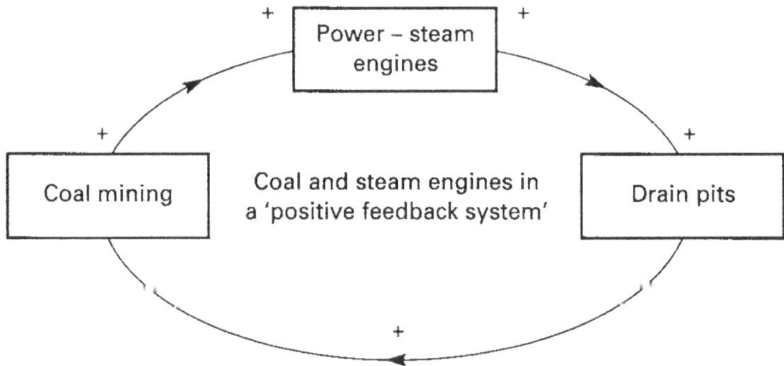

Figure 2.8 Industrialisation as a system of positive energy feedbacks

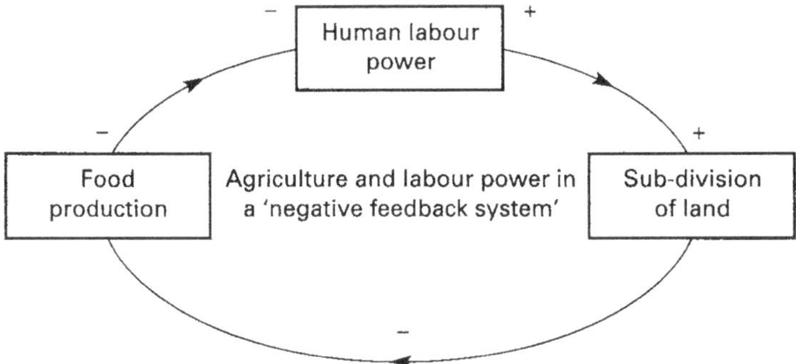

Figure 2.9 Pre-industrialism as a system of negative energy feedbacks

Wrigley's analysis of industrialisation has been subject to some detailed criticism by Gregory (1990). One of Gregory's arguments is that Wrigley overemphasises the role of coal and steam engines and any associated positive feedback loops. In particular, Gregory points to the situation in the cotton industry, which was the major industry to begin adopting steam engines in the eighteenth and early nineteenth centuries and was also the principal sector of economic growth at this time, and yet which in 1800 had only 25 per cent of its mills powered by steam (Gregory, 1990, p. 371). The rest were still powered by a source of the organic economy, that is through water power. Cotton and thereby factory production, which are so often seen as requiring the energy inputs of coal, were hence generally powered by water-mills. For Gregory, therefore, the expansion in industrial production was associated much more with an '*advanced organic economy*' than with any mineral economy. Even during the second half of the nineteenth century the organic economy remained highly significant for industrial production: Thompson (1985) has estimated that in 1850 about a third of the power for industry came from water-mills, although by the 1870s this had declined to some 10–12 per cent. Gregory argues that rather than seeing the industrial revolution as signalling a replacement of the organic economy by the mineral economy, one should much more see it as a process of '*combined development*', in which an advanced organic economy and a mineral-based one effectively fed into one another: 'feeding one another's achievements, endorsing one another's effects' (Gregory, 1990, p. 370).

One instance of such combined development is the impact of industrialisation on agricultural development. Industrialisation not only involved the exploitation of non-agricultural resources, but also widened agricultural production by creating demands for new agricultural products. Cotton production, for instance, was greatly expanded to meet the demands of the textile producers. Furthermore, agriculture was also a major market for the products of industrialisation. In particular agriculture absorbed large energy inputs through industrial products (Figure 2.10). For Simmons (1989; 1990) the use of fossil fuel powered machinery within agriculture marks the beginning of a process of the '*industrialisation of agriculture*'. This process began with the introduction of steam engines for threshing and ploughing, although it was rapidly accelerated by the development of the tractor which became widely adopted in western Europe and the United States in the period after the Second World War.

The application of fossil fuel powered machinery is not, however, the only energy input from industrial production. A second source of energy inputs is through chemical fertilisers and pesticides, both of which represent a large degree of embedded energy. The input of energy via machinery, fertilisers and pesticides is now of such a scale that Simmons (1990, p. 185) talks of agriculture being 'transformed from a solar-based process to one which is underlain by large quantities of fossil fuel', just as in industrial production.

One can suggest that one impact of industrialisation is its narrowing of geographical distinctions, or 'areal differentiation'. The world can be said to be becoming a more uniform place: agriculture is effectively industry, rural societies are urbanised, places are integrated and indistinguishable. One of the key features of industrialisation has been the ability to move energy extracted from natures away from its site of production and into a whole range of other activities and areas. Simmons (1989, p. 202), for example, argues that railways effectively acted in the early nineteenth century as a way of channelling the energy produced from hydrocarbons into other areas. Trains used coal: they in effect became a point where energy was invested. They also spread energy by distributing resources, including coal, to other areas. Newer forms of energy have heightened such processes: oil, for instance, is relatively easily transported and electricity is a very transportable form of energy. From Simmons' and Wrigley's essentially 'system' perspective on society and nature interaction, industrialisation involved both heightened extraction of energy and greater diffusion of such energy to create what Gregory terms 'combined development'.

On the other hand it can also be suggested that industrialisation has increased geographical differentiation, for instance by increasing distinctions between urban/industrial society and rural/agricultural society. In the pre-industrial era and the early stages of industrialisation both agricultural production and industrial production were located in what one might term 'rural locations'. Nef (1964), amongst others, has attributed this to the location of the raw materials and energy supplies used in early industry although, as will be discussed shortly, others have suggested a variety of more social factors. With the rise in fossil fuel extraction

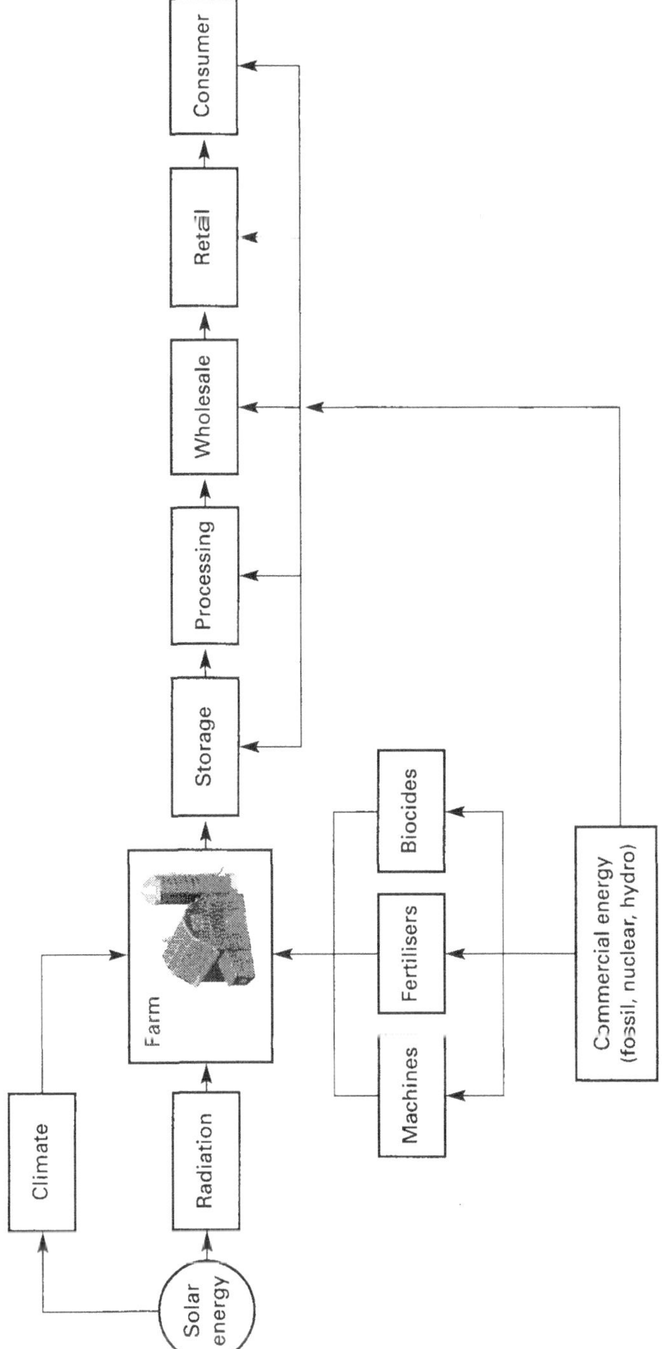

Figure 2.10 Energy inputs into modern agriculture (Source: Simmons, 1990a)

industry became highly concentrated, with the emergence of 'industrial regions' such as South Wales, Central Scotland, Lancashire, Yorkshire, the North East and the Midlands (see Figure 2.11). People migrated to these areas of industrial expansion and with an expansion in numbers came a series of changes in the way people organised their life: '*urbanisation*' (an increase in the number of people living in urban areas) is seen to have led to '*urbanism*' (a rise in a distinctive urban way of life). There was also a transformation in the appearance and environmental quality of many areas. As Mannion (1991b, p. 141) notes, cities can be seen as 'fuel-powered systems' which, as well as providing concentrations of human activity and

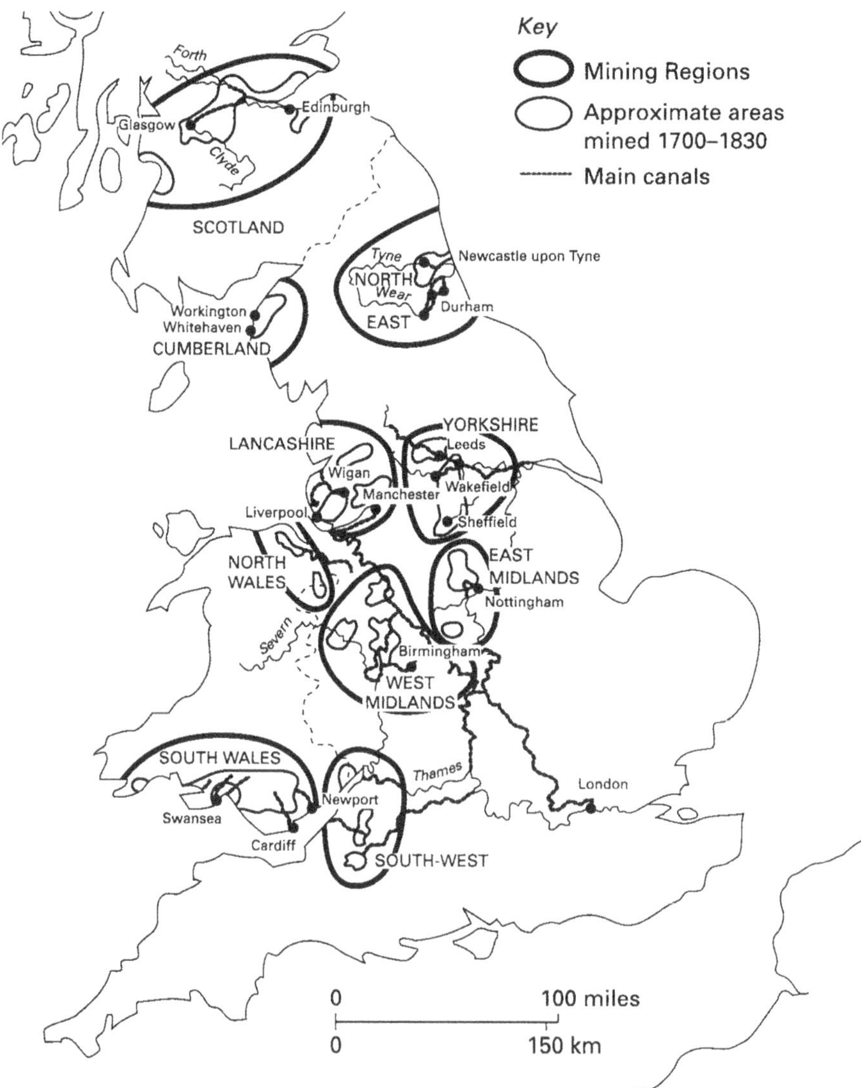

Figure 2.11 Coal mining districts in Britain, c. 1700–1830 (after Gregory, 1990)

interaction, 'also produce large quantities of waste material ranging from toxic metals to domestic rubbish'. The extent of these wastes transformed local environments (see Box 2.7) and also had more widespread impacts through the release of materials into atmospheric and hydrospheric circulatorary systems. Current examples of these regional and globalised environmental impacts will be discussed in detail in Chapter 3. It is, however, important to note that many of the environmental changes that form the subject of current concern have a history that may be quite closely tied to the emergence of industrialisation. Mannion (1991a, 1991b), for example, has amassed some evidence that waters in lakes in Britain started to become more acidic from about the 1840s, indicative that the phenomenon of 'acid rain' is not simply a product of contemporary industrial production. Atmospheric concentrations of carbon dioxide and methane also began to increase from around the 1750s (see also Chapter 3).

Box 2.7 The environmental impacts of industrialisation: the example of the Swansea Valley

The city of Swansea in Wales experienced rapid industrialisation in the course of the late seventeenth, eighteenth and nineteenth centuries and its main area of industry, the Tawe Valley, went from being described as an area of 'verdant valleys and shady groves' in the seventeenth century to being described in the 1880s as containing the 'ugliness of grime, and dust, and mud, and smoke and indescribable tastes and odours' (Bridges and Morgan, 1990, pp. 277–99). The seventeenth century saw the emergence of small-scale workshops, collieries and mills within the Tawe Valley and by the eighteenth century the area had become a major coal production centre within the South Wales coalfield. In 1717 the first of a series of metallurgical industries was established in the valley when the Llangyfelach Copper Works was started. Copper ores were imported into the area up the Tawe to be smelted using local coal. By the end of the eighteenth century a series of metal smelting industries had been established and by the beginning of the nineteenth century the valley had 'entered the major phase of industrialisation' (Swansea City Council, 1978, p. 5). As well as copper, other metals such as zinc, nickel, iron, lead, cobalt and silver were also produced in the valley. Swansea became a point of wealth accumulation, being described as 'the non-ferrous metallurgical centre of the world' (Swansea City Council, 1978, p. 6). The city, for example, accounted for some 85–90 per cent of the copper production in Britain in 1800 (Bridges and Morgan, 1990). This dominance declined, however, from the 1840s with the expansion of copper mining in such places as Cuba, Chile, Spain, Portugal and North America. Swansea however still remained a site of industrialisation with the emergence of tin plating and steel manufacturing, which remained highly significant employers and centres of wealth accumulation until after the Second World War. By the 1960s, however, metallurgical working had largely ceased in the Swansea area: the last copper works in the Lower Tawe Valley was closed in 1974 and the last tin plate works in 1961.

▶

The metalworking industries had a profound, albeit relatively short lived, economic and social impact on Swansea. They also had rather less spectacular but still persistent environmental impacts. A series of studies (see Bromley and Humphrys, 1979; Bridges and Morgan, 1990) have shown the continuing presence in the soils and waters of the valley of a series of hazardous and toxic pollutants including arsenic, cadmium, chromium, copper, lead, nickel, sulphur dioxide and zinc. In addition to this, the decline of the metalworking industries led to large areas of the valley being abandoned for economic activity and becoming areas of dereliction. From the early 1960s a series of attempts have been made to improve the economic fortunes of the area and to improve its physical appearance. However, as the maps in Figure B2.7 show, levels of several pollutants produced during Swansea's period of industrialisation remained significantly high in some areas some two centuries after the first copper works started production in the Tawe Valley.

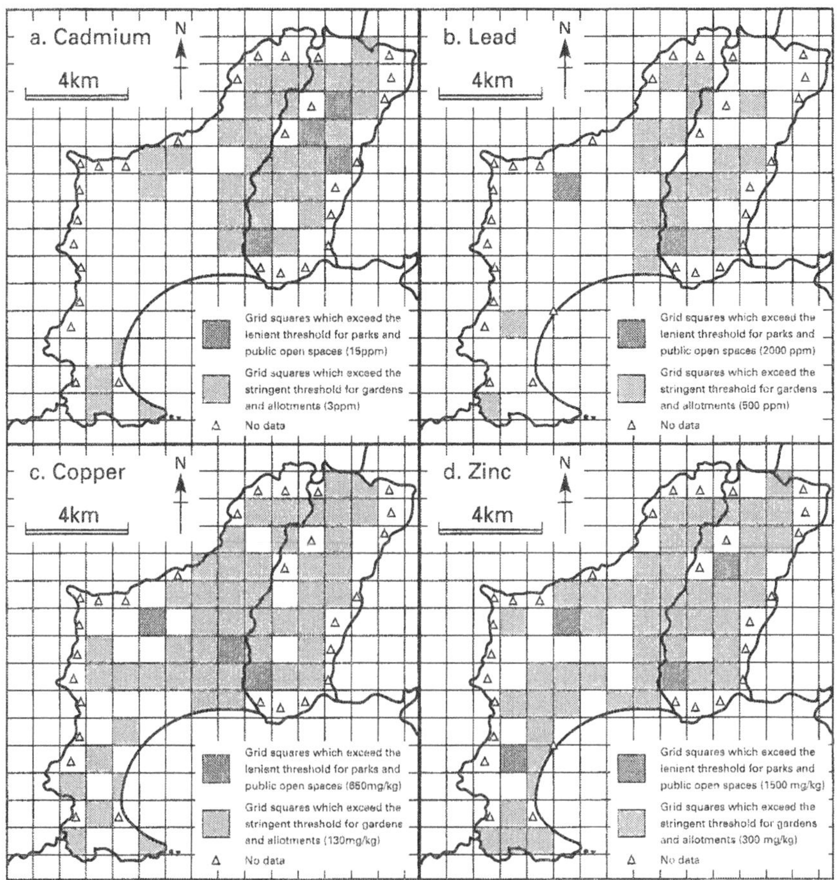

Figure B2.7 Industrial pollution in Swansea (Source: Bridges and Morgan, 1990)

Not only did industrialisation contribute to the emergence of urban life, it can also be said to have led to the creation of what is now frequently taken to constitute 'rural society', namely a community of agriculturalists. In the course of industrialisation agriculture, which had been the principal economic activity in rural areas, became in most virtually the sole economic activity. As Lash and Urry have remarked (1987, p. 94), 'those regions which did not experience nineteenth century industrialization ... were ironically more agriculturally based in 1900 than they had been in the previous couple of centuries'. Prior to the onset of large-scale industrialisation, many rural areas were sites of industrial activity. Much of this industry was unable to compete successfully with newer, more urban-centred forms of industry and, as a result, there was a de-industrialisation of the countryside at the same time as there was urban industrialisation.

Given these two contradictory processes one can perhaps agree with Gregory's characterisation of industrialisation as a process of combined *and* uneven development. This process can be understood from an 'ecological' approach as reflecting concentrations and diffusions of energy extracted by society from nature. It can, however, also be understood from a more sociological perspective.

2.3.3 Exploitation of nature and changing social organisation

For Gregory, and many others historians and geographers, the industrial revolution involved much more than simply the 'discovery' of coal and the invention of the steam engine, although both these were of importance. Their importance, however, lay primarily in the way that they help change the society and economy of Britain. In particular it has been suggested that the industrial revolution was first and foremost a change in the way people were organised to work. At least three significant transformations in the organisation of work have been identified as being associated with industrialisation. First, machinery reduced work to essentially a task of supervision; no longer was work associated with the expenditure of physical muscle power, but was much more associated with the supervision of a machine. Second, working with a machine altered the pacing of work, a point nicely illustrated by Thomas Hardy's description of the controller of a threshing machine in his book, *Tess of the D'Urbervilles*: 'he [the threshing machine controller] was in the agricultural world but not of it. He served fire and smoke; these denizens of the field [the farm labourers] served vegetation, weather, frost and sun' (Hardy, 1978, p. 405).

In other words the pace of work of the threshing operator was dictated by the motions of the steam engines, which were regular and largely unaffected by the vagaries of the weather and even the distinction between night and day. The rate at which agricultural labourers worked was, however, very much conditioned by these environmental factors, as well as the personal vagaries of the labourers. As Thompson (1967) has put it, industrialisation involved a transition in work practices towards the rhythms of mechanisation and abstract time, rather than 'natural seasons' and tasks to be achieved.

A third link between industrialisation and transformations in the organisation of work is that industry permitted the employment of people being displaced from agricultural production, in part because of the adoption of industrial technology within agriculture. This displacement of workers allowed production to expand across a whole range of sectors; in effect the displacement of workers and their labour power was perhaps as significant, or indeed more significant, than the diffusion of power produced from fossil fuels.

These three transformations, and a variety of others that have also been suggested, are seen to be of significance in that they were an extension of processes already in existence that were related to a capitalist form of economic organisation (see Table 2.2). The transformation of work into supervision can be seen as a continuation of the process of *'proletarianisation'*, whereby workers were being removed from resources through which they could feed themselves. The imposition of a mechanised time discipline can be seen as an expansion in the control capitalists were exerting in early forms of production, such as *'proto-industrialisation'*. This latter term is one coined by Mendels (1972) to describe forms of economic production whereby the efforts of a series of quite disparate and relatively autonomous workers, often living in rural locations, were brought together by merchants in order to create production for markets. Mendels argues that such forms of production both preceded and acted as a preparation for the industrial revolution. Gregory (1990) has provided a very full discussion of how proto-industrialisation can be seen as a precursor to industrialisation and examines it both through an 'ecological perspective' and a more sociological (or to be slightly more precise a 'political economy') perspective. In the latter, he pays particular attention to the ways in which rural industrial production in woollen textiles involved its own internal discipline: 'different to that of the factory system, to be sure, but none the less decisive' (Gregory, 1990, p. 356). Finally, one can see the displacement of workers from agriculture into industry as both a continuation of the processes of 'proletarianisation' and as a necessary step to permit the *'expanded accumulation of capital'*, whereby an increasing range of activities are required to sustain the possibility of profitable production.

Table 2.2 Transformations in the labour process under industrialisation and within capitalism

Under industrialisation	Within capitalism
Reduction of labour to supervision of machines	*'Proletarianisation'*: the creating of a waged labour force
Alteration of pace of work: creation of time discipline	*'Subsumption of labour to capital'* – eg 'proto-industrialisation'
Displacement of agricultural work-force	*'Proletarianisation'* and *'expanded accumulation of capital'*

2.3.4 Summary

In this section two distinct interpretations of industrialisation have been discussed: an ecological analysis that focuses on energy production and diffusion and a political economy perspective that focuses on the organisation of the labour process. Each interpretation asks somewhat different questions and produces rather different interpretations as to what industrialisation is, when it occurred and what were its consequences. The ecological interpretation, for instance, sees industrialisation as representing a movement from an organic to a mineral based economy and places great stress on technological changes in the eighteenth and nineteenth centuries, although there is also some recognition of precursor forms of mineral exploitation. The political economy perspective, on the other hand, sees industrialisation in terms of changes in the social organisation of work related in part to the introduction of machines but also stemming from rather longer-run changes related to the emergence of capitalist social relations. For the ecological approach heightened exploitation of nature in the form of animate and inanimate energy and materials is a fundamental aspect of industrialisation, while in many political economic interpretations it does not figure at all. However, it should be noted that the exclusion of nature from political economic accounts is not a necessary feature, in that industrialisation as a process of social reorganisation did not occur in an environmental vacuum. Changes in social organisation, such as the emergence of the mechanisation of agriculture and the emergence of factory production, had clear and lasting environmental impacts. Indeed many contemporary 'environmental problems' such as acid rain, atmospheric, marine and terrestrial pollution and climate change may have roots that can be traced to these forms of social organisation, as will be discussed in Chapters 3 and 5. Many of the transformations of nature discussed here as part of an 'ecological' interpretation of industrialisation can hence be equally well interpreted from a political economic perspective. Indeed there have been calls to add an ecological dimension to political economy perspectives, with people such as Blaikie and Brookfield (1987b), Redclift (1987) and Rees (1990) arguing for the development of what has been termed a 'political ecology approach' (see Emel and Peet, 1989).

2.4 Imperialism

The previous section looked at industrialisation, particularly as it occurred in Britain, although fleeting references were made to the situation in other parts of the world. The processes of combined and uneven development identified as being associated with industrialisation can be seen to have operated at global, as well as national and regional scales. Indeed while industrialisation in Britain has frequently been portrayed as a process of internal 'national development' such interpretations are open to what Taylor (1989) has termed the 'error of developmentalism', in that they see the world being composed of large numbers of autonomously functioning national societies, each of which changes or 'develops' largely through processes operating within each society. Taylor (1985, 1989) likens this view to 'a world of ladders': each national society is seen to be seeking to climb a common set of steps to development but, at any one point in time, some societies may be further up the ladder than others. By contrast, Taylor highlights how the development in one country is connected to the development in others through the operation of what he terms, after Wallerstein (1979) a 'world system' or 'world economy'. While not fully adopting this world perspective we will in this section geographically widen the focus beyond Britain to explore the processes of what has been termed 'imperialism' and examine how they were associated with the changes in the exploitation of natures that spanned vast areas of the globe.

2.4.1 Interpretations of imperialism

Imperialism has been defined by Smith (1986, p. 216) as 'an unequal territorial relationship, usually between states, based on domination and subordination'. The domination of territories is generally seen to be achieved by economic or political means or by a combination of both. One of the most common uses of the term imperialism is hence as a reference to the economic and political relationships between advanced capitalist and other countries, variously described as un- or underdeveloped (see Box 2.8). Imperialism is a word that has become synonymous with the oppression and 'exploitation' of weak, impoverished countries by powerful ones.

Box 2.8 Development and underdevelopment

People and places in the world are frequently described as being developed or undeveloped or less developed or underdeveloped or overdeveloped. These terms are used in academic, popular and political discourses and yet they are also terms that have been repeatedly questioned (see for example, Brookfield, 1975; Mabogunje, 1980; Unwin, 1983; Larrain, 1989; Taylor, 1989; Crush (ed.), 1995; Brown, 1996). As Thomas and Potter note, although there are a variety of different definitions of development and its associated terms, commonly the term development implies 'a process of change from a less desirable to a more desirable kind of society' (Thomas and Potter, 1992, p. 116). Indicative of this is Lee's (1994, p. 128) definition of development as: 'A process of becoming and a state of being ... [that] would enable people in societies to make their histories and geographies under conditions of their own choosing'.

Such definitions clearly raise a series of further questions concerning to what extent there is any consensus over what might make a society more desirable to people, and also to other 'natural' constituents of this world. To some people the term development is problematic in that it remains an abstract ideal with no clear substance. On the other hand the term is often given very specific but unacknowledged meanings. Mabogunje (1980), for example, has argued that the term is commonly used to refer to one or more of the following: i) economic growth; ii) modernisation; iii) distributive justice; and iv) socio-economic change. In the first development is taken to mean increases in the production of goods, services and money. Development is often discussed through such measures as gross national product (GNP) or gross domestic product (GDP), often expressed per capita. Modernisation is seen as a more social phenomenon in that development is seen to imply changes in the way people act or behave or in the ways societies are organised. Again this notion of development is often discussed through particular measures: levels of education or the consumption of particular commodities are, for example, often seen as indicators of the adoption/non-adoption of particular ways of behaving. This approach can, however, be criticised for privileging particular ways of behaving and social institutions – often those prevalent in western, industrial, capitalist counties – over others, often with no justification. This approach can be seen to underlie such classification as 'First', 'Second' and "Third World' in that in each there are seen to be distinct forms of social organisation and ways of acting. It also led to the classification of areas and people as being 'developed' and 'un-' or 'underdeveloped': developed societies were seen to have undergone modernisation while un- or underdeveloped countries were seen to be non-modern or relatively unmodernised respectively.

Socially distributive approaches to development are relational in focus and tend to classify people and places into a gradational hierarchy based on levels of control of economic or social resources. Terms such as 'more' or 'less' developed are often used in this approach. Less quantitative differentiations are often used in socio-economic perspectives on development, where this term is used to imply change from one form of society to another, often without any particular notion of progressive change. Hence one has the 'development' of agricultural societies into industrial ones, or feudal societies into capitalist ones, or nations into empires, or firms and institutions into world economies.

While there are important differences between these interpretations of development it is possible to identify two common and rather problematic aspects of them. First, it has been argued by Mabogunje (1980), and also Unwin (1983), that the latter four all tend to draw upon and become dominated by the first. In other words discussions of development there tends to be of a high degree of 'economism' whereby development is seen to involve 'to a greater or lesser extent ... increased productivity, higher levels of consumption per capita and a shift from primary to secondary and tertiary economic activities' (Unwin, 1983, p. 236). Unwin (1983, p. 237) suggests that there is a need to question this focus, not least because it tends to divert attention from the making of 'uncomfortable decisions relating to social change' – what might be seen as development in Lee's sense of people choosing the histories and geographies they make – although he does also concede that 'the term "development" is probably too deeply embedded in our vocabulary for it ... to be abandoned' (p. 241).

In addition to debates as to what might or might not constitute development there has also been an important debate as to the spatial scale on which development takes place. As noted in the text, people such as Taylor (1989, 1992) have been highly critical of adopting a national scale of analysis. Taylor argues that while many discussions of development assume, albeit often implicitly, that countries or nation states constitute the boundaries of what is taken to be an economy or society this is a flawed assumption, particularly in the contemporary period when there is a global flow of people and products between places that are quite distant spatially. For Taylor societies and economies are neither autonomous national units or spaces not geographically undifferentiated but are rather integrated into and constituted by a 'world economic system'. He further argues, drawing on the work of Wallerstein (1979, 1983), that the development of some countries proceeds at the expense of other countries: '[t]he world economy cannot sustain all countries rising simultaneously ... [and] the very success of any one country lessens the opportunities for its rivals' (Taylor, 1989, p. 312). This idea is very closely connected to the writings of what is commonly referred to as '*dependency theory*', which argued that the development of some societies was the very consequence of the 'underdevelopment' of others. Griffin provides a clear illustration of this view:

> Nearly all of the people in today's underdeveloped areas were members of viable societies which could satisfy the economic needs of the community. Yet these societies were shattered when they came into contact with an expanding Europe. Europe did not 'discover' the underdeveloped countries, on the contrary it created them (Griffin, 1969, p. 41).

Dependency theory and the world systems analysis of Wallerstein have both been criticised for portraying development as a 'zero-sum' game: it appears that countries can only 'develop' at the expense of others, rather than perhaps establishing new ways of utilising their own resources. Critics such as Brenner (1977), for example, argue that dependency and world-systems theories focus solely on the distribution of commodities and money and fail to consider the modes by which these are produced. This is an important argument to make but, as shown in the following section of this chapter and also in Chapter 5, it is also clear that from at least the sixteenth century the development of some peoples and countries has been closely linked with the demise of others. For this reason in this book we follow the dependency theorists in describing countries as being 'developed' and 'underdeveloped' to highlight the at least potential interrelationships between these states of development. It might also be reasonable (given the arguments over the environmental problems associated with western modes of production and consumption that will be discussed in Chapter 3) to suggest that even more appropriate terms might be 'overdeveloped' and 'underdeveloped'.

The word imperialism has, however, been used in rather more precise ways. The general definition of imperialism as unequal relations between countries based on domination and subordination has been narrowed down in a number of ways. Two forms of definition have been particularly significant. First, imperialism has been defined in terms of relations of domination and subordination based on 'political-territorial control'. Imperialism here becomes the domination of some states by other states and can be linked, amongst other things, to the process of 'colonialism', which is the movement of people from one country to another to control that country. Second, imperialism has frequently been used to refer to relations of domination and subordination based on economic, as opposed to political-territorial, mechanisms. Three distinct economic motivations for imperialist domination and subordination have been identified: foreign markets, investment of capital and low-cost raw materials (see Box 2.9).

Box 2.9 Economic theories of imperialism

Imperialism has often been seen in terms of relations of domination and subordination based on economic mechanisms. Since the early twentieth century three distinct economic mechanisms leading to imperialism have been identified: i) a quest for foreign markets; ii) a need for the export of surplus capital; and iii) the extraction of raw materials at below 'economic' prices. The last of these can be seen to be the mechanism most directly linked to the exploitation of nature, in that European conquest is seen to be connected to a desire to obtain access to 'natural' resources of other countries in order to either supplement existing internal (and perhaps dwindling) resources or else because they can be obtained for lower costs. A particularly significant theorist of imperialism who makes much of this form of exploitation was Rosa Luxemburg (see Luxemburg, 1951 and Luxemburg and Bukharin, 1972).

Luxemburg also emphasised the second of the three economic motivations for imperialism, namely external markets. This view was also highlighted by Marx, who argued that capitalist economies had an inherent tendency to expand spatially in order to find new markets in which they could keep selling their products (see Marx, 1976; Marx and Engels, 1971; and also Harvey, 1982).

A third set of economic interpretations of imperialism have focused on the export of capital to the colonies. This idea is often traced back to the writings of Hobson (1902) and Lenin (1915) who, although taking two quite different political perspectives on imperialism, both argued that European imperialism stemmed from a shortage of internal investment opportunities for capitalists, which led to them looking outwards towards the colonies for investment. Hence in their view imperialism was essentially about the 'export of capital' rather than the export of manufactured products or the import of cheap raw materials.

There has emerged a series of studies and debates relating to the relative importance of these three mechanisms at different points in time and in relation to differing imperial and colonised countries. General reviews and critiques of some of these can be found in Kemp (1967); Barratt-Brown (1974); Brewer (1980); Elson (1984) and Smith N. (1984).

In addition to these two commonly used interpretations of imperialism there are two other ways of looking at imperialism that are of significance when considering society–nature relations. First, Crosby (1986) has argued for the significance of '*ecological imperialism*', which he takes to refer to the biological impacts of European expansion overseas. Second, writers such as Pawson (1990) have referred to a '*culture of imperialism*' and argued that particular cultural beliefs and value beliefs were bound up with political and economic expansion.

A key question in understanding imperialism is whether one uses one or all of these ideas of imperialism. One way of linking some of the ideas of imperialism is historically, that is by suggesting that at various times through history one or more of these understandings of imperialism have been relevant and that there has been movement from one type of imperialism to another. This is broadly the perspective adopted by Knox and Agnew (1989), who broadly identify four periods of imperialism that might be described as an age of plunder, an age of colonialism and imperialism, a period of post-colonial modernisation, and a period of multinationals and a world capitalist economy – each of which has distinctive processes or sets of processes.

The age of plunder

Knox and Agnew (1989) argue that in the fifteenth and sixteenth centuries European merchants, adventurers and government officials extracted 'tributes' from people living in non-European countries. Tributes, as Wallerstein (1974) has outlined, were payments said to be made in return for military protection. The payments were, Wallerstein argues, always intended to be in excess of the costs of providing any military protection and in this way acted as a mechanism of economic exploitation.

Wallerstein argues, however, that tributes eventually became unsustainable as mechanisms of exploitation in that the costs of providing military protection and extracting tributes tended to rise above the value of the tributes. Hecht and Cockburn (1990), for example, observe how Portuguese and Spanish exploration and plunder of South America became increasingly weighed down by military costs in a vicious cycle, whereby political expansion became necessary simply to recover the resources depleted by war rather than increase the wealth of the home country. Wallerstein (1974) suggests that the breakdown of the tribute system marked a fundamental socio-historical change from 'empire building' towards the building of a 'European world economy'. He argues that up until the sixteenth century countries tended to build up power and wealth through military conquest and political domination. From the sixteenth century a country's fortunes were, so Wallerstein argues, principally determined economically. This does not mean that there was an end to overseas expansion: rather from the sixteenth century the motives and mechanisms for this expansion became, so Wallerstein argues, principally economic. This change in the motives and mechanisms for overseas expansion involved the movement from the 'age of plunder' to the 'age of colonialism and imperialism'.

Age of colonialism and imperialism

According to Knox and Agnew, from the eighteenth century onwards European countries exercised 'political-territorial control' in association with the 'economic' processes of expanding markets, exporting surplus capital and cheapening raw material supplies, and with the 'social' colonisation of areas by people from the colonial powers. These three sets of processes produced what Knox and Agnew (1989, p. 240) term a 'new colonialism' that can perhaps more correctly be seen as the age of imperialism *per se,* in that exploitation of nature and other societies was achieved through economic mechanisms that were not totally reliant upon political-territorial control.

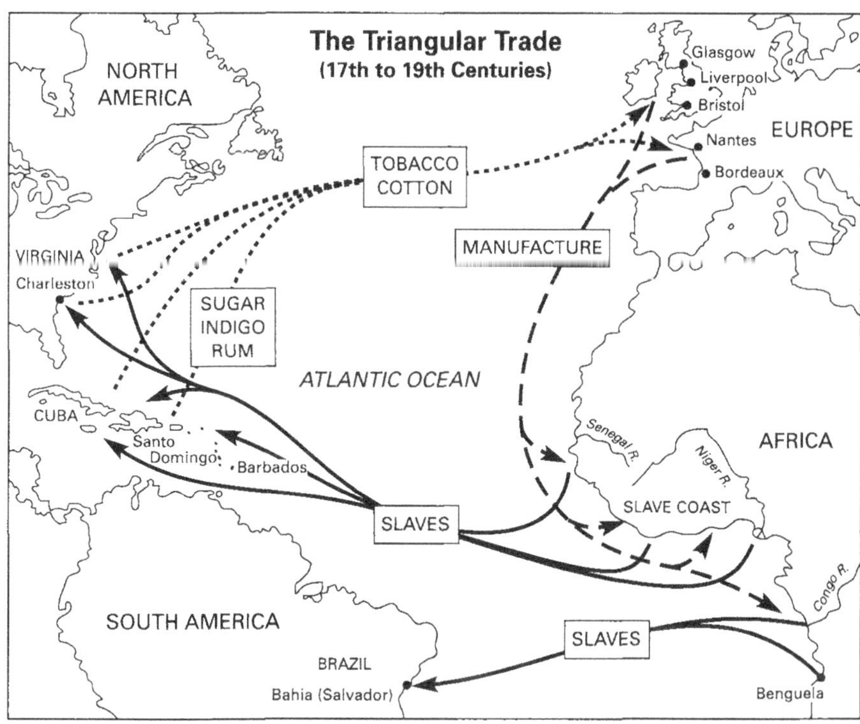

Figure 2.12 The triangular trading system (Source: Knox and Agnew, 1989)

This is not to say that there were no links between political-territorial control and economic exploitation. The eighteenth century saw Britain expanding its political-colonial control and reaping considerable economic benefits. For example, by the 1790s the West Indian and American colonies bought half of Britain's exports, which were mainly manufactured goods, and supplied a third of Britain's imports (Pawson, 1990). Both colonial areas were integrated in what became known as the 'Triangular Trade' that operated between Europe, the Americas and Africa (see Figure 2.12). Another trading system operated between Europe and India and colonies in the Far East. Political expansion would hence seem to have supported both desires for expanding markets and cheap raw materials. In addition one can also find empirical support for the theory that imperialism involved the exportation of surplus capital: by 1914 Britain had become the world's largest lender of finance (Pawson, 1990, p. 522). There was also a close connection to the social process of colonisation: Pawson (1990, p. 524) estimates that two-thirds of the capital invested overseas between 1865 and 1914 went into areas of recent European settlement.

Post-colonial modernisation

While in the nineteenth and early twentieth centuries European political expansion, economic exploitation and colonisation frequently acted in a mutually reinforcing manner, by the second half of the twentieth century the mutually supportive processes of imperialism had begun to disintegrate. Political-territorial control began to break down with the emergence of various independence movements and economic relationships began to change. In particular processes of industrialisation spread to some of the colonial areas. During the 1960s exports of manufactured goods from 'less developed countries' (LDCs) grew from about $3 billion in 1960 to over $9 billion in 1970, while by 1980 they had reached over $80 billion (Knox and Agnew, 1989, p. 262). It must be stressed, however, that this process of 'modernisation' has been very limited in its geographical extent: more than 80 per cent of the total manufactured output of the LDCs comes from only ten countries: Argentina, Brazil, Hong Kong, India, Malaysia, Mexico, Pakistan, Singapore, South Korea and Taiwan (Knox and Agnew, 1989, p. 263).

Multinationals and the world economy

Industrialisation was, however, seen in many of the newly politically independent countries as being a means of achieving economic independence. Strategies such as 'import substitution', where industries were developed to manufacture goods that had formerly been imported from western industrial countries, and 'protection' for domestic industries were adopted in the endeavour to establish independent industrialisation. Subsequent experience has, however, shown that much of the industrialisation, far from creating economic independence, has produced economic interdependency. In particular, multinational firms have come to dominate many of the economies of the 'newly industrialised countries' (NICs) and exploit them for their reserves of 'natural resources' and 'cheap labour' (see Fröbel *et al.*, 1980). In addition multinational companies have sought to increase sales of their products within these and other underdeveloped countries, even when the products are ill suited to these countries (see Chapter 5). It can be seen that even though political-territorial control may have been ceded by European states, and large-scale migration of European settlers has ended, the processes of capitalist penetration through the creation of new markets, capital investment and exploitation of human and natural resources has continued. In a sense the processes of imperialism continues, although less through the mechanisms of domination of states by another, more powerful, state and rather more through the mechanisms of multinational corporations that are effectively oblivious to nation states.

2.4.2 Imperialism and nature

The preceding section has presented an extremely brief and simplified interpretation of imperialism that stresses both historical change and continuity into the present. It is now possible to consider the roles of nature within imperialism. Attention in this discussion will focus principally on the phase described as 'imperialism and colonialism'. In other words, it will focus on the impacts of European political-territorial control of colonies, the three economic processes of imperialism and the impact of European settlement within colonial areas as expressed between the eighteenth and early nineteenth centuries. However, as the preceding discussion of the history of imperialism has highlighted, many of the features of imperialism and colonialism have some applicability to the other phases of imperialism, including that of the present.

Imperialism and the exploitation of nature as a resource

One of the most direct connections that can be drawn between imperialism and the natural is the appropriation by European states of the 'natural resources' of other countries. This appropriation and exploitation of elements of nature occurred in four principal ways, namely through: i) the establishment of plantation economies; ii) the creation of settler economies; iii) the 'modernisation' of indigenous agriculture; and iv) the establishment of industrialisation.

The establishment of plantations was one of the most observable features of imperialism: plantations were large agricultural productive units using large labour inputs. They were established by colonial powers to supply themselves with cheap agricultural products such as cocoa, coffee, sugar, tea, rubber and tobacco. They were established in tropical and subtropical areas such as the West Indies, southern America and the Far East, and generally were used to grow one major crop. Labour was frequently transported to the area of production, through either slave trading or the use of indentured or captured indigenous people.

Plantations could produce high output for the colonial powers. They were, however, highly susceptible to disease, particularly through their use of monoculture. Insecticides and pesticides were increasingly applied to such agricultural systems but the effectiveness of these frequently diminishes over time as pests develop immunity to the chemicals. Pests were one reason for plantation economies experiencing what Redclift (1987, pp. 82–3) has termed a '*cycle of monocrop cultivation*', in which one monocrop becomes replaced by another monocrop cultivation. Examples include Sri Lanka (or Ceylon as it was under British colonialism), where plantation agriculture changed crops from cocoa to coffee to cinchona to tea and rubber; and Cuba, where the sequence was tobacco, sugar, cotton, coffee and then back to sugar. Another reason for the cycle was the susceptibility of monoculture to disease: countries would grow one plantation crop until production was hit by disease, at which point another crop would be introduced. A third reason, however, was the intense competition between plantation economies, competition that allowed colonial powers to drive down prices until it became uneconomic for some countries to produce the crop, at which point production shifted to another crop.

Although plantations were a clear manifestation of European imperialism there were also a series of other manifestations, including the creation of settler economies. These economies were associated with the movement of people from the colonial powers. The migration of people from Europe tended to be towards areas with a temperate climate. Crosby (1986, p. 5) argues that between 1820 and 1930 well over 50 million migrated into what he termed 'neo-European lands'. These areas of settler economies were also, in general, in areas with relatively sparse indigenous population and settler economies hence tended to be less labour intensive than plantation economies and to emulate the methods adopted within European agriculture. The result of such emulation was not always successful although, on the whole, as Redclift argues, settler economies encouraged a rather different form of resource use than did the plantation system:

> The principal difference in resource use between 'settler' and 'plantation' societies is that in the former the local environment was looked upon as a means of support by a privileged European elite. In 'plantation societies' the integration with metropolitan capital was closer, and the renewability of resources correspondingly less important (Redclift, 1984, p. 16).

Renewability had long been a concern of the indigenous people in these areas for they, even more than the European settlers, relied upon the local environment to sustain them. Imperialism however not only *implanted* new forms of agricultural production in the colonies, it also led to the *transformation* of indigenous agricultural production. Mabogunje (1980), for example, has identified six effects of imperialism and colonialism on indigenous agricultural systems:

(1) the creation of a market in land;
(2) the fragmentation of land holdings;
(3) over-use of the environment;
(4) increasing rural indebtedness;
(5) dispersal of the rural population;
(6) collapse of traditional labour organisation.

The impact of these changes was, Mabogunje argues, the collapse of 'traditional rural structures' and an increasing dependency upon the capitalist world economy. Similar arguments have been advanced by a variety of writers including Redclift (1984; 1987), Blaikie and Brookfield (1987a) and Watts (1983), who all note how the social changes associated with the transformation of indigenous production had a series of environmental impacts, many of them adverse. These issues will be discussed further in Chapters 3, 4 and 5, but it is important to note that for some people the principal feature of imperial modernisation of agriculture was environmental and ecological degradation rather than resource exploitation.

Considerable attention has been paid to the impact of imperialism on the exploitation of environmental resources through agriculture, but it is also important to recognise that industrialisation was also a product of imperialism, at least in some places. As noted in Box 2.9, one of the economic interpretations of European imperialism is that it was driven by a search for raw materials. Many of these raw materials were agricultural in nature, although destined for industrial production within Europe. Probably the most significant example of this was the flow of cotton to Britain as part of a 'triangular system of trade', whereby slaves were transported from Africa to work on plantations in North America and the Caribbean growing cotton; the cotton was then transported to Britain for spinning and weaving into cloth, some of which then returned to Africa for sale (see Figure 2.12).

As well as there being a flow of agricultural products to the imperial core countries, there was also a similar movement in mineral products, although Rees (1990) argues that their significance has often been over-played. Minerals such as gold and silver certainly appeared to play an important role in the age of plunder, in that these were the resources most prized by many Europeans. Initially much was appropriated via tributes or as the spoils of military conquest but soon attention was paid to the mining of gold and silver, particularly in South America (see Figure 2.13). Major silver mines included those of Zacatecas and Guanajuato in present day Mexico and Potosí in the Peruvian Andes. The latter area was known as the 'silver mountain' and was worked through a form of slave labour known as the *mita*

Figure 2.13 Colonial mining areas in Latin America (after Newson, 1996)

(Bernstein *et al.*, 1996; Newson, 1987). In the *mita* system, which was originally established by the pre-colonial Incas, each village was required to supply one seventh of its adult male population to work in the mines each year. At its height it is estimated that some 13,500 men worked in the Potosí mines as forced labour (Newson, 1986, p. 24). In other areas of South America, which were not so densely populated and where labour was hence not so plentiful, mining was carried out through the importation of slave labour as in many Brazilian and Colombian gold mines, through the employment of paid labourers as in northern Mexico (Newson, 1987) or through self-employed, *petit bourgeois*, prospectors such as the *garimpo* in Amazonia. In all these forms of mining there have been considerable environmental impacts, some of which have had serious effects on human health (see Hecht and Cockburn, 1990; Veiga *et al.*, 1995; and Chapter 5)

Ecological imperialism

In the preceding section we looked at how humans, and principally Europeans, sought to exploit nature as a resource. However, as already mentioned, Crosby's (1986) notions of 'ecological imperialism' suggest that there might be a series of other, perhaps less acknowledged, connections between imperialism and nature. Simmons (1989) has argued that it is possible to classify human impacts on the environment in terms of domestication, simplification, obliteration and diversification (see Table 2.3). European colonialism and imperialism can be seen to have led to all four types of changes.

Domestication was practised by many of the indigenous populations in the areas of European colonisation: Fiedel (1987), for example, has argued that over 100 species of plants were domesticated by the indigenous people of America. However, Europeans often took domestication to new levels: indeed many of the crops of the plantation system were indigenous plants that had been domesticated. Examples of this include bananas, tobacco, tomatoes and also maize, which was important in many settler economies.

Table 2.3 Simmons' typology of human impact on the environment

Type of impact	Description
Domestication	The reshaping of a plant or animal species in the mould of what a particular human culture wants.
Simplification	Reducing the variety of animal and plant species. Often a by-product of domestication in that when some parts of the environment become labelled a resource other parts come to be seen as a hindrance (eg when some plants become 'flowers' others become 'weeds').
Obliteration	Removal of virtually all forms of life. This is generally an *unintended* consequence of human activity, such as through the escape of radiation as occurred at Chernobyl.
Diversification	An increase in variety of species in an area. Occurs when there is an introduction, either deliberately or accidentally or 'naturally', of 'exotic' species into an area.

(Source: after Simmons, 1989)

Myers (1979) has shown how imperialism has led to an obliteration of species and a simplification of natures at an increasing rate: in the period between 1600 and 1900 some 75 known mammal and bird species were made extinct, a rate of loss of over one species per year. Some of this, as Arnold (1996) makes clear, was very much a direct consequence of imperial exploitation of nature. He cites as examples the fur trade and timber trades and provides a particularly graphic statistical description of the former:

> The opening up of Canada by the French and English via the Hudson Bay, the St Lawrence River and the Great Lakes held out the prospect of a lucrative and seemingly inexhaustible new source of furs, especially much prized beaver skins. The resultant trade was massive and horrific. In just one year, 1742, Fort York handled 130,000 beaver skins and 9,000 marten pelts; in the 1760s, one of the posts of the Hudson Bay Company supplied nearly 100,000 beaver skins ... In 1743 the single French port of La Rochelle imported 127,000 beaver skins, 30,000 martens, 12,000 otters, 110,000 raccoons and 16,000 bears ... By 1831 the beaver was extinct in the northern Great Plains, and by the 1830s, after barely two hundred years of hunting and trapping, North America's fur trade was effectively over. There were few animals left to trap, and a whole rich spectrum of American wildlife had disappeared (Arnold, 1996, p. 123).

As well as simplification and obliteration as a direct consequence of European imperial exploitation there were, as both Arnold and more particularly Crosby (1986) have documented, other less obvious impacts on nature. Crosby in particular has stressed that imperialism involved not only the movement of people and products, but also involved the movement of parts of nature, most particularly plants, animals and diseases. As Johnston (1989, p. 89) records, the arrival of Europeans led, in the case of New Zealand, to an increase in the diversity of species with the introduction of 130 species of birds, 40 species of fish and 50 mammal species (including the rabbit, which was to create an ecological crisis by creating extensive soil erosion). In other countries the impact was more negative, particularly when disease as well as people, plants and animals were moved around the globe. European imperialism was indeed often associated with a simplification of nature as non-European flora and fauna were hit by disease for which they had no immunity or else were overrun by European species (see Crosby, 1986; Arnold, 1996).

The impact of disease was not just restricted to a simplification of the environment; in many areas it led to the complete obliteration of some groups of people. Smallpox, for instance, was taken to South America by the Spaniards and decimated the indigenous population. Arnold (1996, p. 78), for example, records that while the estimated population of Latin America prior to the Spanish conquest was somewhere between 75 and 100 million, by the eighteenth century it had declined to some 250,000. Stannard (1992, p. 146) has described the impact of European colonisation on the indigenous population as being effectively the 'worst human holocaust the world has ever witnessed'. European colonisation and associated movement of disease also led to the deaths of North American Indians, although the precise numbers affected is the subject of some debate. In 1798 smallpox was transmitted to Australia and killed about a third of the Aborigine population, while in

New Zealand the Maori population was similarly decimated, declining from 120,000 in 1840 to only 56,000 in 1858 (Pawson, 1990, p. 537). Decimation of indigenous populations also occurred through more directed human activity, as Arnold (1996) emphasises. An estimated 10 million Africans were taken from their communities by Europeans to work, and often die, as slaves in the plantations of America or the Caribbean or elsewhere in the imperial empires. As Bernstein *et al.* (1992) note, slave trading not only had profound impacts on the social and cultural composition of recipient areas but it also produced social, economic and political dislocation in the communities from which the slaves were taken. Through both direct exploitation and via the agency of disease, imperialism sometimes resulted in not only the destruction of previously sustainable 'natural' economies, but also at times to the wholesale destruction of indigenous populations.

The impact of imperialism and colonialism in terms of society and nature relations was clearly in many instances destructive. However, as Blaikie and Brookfield (1987a) note this was not universally the case – there are some instances where colonial authorities had strategies to improve the conservation of 'natural resources', and much more generally there was within European imperialism a 'civilising, even missionary ideology' focused on 'improvement', which might seem at odds with the destructive effects detailed above.

Imperialism, culture and nature

So far imperialism has been described in economic, political and to some extent ecological terms. Pawson (1990) has, however, argued that a necessary component of European political and economic expansion was an 'ideology of imperialism': that is 'a cultural perspective' that served to justify 'the subjugation of peoples and lands beyond Europe' (Pawson, 1990, p. 527).

Sometimes this perspective was quite blatantly expressed. Godlewska, for example, has argued that the text of the *Description de l'Egypte*, which was a survey of Egypt produced by a team of French scholars and travellers, displayed a Messianic desire for conquest and control and portrayed Egypt as 'a country blessed by nature' but in need of 'French law and ... technology to realise its full potential' (Godlewska, 1995, p. 5). Egypt was invaded by Napoleon in 1798 but remained under French rule only until 1801, when it came under British control. There were similar clear expressions of an imperialistic desire with the British Prime Minister Arthur Balfour, for example, suggesting than the Egyptians had under British rule gained a 'far better government than in the whole history of the world they ever had before' and that 'We are in Egypt not merely for the sake of the Egyptians, though we are there for their sake' (quoted in Said, 1992, p. 33).

A key element of these expressions of imperialist supremacy was conceptions of the natural environment and the relationship various people held to it. Three distinct, albeit overlapping, conceptions were commonly held. First, as Hecht and Cockburn (1990) remark, the extent and diversity of newly 'discovered' areas such as the Amazon led to a firing of ambitions for imperialist expansion and exploitation. They remark, for instance, that the extent and diversity of the Amazon rain forest led European colonisers to view it as 'virgin soil' that awaited the seed of 'civilisation' in order that its chaotic, untamed nature could be tamed, could 'come into the reach of human history and the realm of profit' (Hecht and Cockburn, 1990, p. 4). The physical environment of the colonies was seen by some colonialists as a challenge to human ingenuity and power on such a scale that only the power and knowledge of European society could conquer it.

This bolstering of the self-image of imperialism was associated with a denigration of the achievements of other societies and in particular the role of the indigenous people of the areas colonised. Again Hecht and Cockburn are instructive on this point, commenting on how the European colonisers of Amazonia were constantly remarking on both the region's richness in vegetation and minerals and the poverty of the indigenous people. The Europeans generally squared these observations by suggesting that the Amazonians were a lazy, inept people. However, as Hecht and Cockburn demonstrate, the Europeans were mistaken on two fronts. First, the Amazon although rich in vegetation has very poor soils and has many pests and diseases that make agriculture in the area difficult to sustain, particularly on any European model of agricultural development. Second, the Amazonian people had adopted some highly complex and successful agricultural practices in the face of such environmental problems. Research has found, for instance, that Amazonian Indians, such as the Kayapo, have manipulated the tropical rain forest in significant ways, including 'slash and burn' (which makes nutrients stored in the vegetation available for crop growth and utilises 'natural' successional processes of growth to keep ground covered to prevent soil erosion) and the planting of useful species, such as Piqui tree, along their trekking trails, camping sites and hunting areas. The exploitation of nature by these Indians was hence undertaken on a very spatially extensive scale (the lands of the Kayapo, for example, are equivalent in size to the area of France). Such spatially extensive practices do not accord well with the European concept of agriculture as involving intensive, localised, permanently located exploitation of nature within a clearly identified 'farm'. Certainly the existence of any agricultural practices in the Amazon by indigenous people was consistently overlooked by European colonisers. Even today the Amazon is portrayed as a 'natural forest', a forest external to human activity. However, Hecht and Cockburn note that even today some two million people make their living from the forest and that wherever one looks in the Amazon forest 'the landscape almost invariably bears the imprint of human agency'. They add that the imperialist's 'patronizing eye', however, 'sees nothing of this and the image of the forest as a wild biological entity remains, making it easier to envisage forest clearance as the only rational form of development' (Hecht and Cockburn, 1990, p. 37).

A third common element in the construction of an attitude of imperialist supremacy was the influence of various theories about the relationship between society and nature and in particular those philosophies that attempt to draw a dividing line between people and the other elements of nature. As we discussed in Chapter 1, a series of distinctions have been drawn between society and nature; between human and non-human elements of the world. In the age of imperialism, and since then too, distinctions between human and non-human have been applied to establish social divisions between people. Pawson (1990, p. 528), for example, remarks how '[w]omen, children, the socially marginalized – even the Irish' were frequently described as being 'closer to nature, some or all of their characteristics being held to reflect animal qualities'. Likewise other cultures 'particularly those without accumulated wealth or skills respected by Europeans' (Pawson, 1990, p. 528) were assigned a place nearer the animals than to the so-called 'civilised' European. In this manner, distinctions drawn between society and nature became used to justify social discrimination and inequality: as Pawson puts it 'an hierarchical conception of the natural world' became a 'cultural pillar of historical capitalism'. Distinction between society and nature became used to 'naturalise', to make universal and unexceptional, social hierarchies based on processes of exploitation.

2.4.3 Summary

In this section we have discussed the concept of imperialism and noted how it has been interpreted in a variety of different ways. We have outlined how the history of imperialism, in the general sense of European domination of other countries, can be classified into four stages – mercantile plunder, colonialism and imperialism, modernisation and the contemporary stage of multinationals and the global economy – but have focused our attention particularly on colonialism and imperialism. We have highlighted how colonisation and imperialism were connected into the exploitation of natural resources, and also how elements of nature had their own agency within imperialism, as in the spread of animals, plants and disease as an unintended by-product of the movement of people and products. Attention has also been drawn to how nature acts as a symbolic resource and how within imperialism it was used to bolster the imperialist at the expense of the colonised. Hence as well as the social exploitation of nature there was also social exploitation through nature.

2.5 Post-industrialism

The final form of society–nature relations to be discussed in this chapter is that of 'post-industrialism'. A wide range of authors (e.g. Bell, 1973; Tourraine, 1974; Ley, 1980, 1987; Harvey, 1989; Soja, 1989) have argued that some time after industrialisation there emerged a number of 'post-industrial' or 'postmodern' societies or social spaces. The precise meaning of the terms 'post-industrial' and 'postmodern', and their interrelationship, has been the subject of considerable debate. One of the most direct definitions that is of value here is that of Lee (1986, p. 362), who states that a post-industrial society is 'a society in which manufacturing no longer dominates economic activity'. Under such a definition, countries in which the number of people employed in primary and secondary activities is exceeded by the number of people employed in 'tertiary' or 'service' occupations could be said to be 'post-industrial'. Table 2.4 shows countries that did, and did not, fall into such a definition in 1998.

Considerable attention has been paid in the literature to the formation of such societies and the social and economic consequences of having a post-industrial occupational structure (e.g. Tourraine, 1974; Thrift, 1987; Urry, 1995a). In the present context, however, what is of principal concern is the use of nature in societies where industry no longer dominates economic activity. If one adopts the

Table 2.4 Post-industrial and non-post-industrial societies in 1996

Country	Percentage of working population employed in service sector
The Netherlands	78
Hong Kong	74
Canada	73
USA	73
Japan	61
United Kingdom	55
Argentina	47
Bolivia	43
Portugal	42
Poland	34
China	16

Notes: Service sector defined as categories 6–9 of International Standard Industrial Classification of all Economic Activities, revision 2 and G–Q of the International Standard Industrial Classification of all Economic Activities, revision 3.
(Source: International Labour Office, 1997)

perspective of people like Simmons and Wrigley, that industrial societies are fundamentally linked to heightened exploitation of the resources (and particularly energy supplies of the physical, 'natural' environment) then the suggestion that this form of society may no longer be present within particular areas is of considerable environmental significance. One view espoused, for example by MacCannell (1989), is that post-industrialism signals a reversal of many of the trends that characterised industrial society, including 'mechanisation', in which technology is used to replace people in the production process (and hence their organic energy is replaced by largely mineral energy), and 'material'-centred production. With reference to the latter, the products of the service industries that are a central component of a post-industrial economy are such 'intangibles' as information, credit and personal service. A key question that this section will seek to address is whether the movement from mechanisation and material production can be said to mark the end of the 'exploitation of nature', or whether this exploitation is merely continuing (and even expanding) in new forms. This question will be addressed through an examination of three important elements of post-industrial society: the tourist industry, the mass media and a counterurbanising population.

2.5.1 Uses of nature in post-industrial society: case study 1 – the tourist industry

For MacCannell (1989) tourism is central to a post-industrial economy, and one can certainly see in tourism many of the distinguishing features of a service economy. For example, tourism is very labour intensive: it has been calculated that tourist-related services now employ some 60 million people world-wide (Urry, 1992, pp. 1–2), while in the UK alone the figure is 1.5 million people, said to be increasing at a rate of some 1000 per week (Urry, 1990, p. 5). Many of these jobs are, in contrast to traditional industrial jobs, part time in character. They are also often seasonal and may involve unusual hours. In these senses tourist jobs exhibit flexibility. Furthermore, tourism is an 'intangible' product and in very many instances does involve highly personal interactions: the quality of service, for instance in a restaurant or hotel, may critically influence whether a holiday is a pleasurable experience. It has also been suggested that tourism is a 'natural', renewable resource industry in that people simply gaze on resources rather than actively consume them (Murphy, 1985).

The term 'tourist' is derived from the word *tour*, which means 'a journey at which one returns to the starting point' (Murphy, 1985, p. 4). For the tourist, the point of departure and return is generally that of the home. Tourism can, hence, be said to be the leaving of or departing from the home environment, with its routines of work and everyday life. The practice of taking such a break from everyday practices and environments has become a normal feature of most people's lives in modern western society. As Urry has put it:

> To be a tourist is one of the characteristics of the 'modern' experience. Not to 'go away' is like not possessing a car or a nice house. It is a marker of status in modern societies and it is also thought to be necessary to health (Urry, 1990, p. 4).

The general practice of taking a holiday, of taking a break from the normal routines and environments of everyday life, is a relatively recent phenomenon. Before the nineteenth century it was only members of the upper classes who travelled away from their home environments for any reason apart from work or business. During the nineteenth century tourism emerged as a mass phenomenon. People from a variety of classes began to travel 'on holiday', with, for instance, Thomas Cook running the first 'package' trip in 1844. Mass tourism gradually expanded through the second half of the nineteenth century and into the twentieth. The second half of the twentieth century has seen a rapid expansion in mass tourism on a global scale. The number of international tourist air trips in the world, for example, rose from 25 million in 1950 to 183 million in 1970 and 300 million in 1982 (figures from Murphy (1985, p. 3) and Urry (1990, p. 47)). In 1981 there were an estimated two billion domestic tourist trips in the world, representing a 240 per cent increase on the 1975 figures (Murphy, 1985, p. 3), although it should be noted that domestic trips, that is trips that do not cross an international boundary, are very difficult to quantify accurately. Other measures of tourist growth include the amount of money spent on tourism (said to be $488 billion in 1978 and $919 billion in 1981) and the number of people employed within the tourist industry (in 1975 an estimated 1.5 million people were employed in tourist or related employment, by 1989 this figure had risen to 60 million) (figures from Murphy (1985) and Urry (1992)). This globalisation of the tourist industry has, in part, been built upon transformations in technology: the nineteenth-century growth of package holidays and seaside resorts, for instance, was dependent upon the expansion in railway travel. Similarly the post-war growth in international travel has been achieved through the use of air travel. The globalisation of tourism has, however, also been based on first, changes in the social organisation of travel and tourism as emphasised by Urry (1995d), and second, exploiting nature in new ways, and in particular through exploiting the perceptions and feelings that people have about elements or features of the 'natural world'. It is this latter aspect of contemporary tourism on which we will focus particular attention, although we would add that it is closely interlinked with issues of technology and social organisation.

As discussed in Chapter 1 people's perceptions and feelings about nature are varied, ranging, for instance, from placing 'nature' at the centre of one's life within extreme ecocentrism to denying it virtually any existence within extreme technocentrism. People such as Urry (1990; 1992) and Harrison (1991) have suggested that a particularly important view within contemporary tourism has been a 'romantic view of nature': that is a view that places great stress on the value of experiencing parts of a physical environment that stands, as far as is possible, outside society. Indeed for Urry (1990) the 'romantic gaze' has been a major component within the globalisation of tourism. In particular he argues that a change from a 'collective' to a 'romantic' attitude to tourism was an important element in the demise of one of the central tourist attractions within nineteenth-century British tourism: namely, the seaside resort. The collective view or 'gaze' of tourism sees tourism as an experience that should be conducted in the presence of others. People are necessary to add a sense of fun or 'carnival' to the holiday experience. Up until the 1960s British seaside resorts were seen to be the place where such collective experiences could be gained, but since then many of these seaside resorts have been in decline. This is partly through competition from new locations for the collective experience, such as 'theme parks' like Disneyland and Alton Towers, 'foreign resorts' such as the Balearic Islands and the coastal resorts of Spain, and cosmopolitan global cities such as London, New York, Paris, Sydney, Hong Kong and Bangkok.

While theme parks, foreign resorts and cities can be seen as extensions in the localities for the collective experience of tourism there has also been a movement of tourism away from such sites and into a more dispersed and global pattern. This movement has, Urry argues, been very much influenced by the rise in the romantic view of tourism in which emphasis is placed on 'solitude, privacy and a personal, semi-spiritual relationship' (Urry, 1990, p. 45) with the tourist 'attraction', which is frequently seen to be an object of the 'natural world'. Such natural objects include 'natural landscapes' and, within safari holidays, the 'animals of nature'. There has also been since the 1980s an emergence of so-called 'green' or 'eco-' tourism (see Jones, 1987; Goodwin, 1996) in which the conservation of nature also become a tourist object. According to Jefferys (1998) ecotourism is the most rapidly growing form of tourism, worth some $175 billion in 1995, a figure that is expected to double by the year 2000. Tourists seeking these various 'romantic' experiences of nature have shunned traditional tourist sites and turned to new areas. At the international level this has led to the 'globalisation' of tourism as the search for the romantic experience, an experience of nature largely devoid of the presence of other people, has led to virtually every country in the world becoming a tourist destination for someone. Each year a new holiday destination becomes *the* place to go to: a few years ago it was the West Indies, then it was Turkey, then The Gambia and Mauritius, this coming year – who knows! As Briguglio *et al.* (1996) note, many of the most rapidly growing tourist destinations appear to be 'nature oriented'. So, for example the Galapagos Islands, on which giant tortoises live, had virtually no tourist industry in the 1970s but now receive over 36,000 visitors per annum, while in Manaus in the centre of the Amazonian rain forest the number of tourists increased from 12,000 in 1983 to 70,000 some five years later (Warner, 1991).

The continual 'discovery' of a new tourist destination is an inevitable consequence of the growth of a romantic view of tourism; as Urry (1990, p. 46) puts it, the more people seek a romantic experience through tourism the more the conditions for finding a romantic experience are undermined. It is, for instance, hard to find a solitary experience of nature if other people are at the same location in search of a similar experience. Hence the romantic tourist tends to move on to a new location once their favoured location becomes popular.

Shields (1991) has provided a useful study of the undermining of a 'romantic experience' with the arrival of mass tourism in his study of Niagara Falls. The Falls were 'discovered' in 1697 by Father Hennepin, who was the first non-indigenous person to see them and who described them as 'a Prodigy of Nature', 'an incredible Cataract or Waterfall, which has no equal' (Hennepin, 1698, quoted in Shields, 1991, p. 117). Niagara, Shields argues, became a symbol of a remote, awesome 'nature' that could never be tamed by society. Indeed it was argued that the scale of the Falls was so great that they could never be captured, even in terms of an artistic representation; in 1806 for instance, it was claimed that the Falls 'will forever scorn the confinements of art' (*Monthly Anthology and Boston Review*, quoted in Shields, 1991, p. 119). The only way to experience the wonder of the Falls was said to be to visit it: a journey that was, up until the construction of the Erie Canal in 1825, difficult and somewhat hazardous.

Niagara Falls up until the early nineteenth century fulfilled many of the conditions of the romantic tourist in that it was: a) a journey that took the tourist beyond the geographical and experiential confines of mass society and its technologies of reproduction; b) a journey that only a few people could undertake; and c) a journey to an object that demonstrated the power of 'nature' and the impotence of human agency. Niagara could become an object outside society, within nature, and around which the 'tourists' could re-evaluate and re-order their lives. Niagara was, so Shields argues, a shrine of nature to which people in the eighteenth and nineteenth century effectively made a pilgrimage.

This iconic status was, however, undermined during the course of the later nineteenth century and in the twentieth century. Mention has already been made of the Erie Canal, and Shields (1991, p. 124) suggests that the opening of this: 'proved to be an abrupt turning-point for Niagara. Suddenly it was opened to trade, industry and tourism on such a scale that it could no longer remain a sacred separate landscape'.

The Falls quickly became transformed from a personal experience of nature to a communal experience of spectacles: two years after the opening of the Canal, for example, an estimated crowd of 15,000 people assembled to watch a condemned schooner, the 'Michigan', go over the Falls. Soon there were stalls selling souvenirs of the Falls and exhibitions of exotic animals. Rather than being an unrepresentable phenomenon of nature, Niagara became a widely reproduced sign, appearing, for example, on a whole host of commodities and services. In the process, awesome Niagara became a demonstration of the mastery of society over nature, most particularly the powers of technology and commerce to represent and exploit nature.

This twin usage of Niagara has continued at an ever-increasing rate within the twentieth century, where the Falls have been used to portray many products, people and forces within contemporary society. Shredded Wheat cereals, for example, make use of a stylised image of the Falls as its brand logo, while Niagara became the setting for numerous film and television shoots, including a 1952 film entitled *Niagara* starring Marilyn Monroe and Joseph Cotton. In the promotional posters for this film Marilyn Monroe was likened to Niagara in a way that clearly resonates with the naturalisation of gender difference mentioned in the context of imperialism and in Chapter 1: '*Marilyn Monroe and Niagara:* a raging torrent of emotion that even nature can't control' (Shields, 1991, p. 143). Niagara has also become a frequently used image in the promotion of the nation states of both Canada and the United States. With such commercialisation and reinterpretation the Falls have become ill suited to a tourist in pursuit of a romantic experience of nature and increasingly tourists are said to have sought other sites, initially wilderness areas such as Yellowstone and Yosemite and increasingly within the contemporary period through travel to foreign countries.

Similar processes of 'discovery' and 'destruction' of romantic tourist attractions can be found across the globe. Within Britain the romantic view of nature has been particularly significant in the development of countryside recreation. Harrison (1991), for instance, has argued that a 'countryside aesthetic', a belief that the value of the countryside to society lies in its 'landscape' and that its appropriate use lies in the pursuit of solitary and quiet experience of this landscape, has dominated policies on recreation in the countryside throughout post-war Britain. This argument is supported by MacNaghten and Urry, who examined a range of policy documents of the *Countryside Commission*, which was concerned both to conserve the English countryside and to promote its recreational use, and found that there was:

> a repetitive emphasis on the 'beauty of the English countryside' [which] particularly embodies the 'romantic' gaze. Whether alone, or in company, the model of the person presented is of a privatised individual experiencing and consuming qualities associated with a natural beauty ... Linked to this 'gaze' in such documents are pictures of 'unspoilt' countryside – usually unpeopled, majestic and awe inspiring. The message implies a ... method of engagement ... of quietly and unobtrusively acquiring these enduring qualities. Farmers are either hidden from the tourist's gaze *or* simply referred to as 'markers of the land' (constructing them symbolically as integral to the romantic gaze), while the significance of the vernacular, the familiar and the locally distinctive are down played. Similarly the leisure subject is usually hidden from the texts (since the tourist has no place in the romantic gaze) and no space is allocated for a more intrusive, active or participatory type of engagement in the countryside (MacNaghten and Urry, 1998, pp. 187–8).

As Harrison has outlined, amongst the consequences of this aesthetic has been the expansion of recreational provision for the quiet enjoyment of the countryside: one can include here, for instance, the development of national and country parks and the protection and expansion of public footpaths, including long-distance footpaths such as the Pennine Way. Other consequences which she, and also MacNaghten and Urry (1998) have identified, include the 'institutionalisation' of countryside recreation whereby membership of an organisation becomes the means of gaining access to the countryside. Organisations such as the *Camping and Caravanning Club*, the *Ramblers Association* and the *Youth Hostels Association* (YHA), for example, have long been granted preferential access to parts of the countryside, for instance by being allowed dispensations from some planning restrictions or requirements. Closely connected to this institutionalisation of countryside recreation has been both *'social exclusion'* and *'social disciplining'*. The former refers to the way some people have been effectively denied access to the countryside, a process that is in part the direct result of the institutionalisation of countryside recreation, in that by privileging some uses and users of the countryside one is by necessity curtailing the relative accessibility of the countryside to other people. Harrison (1991), for example, argues that people who do not seek solitary and quiet countryside recreation have effectively been denied access to the countryside. Social exclusion from the countryside may also be a direct consequence of the adoption of the countryside aesthetic, in that it is seen to be necessary to restrict the number of people visiting the area because otherwise the very quality of the area that attracts people is seen to be lost. Many governmental agencies responsible for areas of the countryside have adopted a range of people management strategies in order to 'conserve' the natural and amenity value of popular areas of the countryside. These strategies run from restricting the number of car-parking spaces and the width of roads in popular areas, through the development of counter-actions to divert people away from particular 'honey-pot' areas through to charging for access. All these strategies are exclusionary, and the last has very significant social implications in that it is most likely to restrict access by lower-income groups. For this reason there has been a tendency for public bodies in Britain, and indeed in many other countries, to avoid charging for access to the countryside. However, recent years have seen an increasing privatisation and marketisation of the countryside, processes that are seen to have led to an increasing 'commodification' of the countryside (see Cloke, 1992; 1993).

As well as contributing to social exclusion the countryside aesthetic and romantic gaze can be seen to have been connected to what may be described as 'social disciplining' or the 'ordering of ways of behaving' (see Foucault, 1977; Matless, 1994; Driver, 1997; MacNaghten and Urry, 1998). As MacNaghten and Urry (1998, p. 188) note, this sometimes occurs through overt legislation but more usually through 'codes of self-discipline'. A clear example of the former mode of social discipline is the 1994 *Criminal Justice and Public Order Act* which, for instance, allowed police to prevent people from assembling on land where such assembly might cause serious disruption to local communities or damage 'to sites of historical, architectural, or scientific importance', or where people were meeting for the purpose of 'attending or preparing for a *rave*', where this was defined as listening to music that 'includes sounds wholly or predominantly characterised by the emission of a succession of repetitive beats' (quotes from the Act in Halfacree, 1996; McKay, 1996). Examples of less legalistic social disciplining can be seen to be associated with the institutionalisation of countryside recreation in that, as part of their being granted preferential access to countryside, members of organisations such as the Camping and Caravanning Club, the Ramblers and YHA have to agree to abide by particular 'codes of behaviour' that enact the countryside aesthetic. Even people outside these organisations are entreated by various agencies to follow particular codes of conduct, most notably the 'Countryside Code', which was established in 1957 by the *National Parks Commission*. As MacNaghten and Urry (1998) argue, this code both ensures that recreationalists behave in a manner that is in line with the interests of private landowners – great stress is, for example, placed on not trespassing on land, damaging crops or disturbing livestock – and also helps sustain the notion that the right way of enjoying the countryside is to be involved in 'quiet, orderly forms of behaviour'.

It is, therefore, possible to suggest that a 'romantic' or 'countryside aesthetic' has been an important element in determining the course of both international tourism and more local countryside recreation, although it must be stressed that a variety of other factors, such as technological, social organisational and political economic changes have also played a part. One question that is raised by the significance of the romantic gaze in contemporary tourism and recreation is why it is so important to so many people. One reason may be the way the countryside and environmental issues have been an important subject within the media.

2.5.2 Uses of nature in post-industrial society: case study 2 – the media

The media is another prime example of a service industry: it produces an extremely intangible product (an image, sound, impression or message) and it is frequently portrayed as a highly creative, personal activity. It is also quantitatively a highly significant element in contemporary society. In the United Kingdom, for example, 99 per cent of households had a television in 1995 and 80 per cent a video recorder (Office for National Statistics, 1997; see also Burgess, 1990). In the United States just over 98 per cent of households had a television in 1995 and the average number of television sets per household was 2.3 (US Bureau of the Census, 1997). Globally it has been estimated that there are some 135 million television sets in use, a figure that has tripled over the last 20 years (Rayner *et al.*, 1998).

The media industry produces, or creates, a wide range of images or messages, or 'symbolic products' as they are sometimes called (see Burgess, 1990). These symbolic products are transmitted by a variety of media – newspapers, magazines, books, television, radio and cinema. They are also produced by a wide range of organisations, ranging from the explicitly commercial products of advertising agencies, through commercial media producers (such as commercial television and radio companies, newspaper and book publishers), through public service broadcasters such as the BBC and into the local productions of community groups and individuals. These media organisation produce a multitude of images, some of which are mutually supporting, some of which are contradictory. Amongst this myriad of images it is, however, possible to discern a plethora that would appear to foster particular images of nature, including many that connect to a romantic image of the countryside.

Romantic imagery for example is frequently used in advertising, in particular what has been termed 'lifestyle advertising' (Thrift, 1989, p. 18). The lifestyle advert is one that tries to sell a product by placing it in a particular setting and producing associations between the product to be sold, other products and settings. Romantic nature is often used as a preferred setting.

Such images are not just used in advertisements. Phillips *et al.* (forthcoming), for example, highlight the popularity of rural television drama while Thrift (1989) points variously to the 'flood of books on the country'; to the development of 'country style' clothing, notably the Barbour jacket, the 'green wellingtons' and the tweed jacket; the creation of 'country look' house interiors; and the publication of a rising number of rural magazines. Thrift, and also Urry (1990), link these developments, and the other manifestations of a concern with a romantic countryside, with the growth of a so-called 'service class'. The service class is a term used to describe the people who work in the service industries or undertake service functions: it is composed of managers, professionals and creative service producers. It therefore includes many of the people responsible for creating images like those in lifestyle advertisements (see Thrift, 1989). It is also the group that participates most in countryside recreation: in Britain it was calculated that in 1986 one-third of socio-economic groups A and B, that is those in professional and managerial occupations (or the service class), made at least five visits to the countryside in a year, a figure twice that of semi- and unskilled workers (Thrift, 1989, p. 35; Urry, 1990, p. 97). Indeed Thrift (1989, p. 31) has suggested that the service class has taken the countryside 'to its heart'. It is also, and perhaps relatedly, the group that has participated most in one of the most significant ways in which the resources of the countryside, and particularly land, have been used in recent years: namely counterurbanisation.

2.5.3 Uses of nature in post-industrial society: case study 3 – counterurbanisation

Counterurbanisation can be seen to refer to a process of population deconcentration through the growth in populations living in 'non-urban' areas and an associated decline in the population living in urban areas. Moreover, as Halfacree (1994, p. 164) notes, the growth in the non-rural population is generally explained as a consequence of 'net migration to rural areas, rather than by adjustments in the rate of 'natural change', where this last term is seen to be balance of birth- as against death-rates. Counterurbanisation is seen to have emerged in western 'developed' countries in the 1960s and continued at an increasing rate through the 1970s and 1980s, although there is some suggestion that its rate of growth has fallen during the 1990s (Champion, 1989; 1992). By the start of 1988 more than half of the new houses being built in Britain were in rural or semi-rural locations (Urry, 1990, p. 97).

One explanation of the process is that it reflects the rising importance of the service economy within the core countries. Service activities have rather different, and rather less severe, locational requirements than industrial and manufacturing activities. It was therefore possible for service firms to become more geographically dispersed. As a result their was a deconcentration of employment and people followed the jobs into the countryside. This dispersal was encouraged, some people argue, by the preference of members of the service class for rural residential locations, or more precisely one should say *certain* rural residential locations. Thrift (1987; 1989), for example, has argued that service-class members are attracted to rural locations that fulfil a particular expectation of the countryside: an expectation that is in part informed by the romantic notion of a natural landscape outside modern society. As Urry (1990, p. 97) puts it: 'The countryside is *thought* to embody some or all of the following features: a lack of planning and regimentation, a vernacular quaint architecture, winding lanes and the lack of social intervention'.

McLaughlin (1986) has similarly talked of a 'village England ideology' that draws upon a naturalist, romantic aesthetic to portray the countryside as a model site for living: its natural beauty being taken to imply that it is necessarily a nice, wholesome place to live. McLaughlin suggests that this image is highly misleading and, particularly when incorporated into government and local authority policy, has socially negative consequences. In particular McLaughlin argues that the image of the countryside as an idyllic place to live has led to the perpetuation of rural deprivation: policy makers have been taken in by the image of the countryside as a place outside the ravages of modern society and have therefore assumed, wrongly, that there are no social, or indeed environmental, problems in these areas. The images of the countryside and the natural environment created by the media industry and so influential in both the tourist industry and the spread of the service class into rural areas have tended to ignore some of the environmental and social problems of these areas, and arguably have helped exacerbate these problems. Indeed, while the shift to a post-industrial economy and society is often portrayed as being environmentally and socially advantageous it may well have serious negative impacts. It is to outlining some of the impacts of the post-industrial economy on parts of the 'non-human' or 'external' nature that we shall now, briefly, turn. We will then outline some discussions that have emerged over the 'social significance' of the use of nature in a post-industrial society; in particular we will explore how people's use of external nature may have impacts on their 'internal nature'. Finally we will briefly raise some issues about the validity of the concept of post-industrial society that stem from its continuing use of external nature.

2.5.4 Impacts of post-industrial economies on nature: on external nature

As discussed at the beginning of this section on post-industrial society, the onset of such a society is often portrayed as being environmentally advantageous, largely because it is centred around the production of 'non-material' images and services rather than material products extracted from nature. Post-industrial societies may be seen as involving a shift from the consumption of material nature to the consumption of nature as sign. Tourism, for example, particularly in the form of the aesthetic appreciation of nature, has been portrayed as being a 'natural' renewable resource industry' (Murphy, 1985, p. 1), in which people simply come to gaze upon a landscape, historic monument or some other tourist attraction. No resources are consumed, there is therefore no possibility of a resource crisis.

However, as the tourist industry has expanded it has become apparent that tourism involves rather more than simply looking at a tourist attraction, and that it can have considerable impacts upon the resources on which it is centred. Mishan, for example, has argued that:

> The tourist trade, in a competitive scramble to uncover all places of once quiet repose, of wonder, of beauty and historic interest to the money-flushed multitude, is in effect quite literally and irrevocably destroying them (Mishan, 1969, p. 141).

Within Britain it has become clear that countryside recreation, even when in accord with a heavily managed and institutionalised 'countryside aesthetic', can lead to the destruction of resources. Duffus and Dearden (1990), for example, have argued that recreational activities can be classified into those that are 'environmentally consumptive', in that they remove something from an environment that can never be replaced, and those that are 'non-consumptive', in which elements of an environment are not 'deliberately' removed by people. However, while people undertaking activities such as walking may not deliberately seek to remove or destroy a part of the environment, in many instances this is the consequence of their actions. A series of studies have suggested that 'human trampling' quite literally impacts on nature, in that the impact of, often heavily clad, feet can lead to: a) soil compaction; b) an increase in the distribution of bare ground by removing vegetation from the area; c) an increase in soil pH, through either mechanical compaction or vegetation removal; and d) reduced water infiltration and increased surface runoff, thereby leading to accelerated erosion, and gullying (see Cook-McGuail, 1978; Cloke and Park, 1985; Lance *et al.*, 1989). Soils such as peat with a high 'void capacity', or space within the soil, can be particularly impacted, and serious erosion problems have been identified in peat moorland areas traversed by popular footpaths such as the Pennine Way within Britain. People walking along the Pennine Way can, hence, be said to be 'consuming the peat' by walking on it, as much as if they were going on to the moor to physically dig up the peat.

It has similarly become increasingly evident that the global search for a romantic experience of nature has led to considerable environmental destruction. For instance, the onset of tourist development in the Algarve, Portugal, has been compared, in terms of its impact on the relief of the area, to the onset of coal mining in South Wales at the beginning of the industrial revolution (Barrett, 1989). Even destinations seemingly remote from mass tourism have experienced considerable environmental impacts. In Chapter 1 (Box 1.1), for example, mention was made of the impacts of tourism on Antarctica. Another revealing case is the mounting of a special expedition to Everest Base Camp in 1989 simply to clear up the litter left by tourists and climbers (Ryan, 1991, p. 96). Many areas in the Himalayas are experiencing deforestation as a result of providing tourists with hot water and fuel (Kohl, 1989). This is at the same time as local people are forbidden from using wood to heat their homes. Local people are also frequently drawn into the destruction of natural resources. In Barbados, for example, the ecology of the coral reefs is being destroyed by a combination of local people removing plants and fish from the coral and the pumping of raw sewage from beachside hotels (Urry, 1990).

Such environmental destruction can frequently be avoided by adopting various management strategies, many of which have been described as constituting a form of 'green', 'eco-' or 'sustainable' tourism. Goodwin (1996), for example, defines ecotourism as:

> low impact tourism which contributes to the maintenance of species and habitats either directly through a contribution to conservation and/or indirectly by providing revenue to the local community sufficient for local people to value, and therefore protect, their wildlife heritage area as a source of income.

Examples of this form of tourism include the Huaorani Indian community of Quehueire, Ono in the Ecuadorean rain forest, that has set up a self-managed ecotourism programme that aims to create a sustainable tourist economy and prevent oil companies from clearing the forest (Holmes, 1997), and the *Dolphin Fleet* in Provincetown, USA and the *Seal Sanctuary* in Gweek, Britain, which both fund marine research and conservation activities out of receipts from tourist visitors. On the other hand, it is also clear that many ecotourist iniatives do not live up to all their claims. Jefferys (1991, p. 27), for example, argues that 'ecotourism has become something of a "bandwagon" term' essentially just used by tour operators to attract customers by adding an 'eco-prefix' (see also Goodwin, 1996; Buckley and Araujo, 1997). Indeed it is possible to argue that there is an inherent tendency, given the way tourism is priced, for environmental damage to be neglected until it has *seriously* damaged an environment. The 'natural resources' used by tourism are frequently seen as public or common goods: goods for which no one pays. There is a strong incentive for people to use these resources as soon as possible – to enjoy the unspoilt view before anyone else does. Delay in use will not increase the value of the good and will in all likelihood see its value diminish as other people make use of the resource. Such factors make controlling resource use highly complex. The problems associated with managing resource use will form the subject of the next two chapters of this book. For the moment it is only necessary to note that these problems remain highly significant within at least the tourist sector of a post-industrial economy.

2.5.5 Impacts of post-industrial economies on nature: on human nature

The impact of tourism on the physical world may be considerable; it has also been suggested that tourism has had profound impacts on the nature of people's relationship with their environment. The precise nature of the impact of tourism on people's experiences is a matter of some debate, with at least three positions having been advanced (Urry, 1990). First, for some people such as Boorstin (1964), contemporary tourism has seen people becoming divorced from real, direct experiences of the natural world. Tourism is seen as an artificial experience in which people are transported across the world within an 'environmental bubble' that isolates them from real experience and within which the tourist watches stage-managed spectacles or 'pseudo-events'. This form of tourism is exemplified in the package holiday where it has been suggested that travel agents, couriers and hotel managers act as surrogate parents, relieving the tourist of problems and protecting them from any harsh reality.

Some people have objected to this portrayal of tourism and suggest that, rather than being the epitome of modern commoditisation, tourism reflects a contemporary manifestation of a search for an authentic experience of nature. This is the view espoused by MacCannell (1989) and also appears to be the view held by Shields (1991) in his discussion of trips to Niagara Falls. Contemporary tourism according to this view reflects a 'democratisation' of the search for the authentic although, as the quote from Mishan reveals, for some people the arrival of mass tourism leads to the inevitable destruction of the authentic. Urry (1990) has, however, argued that Mishan's arguments are essentially expressions of middle-class anxiety about the fate of they resources they use: the poor still do not have the opportunity to indulge in a search for the authentic and the really rich are able to retain exclusive access to environmental resources within expensive resorts, or on their private yachts or secluded islands.

Very similar comments may be applied to counterurbanisation: one motive for moving to rural areas may be that they are seen to be places where there are more original, more authentic relationships with a natural world (see Bell, 1994; Halfacree, 1997). Another motive may be, as noted earlier, the desire to live in a quiet, unpopulated, spacious landscape. In either case, by contributing to the process of counterurbanisation, migrants to the countryside are arguably destroying the very object they desire. Those who are rich enough may be able to protect their desired object, for instance by retreating into a 'private estate' or by moving further from the city, while those with less wealth may be forced to accept the alteration of their environment: to live in what they know to be an 'inauthentic village'. Evidence of such an acceptance of 'pseudo' rurality is the growth of housing estates that incorporate elements of 'vernacular' or 'rustic' design consciousness, that is they are fairly standardised houses that incorporate selected elements of local or rural housing styles (Cloke and Thrift, 1987). A clear instance of this is the Watermead development near Aylesbury, Buckinghamshire, which has been studied by Murdoch and Marsden (1991; 1994). As Plate 2.1 indicates, the design and promotion of housing at Watermead makes clear use of rural and rustic imagery, not least with the inclusion of ducks that are arguably suggestive of a traditional village pond. As Cloke and Phillips (forthcoming) and Cloke *et al.* (forthcoming) document, to some people such estates are merely a stepping-stone towards their ideal of a real country home, while for others their 'inauthenticity' does not matter.

Similarly, within tourism it has been suggested that the quest for authenticity is meaningless. Urry (1990), for example, argues that it is incorrect to see the search for authenticity as the basis for tourism. Rather tourism, and perhaps by extension rural living and countryside recreation, acts merely as a means of expressing difference from, indeed distance from, the realm of organised work and everyday experience. Within tourist research Feifer (1985) has suggested the term 'post-tourists' is applicable to some travellers:

> Post-tourists ... almost delight in the inauthenticity of normal tourist experience. 'Post-tourists' find pleasure in the multiplicity of tourist games. They know there is *no* authentic tourist experience, that there is merely a series of games or texts to be played (Urry, 1990, p. 11).

Tourism to such 'post-tourists' becomes simply a reading of the various signs or images that are presented to them, both within the tourist experience and more widely through the media industries.

These distinctions between forms of tourism show some parallels with the philosophical positions that have been advanced about the concept of nature (see Figure 1.1). The search for the authentic can be seen as a 'naturalist' search for social order from nature. Tourism as 'pseudo-event' also suggests that nature may provide a grounding for an authentic, fulfilling way of life. It suggests, however, that this way of life is not possible within the structure of contemporary society: the post-industrial society, in this interpretation, is forever alienated from a natural way of life. This viewpoint may be seen as a dualistic construction of

Plate 2.1 New constructions of rusticity: the case of Watermead

society–nature relations. The 'post-tourist' adopts a more social constructionist position and rejects notions of an independent nature or natural state. The 'post-tourist' cannot be alienated from anything, given that for them there is nothing outside society to be alienated from.

2.5.6 Is post-industrial society really post-industrial?

In this section we have sought to outline the ways in which forms of nature have been used in a variety of elements of the so-called post-industrial economy. It has been noted at several points that the use of nature within tourism, the media and by rural residences can perhaps be seen as manifestations of a non-exploitative relationship with nature, in that the nature used is not consumed in the same manner that it is within, say, coal mining and the consumption of fossil fuels in power stations. Clearly there may be some instances of localised environmental degradation, but one can perhaps see these as merely instances of poorly managed post-industrial exploitation of nature. However, some people have objected to the term 'post-industrial society' and the suggestion that contemporary societies are seeing nature in less exploitative ways. Indeed it can be argued that each of the phenomena we have been looking at – the tourist industry, the mass media and counterurbanisation – rely on increased, not lessened, exploitation of natural resources. While the global tourist industry may have received considerable impetus from an ever-widening search for a romantic experience of nature, it has been fuelled by the consumption of petroleum products to power the aircraft. Similarly, the growth of the media has relied on electronic power, produced largely from either fossil or nuclear fuels. The growing decentralisation of population described as counterurbanisation has likewise seen the increased consumption of fuel: this time in the form of petroleum for cars. Between 1965 and 1985, for instance, it has been estimated that there has been a 60 per cent increase in the total passenger mileage within Britain. As Wallace remarks:

> however much some aspects of core societies appear to have become 'post-industrial', the interactions between people and their natural environment have a fundamental bearing on the course of economic development and the range of economic opportunities in particular places (Wallace, 1990, p. 61).

The new, more symbolic uses of nature within post-industrial societies do not, as some people have assumed, mark the end of industrial ways of using natures. Instead it may be argued that they add new symbolic uses to more long-established material ones and these new uses may well be instrumental in increasing society's material demands upon natures and may indeed be leading to a series of quite serious environmental problems.

2.6 The narratives of the social exploitation of nature: some common threads?

This chapter has explored four episodes in the history/prehistory of society–nature relations. Particular emphasis has been placed on the extent and form of social exploitation of nature, where this is understood as the instrumental use of 'non-human' or 'external' natures. It has been noted that while many environmental histories have adopted a broadly similar focus many of them have constructed this is an overly reductionist and linear way. Events such as the emergence of agriculture represented an expansion in the way in which societies exploited natural resources such as soils, plants and animals. Similarly, the emergence of industry, both in prehistory and during the so-called industrial revolution, involved, at least in part, an expansion of the way in which societies exploited natural resources such as fossil fuels. However, we have emphasised that social exploitation of nature is not the only society–nature relation, and indeed at various points we have identified other types of relationships. In the discussion of imperialism, for example, we highlighted through the notion of ecological imperialism how some constituents of nature had an agency or effect that, although poorly recognised or understood by European imperialists, was of considerable significance. We also highlighted how nature can act as symbolic resource and how it was frequently used within imperialism to establish social divisions and relations of exploitation between people, as opposed to, or running alongside, relations of exploitation of nature.

In the section on post-industrial societies we highlighted once again how nature can be conceived as a symbolic as much as a material resource and we explored whether the exploitation of the former set of resources has become of primary significance in what might be described as post-industrial or postmodern societies. We argued that while the symbolic use of nature is of great importance in such societies, there are often connections into the heightened, not lessened, exploitation of material natures. We also mentioned that both forms of the social exploitation of nature are often connected into wider social organisational or political-economic changes. The importance of recognising such changes was also highlighted in the discussion on industrialisation, where it was effectively argued that caution needs to be exercised when ascribing things as being related to society–nature relationships. It was argued, for instance, that while industrialisation might be seen as constituting a new and heightened form of social exploitation of nature it can also be viewed equally, if not more, as being about the constitution of new, and heightened, forms of social exploitation of people. Overall, we would therefore suggest that the historical geography of society–nature relations is complex, multidimensional and probably far from linear.

Chapter 3

Limits to Exploitation? Problems of the Contemporary Global Environment

3.1 Introduction

This chapter aims to examine the relationships between societies and natures in the contemporary world. It will draw on the idea of increasing rates of exploitation explained in Chapter 2 and then seek to discuss whether or not the level of exploitation is approaching its maximum limit. The chapter focuses on the ideas presented by 'neo-Malthusians' and discusses the counter view of the 'cornucopian' school of thought. This will be accomplished by reviewing how modern societies, through essentially anthropocentric forms of exploitation (namely agriculture and industry) on global, regional and local scales cause environmental problems.

In particular the discussion will focus on two key issues: first, whether an increasingly large human population can continue to produce enough food and generate enough energy without critically depleting the earth's resources; and second, at what cost to the physical environment. In doing so, three other themes will also form part of the discussion. In particular the chapter will outline the contribution of the developed and underdeveloped worlds to the unsustainable use of the earth's resources. Those factors that enforce the exploitative relationship between the social and the natural will be reviewed briefly, although a detailed discussion on exploitation in a social context forms the main theme for Chapter 5. Finally the role of so-called 'natural agents' of environmental change, which can cause or excerabate environmental problems, will be compared with that of human actions.

The debate about over-exploitation of the earth's resources often focuses upon the ability of a rising human population to continue to support itself indefinitely. Concern over the well-being of Planet Earth centres on the way in which societies exploit the earth's resources. Resources can be split into three categories. Non-renewable resources are those that cannot be regenerated by natural processes except over extremely long time-scales. Examples are fossil fuels and minerals. In human time-scales these resources have a fixed supply. Renewable resources regenerate over much shorter time-scales, either by reproduction (plants and animals) or by recycling (clean air and water). Continuing resources are inexhaustible and their supply is unaffected by human actions. They include resources provided by the sun (solar energy) and geothermal energy (Jacobs, 1991).

Societies exploit natures as resources in numerous ways: for food, energy, shelter, recreation and wealth. Such actions can create problems. Jacobs (1991, pp. 3–5) states that the biosphere performs three principal functions for the economic activities of humankind: the provision of resources, the assimilation of waste and the provision of environmental services (life support and amenities). Excessive use of these functions can lead to environmental problems. Sloep and van Dam-Mieras (1995, p. 42) define an environmental problem as 'any change of state in the physical environment which is brought about by human interference with the physical environment, and has effects which societies deem unacceptable in the light of its shared norms'.

3.2 Limits to growth

Since the days of Socrates people have been debating the earth's carrying capacity (see Box 3.1 on page 111) (Reid, 1998). Many experts have long predicted that if population growth continued at its present rate the planet would become overpopulated such that the earth's resources would not sustain human population. These 'doom and gloom pessimists' who warn of a global human disaster have become known as 'neo-Malthusians' (see also Chapter 1). The name derives from Thomas Malthus, who warned of the dangers of population growth exceeding food supply in 1830. Malthus argued that because food production grew arthimetically and population growth, when unchecked, increased geometrically, demand for food would inevitably at some point outstrip supply. Thus, an imbalance between population and available physical resources would occur and population levels would readjust to balance with food supply through negative feedbacks, such as famine, disease and war. Malthus suggested societies should take preventive measures such as abstinence from marriage to reduce fertility rates.

A rise in the prominence of environmental movements during the 1960s and early 1970s led to widespread concern about population growth and resource depletion. Once again experts warned of impending disaster and argued that societies were reaching the limits to growth. Ehrlich warned societies that they were on the verge of unbelievable famines, and that feeding six billion people would be impossible (Reid, 1998). In 1972 Meadows *et al.* published the results of a computer model called 'limits to growth'. The model aimed to predict the future prospects of societies based on current trends in population growth, industrial capital, food production, consumption of resources and pollution. The results were bleak, with Meadows *et al.* suggesting that if current trends continued unchallenged the limits to growth of the earth would be reached within the next 100 years (Figure 3.1). They argued that if societies wish to avoid such an outcome they need to alter growth trends and embrace sustainable development immediately. Errors and uncertainty about the data in the original model questioned the validity of the results (see Pepper, 1996 for a summary of the main criticisms) so Meadows *et al.* (1992) updated and revised the limits to growth model and still reached similar findings, predicting the earth would not sustain a population of ten billion people beyond the next hundred years if current trends continued. Other work such as *Blueprint for survival* (Goldsmith *et al.*, 1972) and Schumacher's (1973) *Small is beautiful* reaffirmed the need for people to review society's pursuit of economic and social profit and progress at the expense of economic inefficiency, pollution and environmental degradation.

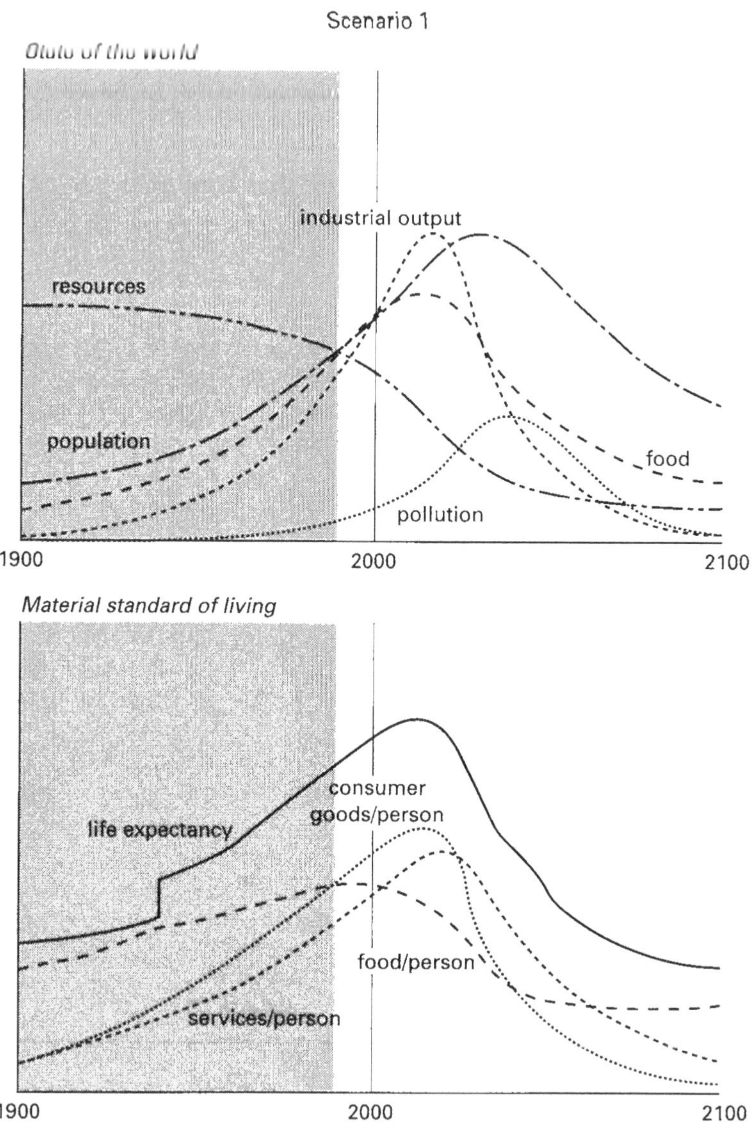

Figure 3.1 The standard run of the limits to growth model (source: Meadows *et al.*, 1992)

To date all the predictions of global disaster have been incorrect. Malthus, for example, failed to recognise the role of technology in improving methods of food production. Societies have not suffered famines on an unprecedented global scale (although they have occurred locally, as discussed later and in Chapter 5), but the shape of an environmental debate had formed. The relationship between population growth and resource depletion became established as a dominant mode of thought amongst many, but not all, scientists and environmentalists. This argument continues between the neo-Malthusians and the opposing cornucopians.

Recently a series of volumes have been published arguing why societies will never reach beyond their capacity to grow (Simon, 1981; Simon and Kahn, 1984: Simon, 1995). These arguments are summarised in the next section.

3.3 Neo-Malthusians and cornucopians

Founded on the views of Thomas Malthus, neo-Malthusians believe that the earth is becoming more crowded and more polluted as its resources are consumed by societies, and that the Earth cannot continue to support an increasing human population unless societies change the way in which they exploit it. Ehrlich and Ehrlich (1990) argue that societies continue to exploit the earth in an unsustainable way. They suggest that the environmental cost of sustaining the world's population is seriously undermining the earth's life support systems. Global warming, ozone depletion, acid rain, increasing rates of soil erosion, land degradation, pollution of hydrological, atmospheric and terrestrial systems, lower supplies of groundwater and loss of biodiversity provide evidence of a system moving towards collapse. Over-consumption by the affluent, over-exploitation, population growth, poverty, trade and the uneven distribution of wealth are cited as the principal reasons for environmental problems (Harrison, 1990). Terhal (1992) and Blowers and Glasbergen (1996) argue that the demands of the richest countries have placed great pressure on the world's non-renewable resources. The top 20 per cent of the wealthiest nations consume 80 per cent of the earth's natural resources and by doing so degrade the environment; for example, the US per capita energy use is 20 times greater than India's and the *National Geographic* (1998) argues that if the entire world consumes resources as profligately as the industrial world has done, and continues to do, the strain on renewable and non-renewable resources may become overwhelming.

Terhal (1992) argues that poverty also forces people to exploit their environment in an unsustainable manner and, in order to raise the level of per capita income in poor countries with high population growth, Daly (1993) argues that their use of resources will have to increase by a factor of approximately 36 because as a country develops and industrialises its population generally becomes more affluent and consumes more resources. If consumption in developed countries remains the same but increases as developing countries industrialise the pressure on resources will become even greater.

Neo-Malthusians advocate that societies should work with natures to promote economic growth that sustains the earth's life support systems and limit population growth (Miller, 1990).

Ehrlich and Ehrlich (1990) argue that every person added to the planet consumes more resources, placing more pressure on the environment – assuming current methods of resource exploitation remain relatively unchanged. Human population has grown at a tremendous rate. In 1650 world population reached 500 million people. From 1800 to 1900 it doubled to two billion and in 1990 it had grown to approximately five billion (Sage, 1996). By 1996, 90 million people were added to the earth's population each year (Lean, 1996) and now it is around 137 million (Reid, 1998). However, current predictions of global population growth are less dramatic. Lutz *et al.* (1997) suggest that the odds of world population doubling by 2050 are less than one-third and they argue that even underdeveloped countries in sub-Saharan Africa have begun the transition to low fertility rates.

Although it can be argued that lower rates of population growth will place less pressure on nature in the long term as resource consumption rates fall, in absolute terms population levels will remain high, especially in underdeveloped countries, until the 'peak number' have passed through their reproductive years (see Table 3.1). India, for example, could still have a population of around two billion even if family size dropped from its 1990 level of 4.3 to 2.4 and remained steady (Ehrlich and Ehrlich, 1990).

Table 3.1 Population growth rates for selected countries

Country (by income group)	Average annual growth rate (%)		Total population (millions)	
	1980–90	1990–95	1980	1995
Low income economies				
Mozambique	1.6	2.6	12	16.2
Ethiopia	3.1	1.9	38	56.4
Tanzania	3.2	3.0	19	29.6
Chad	2.4	2.5	4	6.4
Madagascar	2.9	3.1	9	13.7
Kenya	3.4	2.7	17	26.7
India	2.1	1.8	687	929
China	1.5	1.1	981	1200
Middle income economies				
Bolivia	2.0	2.4	5	7
Philippines	2.4	2.2	48	69
Bulgaria	–0.2	–0.7	9	8
Romania	0.4	–0.4	22	23
Algeria	2.9	2.2	19	28
Brazil	2.0	1.5	121	159
Saudi Arabia	5.2	3.7	9	19
High income economies				
UAE	5.7	5.8	1	2
UK	0.2	0.3	56	59
Canada	1.2	1.3	25	30
Sweden	0.3	0.6	8	9
USA	0.9	1	228	263
Japan	0.6	0.3	117	125

(Source: World Bank, 1997)

Two polar views have developed about the role of population growth and density in the limits to growth scenario. Parenti (1996) argues that in a global context higher rates of population growth in underdeveloped countries become less significant because the population density of many countries is low when compared to developed nations such as The Netherlands or Japan. Harrison (1990) is less optimistic and argues that population growth in underdeveloped countries is responsible for two-thirds of deforestation. He suggests that population growth places excessive stress on resources. For example, the demand for fuel wood promotes deforestation and exacerbates problems such as flooding in Malaysia and Bangladesh and forces subsistence farmers to exploit marginal lands, therefore higher numbers of poor people will place a great strain on the environmental resources of each country. Western (1988) also presents an example that appears to support the neo-Malthusian school of thought. He argues that an upsurge in population growth, combined with land tenure problems on the island of Palawan in the Philippines, has resulted in the acceleration of migrants into the steep mountainous areas where they practise shifting cultivation, destroying the forest cover. A sizeable proportion of deforestation has also occurred in Amazonia as a result of government policies encouraging migrants from over-populated areas to clear forests for ranching and farming (Fearnside, 1989; Treece, 1989; Foresta, 1992), although it will be argued in Chapter 5 that population growth might not be the major reason for the depletion of these resources.

Ehrlich and Ehrlich (1990) argue that it is worrying when countries like Japan can only support high population densities through importing resources from elsewhere and they could not sustain their population solely by exploiting their own land. He asks what would happen if the countries than import from reach a point when their own population is as high and then they have not got the resources to export? For example, China has a massive imbalance between population and resources: a population of 1192 million in an area with 10 per cent of global freshwater, 5 per cent fossil

Box 3.1 Carrying capacity

Postel (1994, p. 3) defines carrying capacity as 'the largest number of any given species that a habitat can support indefinitely'. The concept of carrying capacity is often applied by biologists (Ehrlich, 1982) but has also been used for leisure and recreational planning (Stanley, 1980), to determine when an area is overpopulated (Ramanaiah and Reddy, 1983), rangeland management (Walker, 1993) and to regulate access to fragile environments (Kliskey, 1994).

Over-population may be defined as 'too many people in a given area, too high population density' (Ehrlich and Ehrlich, 1990, p. 38). Determining when an area or country is over-populated is, however, difficult. Ehrlich and Ehrlich (1990) suggest that this threshold is passed when a population of a given area cannot support itself without depleting or degrading its natural resources. When this threshold is reached the carrying capacity of the land has been exceeded. If this definition is accepted when considering the impact of population on resources, then many areas of the world have exceeded their carrying capacity and are over-populated. Examples include Africa because it cannot feed its population and the United States because it is eroding its soils and consuming its water resources (Ehrlich and Ehrlich, 1990), but it is worth remembering that other factors possibly cause famine (see text and Chapter 5).

fuels and 4 per cent non-mineral fuels and only 10 per cent and 50 per cent of its land is suitable for arable and pastural agriculture (Royle and Phillips, 1997). Ehrlich and Ehrlich (1990) argue that a country that cannot rely solely on its own resources to meet the needs of its population is in danger because it exceeds the carrying capacity (see Box 3.1) of its land and this often results in environmental problems.

Neo-Malthusians argue that resource consumption is not the only problem, but that the environmental costs of the development of the populous modern industrial, urbanised and mechanised agricultural societies has also been high. Harrison (1990), for example, argues that population growth accounts for two-thirds of the increase in carbon dioxide emissions between 1950 and 1985. If current growth trends continue people in underdeveloped countries could produce 5.75 billion tonnes of carbon dioxide alone. Population growth is also linked to methane emissions, produced by livestock and decomposition in irrigated fields, and neo-Malthusians argue that high population growth means that more land is needed for food production at the expense of biodiversity. For example, 125 million hectares of natural forests were converted for agricultural and non-agricultural uses between 1971 and 1986. Harrison (1990) also argues that an extra 280 million hectares of land will be needed for non-agricultural purposes if a further five billion people are added to the planet as urban population growth continues. A dramatic rise in urban air pollution has paralleled urban population growth and car use: 500 million cars by the beginning of the 1990s (Keyfitz, 1989).

In contrast to the neo-Malthusian school of thought, cornucopians (derived from the term *cornucopia*, meaning the horn of plenty and a symbol of abundance, (Miller, 1990)), do not share the view that societies are heading for a global disaster. They argue that the warnings of an impending global disaster in the late 1960s and early 1970s never materialised (Simon, 1980; 1994; 1995). Ehrlich (1968), for example, predicted that 'the battle to feed all of humanity is over. In the 1970s the world will undergo famines – hundreds of millions of people are going to starve to death.' Cornucopians such as Simon argue that there is plenty of evidence to suggest the earth's systems will not collapse under the pressure of higher population and that societies can solve environmental problems, many of which are short-term phenomena. Findlay and Findlay (1984) also argue that populous countries like India have managed to avoid famine despite large population growth since the 1940s.

Cornucopians believe that if present trends continue, economic growth and technological advances will create a less crowded, less polluted and more resource-rich world (Miller, 1990, p. 22). They tend to advance the following arguments: first, that global resource scarcity is not a serious problem. Second, there is no historical evidence to suggest that societies will run out of resources, despite higher levels of population growth. Third, more resources, including substitutes for existing ones, will become available in the future as a result of technological innovation. Simon (1995) recalls that in England in the 1600s there were concerns about timber shortages. This arguably provided the stimuli to solve the problem by the development of coal. There is, according to cornucopians, also no reason to suggest that this pattern of procuring and discovering new resources from the earth will not continue. Simon (1991a; 1994) suggests that the neo-Malthusian view does not take account of an improvement in technology and technological efficiency to discover and recover existing resources, as well as developing ways of exploiting alternatives.

Fourth, population and economic growth have also stimulated advances in technology. Simon (1995) argues that more people do cause problems but people are needed to solve those problems. This makes people the ultimate resource. Simon (1994; 1995) suggests that population size and growth is advantageous because it adds to the stock of useful knowledge and more people will increase the rate of cultural and material progress over the long term. These ideas had previously been developed by Boserup (see Box 3.2). Simon (1995) does accept that problems such as pollution will occur, but he suggests that these are short-term problems and a solution will be found. All these views are explored in more detail in Simon (1981), Simon and Kahn (1984) and Simon (1995).

Box 3.2 Boserup

Boserup (1981; 1965/1993) disagrees with the Malthusian view and argues that population growth can be a positive stimulus for agricultural development because higher population densities lead people to develop more intensive agricultural systems, in other words 'necessity is the mother of invention'. However, Boserup admits that over-population growth can lead to inappropriate methods of increasing productivity, which results in land degradation, but argues that other factors, besides population growth usually impede agricultural development, an issue discussed in Chapter 5. There is evidence to suggest that Boserup's hypothesis is correct. One example has been published by Tiffen *et al.* (1994), who present a case study of agrarian practices in Kenya where food output has been increased with additional inputs of labour, suggesting that population growth can promote sustainable land management.

An analysis of measures of human welfare suggest that the exploitative relationship between societies and natures has continued to be beneficial to people and that the quality of life is improving (Table 3.2). Simon (1994; 1995) argues that despite a huge growth in human population during the past 200 years, there has been an unprecedented improvement in the measures of human welfare, including lower death rates, better health and longer life expectancy (see also Simon and Boggs, 1995; Fogel, 1995; Haines, 1995; Hill, 1995). Evidence from statistics indicates that human welfare is improving globally. From the 1960s per capita income, school enrolment, literacy, life expectancy and other measures of human welfare all increased in low, middle and high income countries (Repetto, 1987). These trends have continued in the 1990s. Maps displaying 'the physical quality of life index' (PQLI) suggests that the quality of life for people between 1990 and 1993 in poor nations is improving despite high levels of population growth. The index is based on life expectancy at age one, literacy rates and infant mortality. However, 13 countries in sub-Saharan Africa went against the trend, showing a decline in the PQLI score as a result of increasing infant mortality and decreasing life expectancy (Doyle, 1996a). It is unclear whether population growth and poverty have slowed down such progress but

Repetto (1987) believes the balance of expert opinion would agree that they have impeded further improvement. Simon (1994) argues that there is no persuasive reason why an improvement in these indices will not continue indefinitely.

Table 3.2 Some human welfare indicators for selected countries

Country	Infant mortality rate (per 1000 births)		Life expectancy (years)	
	1980	1995	1990	1995
Mozambique	145	113	47	47
Ethiopia	155	112	48	49
Chad	147	117	47	48
Burkina Faso	121	99	48	49
Bangladesh	132	79	52	58
Kenya	72	58	59	58
India	116	68	59	62
China	42	34	70	69
UK	12	6	76	77
USA	13	8	76	77
Japan	8	4	79	80
Sweden	7	4	78	79

(Source: World Development Reports, 1992; 1997)

The limits to growth depend upon not only the number of people on the planet but how much they consume and produce, and how efficiently, in terms of maximising the use of resources and minimising the environmental cost. The next section debates whether one of societies' most important resources, food, is reaching its limits to growth through over-exploitation of soil and water.

3.4 Can society feed itself?

There has been a considerable amount of debate as to whether a growing human population can feed itself (Rhoades, 1991). Neo-Malthusians argue that the earth cannot continue to support a larger human population and that food production has a high environmental cost on local, regional and global scales. Between 1950 and 1997 there has been unprecedented economic growth and this is evident in the growing pressure on the earth's natural systems. During that period use of timber tripled, paper use increased sixfold, grain consumption nearly tripled, fish catches increased nearly fivefold (Brown, 1998). These statistics lead neo-Malthusians like Brown and Ehrlich to argue that societies cannot continue to meet increasing demands for food because the methods that have been used are making soils unproductive. Cornucopians dismiss these concerns and suggest that technology, human ingenuity and management can meet future food demands.

3.4.1 Is global food production increasing or decreasing?

Cornucopians like Simon are correct to point out that societies have developed agricultural techniques that have had an amazing capacity to produce food. Statistics show that global food production has increased dramatically over the last 40 years. For example, world grain production showed an 2.6-fold increase in output between 1950 and 1984 and world fish catches have increased fivefold since 1950. Crop yields have increased by four-fifths on already cultivated land and between 1950 and 1993 irrigated land increased by 154 million hectares to a total of 248 million hectares (Brown 1997). Even in underdeveloped countries farm output rose by 14.4 per cent between 1985 and 1990 (Avery, 1995). Thus, despite neo-Malthusian claims of food shortages, *globally* food production has kept up with population growth and Avery (1995) claims that this has been achieved without dramatically increasing the agricultural land base.

However, neo-Malthusians like Brown (1994; 1997; 1998), Postel (1994) and Lean (1994) suggest that the trends of rising *global* food production are coming to an end. They argue that societies may be facing a crisis because *global* food production is showing signs of slowing down, with a dramatic decline in both food stocks and per capita food supply. Brown (1994; 1998) suggests this dramatic slowdown commenced in the 1980s, stating that since 1984 there has been no significant increase in grain production: the 1.67 billion tons produced in 1989 was only 1 per cent higher than the 1984 figure, between 1990 and 1996 it increased by 3 per cent (see Figure 3.2) and the World Bank stated that 1995 was the third successive year grain crop production fell. Food stocks in developed nations, the main food producers, have also fallen significantly. For example, Europe's surplus fell from 33 to 5.5 million tonnes between 1993 and 1995 (Lean, 1995a). From 1995 to 1996 world wheat and corn prices doubled in price as a result of production falling behind demand. The fact that total world fish catches peaked in 1989 at just over 100 million tonnes, while marine catches have fallen from 82 million tonnes to 77 million tonnes by 1991 leads Brown (1994, 1997) to predict that per capita seafood supply will continue to fall into the next century. Brown argues that all the figures cited above indicate that the food production patterns of the previous 40 years have come to a halt and societies have reached the point where future growth in *global* food will be much slower, with little evidence to suggest the process will be reversed. These changes are summarised in Table 3.3.

Neo-Malthusians are more alarmed by figures suggesting that, when the rise in the world's population is taken into account, there has been a 7 per cent decline in food availability per person and global food demand could double by 2030 with demand likely to be greatest in underdeveloped countries (Crosson, 1997). They suggest that high population growth rates and the need to improve nutrition levels mean that demand for food could increase by around 2.7 times, and increased affluence of people in underdeveloped countries could also place pressure on future food stocks. Brown (1997) argues that as people become more affluent they consume more food. Throughout Asia use of food grain is increasing and many countries like Japan, Taiwan and South Korea import about 70 per cent of their grain. China will need to import about 200 million tonnes of grain by 2030 (see Table 3.4). If these estimates are realistic and the short-term drop in the productivity of grain-producing nations becomes a long-term trend there may be insufficient grain on the international market to meet the demands of importing countries.

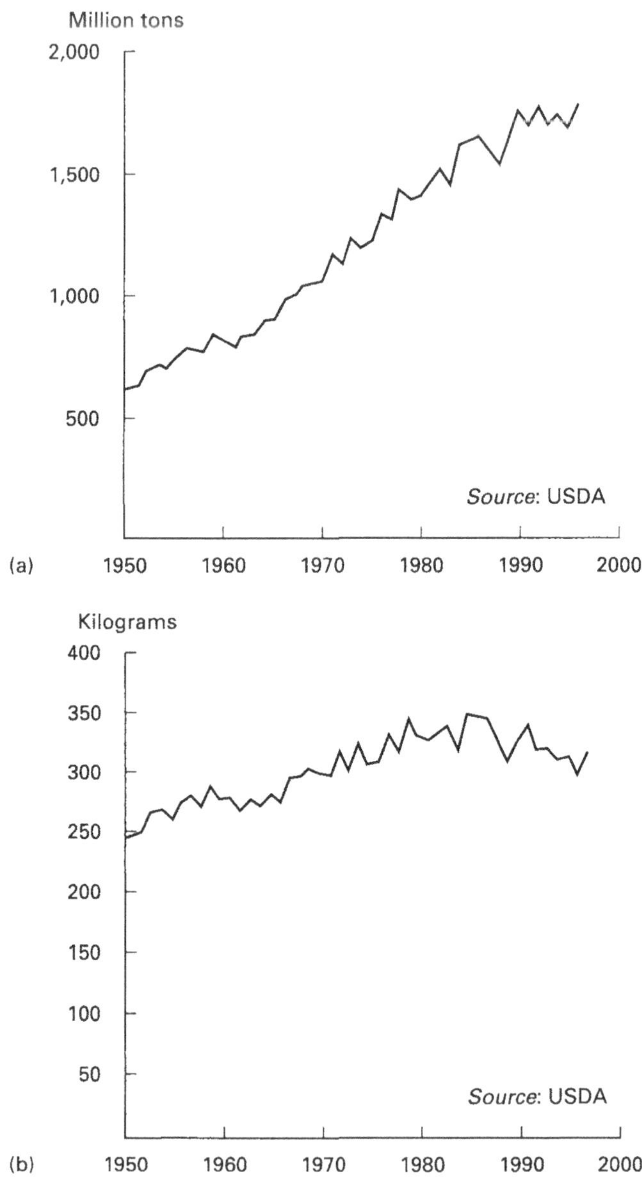

Figure 3.2 (a) Annual world grain production between 1950 and 1996. (b) Annual world grain production per person between 1950 and 1996 (source: Brown, 1997)

Table 3.3 Comparison of key global indicators in the economic and environmental eras

Indicator	The Economic Era: 1950–90	The Environmental Era: 1990–2030
World population	More than doubled from 2.5 to 5.3 billion, adding 2.8 billion (or 70 million per year) and slowing progress	Projected to increase from 5.3 to 8.9 billion, adding 3.6 billion (or 90 million per year). For much of humanity, this may reverse progress
Grain production	Nearly tripled from 631 million tons to 1780 million tons, or 29 million tons per year	Expanding by 12 million tons per year (rate of last eight years) may be best that can be expected
Beef and mutton production	Increased 2.6 times, from 24 to 62 million tons	Little growth expected
Fish catch	From 22 to 100 million tons; per capita up from 9 to 19 kilograms	No growth expected; per capita dropping from 19 kilograms to 11 kilograms
Economic growth	Economy expanded 4.9-fold from nearly $4 trillion to $19 trillion, an annual gain of 4.2 per cent. Growth is focus of national economic policy making	Averaging even half the 1950–90 rate may be difficult. Focus will shift from growth to sustainability and distribution
Growth in grain demand	Two-thirds from population growth; one-third from rising incomes	Nearly all expansion will be needed to sustain population growth
National security	Largely ideological and military in nature; defined by the cold war	Food and job security will dominate, often driving hungry and jobless people across national borders

(Source: Brown, L.R., 1994)

Table 3.4 Projected increase in population of selected regions/countries that may need to import grain (in millions)

Region/county	1995	2030
Middle East	243	544
North Africa (Morocco–Egypt)	136	239
Sub-Saharan Africa	585	1450
India	c. 950	1450
Pakistan	132	312
Mexico	94	150
Brazil	161	210

(Source: Brown, L.R., 1997)

3.4.2 Will water scarcity affect food production?

Neo-Malthusians also argue that societies will not be able to continue to feed a growing human population because there will be insufficient supplies of water. They suggest that water scarcity is already a problem, and certain regions appear to be experiencing shortages as clean, safe, fresh water supplies are diminishing because water is being extracted from groundwater supplies faster than natural processes can replenish them. Forestier (1989), for example, claims that water shortages are expected in 450 of China's 644 major cities by the year 2000. Ehrlich and Ehrlich (1990) argue that water supplies in developed nations are also under threat from industrial and agricultural exploitation. In the USA, for example, accessible portions of the Ogallala aquifer and the San Joaquin valley are being depleted to irrigate cropland (also see Parfit, 1993). Water tables are falling in many areas of the world including most of Africa, the Middle East, north-west Mexico, parts of mid-west USA, Chile, Argentina and Australia. In Libya, for example, which is reliant on groundwater and aquifers to supply water to agricultural areas, withdrawals now exceed natural replenishment by 50 per cent, causing groundwater levels to fall and leading to the intrusion of sea water into coastal aquifers (El Asswad, 1995). Brown (1998), Hinrichsen (1997) and Lean (1996) imply that water shortages are already leading to a neo-Malthusian disaster. They claim that water shortages may plague 52 countries and affect 3.2 billion people by 2025 and conflicts over water supplies already exist in the Middle East and north Africa (Gleick, 1994). Swain (1997) believes that Egypt has used the threat of war to prevent Sudan from taking any action that would adversely affect water supplies. Lean (1996) estimates that 20 nations cannot provide sufficient water to meet their basic needs and Postel (1994) believes that regional water shortages are beginning to slow food production, while millions of people are being forced to utilise water that contains large quantities of pollutants and high levels of bacteria (Witter and Carrasco, 1995). Industrial pollution has contaminated virtually all the water supply in Poland, forcing people to purchase bottled water (Fischhoff, 1991); one billion people lack access to safe water and 1.8 billion do not have adequate sanitary facilities (Doyle, 1997a). Anderson (1995) suggests that these trends are cause for concern because water availability is controlled by the hydrological cycle and is fixed, whereas consumption has been steadily rising since the 1960s.

Cornucopians reject this pessimistic view that societies will not be able to provide enough clean water. Anderson (1995) points out that doomsday predictions that America would have water rations and the oceans would be as dead as Lake Erie by the late 1970s have proved to be extreme and untrue. Cornucopians suggest that societies have many ways in which they can maintain clean water supplies.

Water supplies can be augmented through structural adjustments (build reservoirs and reroute rivers), but there are environmental problems with reservoir construction. Swain (1997) points out that to bolster agricultural production countries such as Ethiopia, Sudan and Egypt have turned to large-scale water projects. For example, Egypt constructed the Aswan high dam and proposes to build the Jonglei canal to feed its irrigation systems. El Asswad (1995) suggests that other complementary measures such as reducing population, encouraging careful use of water, using salt resistant crops and limiting availability may also stop wastage, reverse the idea that water is 'cheaper than dirt' and lower water consumption rates.

Desalinisation plants can convert sea water into drinkable fresh water. Desalinised water forms a major water supply in Kuwait and is used in arid regions of Spain and Australia or areas with limited groundwater supply such as Jersey. By 1989, 7536 plants across the globe produced 13 billion cubic metres of water a day, but 60 per cent of the capacity is based in the Middle East. The cost of producing desalinised water, between $1.60 and $2.20 compared with 50c for UK tap water, is expensive but it is another technological solution that could meet future water supplies (Coghlan, 1991).

Water can also often be used more efficiently, which will help to sustain agricultural production. In Israel, for example, highly efficient water irrigation systems have reduced the amount of water applied to crops by 36 per cent and tripled the area under irrigation. Sometimes emphasis is given to large capitally intensive schemes, but quite often low technology solutions can have important results. For example, construction of stone lines in fields in the province of Yatenga, Burkino Faso, retained soil moisture and reduced soil erosion, and these techniques can raise yields by 50 per cent (Postel, 1994). Soil and water can also be improved by better cropping practices. In one example, intermittent flooding of rice paddies can keep the soil saturated, saving water by as much as 40 per cent. Furthermore, rotating rice with other crops such as soyabeans, sugar cane or sweet potatoes replenishes the soil with nitrogen and reduces fertiliser use by up to 30 per cent (Rosegrant and Livernash, 1996).

3.4.3 Are we running out of fish? The state of world fish stocks

Fish provide a substantial part of human nutrition. Neo-Malthusians argue that ocean fishing is fast reaching its limits to growth. They point out that regionally commercial ocean fish stocks are declining to the point of exhaustion as a result of over-exploitation and argue that the widespread belief that there is an unending supply of fish is fast becoming a myth. Many researchers now believe that the fishing industry is extracting fish from the oceans faster than the rate of natural replenishment and blame poor management, the efficiency of technology and high fish consumption (see, for example, Cook *et al.*, 1997; Gwyer, 1991; Parfit, 1995; Safina, 1995, 1998; Russell, 1996). These authors point out that the world's fish catch peaked in 1989 at 82 million metric tons and the figure has either fallen or remained constant.

In 1995 the UN Food and Agriculture Organization (FAO) claimed that nine of the 17 major global fishing grounds are in serious decline (Russell, 1996), including Peruvian anchovy, Californian sardine and redfish, Atlantic bluefin tuna, pollock in Russia and mackerel off the south-west coast of England (see Harris, 1986; Safina, 1995; Pearce, 1996; Russell, 1996). The fish that were caught off the Grand Banks of Newfoundland, Canada, and Georges Bank off the coast of New England are now considered to be commercially extinct and Cook *et al.* (1997) believe that annual cod catches in the North Sea now exceed the sustainable yield and have warned fisherman that their industry will collapse. If current trends continue, Russell (1996) states that a shortfall of 30 million tons of fish for human consumption will occur by the year 2000.

Safina (1995) identifies the four factors that he believes have undermined fish stocks: first, expansion of the fishing industry, with the number of fishing vessels trawling the world's oceans having doubled to approximately one million from 1970 levels (Russell, 1996). Second, improved technology has played a major role by increasing the efficiency of catching fish. Recent developments include sonar and radar, which have improved navigation, while satellite data can relay weather information and help fishermen locate potential fishing areas. The introduction of larger nets and factory ships, which annually harvest up to 90 per cent of the fish in some populations, have also helped to increase catches (Safina, 1995). Third, governments have helped revolutionise fishing technology by providing fishermen with grants and subsidies (Gwyer, 1991). In underdeveloped nations, where the majority of fishing is done at a subsistence level, the introduction of technology can have a detrimental impact on fish stocks. The introduction of modern trawlers and fish processing technology in the coastal fishing grounds of Kerala, south India has seriously depleted fish numbers. Since the 1980s government subsidies have helped local fishermen mechanise their fishing boats to increase their harvest (see Mitchell and King, 1984; Kurien, 1993; Myers *et al.*, 1996). Fourth, Gwyer (1991) argues that governments have encouraged excessive fishing by allowing unrestricted access to fishing areas to meet international demand for fish (an issue that will be further discussed in Chapter 4).

However, cornucopians such as Wise (1995) point out that world fish landings between 1980 and 1990 increased by over 3 per cent per year and fish catches have always kept pace with human population increases. He argues that fish are not a scarce commodity because their price has not increased and believes that it is reasonable to expect a tripling or quadrupling of fish catches in the future through appropriate management and alternatives to deep sea fishing, including aquaculture and fish farms.

Aquaculture now provides approximately one-fifth of the human fish consumption and its contribution has doubled by adding ten million metric tons in mainly freshwater fish to global fish markets. According to the World Bank aquaculture supplied 22 per cent of 86 million metric tons caught, making fish farms the fastest growing source of fish (Plucknett and Winkelmann, 1995). At present aquaculture is restricted to certain species, for example the saltwater salmon and shrimp, but this form of fishing provides a sizeable proportion of their market share (Safina, 1995). But some analysts believe that aquaculture cannot be seen as a viable alternative to ocean fishing in terms of meeting market demand or as a sustainable way of procuring food. Abramovitz (1997) argues that the construction of fish farms has severely damaged mangroves, coastal areas known to protect juvenile marine fish, and that fish farms require high financial investment. After the introduction of aquaculture operations in India and Bangladesh wild shrimp catches dropped by up to 90 per cent and most operations have a life expectancy of only five to ten years. Furthermore, fish farms can suffer from similar problems to land-based farms, namely bacteria, viruses and pollution (Plucknett and Winkelmann, 1995). A comparison of the tonnage of fish produced by aquaculture and conventional capture methods (Figure 3.3) reveals that aquaculture presently accounts for only a minor proportion of the global fish market, so without major expansion it looks unlikely that aquaculture can be a reliable substitute.

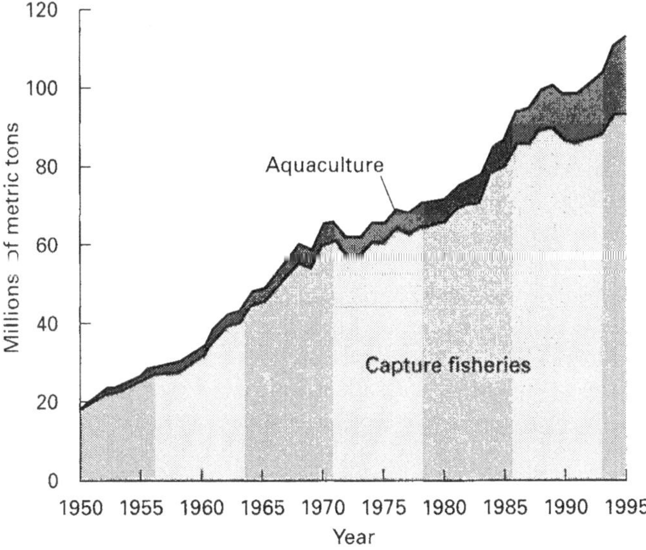

Figure 3.3 Global fish supply from capture fisheries and aquaculture (Grace, 1998)

3.4.4 Does environmental degradation limit food production?

Neo-Malthusians argue that environmental degradation has contributed to the decline in world food production. The World Bank concur with this assessment, stating that 'most new lands brought under cultivation are marginal and ecologically fragile and cannot make up for the land removed from cultivation each year due to urbanisation and land degradation' (Lean, 1996, p. 12). Postel (1994) blames poor agricultural methods for the 550 million hectares suffering from degradation.

Brown (1994; 1998) suggests that if this is going to reflect a long-term pattern of diminishing resources, increasing population and demand, societies will reach a point when they cannot support themselves. He states that at 'a global level population growth is outrunning production of oceanic fisheries, rangelands and croplands'. Brown forecasts that the pattern of growth in food production will not be sustained over the next 40 years because environmental trends will shape economic trends. The environmental cost of food production, therefore, has become an important issue. Conversion of forests or other land uses to agricultural production, soil erosion and excessive use of fertilisers and pesticides carry both environmental and economic costs. This section outlines the problems caused by agriculture and how these problems will affect the ability of societies to sustain food production.

Soil erosion and degradation

Brown (1991) argues that the world has lost nearly one-fifth of its topsoil from agricultural land. Soil degradation includes erosion, compaction, water excess and deficit, acidification, salinisation and the toxic accumulation of agricultural and industrial pollutants. All of these impacts can seriously reduce the quality and productivity of the soil (Ellis and Mellor, 1995) and soil erosion can produce indirect effects (including the siltation of rivers), increase the incidence of flooding and cause eutrophication (Crosson, 1997). The extent of soil degradation is influenced by several interrelated factors. These include climate, relief, land use and socio-economic and political controls. Millington *et al*. (1989), McGregor and Barker (1991), Brookfield and Padoch (1994) and Ellis and Mellor (1995) suggest that population pressure, skewed land resource distribution, poverty, marginalisation, fuel wood demands, inappropriate land tenure, size of landholdings and poor infrastructure exacerbate the processes that lead to accelerated soil erosion and land degradation. These issues will be discussed in Chapter 5.

Hinrichsen (1997) argues that overgrazing and deforestation cause excessive soil erosion, especially in the underdeveloped world. Land degradation of drylands affects approximately 60 per cent, or 2.6 billion hectares, of rangeland, and agricultural and forestry losses amount to over $40 billion. In the sub-Saharan region of northern Africa compaction and reduced vegetation cover by grazing animals exposes the soil to erosion. Deforestation removes trees that protect the soil from rainsplash erosion and bind the soil particles together to increase the soil's structural stability (Aweto and Adejumbobi, 1991). This process is described in detail by Sioli (1985) and is summarised in Figure 3.4. Certain forms of shifting cultivation also degrade the soil, for example when forest or scrub is burned to make swiddens that are cultivated for one or two seasons until the soil becomes infertile. The plot is abandoned and a new area of forest is then cleared (Brookfield and Padoch, 1994).

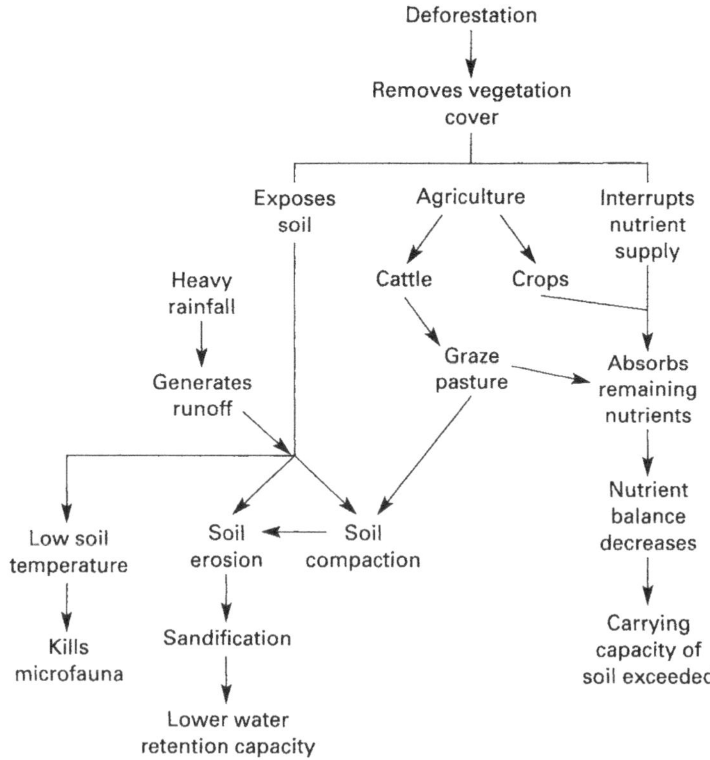

Figure 3.4 Soil erosion and deforestation (source of information: Sioli, 1985)

Nearly two billion hectares of crop and grazing land is suffering from soil degradation. Hinrichsen (1997) estimates that soil erosion and other forms of degradation annually ruin five to seven million hectares of farming land. Brown and Wolf (1984) estimated excessive soil erosion on a global scale to be 25.4 billion metric tonnes each year, but Pimental *et al.* (1995) suggest that figure is now closer to 75 billion metric tonnes. If these figures are correct and soil erosion continues at this rate food production could be seriously threatened. Cropland in the former Soviet Union declined from 123 to 99 million hectares between 1977 and 1993 due to soil erosion, and an estimated 14 million hectares of erodible land in the USA has been converted to grassland or planted with trees because the soil was incapable of sustaining cultivation (Brown, 1994; Doyle, 1996b). Soil erosion is exceeding natural rates of soil formation in many places (Table 3.5); for example erosion rates exceed 200 tonnes per hectare in mountainous areas of Nepal and on the loess plateau of China (Ellis and Mellor, 1995).

Table 3.5 Rates of soil erosion and formation

Soil formation	1 cm/100–400 years	c.0.3–1.3 t/ha
Natural erosion		<1.0 t/ha
Accelerated erosion		>10 t/ha–100 t/ha

(Source: Ellis and Mellor, 1995)

Cornucopians argue that soil erosion and land degradation are short-term and remediable problems that will not affect people's ability to produce more food. They counter claims suggesting that food production will be restricted by soil erosion and argue that existing and new farming practices will become more efficient and less environmentally damaging. They also argue that land degradation is not considered a major obstacle to increasing food production because there is potential to convert more land into agricultural production. For example, only 40 per cent of potential arable land is yet to be cultivated, and eight and three million hectares of grainland are currently out of production in the USA and Europe respectively because of over-production in the developed world. This land can be returned to active farming. For example, idle cropland in the European Union in 1996 could provide an extra 18 million tons of grain, while in the USA approximately seven million hectares could be farmed more sustainably, providing an additional 28 million tons of grain (Brown, 1997). Prosterman *et al.* (1996) argue that even populous countries like China could dramatically improve their food production if land rights were secure enough to give farmers an incentive to farm more assiduously. China still underutilises two-thirds of its farmland and crop yields could increase two to three times, but many farmers are reluctant to invest in land because they fear it will be taken from them by the government in the future. Secure land rights throughout the developing world are vital to increase food production and protect the soil. Optimistically, Avery (1995) argues that societies could feed an extra billion people without using any fragile areas or using heavy doses of chemicals by re-establishing land diverted from agricultural production in the USA and converting the Argentine pampas grasslands into wheat production. Brown, however, doubts whether reutilising former cropland will expand grain production area by more than 1.6 per cent, and suggests that the amount of cropland is unlikely to increase dramatically unless crop prices rise to keep pace with population growth. Historically, however, this has been accomplished, and there is no reason why it should not continue according to Simon (1995).

Because the amount of land used for agricultural production is not reaching its limits, Crosson (1997) argues that land degradation is unlikely to threaten *global* food production. It may, however, present a serious threat in *regions* of the underdeveloped world that have high numbers of people unless soil conservation measures are implemented. Examples include the Nile Delta, the Tigris river basin and the steeply sloped areas of China and south-east Asia.

Other researchers also question the accuracy of soil erosion estimates outlined in the previous section. Oldeman *et al.* (1992) concluded that 1965 million hectares were degraded to some extent but only 1 per cent extremely. Water and wind erosion accounted for 85 per cent of this figure, while chemical and physical degradation (such as salinisation or machinery compaction) caused 15 per cent (Crosson, 1997) and, more significantly, 6770 million hectares were considered to be suffering very little or no degradation. Crosson (1997) has calculated estimates of the impact of soil erosion on productivity to be in the order of 0.1 to 0.2 per cent per year between 1945 and 1990. Such low estimates suggest that soil erosion may be worsening but it is not a serious threat. Furthermore Crosson (1997) also argues that not all eroded soil is immediately lost to agricultural production as it can be redeposited on alluvial plains that provide nutrient-rich sediments for farming.

Protecting the soil from erosion and degradation will help maintain soil fertility and/or increase food production. Measures to control soil erosion can be effective; for example the US Soil Conservation Service estimated that the Conservation Reserve Program instigated by the Department of Agriculture reduced soil erosion rates by one-third or half a billion tons (Ehrlich and Ehrlich 1990). Three strategies can reduce soil erosion and degradation: soil agronomic measures, soil management practices and mechanical remedies (Morgan, 1995). Agronomic measures, including strip cropping and agroforestry, minimise the time the soil is exposed to rainsplash. Forest farming and crop rotation schemes, such as planting corn, beans and squash in southern Mexico, provide a continuous soil cover and reduce the need for pesticide and fertiliser. Soil management practices increase the resistance of the soil to erosion and include mulching, conservation tillage practices and soil conditioners. Mulching or crop residue management leaves the remnants of the previous harvests on the soil to provide cover while conservation tillage techniques plant seeds without mechanically disturbing the upper part of the soil. Conditioners enhance the structural properties of a soil to increase its resistance to erosion, and windbreaks, grassed waterways, the creation of buffer strips and contour farming are mechanical techniques that lower erosion rates by reducing the velocity of runoff and acting as sediment traps. Morgan (1995) provides an excellent review of soil conservation techniques.

The incorporation of appropriate soil management practices into agricultural systems could greatly assist future food production and conserve the soil. The advantage of having numerous strategies to conserve the soil is that they can be adapted or chosen to fit in with traditional agricultural practices in developed and underdeveloped countries. Soil agronomic techniques are also inexpensive. However, erosion control practices must be used in the correct socio-economic and political circumstances to work effectively (Ellis and Mellor, 1995). Learning from past mistakes, such as the transplant of temperate agricultural systems to underdeveloped countries, use of inappropriate technology and neglect of indigenous crops, is also necessary.

Food production can be increased by improving land productivity. Scientific advances in food production (eg the 'green revolution': see Box 3.3) increased world grain production by 169 per cent between 1950 and 1990, while the amount of land harvested only rose by 17 per cent (Postel, 1994). Even today yields of major grains are significantly below their genetic potential and people have suggested that biotechnological advances will increase food production in line with population growth. Advances in genetic engineering offer great potential to develop new disease-resistant strains of rice. One example is the recent cloning of gene Xa21, which offers resistance against blight, into popular rice varieties IR64 and IR72 that are cultivated in over 22 million hectares in Africa and Asia (Ronald, 1997). Development of hybrid rice has produced promising results in China, Korea, Vietnam and India, where yields exceed conventional high-yielding varieties by approximately one ton per hectare (Rosegrant and Livernash, 1996). Genetic engineering and the gene pool contained within the world's flora provides an opportunity to produce more resistant, high-yielding crops.

Box 3.3 The green revolution

In the 1940s a group of US scientists developed new strains of high-yielding crops to help Mexico produce more food and new strains of maize, wheat and rice were successfully introduced. The term 'green revolution' was used to describe the beneficial effects of these new high-yield varieties as they were introduced across most of the underdeveloped world (Scoging, 1991). The ideas behind the green revolution continue today with biotechnology and genetic engineering.

As well as developing new genetically engineered crops, Vietmeyer (1986) argues that there is tremendous opportunity to exploit the vast majority of the world's edible plants to meet global food demands. Throughout history humans have procured food from about 3000 plants, but recently have relied upon approximately 20. Vietmeyer provides a good review of some of the plants that could form a more substantial proportion of human food. If people wish to do so, societies must preserve the biological diversity of the earth's biomes. It is also possible to manipulate the genes in animals for the benefit of societies. Velander *et al.* (1997) describe such benefits and work of this kind might help to produce food from livestock in the future. Genetically engineered bacteria could also be deployed in other areas, for example in cleaning up oil spills (see Miller, 1998).

Rosegrant and Livernash (1996) argue that improvements in pest management would lessen the environmental burden of excessive pesticide and herbicide use and help to maintain food production. For example, the inapproriate use of an insecticide failed to control the brown planthopper, which feed on rice in Indonesia, and resulted in the loss of more than one million tonnes of rice in 1977. The insecticide also killed off the natural predators of the planthopper. Once insecticides were banned and predators were encouraged, rice production has continued to grow and pesticide use declined, saving $120 million on subsidies (Rosegrant and Livernash, 1996). Another alternative is organic farming, which is more environmentally friendly because it is less mechanised and uses fewer chemicals. Results presented by Lockeretz *et al.* (1981) from organic farms in the corn belt of the USA suggest that production of crops and profits are not significantly different from their conventional counterparts. Operating expenses incurred by organic farmers and the value of production were considerably lower than conventional farms, with the result that the net returns showed little difference in monetary terms. Crop yields generally were 10 per cent lower on organic farms, although the figure varied with crop type – wheat yields, in particular, were much higher using conventional farming methods. Moreover, Lockeretz *et al.* (1981) report that over a five-year period organic farms consumed less fossil energy and suffered lower rates of soil erosion but had mixed effects on the nutrient status of the soil. The results suggest that organic farming has less of an impact on the environment, and this is important when replacing nutrients and water from eroded land costs about $196 per hectare, with world-wide costs approaching $400 billion a year (Abramovitz, 1997).

Neo-Malthusians claim that growth in food production has relied on the use of fertilisers to maintain soil fertility. They argue that fertilisers will not have so dramatic an impact on crop yields in the future. Ehrlich and Ehrlich (1990) argue that fertiliser application may have passed the point of diminishing returns as gains in crop yields from additional fertiliser use are falling on land that has been cultivated over long periods of time. For example, in the US corn belt one ton of extra fertiliser increased grain yield by 15 to 20 tons in 1970, but only five to ten tons in 1990. They also suggest that fertiliser is expensive and this restricts its use in underdeveloped countries.

However there is a strong argument that the impact of fertilisers and pesticides on the physical environment and human health has been blown out of proportion and may still have a role to play in sustaining food production. Avery (1995) suggests that improvements in farm technology and careful management will reduce the environmental impact of fertiliser, but this assumes that farmers will be persuaded to adopt such techniques. He also claims modern pesticides are now far less persistent, are much narrower in toxicity, and application rates are falling. Ames (1995) argues that the links between cancer and pesticide use are exaggerated and the effects of synthetic pesticides in food products are no more threatening than natural ones. Whelan (1995) also suggests that the benefits of using pesticides like DDT may outweigh the dangers. She argues that DDT helped to eradicate malaria and therefore dramatically cut the human death rate from the disease, and there is no strong scientific evidence to support the claim that trace levels of synthetic chemicals like PCBs and DDT have an adverse effect on wildlife or humans.

Irrigation and salinisation

One environmental cost of farming in semi-arid and arid areas is salinisation. Salinisation describes the accumulation of salts in a soil (Ellis and Mellor, 1995). The FAO estimates salt accumulation has severely damaged 30 million of the world's 240 million hectares of irrigated land (Hinrichsen, 1997). Irrigation promotes salinisation because when water evaporates it concentrates salts in the upper part of the soil. Eventually salt concentration exceeds the threshold tolerance of plants, rendering the soil useless for cultivation unless the crop is salt tolerant (Szabolcs, 1986). According to Ehrlich and Ehrlich (1990) it is unlikely that yields will continue to rise as croplands degrade and the expansion of irrigation slows down. Since 1979 the growth in irrigated area per person has shrunk by about 7 per cent (Brown, 1997). Many areas are threatened by salinisation, including Imperial and San Joaquin Valleys in California, inland areas around the Aral Sea in the former USSR and 440,000 hectares of former cropland and pasture are salt affected in Australia, a 550 per cent increase since 1955 (Conacher, 1990). India has ten million hectares unusable with a further 25 million threatened by salt accumulation. Salinisation is a persistent problem all over the world: Pakistan, China, central Asia, south-east Asia, the Middle East and north Africa (Repetto, 1987). Crop yields on irrigated land have fallen by half in recent years due to the poor design of irrigation schemes (Hinrichsen, 1997).

So far the discussion has largely focused on *global* food production. Most of the evidence presented here suggests that food production can be sustained to meet population growth in global terms despite some environmental problems, but this does not mean that everybody will receive enough food. This argument is demonstrated by the recurrence of famine in Africa.

3.4.5 Problems of regional and local food production: famine in Africa

In November 1996 the World Bank released figures on food production and warned of an impending shortage of food that threatened to create a Malthusian-type disaster. However, concern was expressed about *regional* food scarcity rather than *global* shortages. The figures suggest that the problem is more acute for underdeveloped countries, especially in Africa. Around 800 million people do not receive sufficient nourishment to meet their basic needs and 82 countries (50 per cent of them in Africa) cannot grow or import enough food to sustain their people. The World Bank predicts that the number of children facing starvation will double its present figure over the next two decades unless more food is produced (Lean, 1996).

Regionally food shortages in Africa have led to famines that superficially appear to be Malthusian-type scenarios. Recently Ethiopia, Sudan and Somalia all suffered regional famines (Suau, 1985; Ellis, 1987; Caputo, 1993) (see Box 3.4), and an estimated 350,000 people have died as a result of famine and war in Somalia since 1991. A review of the causes of famine in Africa not only provides evidence for the problems of regional food production but suggests that environmental change and problems may be socially rather than naturally driven.

Box 3.4 The terminologies of hunger, famine, drought and desertification

Terms such as hunger, famine, starvation, malnutrition, drought and desertification are often bound tightly together in everyday discussions and interpretations, so one might hear people saying that famine occurs when people go hungry and starve and that this is created by drought that causes areas to become deserts. However, as Crow (1992) argues, common usage of these terms may well be unhelpful because 'they imply connections ... which may not further our understanding'. He argues for instance that there is a need to distinguish between *famine*, which he defines as 'acute starvation associated with an increase in mortality', and '*chronic hunger*', which is described as 'sustained nutritional deprivation' (Crow, 1992, p. 15). Chronic hunger is associated with famine, but one can also have a widespread presence of chronic hunger that, although leading to a large number of deaths, does not lead to an actual *increase* in death rates (and thereby, according to Crow's definition, famine). Instead, because of the continual persistence of chronic hunger, death rates in many areas are constantly higher than they would be if there were adequate levels of food and thereby nutrition. Famine can be seen as a particularly dramatic, albeit often quite short lived instance of hunger, while chronic hunger is a more long-term, and thereby often quite 'naturalised' form of hunger (see Chapter 1 for a discussion of the impacts of naturalisation).

Both famine and chronic hunger are often seen to be associated with drought and desertification. Drought refers to a period of time with unexpectedly low precipitation, or when precipitation 'is made less effective by other weather conditions such as high temperature, low humidity and strong winds' (Kemp, 1990, p. 38). Such a definition, however, begs two key issues. First, after what length of time should one define a drought? Is it low precipitation over weeks, months or years? The second question asks what constitutes low? Often the seasonal or annual norms are used as the definitive rainfall: hence Hulme and Kelly (1993) define drought as a period of two years or more with below average rainfall. Answers as to appropriate periodicity and quantity of rainfall are, however, never definitive, in the sense that what matters is the context of low rainfall. In wet, humid areas, below average precipitation may not create any particular problems, while in arid areas with generally low levels of precipitation a slight fall or delay in rainfall can have critical consequences, including that of famine and also, arguably, desertification.

Desertification refers to the creation of desert-like conditions. Scoging (1991) states that desertification affects over 180 million people and approximately 30 million square kilometres of land. As Wellens and Millington (1992, p. 245) note, it is a highly evocative term that in many people's minds 'conjures up a picture of encroaching desert – of sand dunes advancing over agricultural land around the edges of deserts'. Climate change and droughts are often caught up in this imagery. For example, Bryson (1973) suggested that drought in sub-Saharan Africa was a result of climate change. Desertification of the Sahel has also been attributed to climate change and drought. Kemp (1990) describes how the Intertropical Convergence Zone (ITCZ) moves south during the southern hemisphere's summer, reaching its southernmost extent in January or February when it remains above 8° north of the equator in west Africa but lies further south in east Africa. During this period dry, hot conditions prevail over the Sahel. When the ITCZ moves northwards in the winter months bringing hot, moist, maritime tropical air with it, it results in a rainy season (but precipitation patterns are complex) over the Sahel. The northward extent of the ITCZ varies and the rainfall associated with it is also irregular, and when it fails to reach about 20°N the Sahel experiences drought that can last over several years, promoting increased aridity and reduced vegetation growth resulting in desert conditions (Kemp, 1990). Some people interpreted these changes as being irreversible and there was desertificiation in the sense of a 'permanent decline in the potential of the land to support biological activity and, hence, human welfare' (Hulme and Kelly, 1996), together with a rise of chronic hunger and a series of major famines. This notion of irreversible change was frequently conjoined with notions of spatially contiguous change, with it being argued that the desertification in the Sahel effectively involved the expansion of the Saharan desert southwards (see Binns, 1990) However, the imagery of the expanding desert and the interpretation of desertification and famine as exclusively natural processes have come to be challenged in recent years, as discussed in, for example, Kemp (1990), Scoging (1991), Wellens and Millington (1992) and in the text of this chapter and in Chapter 5.

Dreze and Sen (1993) suggest that famine can occur in three ways: (i) a dramatic decline in food availability; (ii) an increase in food prices; (iii) people are unable to grow or buy food due to a natural disaster. A review of the causes of famine in Africa suggests that these three situations occur as a result of the interaction of social, political and natural factors, exploding the myth that natural desertification is the sole cause of famine. Civil wars, ineffective government, inadequate distribution of food, poverty, mismanagement of the land (including cash crop production), as well as drought and climate, all cause famine (Sen, 1981; Findlay and Findlay, 1984; Kemp, 1990; Macmillan, 1991).

It was formerly widely believed that natural desertification and droughts caused famine. Hellden (1986, p. 426), for example, suggested that 'the impact of man on the environment seems to be of minor importance in the semi-arid Sudan compared to climatic impact'. It is true that climate change alters the frequency and severity of drought (Hulme and Kelly, 1993) and that drought can impede food production in semi-arid and arid environments. In the Sahel, a dry, zone that covers many famine-affected countries, rainfall has fallen below its mean annual average over the last 25 years and recurrent droughts are a feature of the Sahelian climate (Nicholson, 1978; 1989). Interestingly, the variation in rainfall and the significant drop in annual precipitation may be the result of natural fluctuations in ocean circulation and climatic variability or an ocean system response to anthropogenic greenhouse gas and sulphate aerosol forcing of climate (Hulme and Kelly, 1993). Warm El Niño–southern oscillation (ENSO) events are also thought to trigger drought in Ethiopia (Seleshi and Demaree, 1995). Other climate changes have also been seen to trigger drought and environmental concern such as El Niño (see Box 3.5).

Box 3.5 El Niño and Southern Oscillation (ENSO)

Furley (1996) suggests that El Niño and southern oscillation (referred to as ENSO) form the largest single global source of inter-annual climatic variation. The southern oscillation is a periodic fluctuation in atmospheric pressure above the southern Pacific Ocean. El Niño is a recurrent, quasi-periodic appearance of anomalously warm surface water that appears along the equator in the central and eastern parts of the Pacific Ocean, especially off the coast of Peru and Ecuador (Kemp, 1990; Betshill *et al.*, 1997); but it is now commonly used to refer to larger-scale ocean–atmosphere variations that occur across the Pacific Ocean. The warming of the Pacific Ocean also heats up the air, encourages cloud formation, and results in abnormal differences in atmospheric pressure and a change in wind direction. The abnormal difference in atmospheric pressure occurs between the south-eastern tropical Pacific and the Australian–Indonesian region. ENSO events occur with varying intensity approximately every 4.5 years (range 2–10 years) and El Niño alternates with 'his twin sister' La Niña, which describes a cooling of the Pacific Ocean and a reversal of the climate changes associated with El Niño (Betshill *et al.*, 1997; Lloyd Parry, 1997).

ENSO causes some dramatic changes in climate and weather across the world, and is thought to result in irregular periods of flooding in Peru, Bolivia and Colombia and aridity across Brazil, India, Indonesia, the mid-west of the USA and Australia (Mannion, 1991b). Although Lockwood (1984) argues that there is no clear evidence to link drought in the Sahel with ENSO events, recent research by Seleshi and Demaree (1995) and Betshill *et al.* (1997) have argued that warm ENSO events triggered drought in Ethiopia and southern Africa.

Climatic variations, including El Niño, are held responsible for droughts and other environmental changes including crop failure and forest fires in Australia and Indonesia, forest fires and hurricanes in Florida, severe flooding in the mid-western USA and ice and winter storms in Canada and the USA (Lloyd Parry, 1997; Miller, 1998). The effects of El Niño are summarised in Figure B3.5.

▶

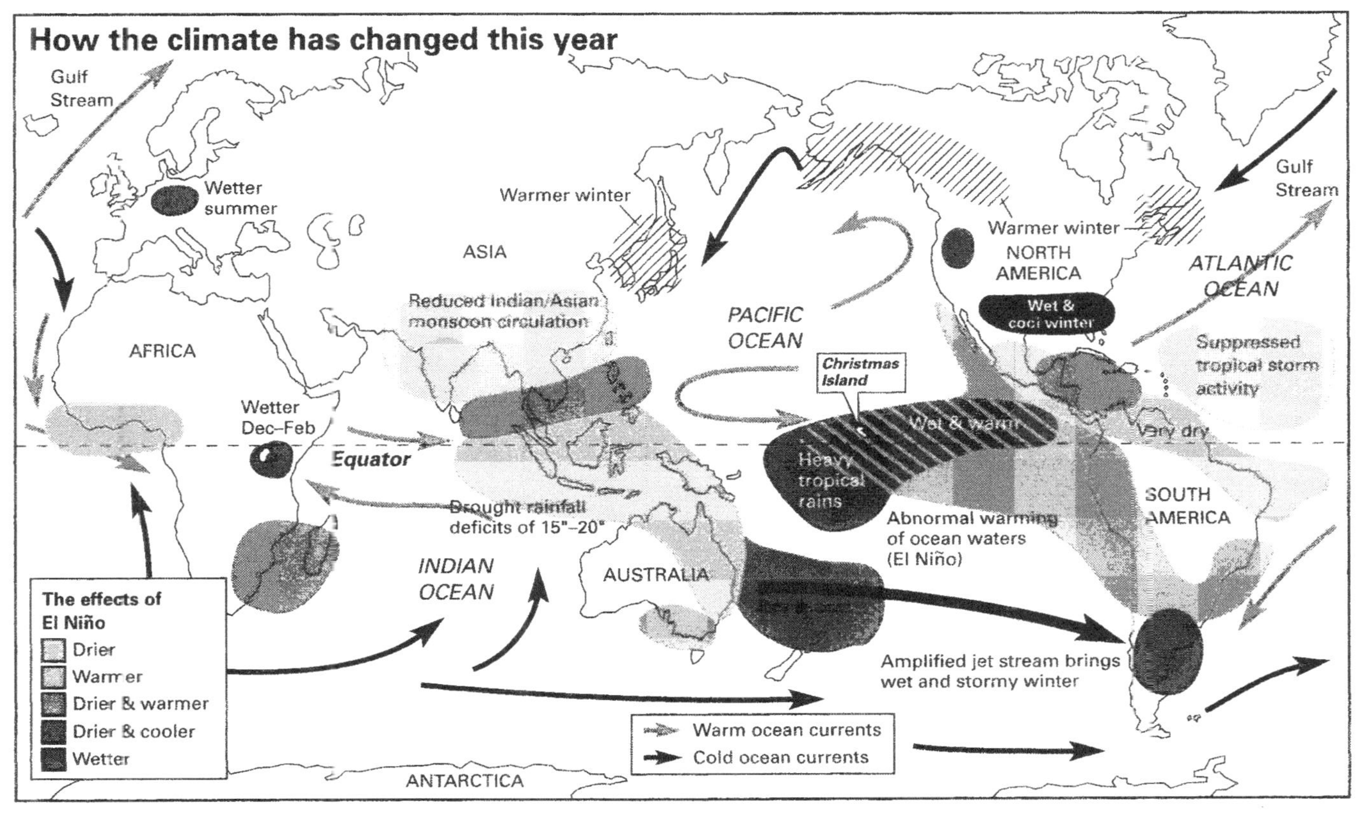

Figure B3.5 The climate effects of El Niño during 1997 (Source: Michael Roscoe IOS, the Independent on Sunday)

However, Hulme and Kelly (1993) argue that drought does not necessarily induce or contribute to desertification. In fact between 1981 and 1990 areas prone to desertification in Africa have decreased by 25.3 million hectares, but the amount of land that can be described as hyper-arid has also increased. They identify land use changes as an important factor in disrupting food production and argue that resource management failure rather than climate often promotes environmental degradation, although climate can aggravate the problem. Such a scenario is presented in Figure 3.5(a).

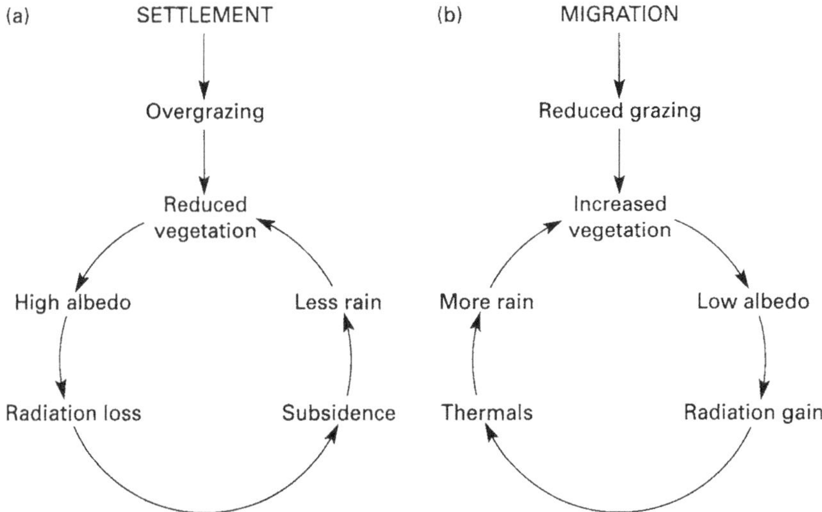

Figure 3.5 (a) Resource management failure often promotes environmental degradation. (b) Climate–land use feedbacks in the semi-arid Sahel (source: Hulme, 1989)

Hulme (1989) suggests that poor agricultural practices cause overgrazing, which initiates a cycle of land degradation resulting in a loss of vegetation cover and an increase in the amount of heat reradiated into the atmosphere from the land surface, thus creating conditions unconducive for cloud formation and producing less rain. Thus, human-induced environmental degradation might cause drought and, to reverse the situation, vegetation cover must be allowed to regenerate, as shown in Figure 3.5(b), which used to happen when the land was used by traditional nomadic herders. However, Scoging (1991) argues that since the 1970s large areas of land have been turned over for settled agriculture, including the production of cash crops, forcing nomadic herders on to more fragile lands that are easily damaged by animals. Land degradation has also resulted from pressure of demand on fuel wood and this process is summarised in Figure 3.6.

Population growth is often cited as the main reason for land degradation and contributes to famine by increasing the number of people who need to be fed. Findlay and Findlay (1984), for example, cite the example of the Ethiopian famine of 1984 and 1985. Prior to the famine food production per head fell by 5 per cent

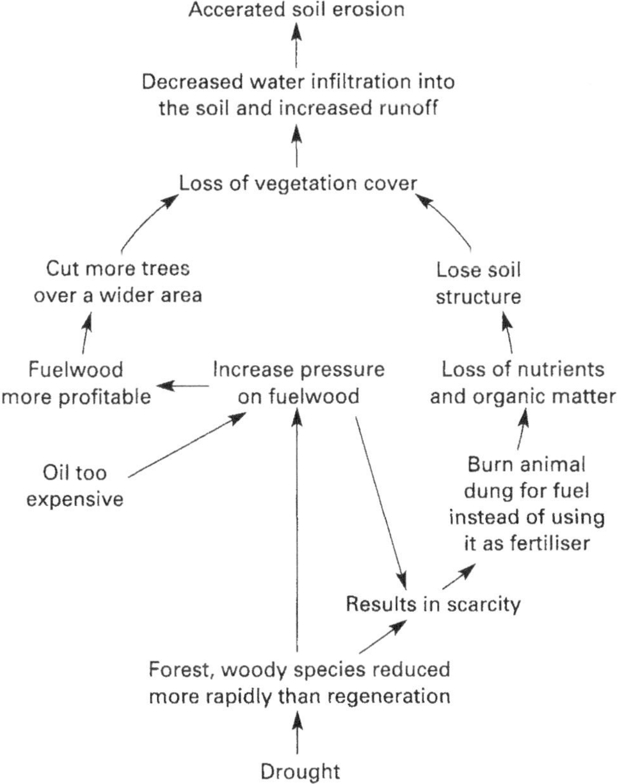

Figure 3.6 Fuel wood crisis: a schematic diagram to show the environmental problems connected with fuel wood collection in underdeveloped countries (source of information: Scoging, 1991)

over a six-year period and the worst affected areas, Wollo, Gondor, Shoa and Tigray, had a population of 20–30 million people. Scoging (1991), however, argues that while population growth will certainly aggravate problems of food production and land degradation it does not cause them, and she forwards the ideas of Boserup (see Box 3.2) and suggests that it is the maintenance of traditional standards of living that is a fundamental impediment to agricultural development. However, the situation is not that simple because other factors affect the ability of people to respond to environmental and social conditions. It has been argued, for example, that famine in Africa has been caused by not only inadequate agricultural development but by civil war. Macmillan (1991) argues that war prevents people from growing food by forcing them to leave their land and homes, while governments spend capital on weapons rather than rural infrastructure and agriculture. At least two million people were internally displaced in Somalia in the late 1980s and early 1990s (Unruh, 1995). Thus when the region experienced a persistent drought in the mid-1980s the problems of food procurement caused by war were exacerbated.

Ineffective government policy can also prevent food and food aid from reaching those threatened by famine. Requests for food aid by the Ethiopian government in

1982 were met with an indifferent response by the international community. There appears to have been a lack of political will, both internationally and nationally, to reduce the imbalance between population and food resources (Findlay and Findlay, 1984). Food production plummeted in Ethiopia when the country's ruler, Mengistu, imposed a new totalitarian regime. People were forced to work in collective villages and on state farms, which appeared to remove any incentive for farmers to work efficiently.

Political rivalry can also disrupt food distribution. Food delivery, for example, was constrained during the Somali famine of 1992–3 by extortion and gunmen and 50 per cent of food donated to Somalia was looted. In Sudan logistical problems and mistrust between rival factions hindered attempts to distribute food aid. Before food could be flown in by aid organisations permits were needed from the government and rebel groups, and only two planes were available to distribute food because war and floods had cut off many remote villages (Caputo, 1993). Once food aid reaches a country poor communications can hamper efforts to distribute it to the areas most affected by food shortages.

Once a conflict ends the situation quickly changes as people return to their homes and begin to cultivate the land. Crop yields doubled after the war ended in Eritrea and the newly formed nation cut foreign food aid requests by half. Ethiopia had its largest harvest in 20 years after the civil war ended in 1991. In Tigray wheat, barley, coffee and teff production resumed. Approximately 250,000 acres of land have been terraced and 42 million seedlings planted to repair the damage caused by decades of poor agricultural practices and neglect (Caputo, 1993). Overcoming the large-scale dislocation of people affected by war and famine is necessary to revitalise food production, but this example does suggest that Africa could feed itself. Unruh (1995) and Cobb jun. (1996) suggest that post-war recovery and agricultural production in Somalia and Eritrea can be accomplished by providing cultivable land, safe water, fuel wood and infrastructure. Drought does impede food production, but it rarely causes mass starvation without human help. The main problem is the inability of a country to cope with drought by, for example, investing in new techniques, plant strains and storage facilities (Mabogunje, 1995).

A range of factors appears to have caused famine in Africa. However, those factors discussed here are internal to Africa and need to be situated in wider contexts, and these may be central in explaining the conditions that lead to famine (see Chapter 5). Finally it must be remembered that hunger affects not only underdeveloped countries – 12 million children and eight million adults were short of the nutrients necessary for growth and good health in the USA in 1987. Poverty and cuts in food stamp programs left many people unable to procure sufficient food (Brown, 1987).

3.4.6 Summary

The evidence presented here outlines two opposing views regarding the ability of societies to meet future food demand. In particular Brown (1998) suggests that the trends shown in Figure 3.2 represent the start of the collapse of our food production systems and a downturn in growth, with resources placed under ever-increasing pressure. Brown suggests that there is evidence that agricultural systems are beginning to collapse. Ocean fish catches have remained fairly constant for the last seven years; regional grain production has fallen due to overstocking and erosion and the lessening effect of fertiliser; regional deforestation has occurred as a result of global demand for paper and regional fuel wood needs.

However, the situation may not be as bad as neo-Malthusians such as Ehrlich and Ehrlich (1990) and Brown (1997) have argued. Bongaarts (1996), Rosegrant and Livernash (1996) and Crosson (1997), for example, believe food production can keep pace with global population growth. The UN predicts that food supply will continue to grow faster than population, at least until 2010 (Reid, 1998). Bongaarts (1996) suggests sufficient food will be produced so long as governments are prepared to pursue policies that facilitate food production and create conditions conducive to maintain soil quality. Thus Brown's (1998) conclusions may be too pessimistic and may represent short-term fluctuations. For example, the downturn in world grain 'carryover stock' expressed as days of consumption (Figure 3.7) sees the food security threshold level broken over the past few years. A similar situation has occurred before, in the mid-1970s for example, and stocks have recovered. It is difficult therefore to predict if the downturns outlined above represent the actual beginnings of system collapse or our ability to produce food has peaked. Given that human population is still expanding it is clearly dangerous to be dismissive about these warning signs and, at face value, it would be sensible to take a cautious approach and address these issues before short-term blips turn into long-term trends.

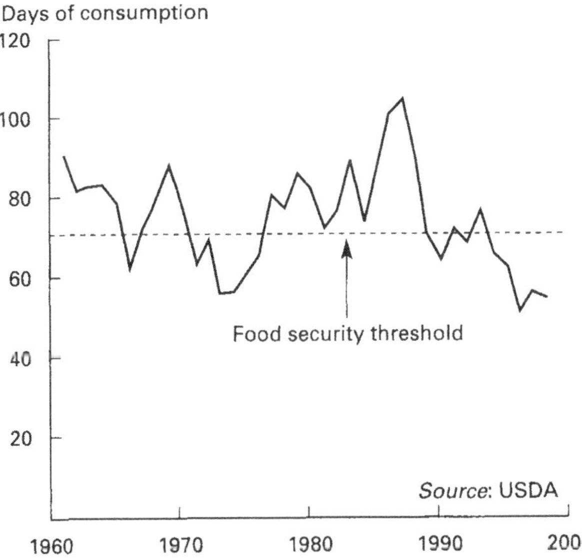

Figure 3.7 World grain carryover stocks, as days of consumption, 1961–98 (source: Brown, 1998)

At present, however, meeting global food demand is not a problem. Reid (1998) argues that food distribution is the main reason why people go hungry and suffer from famine and malnutrition, and that regional disparities must be overcome by providing food, grown in areas where potential arable land now exists, on the international market. Solutions to the current food crisis can be tackled on a number of fronts. First, technological revolutions in food production offer potential to increase yields. Second, better management (in terms of people having access to land, land ownership and how they use the land) can prevent soil erosion and loss of soil fertility. Third, efforts to eliminate malnutrition and population growth will also relieve the situation. Water scarcity, falling water tables and the continued pollution of water supplies are warning signs that local and regional supplies may be exhausted in the short to medium term. The situation, however, may not become critical if these resources can be managed efficiently and technology continues to improve. These issues are explored in Chapters 4 and 5.

3.5 Can societies meet their energy requirements?

Contemporary societies have become dependent on four major fuel sources: oil, gas and coal and more recently, and to a limited extent, nuclear power. By 2025 world-wide demand for fuel is expected to increase by 30 per cent and demand for electricity by 265 per cent. Such a large increase raises the question as to whether or not sufficient energy resources can be found to meet this demand. Neo-Malthusians argue that conventional energy sources will not suffice, because fossils fuels such as oil and gas will run out over the next hundred years or so, and coal and nuclear power will be abandoned for environmental reasons. Neo-Malthusians say that societies will have to turn to alternative energy sources but there is some debate as to whether these forms of energy generation will meet future demand. Cornucopians argue that predictions that fossil fuels will run out will not materialise because fossil fuels will be used more efficiently, new reserves will be discovered and levels of pollution will be reduced. Even if fossil fuels did run out cornucopians like Cohen (1995a, 1995b) and Simon (1995) argue that nuclear power will meet all society's energy needs and we will never run out of energy (Cohen, 1984).

3.5.1 Current levels of nuclear power consumption and of fossil fuel use

Eisenbud (1990) and Pasztor (1991) outline the potential of nuclear power as a major energy source. In 1988 the industry generated 17 per cent of the world's electricity from 42 nuclear power stations and the future looked bright with an additional 105 under construction, and by 1990 uranium was used in over 1000 nuclear reactors as a fuel for nuclear fission to produce energy.

However, the future of nuclear power is a controversial issue. Pasztor (1991) suggests this is the result of a series of nuclear accidents at Three Mile Island and Chernobyl and now the nuclear industry appears to be contracting. Only 16 new nuclear power plants have been commissioned since 1989, Sweden has abandoned nuclear energy altogether, while other countries like the UK have turned to natural gas to supply electricity. It is not surprising, therefore, that global figures for nuclear power are in decline, with the industry contributing 24 per cent of total electricity production in 1988 and the predicted figure for 2005 being 22 per cent. Williams and Woessner (1996), Pasztor (1991), Swinbanks (1997) and Cohen

(1995a) believe that unless new designs, improvements in safety and costs, and convincing assurances over radioactive waste transport and disposal, the decommissioning of existing nuclear reactors, the 'impossibility' of terrorist groups purchasing such substances on the black market (25 nuclear warheads could be constructed from plutonium used in a typical 1 GWe LWR reactor) and nuclear profileration are forthcoming, a revival in nuclear power looks increasingly unlikely. Despite advances in nuclear technology, for example the development of modular high temperature gas cooled reactors, which are considered to be the safest form of nuclear reactor, the industry cannot provide firm assurances over safety.

At present it seems that nuclear power will not provide a substantial amount of energy in the short term. However, Pasztor (1991) and Perera (1993) argue it would be unwise to dismiss completely the future role of nuclear power. France and Japan are amongst a minority of countries that have maintained a commitment to nuclear power. Over the next decade Japan is the only country that will significantly increase the amount of electricity generated by nuclear power, but expansion of the nuclear power industry may occur in former eastern bloc countries (where a total of 49 reactors were under construction in 1991 (Pasztor, 1991)) and in underdeveloped countries such as India and Pakistan. Furthermore future restrictions on fossil fuel use, possible energy shortages and limited access of conventional fuels on the international market, employment opportunities, combined with safer reactors and improved methods of radioactive waste disposal or reprocessing, could raise the profile of nuclear power back to the forefront of national energy policy (Anon, 1995a; Cohen, 1995b).

Currently fossil fuels account for 78 per cent of the world's energy, with renewable energies (including hydropower and biomass) and nuclear power providing 18 per cent and 4 per cent respectively. Fossil fuels, namely coal, oil and gas, are versatile compounds that perform a variety of functions including energy generation, fuel and powering transportation. They are accessible, abundant and cheap and therefore have become a popular conventional method of energy generation for industrialised countries (Charles, 1993).

The world energy market is a dynamic one, the contribution of individual fossil fuel sources has changed over the last 200 years. Coal was the dominant energy fuel until it peaked during the 1920s, when it contributed up to 70 per cent of the world energy supplies. Oil peaked, accounting for just over 40 per cent of the world total, during the early 1970s and the proportion provided by natural gas is currently increasing (Davis, 1990).

Coal generates about 56 per cent of electricity in the USA but accounts for 85 per cent of the potential fossil fuel reserves and therefore is likely to continue to be a central focus of US energy policy (Corcoran, 1991). Demand for electricity in China and Asia has seen coal consumption increase by 5.5 per cent over the past decade. Although fossil fuels are linked to several environmental problems, including the enhanced greenhouse effect, urban smog and acid atmospheric deposition, abundant coal deposits exist in Asia (especially Siberia), North America and Europe (Simmons, 1991), meaning that coal will remain an available energy source for the foreseeable future.

Abundant natural gas supplies make it a viable energy form. Prominent gas fields occur in Siberia, the North Sea, Canada and Alaska. Natural gas exploitation is expanding; for example between 1981 and 1987 gas production in the former USSR increased from 462×10^9 m^3 to 727×10^9 m^3 (Nisbett, 1989). It is estimated that 500 billion barrels of gas are readily recoverable, excluding gas found in coal seams and other geological formations. Exploiting this gas could maintain liquid fuel supplies for a decade after conventional stocks dwindle. At present the cost of producing liquid fuels from gas are approximately 10 per cent higher than crude oil but modest improvements in technology could reduce this figure. Natural gas also contains few substances that form pollutants, namely sulphur, nitrogen and heavy metals (Fouda, 1998). Since privatisation UK companies have opted to use natural gas-fired power stations to provide electricity, a cheaper alternative to coal (Cross, 1993).

The rising demand for energy has been attained by extracting more fossil fuels at ever-increasing rates and they are being depleted 100,000 times faster than they are being formed, raising concerns that supplies of oil and gas will run out in the next 100 years, supporting the neo-Malthusian idea that societies will face resource scarcity. Global oil demand is increasing at approximately 2 per cent per year and growth in energy use in underdeveloped regions (Latin America, Africa and Asia) is up by at least 30 per cent since 1985 (Campbell and Laherrere, 1998). With 65 million barrels of oil produced by the industry every day the output of many of the current oilfields is declining; and global discoveries peaked in the 1960s, leaving only about 1000 billion barrels of conventional oil recoverable.

It is estimated that the world's oil reserves stand at one trillion barrels. Approximately 70 per cent of global oil is associated with areas of plate convergence. Huge reserves exist in the Middle East where the Eurasian and Arabian plates collide and it is anticipated that by around 2002 the Middle East will control the bulk of oil supplies and might restrict availability for political reasons. Oil is found north of the Brooks range in Alaska and east of the Russian Urals close to plate boundaries. The remaining oil deposits occur close to former areas of plate activity such as rift zones. However, the quantity of oil reserves is uncertain because new oil fields remain undiscovered (Howell *et al.*, 1993). Davis (1990) suggests that the amount of recoverable fossil fuels is thought to equal ten trillion barrels of oil, which would last for another 170 years based on current consumption rates. Howell *et al.* (1993) predict 950 billion barrels exist in known oil fields and estimate undiscovered sources could provide between 300 and 900 billion barrels. They argue that oil will run out in about 70 years, while Campbell and Laherrere (1998) suggest that conventional oil supplies will not be able to meet demand in the next decade, but the oil industry reports that proven reserves should maintain supplies for approximately 43 years.

Cornucopians argue that the predictions of oil supplies are misleading. They argue that the life expectancy of fossil fuel sources is dependent upon recoverable stocks, the extent to which advances in technology and economic forces will influence the cost and profit margins of recovering the remaining sources, the environment cost and new oil fields that remain undiscovered. These uncertainties mean that estimates vary, making it difficult to determine if oil will run out. Campbell and Laherrere (1998), George (1998) and Anderson (1998) argue that there is little reason to assume oil and gas supplies will run out in the foreseeable future. For example, the US Energy Information Administration anticipates that oil reserves will continue to increase by almost two-thirds up to 2020 and that the life expectancy of oil may be longer if heavy oils, oil shales and coal can be economically recovered using new extraction technology. Unconventional oil (tar sands, heavy

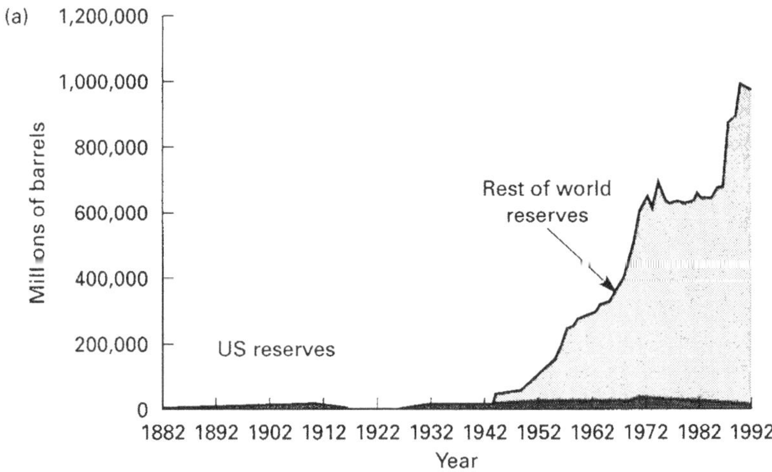

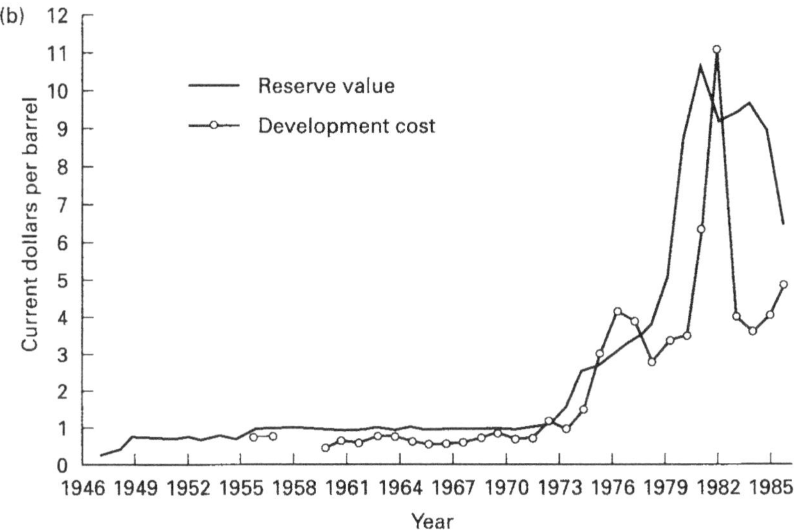

Figure 3.8 (a) Estimated crude oil reserves. (b) Crude reserve value, 1946–86, and post-tax development cost, 1955–86 (source Adelman, 1995)

oils) is abundant and may substitute conventional supply. Recovering 20 per cent of Canada's heavy oil could supply Canada and the USA with oil for the next 100 years, while the Orinoco oil belt in Venezuela contains approximately 1.2 trillion barrels of heavy oil and tar sands and shales in Canada and the former USSR hold over 300 billion barrels. Australia also has at least 28 billion barrels in oil shales. Exploiting this oil will cause a significant amount of pollution and it is uncertain if its recovery will be profitable, although Suncor Energy Ltd are producing unconventional oil profitably in Canada. Other sources extracted from deep water may also become possible with the advent of new technology such as underground imaging, steerable drilling and gas induced extraction techniques (see Anderson, 1998).

Because of the potential to develop alternative sources of oil cornucopians like Singer (1984) argue that concerns over oil scarcity are misplaced, as other sources of energy will provide a low-cost solution and societies will become less dependent on oil and will eventually bypass available supplies. Dusseault (1997) suggests environmental considerations will limit future oil use rather than a lack of hydrocarbons, while Simon (1994) argues discoveries and technological advances arise from human enterprise in time of perceived scarcity and will prevent shortages of all minerals. Relatively low oil prices since 1986 do not support the contention that oil is running out (Stevens, 1996; Adelman, 1995) as high prices have been the product of artificial, short-term fluctuations in supply (such as the OPEC oil crisis, the Iran–Iraq and Gulf Wars and the debt crisis of the early 1980s). Figure 3.8 shows that estimated oil reserves have actually increased and that the value of a barrel of oil has exceeded development costs since the 1970s (Adelman, 1995).

The discussion so far has concentrated on whether societies can continue to meet future energy requirements and to consider the role of fossil fuels and nuclear power. One aspect of this debate is the question of resource scarcity. Neo-Malthusians have argued that societies will run out of resources if we are not careful. So far this view is not fully supported by all the estimates of oil and gas reserves. An analysis of non-fuel minerals in the next section also suggests that these supplies will not become scarce in the foreseeable future.

3.5.2 Non-fuel mineral supplies

An increase in the reserves and a reduction in the price of many key minerals has occurred since the 1950s, providing no evidence that these resources are becoming scarce. The industrial economy of western nations has resulted in widespread demand for metals and minerals. World steel output has increased by 70 times this century and aluminium use by 1700 times, a pattern common to most metals (Repetto, 1987). Despite such demand available metal reserves have increased during the same time period. Technological advances have helped to discover new reserves and allowed societies to exploit previously uneconomic mineral deposits. The price of metals and minerals has remained low, implying that in economic terms exhaustible metals and minerals have not become significantly scarce and show no signs of doing so in the foreseeable future. The price of most non-fuel minerals reveals that prices have remained stable or lowered between 1950 and 1980 (Barnett *et al.*, 1984). Simon (1994) notes that the price of copper is one-tenth of what it was 200 years ago.

Meadows *et al.* (1972) suggested that the world's aluminium resources might be exhausted by 2003 but at present no mineral suffers from absolute resource scarcity: a situation that arises when there are not enough actual or affordable supplies to meet demand (Miller, 1990). Myers *et al.* (1995) reviewed the past trend in availability of 13 major minerals in terms of their price relative to wage rate and to consumer price. The analysis suggests that there was no evidence for increasing costs because mineral scarcity was not a constraint on production. When prices do rise sharply it is normally because of what Simon describes as 'non-physical causes': a dramatic rise in prices rather than a real decline in availability, which is sometimes referred to as relative resource scarcity (Miller, 1990). Goeller (1995) looked at future trends of minerals and claimed it is unlikely that societies will run short of any element before 2050 and will not suffer from total exhaustion of minerals because new technology will almost certainly provide alternatives and substitutes. If, in the unlikely event, certain elements remained constant only a limited number of elements would be exhausted by 2100. Goeller even suggests recycling of minerals is unnecessary as the amount recycled generally only accounts for a minute fraction of the total that is recoverable.

An examination of minerals suggests that societies are not running out of resources. In fact the trend is the complete opposite and mineral reserves are actually increasing despite higher rates of consumption by societies. Technological advances in recovering minerals and the discovery of new sources have paralleled population growth and demand. It seems likely that mineral and metal availability will not limit industrial production and consumption while fossil fuels and nuclear power, if allowed, could provide sufficient energy. However, there is another reason why neo-Malthusians argue that coal and nuclear power will not continue to provide a substantial proportion of future energy demand, and that is because the extraction of resources and the energy generation pollutes the environment. The next section reviews the evidence that suggests that people have paid a high environmental cost for energy generation and the production and consumption of resources.

3.6 Global environmental costs of the industrialised societies

3.6.1 Introduction

Since the mid-1950s there has been an unprecedented quintupling of global economic production (see Table 3.6) and economic trends have shaped the way societies have exploited the earth's resources. Postel (1994) argues that environmental damage has increased proportionately with economic activity: large increases in fossil fuel use, water use and extraction of minerals and timber.

Table 3.6 World economic growth by decade, total and per person, 1950–93

Decade	Annual growth	Annual growth per person
1950–60	4.9	3.1
1960–70	5.2	3.2
1970–80	3.4	1.6
1980–90	2.9	1.1
1990–93 (prel.)	0.9	–0.8

(Source: Worldwatch Institute, Brown, 1994)

Until recently little attention had been given to the process by which economic growth had been accomplished and, more importantly, the environmental cost of achieving growth. Brown (1991) states that behind the 'beautiful mask' of economic statistics lies an 'ugly face': the environmental impact of industrialisation and over exploitation caused by consumption in the developed world and poverty in the underdeveloped world. Neo-Malthusians have therefore argued that during the second half of the twentieth century energy generation and industrial processes have had a high environmental cost at global, regional and local levels.

In this section two of these costs, namely the greenhouse effect and ozone depletion, will be examined. These can both to some extent be described as natural processes, in that they would occur to some degree whether humans were present on Earth or not.

3.6.2 The greenhouse effect

The sun provides the ultimate energy source for weather and climate on Earth in the form of solar (or short wave) radiation. As illustrated in Figure 3.9, approximately two-thirds of solar radiation is absorbed by the atmosphere, land, ice or water. Infra-red radiation is emitted from the earth into the troposphere (the lower part of the atmosphere), where it is absorbed in the form of heat by naturally occurring gases: water vapour (H_2O), carbon dioxide (CO_2), methane (CH_4), nitrous oxide (NO_x), ozone (O_3) and clouds. The trapping of heat in the lower atmosphere by these gases operates in a way similar to a garden greenhouse, hence the process being known as the '*greenhouse effect*' and the gases that trap the heat being described as '*greenhouse gases*'. Without greenhouse gases absorbing this form of heat the lower atmosphere and the earth's surface would be about 33°C colder. The amount of incoming solar radiation at the top of the atmosphere is balanced by outgoing infra-red radiation. Carbon dioxide, methane, nitrous oxides and halocarbons all cause warming in the troposphere and are popularly regarded as the major greenhouse gases because they are readily identified with human activities. However, Elmsley (1996) argues that water vapour accounts for 97 per cent of the natural greenhouse effect and it is the principal greenhouse gas.

The lower atmosphere can be heated up or cooled down by a process known as '*radiative forcing*', defined as 'a change in the net radiation at the top of the troposphere because of a change in either solar or infra-red radiation' (Houghton *et al.*, 1996, p. 15). Thus, radiative forcing upsets the balance between incoming and outgoing radiation. Once this happens the climate system will adjust to a new equilibrium where a negative radiative forcing is likely to result in cooling the system, while a positive forcing is likely to result in warming (Houghton *et al.*, 1996).

Non-human agencies are thought to influence radiative forcing. Observation has shown that total solar irradiance varies during a so-called 11-year sunspot cycle. Changes in solar radiation during a sunspot cycle are estimated to represent about 0.2 $W\,m^{-2}$, although the resultant impact on temperature is thought to be negligible given the variability of solar radiance in one cycle. Volcanic eruptions create a cooling effect by supplying the atmosphere with aerosols that scatter solar radiation in the stratosphere, preventing it reaching the troposphere. Estimates based on the Mount Pinatubo eruption of 1991 suggest the radiative forcing was about –4 $W\,m^{-2}$, dropping to $-1\,W\,m^{-2}$ within two years, and global surface temperatures lowered by 0.3 to 0.5°C in 1992 (Houghton *et al.*, 1996).

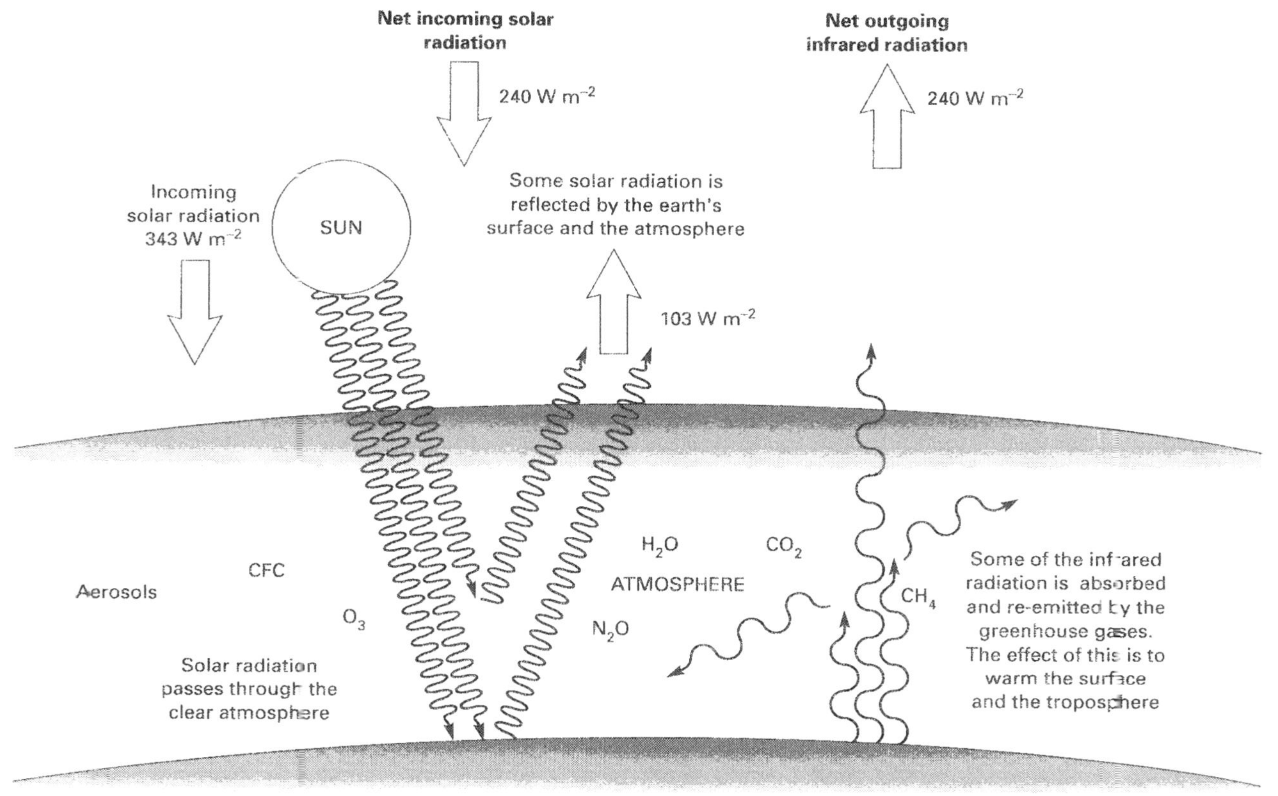

Figure 3.9 The greenhouse effect (after Houghton *et al.*, 1996)

A variety of natural processes are thought to control the concentration of greenhouse gases in the atmosphere. Sources release them into the atmosphere and sinks remove them, creating a system that regulates itself in a way that has similarities to the Gaia hypothesis (see Chapters 1 and 6). Whereas the effects of volcanic eruptions and sunspot cycles are transitory, these natural sources and sinks regulate the atmospheric concentration of greenhouse gases (see Box 3.6). This is evident when looking at the pre-industrial greenhouse gas concentrations that are stable throughout much of the Holocene. Neo-Malthusians, however, argue that anthropogenic factors may now be more influential in radiative forcing and have possibly begun to alter atmospheric greenhouse gas concentrations.

While carbon dioxide, methane, nitrous oxides and halocarbons all cause warming in the troposphere, ozone and aerosols deserve some consideration because they can produce a cooling effect. Ozone is believed to be able to induce both a warming and cooling effect as stratospheric ozone loss over the last 25 years has led to an estimated average global radiative forcing of –0.1 W m^{-2}. Prior to significant stratospheric ozone loss halocarbons are thought to have contributed a net radiative forcing of 0.1 to 0.2 W m^{-2} (Houghton *et al.*, 1996) but it seems that stratospheric ozone loss may offset the positive radiative forcing of tropospheric ozone and halocarbons. Tropospheric ozone also behaves like a greenhouse gas and both vary spatially and vertically in the atmosphere.

Schwartz and Andreae (1996) argue that aerosols (particles that are suspended in the atmosphere) could negate much of the present radiative forcing. Aerosols are formed by volcanic and human activity (eg by fossil fuel combustion, deforestation, cultivation and overgrazing) (Thompson, 1995). Once in the troposphere, aerosols either scatter solar radiation or act as a surface for chemical reactions. The latter is important for cloud formation, which can also reflect solar radiation, but the spatial distribution and concentration of aerosols in the lower atmosphere is extremely variable, therefore it is difficult to estimate their influence on climate change. Radiative forcing by sulphate aerosols is estimated to be between –0.25 and -0.9 W m^{-2} but there is some uncertainty as to the extent that aerosols migitate warming.

Box 3.6 Natural sources and sinks of the major greenhouse gases Sinks (see Tables B3.6a, B3.6b and B3.6c)

The world's oceans hold up to 50 times more carbon dioxide than the atmosphere and they are the largest CO_2 sink (Schimel *et al.*, 1995). Results from several different types of ocean circulation models suggest that the oceans absorb approximately 2.0 ± 0.8 Gt each year of anthropogenic carbon (Schimel *et al.*, 1995) and ocean uptake may also provide a sink for some halocarbons such as carbon tetrachloride and methylchloroform (Prather *et al.*, 1995).

Chemical reactions that take place in the lower atmosphere also control the abundance of most greenhouse gases, for example between 90 and 95 per cent of methane is oxidised by hydroxyl radicals (OH) in the atmosphere (Khalil and

Rasmussen, 1985; Tetlow-Smith, 1995) which also break down or oxidise carbon monoxide, halocarbons and nitrogen dioxide. Some of these chemical reactions can be summarised as follows:

$$OH + CH_4 + O_2 \rightarrow CH_3O_2 + H_2O$$

$$OH + CO + O_2 \rightarrow CO_2 + HO_2$$

OH can be recycled by the following reactions:

$$HO_2 + NO \rightarrow OH + NO_2$$

$$HO_2 + O_3 \rightarrow OH + O_2 + O_2$$

This recycling of OH radicals can be stopped if OH and NO_2 combine to produce nitric acid.

$$OH + NO_2 \rightarrow HNO_3$$

Nitric acid can be lost from the lower atmosphere by precipitation or dry deposition, a major mechanism for NO_x removal, and if the atmosphere continues to be polluted by NO_2 methane concentrations will increase because there will be less OH available to oxidise it. Another knock-on effect of fewer hydroxyl radicals in the atmosphere is that it could increase the concentration of other greenhouse gases. The oxidation of methane in the presence of nitrogen can produce ozone, which acts as a greenhouse gas in the troposphere. If methane reaches and subsequently breaks down in the stratosphere it produces water vapour, another greenhouse gas.

Greenhouse gases are also immobilised by plants and the soil. Carbon is immobilised by afforestation in mid and high latitudes (approximately 0.5 ± 0.5 GtC/yr) and photosynthesis (0.5 to 0.2 GtC/yr) and this is expected to increase as photosynthetic rates increase with higher atmospheric concentrations of CO_2 and Whalen and Reeburgh (1992) show that methane can be oxidised in the upper part of aerobic soils profile by bacteria.

Sources (see Tables B3.6a, B3.6b and B3.6c)

Methane is produced by the process of methogenesis, a process of anaerobic metabolism by methogenic bacteria that are found in the digestive tracts of ruminant animals and other anaerobic environments (Tetlow-Smith, 1995). Methane is also released naturally from northern wetlands, permafrost areas and through the decomposition of methane hydrates in continental shelf areas (Nisbett, 1989). Geological sources of methane, including gas hydrates, release approximately 5 to 15 Tg every year and Whalen and Reeburgh (1992) suggest northern wetlands produce between 20 and 60 $TgCH_4$/yr while tropical wetlands release about 60 $TgCH_4$/yr (Bartlett and Harriss, 1993).

Soils contribute 12 Tg/yr of NO_x with forest soils producing an estimated 4 Tg(N)/yr while other ecosystems and the degassing of groundwater for irrigation are minor sources (Prather *et al.*, 1995). Termites produce 20 Tg of methane each year (Prather *et al.*, 1995) while beavers can also promote methane production by flooding areas of wetlands and forest (Nisbett, 1989).

Oceans release several greenhouse gases including about 3.5 to 50 Tg/yr of methane. It is estimated that the oceans are the largest producer of nitrous oxide, ranging from 1 to 5 Tg(N)/yr, while ocean algae produce a variety of halocarbons (methylhalides and halogenated methanes) but they do not make a significant contribution to atmospheric concentrations. Oceans also contribute about 30 to 70 per cent of methylbromide, an ozone depleting chemical (Prather *et al.*, 1995).

Table B3.6a Annual average anthropogenic carbon budget for 1980–9. CO_2 sources, sinks and storage in the atmosphere are expressed in GtC/yr

CO_2 sources	
(1) Emissions from fossil fuel combustion and cement production	5.5 ± 0.5*
(2) Net emissions from changes in tropical land-use	1.6 ± 1.0†
(3) Total anthropogenic emissions = (1) – (2)	7.1 ± 1.1
Partitioning amongst reservoirs	
(4) Storage in the atmosphere	3.3 ± 0.2
(5) Ocean uptake	2.0 ± 0.8
(6) Uptake by Northern Hemisphere forest regrowth	0.5 ± 0.5#
(7) Inferred sink: 3 – (4 + 5 + 6)	1.3 ± 1.5§

Notes

* For comparison, emissions in 1994 were 6.1 GtC/yr.

† Consistent with Chapter 24 of IPCC WGII (1995).

This number is consistent with the independent estimate, given in IPCC WGII (1995), of 0.7 ±0.2 GtC/yr for the mid- and high-latitude forest sink.

§ This inferred sink is consistent with independent estimates, given in Chapter 9 of Houghton *et al.* (1994), of carbon uptake due to nitrogen fertilisation (0.5 ± 1.0 GtC/yr), plus the range of other uptakes (0–2 GtC/yr) due to CO_2 fertilisation and climatic effects.

(Source: Houghton *et al.*, 1994)

Table B3.6b Estimated sources and sinks of methane in Tg CH_4/yr

Observed atmospheric increase, estimated sinks and sources derived to balance the budget

	Individual estimate	Total
Atmospheric increase	37 (35–40)	37 (35–40)
Atmospheric removal (lifetime = 9.4 yr)		
tropospheric OH	445 (360–530)	
stratosphere	40 (32–48)	
soils	30 (15–45)	
Total sinks		515 (430–600)
Implied total sources (atmospheric increase + total sinks)		552 (465–640)

(b) Inventory of identified sources

Identified sources	Individual estimate	Total
Natural		
Wetlands	115 (55–150)	
Termites	20 (10–50)	
Oceans	10 (5–50)	
Other	15 (10–40)	
Total identified natural sources		160 (110–210)[†]
Anthropogenic		
Total fossil fuel related		100 (70–120)[††]
Individual fossil fuel related sources		
Natural gas	40 (2550)	
Coal mines	30 (1545)	
Petroleum industry	15 (530)	
Coal combustion[†††]	? (130)	
Biospheric carbon		
Enteric fermentation	85 (65–100)	
Rice paddies	60 (20–100)	
Biomass burning	40 (20–80)	
Landfills	40 (20–70)	
Animal waste	25 (20–30)	
Domestic sewage	25 (15–80)	
Total biospheric		275 (200–350)
Total identified anthropogenic sources		375 (300–450)[†]
TOTAL IDENTIFIED SOURCES		535 (410–660)

Total Global Burden: 4850 Tg(CH_4)

Note: The observed increases in methane show that sources exceed sinks by about 35 to 40 Tg each year. All data are rounded to the nearest 5 Tg.

† A pre-industrial level of 700 ppbv would have required a source of 210 Tg(CH_4)/yr if the lifetime has remained constant, and 280 Tg(CH_4)/yr if current tropospheric chemical feedbacks can be extrapolated back. The total anthropogenic emissions of CH_4 based on identified sources, 375 (300–450), is slightly higher than the inferred range from pre-industrial levels, 270–340. but is well within the uncertainties.

†† Fractional source from fossil carbon based on a measure of the atmospheric ratio of $^{14}CH_4$ to $^{12}CH_4$.

††† Judd *et al.* (1993), Khalil *et al.* (1993)

(Source: Houghton *et al.*, 1995)

Table B3.6c Estimated sources and sinks of N_2O typical of the last decade (TgN/yr)

	Range	Likely value
Atmospheric increase	3.1–4.7	3.9†
Sinks		
stratosphere	9–16	12.3
soils	?	
Total sinks	9–16	12.3
Implied total sources (atmospheric increase + total sinks)	13–20	16.2
Identified sources	Range	Likely value
Natural		
Oceans	1–5	3
Tropical soils		
wet forests	2.2–3.7	3
dry savannas	0.5–2.0	1
Temperate soils		
forests	0.1–2.0	1
grasslands	0.5–2.0	1
Total identified natural sources	6–12	9
Anthropogenic		
Cultivated soils	1.8–5.3	3.5
Biomass burning	0.2–1.0	0.5
Industrial sources	0.7–1.8	1.3
Cattle and feed lots	0.2–0.5	0.4
Total identified anthropogenic	3.7–7.7	5.7
TOTAL IDENTIFIED SOURCES	10–17	14.7

† The observed atmospheric increase implies that sources exceed sinks by 3.9 Tg(N)/yr.

(Source: Houghton *et al.*, 1995)

3.6.3 The human-enhanced greenhouse effect

It is widely believed that human activities have the capacity to alter climate. For example, Houghton *et al.* (1995) concluded that 'the balance of evidence suggests that there is a discernible human influence on global climate' in recent centuries and Hansen *et al.* (1988) confidently suggest that major greenhouse climate changes are a certainty. Human activity is believed to be changing the concentration of greenhouse gases in the lower atmosphere by increasing the sources and destroying the sinks of the greenhouse gases. It is believed that this will upset the radiation balance of the lower atmosphere and lead to global warming. This scenario is often described as an 'anthropogenic' or 'enhanced' greenhouse effect.

Neo-Malthusians such as Ehrlich and Ehrlich (1990) argue that the enhanced greenhouse effect is the most serious intransigent environmental problem and suggest that the changes to global climate will have disastrous consequences for societies, including flooding, desertification, water shortage and the spread of disease. They readily accept the data that suggests that global warming is taking place as a result of an increase in the atmospheric concentration of the greenhouse gases. Cornucopians are more sceptical about the evidence for global warming and Simon (1995) believes that societies can intervene to remove any serious threat, while others such as Michaels (1995) question whether the impacts of global warming will ever materialise.

The human and non-human sources and sinks for carbon dioxide, methane and nitrous oxides are summarised in Tables B3.6a, B3.6b and B3.6c. Neo-Malthusians point to the fact that human-made sources now make up a significant proportion of the total. For example, industrial sources include fossil fuel combustion, peat mining and natural gas extraction (Rodhe and Svensson, 1995). Cement manufacture and fossil fuel combustion are the largest anthropogenic contributor to atmospheric CO_2. Between 1980 and 1989 about 5.5 ± 0.5 GtC/yr (GtC – gigatonnes of carbon) of CO_2 was emitted from these sources and since the beginning of the industrial revolution the total amount released approximates 230 GtC (Andres *et al.*, 1994). About 20 per cent of the total annual emission of methane and 24 TgN/yr (TgN – teragrams of nitrogen) of NO_x is also produced from fossil fuel use, which also release aerosols. Natural gas extraction releases about 65 Tg/yr of methane into the atmosphere (Nisbet, 1989). Methane is also leaked from coal mines and pipelines while peat is burnt as a fuel in some parts of the world, including Scandinavia and Ireland, which releases carbon dioxide. For example, use of peat in Sweden currently accounts for 3.5T Wh of CO_2 per year but methane emissions do cease as peatlands are drained prior to extraction (Rodhe and Svensson, 1995). The contribution of the major greenhouse gases to radiative forcing is shown in Figure 3.10.

Agricultural practices and forest destruction also produce greenhouse gases. Deforestation and other land use changes unlock immobilised carbon from biomass and release carbon dioxide and nitrous oxides into the atmosphere. Forest destruc-

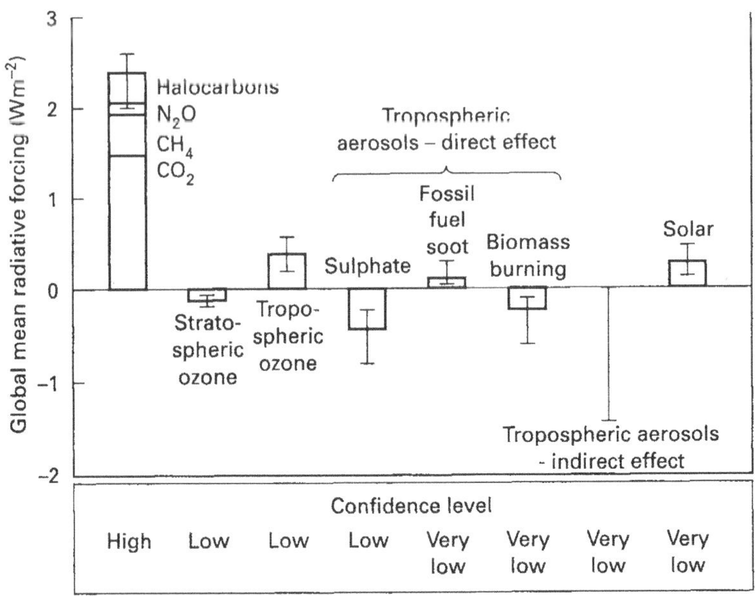

Figure 3.10 Estimates of the globally averaged radiative forcing due to changes in greenhouse gases and aerosols from pre-industrial times to the present day and changes in solar variability from 1850 to the present day (source: Houghton *et al.*, 1996)

tion, especially in the tropics, accounted for *c.* 1.6 ± 1.0 GtC/yr of CO_2 between 1980 and 1989 (Schimel *et al.*, 1995) and approximately 122 ± 40 GtC between 1850 and 1990 (Houghton, 1994). Globally, landfills and biomass burning are also significant human sources of NO_x and aerosols but agriculture represents the largest anthropogenic source for methane production, with rice paddies and ruminant animals annually producing approximately 20–100 and 65–100 Tg of methane respectively, while animal waste produces a further 20–30 Tg.

The effect of anthropogenic sources on the greenhouse effect is evident in historical records of climatic data that provide evidence for recent temperature changes over the past century. Jones and Wigley (1990) and Jones *et al.* (1986a; 1986b) have produced graphs of global temperature variations for 1851 and 1984. Although some anomalies may exist in the dataset as records before 1880 are less reliable, they argue that a linear increase in global temperature of 0.45°C provides evidence of global warming. The Intergovernmental Panel for Climate Change Houghton *et al.*, 1995 shares this view, arguing that the pattern of greenhouse gas concentrations during the Holocene provides a firm link between human action and the 'enhanced greenhouse effect'.

Some people are more cautious about interpreting the data, arguing that scientists do not fully understand how climate change occurs over long, medium and short timescales. Hasselmann (1997), for example, questions whether global temperature increases over the past few centuries have been caused by human activity or whether it is simply an expression of natural climate variability on larger spatial and temporal scales, a point that has yet to be fully resolved. Chambers (1995; 1998) suggests that deciphering an anthropogenic climate change signal from natural climate variability is difficult, as climatic variations of a similar magnitude have occurred over hundreds of thousands of years, primarily caused or triggered by earth-orbital variations (known as astronomical forcing, and/or Milanovitch cycles) and over decadal to century time-scales.

Measurement of carbon dioxide concentration in polar ice cores provides a method of reconstructing atmospheric concentration of CO_2 through the geologically recent past. Results have shown that carbon dioxide and methane concentrations vary with glacial/interglacial cycles. For example, ice core data show that methane and carbon dioxide concentrations have a positive correlation with temperature. During glacial periods atmospheric methane and carbon dioxide values are lower, averaging between 300 and 400 parts per billion by volume (ppbv – parts per billion (10^9) by volume). This data shows that rapid changes in atmospheric concentration of greenhouse gases parallel climatic cycles, but it is unclear whether those changes occurred as rapidly as those in the twentieth century (Prather *et al.*, 1995) or whether current concentrations are outside levels reached as part of natural climatic variability, and Street-Perrott and Roberts (1994) suggest there remains the possibility that atmospheric concentrations are a consequence of climate change rather than a cause of it.

Even the rate at which temperature has increased over the past few centuries appears to have been matched in the past. Suplee (1998, p. 48) suggests that the critical issues that determine anthropogenic influence on climate change are the magnitude and speed of climate change and while 'a number of temperature shifts have occurred since the end of the last glacial ice age, the 20th century warming of 0.5°C is unusually large, abrupt and widespread'. Suplee argues that the rise in greenhouse gases as a result of human activity is responsible, and forms the basis for the argument that this recent upturn in temperature represents a new directional shift in climate. However palaeoenvironmental analyses of sediments have shown that natural rates of climate change have been variable throughout the Holocene and very rapid temperature changes occurred at the end of the last cold stage (Chambers, 1995). Broecker observes:

> the warming ... occurred in a period of no more than five decades. This is amazing; climate warmed at a rate of more than one degree Celcius [*sic*] per decade! This is faster than the most extreme projections for the current greenhouse warming. (Broecker, 1992, p. 42)

Long-term climate trends suggest, therefore, that the recent increase in greenhouse gas might not be anomalous or a significant departure from natural variation. However an analysis and recent trends suggests that there is a link between greenhouse gas concentrations of and human activity. Although water vapour is the most abundant greenhouse gas, with its concentration in the troposphere peaking at 3 per cent, it is not considered to be directly infuenced by human activity (Prather *et al.*, 1995). However, Webster (1994) suggests that levels of atmospheric water vapour could increase if rises in the concentration of other greenhouse gases induce warming. The other greenhouse gases (for example carbon dioxide, methane, nitrous oxides and halocarbons) have increased since the industrial revolution (see Table 3.7). Observations reported by Houghton *et al.* (1996) suggest that tropospheric ozone concentrations have also risen in the northern hemisphere since the early

1900s, and especially during the last 30 years, and compared to pre-industrial times, tropospheric concentration is approximately 25ppbv higher.

Table 3.7 Changing concentrations of some of the major greenhouse gases from pre-industrial times to 1994

	CO_2	CH_4	N_2O	CFC–11	HCFC–22 (a CFC substitute)	CF_4 (a perfluoro-carbon)
Pre-industrial concentration	~280 ppmv	~700 ppbv	~275 ppbv	zero	zero	zero
Concentration in 1994	358 ppmv	1720 ppbv	312 ppbv	268§ pptv±	110§ pptv	72§ pptv
Rate of concentration change*	1.5 ppmv/yr	10 ppbv/yr	0.8 ppbv/yr	0 pptv/yr	5 pptv/yr	1.2 pptv/yr
	0.4%/yr	0.6%/yr	0.25%/yr	0%/yr	5%/yr	2%/yr
Atmospheric lifetime (years)	50–200††	12†††	120	50	12	50,000

± Estimated from 1992–93 data.
§ 1 pptv = 1 part per trillion (million million) by volume.
†† No single lifetime for CO_2 can be defined because of the different rates of uptake by different sink processes.
††† This has been defined as an adjustment time which takes into account the indirect effect of methane on its own lifetime.
* The growth rates of CO_2, CH_4 and N_2O are averaged over the decade beginning 1984, halocarbon growth rates are based on recent years (1990s).

(Source: Houghton *et al.*, 1995)

Atmospheric concentration of the greenhouse gases also varies between the hemispheres. The Intergovernmental Panel for Climate Change (Houghton *et al.*, 1995) interprets these differences as the result of uneven spatial distribution of human-made greenhouse gas sources. For example, methane is concentrated in the northern hemisphere where the land mass is greater and where human sources exceed those in the southern hemisphere by approximately 280 Tg(CH_4)/yr (Prather *et al.*, 1995).

However, other researchers suggest that the short-term trends in global temperatures may be explained by natural rather than human-induced factors. Balling (1994) suggests that the predictions for global warming may be over-exaggerated because of inaccuracies in historical temperature data due to the relocation of measuring stations, urban heat islands, the use of different methods to measure ocean temperature records and the influence of overgrazing and desertification on regional temperatures. Friis-Christensen and Lassen (1991) suggest that over 75 per cent of global warming this century can be statistically explained by variations in the length of the solar sunspot cycle and stratospheric aerosols. The combined influence of overgrazing, desertification, urban heat islands, aerosols and solar variability eliminates most of the linear 0.45°C increase in temperature over the past 100 years and if atmospheric carbon dioxide concentrations double the change in temperature will be less than 1°C. Balling (1994) also argues that the timing of temperature increases and build up of greenhouse gases in the atmosphere are out of synchronisation; 0.32°C of warming took place between 1893 and 1942 before the rapid accumulation of greenhouse gases in the atmosphere.

Cornucopians have argued that pollution is a short-term phenomenon and problems caused by pollution will be rectified by human action. There is some evidence to suggest that the rate of increase of greenhouse gas concentrations is beginning to fall. Dlugokencky *et al.* (1998) report that since 1991 growth rates of atmospheric methane have declined, although the global atmospheric methane burden continues to increase. This is attributed to a decrease in sources, especially lower fossil fuel combustion and tropical biomass burning, by the Intergovernmental Panel for Climate Change (Houghton *et al.*, 1995). Since the introduction of emission controls the growth rate of halocarbons is also slowing down. For example, tropospheric organic chlorine slowed from 2.9 per cent to 1.6 per cent in 1992 (see Chapter 4) and chlorine/bromine loading in the lower atmosphere peaked in 1994 (Houghton *et al.*, 1996). As the use of halocarbons is phased out their presence in the lower atmosphere will diminish, so removing any global warming potential.

It seems that societies can transform natures on a global scale if the evidence for a human-enhanced greenhouse effect is to believed. However, we cannot state that societies control natures because, although people can control the release of greenhouse gases from anthropogenic sources (see Chapter 4), societies have no control over the resultant impact once the gases are in the atmosphere. To cornucopians this fact is of little consequence. They point to the evidence that climate is controlled by innumerable interacting variables including solar input, the atmosphere, the oceans, the water cycle, clouds, ice and snow, the type of land surface as well as human influences (Suplee, 1998). Because of the complexity of climate change scientists who lean towards a cornucopian viewpoint argue that the role of human activity may not be as important as some neo-Malthusians believe, and that the thermal response and the resultant impact of increasing greenhouse gas concentrations might not be as great as some climate models predict. The impact of the global warming on the earth's major systems is reviewed in the next section.

3.6.4 Impact of the global warming

A plethora of studies have been completed to predict the impact of global warming as the earth's system adjusts to the effects of increased greenhouse gas concentrations. These studies use general circulation models to predict how climate and the earth's systems will change in the future (see Box 3.7). For example, Watson *et al.* (1996) consider how all of the major natural ecosystems and managed systems will respond to climate change, and a summary of those predicted changes is provided in Tables 3.8 and 3.9.

Box 3.7 General circulation models (GCMs)

GCMs are computer models that simulate climate by mathematically modelling atmospheric parameters and ocean circulation that cause climate. The earth's surface is divided into grids of 200 miles and for each grid parameters such as air temperature, earth's rotation and surface friction on sea-level and rainfall are calculated. This data is loaded into a computer and used to run the model. Dickenson (1986) and Shine and Henderson-Sellers (1983) review the types of climate models and how they are designed to model climate change, while Kattenburg *et al.* (1996) and Hulme (1997) review the projections of climate change models in more detail. The end result is a prediction of future climate.

A prediction is 'largely based on statistical theory and uses the historical record of past events to estimate the future probability or recurrence of similar events'. Predictions tend to be long term, whereas a forecast is the 'detection and evaluation of an individual event as it evolves through a sequence of environmental processes which are relatively understood' and is short term (Smith, 1993, p.91). Both predictions and forecasts are used to warn society of impending or longer-term hazards and environmental changes; the results can be used to lower vulnerability to a hazard or problem. One example is the predicted temperature increases as a result of global warming from successive modifications of the UK Meteorological Office climate model shown in Table B3.7a.

Table B3.7a Examples of the projected rise in mean global temperature with a doubling of CO_2 concentrations, produced from successive modifications of the earlier UK 'Met Office models'

Model	Cloud representation	Radiative properties of the clouds	Temperature rise (°C)
UKLO (1987)	Empirical, linked to relative humidity. All-water clouds	Fixed	5.2
UKHI (1989a)	Computed liquid water and ice content	Fixed	3.2
UKHI (1989b)	Computed liquid water and ice content	Variable function of water and ice content	1.9
UKTR (1991)	As above, but with deep ocean circulation included		1.6

(Source: Chambers, 1998)

The predicted environmental changes caused by global warming from climate models also vary. One example is the distribution of permafrost in northern regions: climate models simulate a 20–30 per cent increase in active layer thickness for most permafrost regions in the northern hemisphere. Simulations indicate a 25–44 per cent reduction in permafrost cover with 2°C global warming, with areas of continuous permafrost worst affected. The results of four model simulations are shown in Table B3.7b.

Table B3.7b Areas of permafrost ($km^3 \times 10^6$) for contemporary climate and simulated warming (2°C) from four models. Values in parentheses indicate percentage contemporary values

Model	All zones	Continuous	Widespread	Sporadic
Contemporary	25.46 (100)	11.74 (100)	5.59 (100)	8.13 (100)
Holocene	14.21 (56)	3.79 (33)	3.98 (71)	6.43 (79)
GFDL	18.54 (72)	7.84 (67)	4.13 (74)	6.37 (78)
GISS	18.78 (74)	7.71 (66)	4.27 (76)	6.76 (83)
UKMO	19.17 (75)	8.39 (71)	4.24 (76)	6.54 (80)

(Source: Anisimov and Nelson, 1996)

Recent results from GCMs are, however, encouraging. Predictions that include the effect of greenhouse gases and aerosols on climate give a better agreement between observed and predicted temperature patterns (Figure B3.7) (Bennetts, 1995).

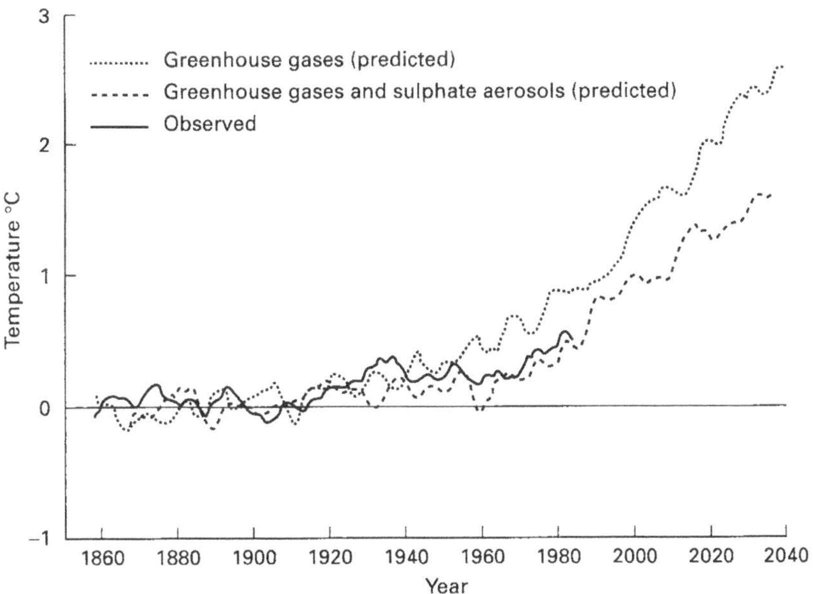

Figure B3.7 Predicted (1860–2050) and observed (1860–present day) surface air temperatures (source: from Bennetts, 1995 in Chambers, 1998)

Table 3.8 Projected impact of climate change on global ecosystems

Ecosystem	Changes
Forests	If CO_2 levels double, a substantial proportion of forest – globally about one-third of existing forests, especially in high latitudes – will undergo broad changes. Changes in physiological processes will occur in response to higher temperatures. Tropical forests are likely to be affected by soil water availability.
Coastal zones and small islands	Expect increased vulnerability of some coastal areas to flooding and erosional land loss. Could affect people as sea-level rises between 15 and 95 cm. Bangladesh, Majuro Atoll, China possibly the worst affected areas.
Rangelands	Increases in CO_2 will reduce palatability and forage quality as C:N ratios and shrub encroachment increase, especially in low latitudes. CO_2 and CH_4 fluxes in tundra will alter with precipitation and temperature changes.
Deserts and degraded land	Deserts will become hotter, altering water balance, hydrology and vegetation. Project increased drought and erratic rainfall distribution. Soil erosion and salinisation will increase. Areas with higher rainfall may experience acidification. Desertification may increase in drought-afflicted areas.
Cryosphere	Extent will be reduced. Increased temperatures will reduce volume of snow cover, permafrost and mountain glaciers, altering hydrological cycle. Implications for ice sheets of Greenland and Antarctic are uncertain.
Mountain regions	Redistribution of vegetation to higher elevations with the loss of some species and ecosystems. Reduction in glacier cover, permafrost and snow. Likely to disrupt economic activities.
Lakes and streams	Altered temperature, flow regimes and water levels. See poleward shift in aquatic species. Biological activity of aquatic ecosystems will occur in response to higher temperature.
Non-tidal wetlands	Altered hydrologic regime (water availability, depth, seasonal flooding), biological and chemical functions. Distribution will shift in response to precipitation and temperature changes.
Coastal ecosystems	Response will be varied. Increase in sea level, storminess could displace wetlands and lowlands. Tidal ranges, salinity of estuaries, rivers and bays likely to change. Possible loss of mangrove swamps, salt marshes. Coral reefs destroyed if temperature increases.
Oceans	Sensitive to temperature. Altered sea-level, circulation patterns, vertical mixing, wave climate and nutrient/biological productivity. Increased input of fresh water could weaken global thermohaline circulation.

Table 3.9 Projected impact of climate change on managed systems

System	Projected impacts
Agriculture	Crop yields and production patterns will change. Some areas will improve productivity and others decrease, eg tropics and subtropics. Possible increased hunger risk in sub-Saharan Africa, south-east Asia and tropics of Latin America. Role of diseases, pests and weeds unclear. Risks may be alleviated by technology and management
Forestry	Climate and non-climatic factors will limit global wood supplies. Boreal forests likely to decline.
Fisheries	Globally marine fish stocks projected to remain about the same. Freshwater and aquaculture production likely to increase at higher latitudes. Possible shift in centres of production.
Industry	HEP, biomass, energy, food mitigation structures may expect some change.
Water	See an intensification of global hydrological cycle: changes in temperature and precipitation patterns. High latitudes experience greater precipitation and runoff; low latitudes see a decrease in runoff. Possible major impact on regional water resources.

(Source: Watson *et al.*, 1996)

Because these predictions are made by modelling climate (see Chapter 4) we do not know for certain that the predicted changes and impacts will actually take place, nor can scientists accurately predict the scale and magnitude of those changes. Notwithstanding these uncertainties, this section briefly outlines the results from some climate change models and their impact on the earth's systems.

Temperature and moisture changes

It is generally accepted that if current emissions of carbon dioxide continue, atmospheric concentrations of CO_2 and other trace gases will double some time in the next century, possibly within the next 40 years (Peters, 1994), and this will result in an increase of global temperature. For example, the Intergovernmental Panel for Climate Change climate models project that global air temperature will increase between 1.0 and 3.5°C compared to 1990 levels over the next century, but temperature or directly related climatic parameters will vary spatially with high latitudes experiencing more marked changes (Watson *et al.*, 1996). This means that future changes in climate will probably affect all aspects of terrestrial and hydrological systems including biogeochemical cycles on global, regional and local scales. For example, temperature changes would trigger widespread changes in rainfall patterns. Suplee (1998) suggests that water vapour concentrations will increase by 6 per cent for every 1°C increase, which will increase rainfall world-wide, but some regions will experience higher rainfall and others drought.

Hydrological changes

Climate change will also affect hydrological systems. For example, global sea level is expected to rise by 15–95 cm by the end of the next century (Watson *et al.*, 1996), with decreases in polar ice and glacier volume (Mannion, 1998). This may contribute around 8.4 cm to sea levels, with a further rise of about 8 cm as a result of thermal contraction of the oceans between 1990 and 2030 according to predictions made by Wigley and Raper (1992). Ray *et al.* (1991) suggest that ocean primary productivity, species abundance, water circulation patterns and temperature will also change. Seasonal ice break up may start to occur much earlier (Ingram *et al.*, 1996) and permafrost distribution could also alter with climate change (Wang and Allard, 1995; Anisimov and Nelson, 1996; Anisimov *et al.* 1997), while Van Blarcum *et al.* (1995) predict that runoff and river flow will increase.

Changes to vegetation

Climate change could have a profound effect on vegetation. Higher temperature and increased levels of carbon dioxide could alter the ecophysiological behaviour of plants, extend growing seasons and shift vegetation zones (see Woodward, 1992; Peters, 1994; Lenihan and Neilson, 1995; Woodward *et al.*, 1995; Plochl and Kramer, 1995; and Pan *et al.*, 1996). For example, growing seasons are expected to change most prominently in latitudes north of 50° and evidence from forests in the northern hemisphere suggests that the peak of the growing season has already advanced by a week since the mid-1970s (Spizzirri, 1996; Suplee, 1998). Woodward (1992) used the results of changes in temperature and precipitation from the GISS general circulation model (Hansen *et al.*, 1984) to map the predicted patterns of vegetation if atmospheric carbon dioxide values doubled (see Figure 3.11). The main differences are a decrease in the areal extent of tundra, broad needleleaved forest, broad evergreen forest and frost-resistant vegetation and the expansion of broadleaved deciduous and evergreen vegetation.

Climate change will alter the geographical distribution, growth rates and metabolic behaviour of individual species as well as global biomes and regional communities. For example, Hughes *et al.* (1995) suggest that the entire present day populations of *Eucalyptus* will be exposed to temperatures and rainfall under which no individuals currently exist, and they predict substantial changes in the tree flora of Australia unless the trees adapt to, or are tolerant of, new climatic regimes. Climate changes may indirectly alter the distribution of plants and animals by influencing rates of predation, incidence of disease, fire frequency, parasitism and new competitive interaction (Peters, 1994).

Results have shown that plant growth could be between 20 and 40 per cent higher (eg Idso and Kimball, 1993) and acclimatory changes in plant metabolism are expected to occur as climate warms and atmospheric carbon dioxide concentration increases. For example, Beerling and Woodward (1996) have measured changes over the length of one growing season for elevated carbon dioxide levels in the boreal forest. Photosynthetic rates increased, leaf stomatal density decreased but leaf photosynthetic capacity remained the same for *Pinus sylvestris*, *Betula pubscens* and *Vaccinium myrtillus*. However, other studies suggest that the impact of higher temperatures and elevated carbon dioxide levels will be restricted (eg Oechel *et al.* 1994; Pushnik *et al.*, 1995), and changes in plant metabolism and soil nutrient–plant interactions in response to climate change over longer time-scales is

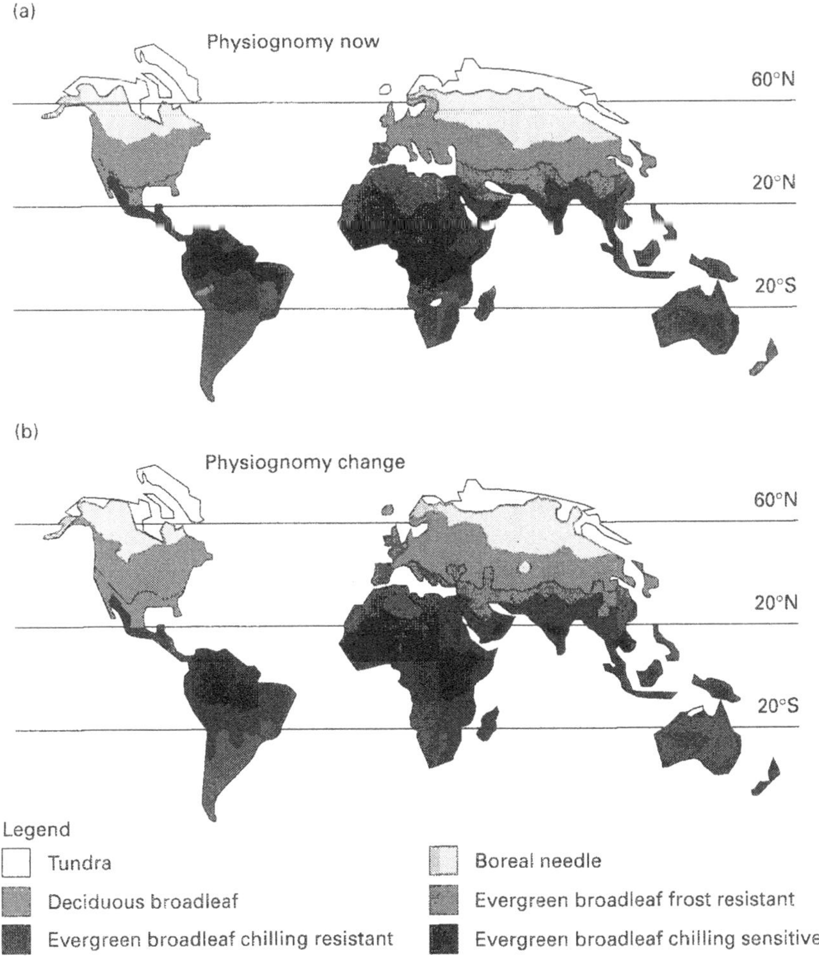

Figure 3.11 Changes in vegetation physiognomy in response to a doubling of carbon dioxide. (a) Global distribution of present-day physiognomic types of vegetation. (b) Predicted global patterns of physiognomy following CO_2-induced climate change (source: Woodward, 1992)

uncertain because the effects of CO_2 fertilisation are dependent upon a number of factors: nutrient availability, soil fertility, water availability, weed competition, insect attacks and the occurrence of extreme climate (drought, hot spells and frost) (Schimel *et al.*, 1995). Oechel *et al.* (1993) argue that plant growth response to increased atmospheric CO_2 may be statistically insignificant over long time-scales.

However, high carbon dioxide levels could have a number of possible positive effects on ecological systems. Some of these are summarised in Table 3.10. But increased carbon dioxide levels may reduce the biodiversity of ecosystems, cause pollinator/plant relationships to become out of synchronisation and result in a decrease in seed production (Fajer and Bazzaz, 1992).

Table 3.10 Potential benefits of CO_2 fertilisation on plants

- Increased production of sugars to feed symbiotic organisms will improve nutrient fixation.
- Rate of leaf water loss will decrease, eg soyabeans by 30%, maize by 50%.
- Increase in vegetative growth (shoots, stems, leaves, roots) by an average of 13%.*
- Increase in reproductive output (flowers, seeds) by an average of 31%.*
- Increase in crop yields by an average of 34%.*
- Increase in photosynthesis. Photosynthetic efficiency improved by c. 50%.
- Additional protection from pollution damage, eg less SO_2 molecules enter leaf interior in CO_2 rich environments.
- Overcome salinity stress in arid, marine and irrigated areas.

* Based on the results from short-term experiments.

(Source: after Fajer and Bazzaz, 1992)

Changes to food production

Neo-Malthusians such as Ehrlich and Ehrlich (1990) and Brown (1997) warn that global warming will have disastrous consequences on food production. Brown argues that extreme weather conditions experienced over the past 15 years could increase with rising levels of greenhouse gases, with droughts, floods, crop-withering heat waves and destructive storms all threatening food supply. Brown cites the heat waves of 1988 and 1995 that lowered grain harvests as an analogue of the possible impact of global warming. Ehrlich and Ehrlich (1990) and Peters (1994) warn that lower soil moisture levels will have an adverse affect on farming in areas like the US mid-west, parts of Mexico and Africa, whereas in areas of increased precipitation soil chemistry will change as leaching and erosion rates increase. Rosenzweig and Parry (1994) warn that climate change in low latitudes will have adverse affects by accelerating crop growth periods and increasing levels of heat and water stress, which could increase crop prices by up to 150 per cent and result in 10 and 60 per cent more people at risk from hunger. Climate change therefore represents a serious risk to food production.

However, the results of models do not fully support this neo-Malthusian assertion. For example, Rosenberg (1982) and Warrick *et al.* (1986) suggest that if problems of crop pests and disease can be controlled, crop yields could increase by 10 to 35 per cent if atmospheric carbon dioxide levels reach between 650 and 700 ppmv (parts per million (10^6) by volume). Most models predict that climatic changes at low latitudes will have negative impacts but at mid- and higher latitudes the impact, including the direct CO_2 effects on crop growth and water use, will be positive. For example, in Finland the length of the growing season is expected to increase by up to ten days as the temperature rises by 1°C, with improved crop productivity as spring cereals could be sown earlier (Kleemola *et al.*, 1995).

Crop growth models have been used to predict the impact of climate change on crops for individual countries and geographical regions, and climate change will affect the distribution of crops and yields as soil properties such as moisture, water retention capacities and temperature change (eg Schulze and Kunz, 1995). For example, Hubbard and Flores-Mendoza (1995) suggest that corn and soyabean production areas may decline by 20–40 per cent while wheat and sorghum production areas will expand by up to 80 per cent in the USA if temperatures increase between 3.5 and 5.9°C and rainfall is 10 per cent lower. Kleemola *et al.* (1995) argue that individual species will respond differently to changes in temperature, moisture and carbon dioxide concentrations. Predictions from general circulation models suggest that precipitation and temperature changes across China will be favourable as cropping patterns diversify, with the areal extent of single cropping agriculture decreasing by 23.1 per cent and triple cropping increasing by 22.4 per cent, while rice cultivation could be prolonged by 30–50 days and allow safe transplanting of rice 20–30 days earlier than at present (Wang and Zhao, 1995; Futang and Zong-Ci, 1995). However, if evaporation rates increase by the predicted amount – 16–40 per cent – compared to a 15–20 per cent increase in precipitation this could lead to water shortages and adversely affect rice production (Ohta *et al.*, 1995). Rosenzweig and Parry (1994) suggest that the impact of climate change on food production can be coped with through farm-level adaptation and future technological yield improvements, and accept that the uncertainty of climate change and its impact on food production regionally and globally makes it extremely difficult to know if societies can increase food production at the same rate as human population growth.

Insect outbreaks and disease

Ehrlich and Ehrlich (1990) suggest that herbivorous insect and disease outbreaks are almost certain to change with climate, which could have an adverse effect on food production. For example there is already evidence that trees are becoming stressed in response to warmer temperatures, causing soil moisture loss and increasing the frequency of insect attacks in the boreal forest/tundra zone (Jardine, 1994; Taubes, 1995). Luo *et al.* (1995) suggest that temperature and UV-B changes in cool subtropical and warm, humid subtropical areas will increase the risk of leaf blast epidemics, resulting in a drop in yields of 9–10 per cent. Williams and Liebhold (1995) modelled the changing geographical distribution and spatial extent of outbreaks of the western spruce budworm (*Choristoneura occidentalis* Freeman (L.)) in Oregon, USA and the gypsy moth (*Lymantria dispar* (L.)) in Pennsylvania, USA. The results for five climate change scenarios are shown in Table 3.11, and reveal that the spatial pattern of defoliation caused by the budworm and the moth could significantly alter. However, the results from two general circulation models contrast sharply.

Past global climate changes have greatly affected the geography of serious diseases and future climatic change could result in the spread of disease and affect large numbers of people. For example, Martin and Lefebvre (1995) suggest malaria could spread from its present distribution in tropical areas to higher latitudes. This movement is caused because the breeding grounds and lifecycle of the malaria parasites (*Plasmodium falciparum, P. vivax, P. ovale* and *P. malariae*) and their carrier, the *Anopheles* mosquito, are controlled by rainfall, temperature and air humidity.

Table 3.11 Percentage and areal changes of defoliation by the budworm and gypsy moth in Oregon and Pennsylvania, respectively, under five climate change scenarios

Scenario	Oregon		Pennsylvania	
	%	km^2	%	km^2
Ambient °C + pptn	19.9	49948	62.8	55182
Increase 2°C	7.7	199263	64.5	56676
Increase 2°C + 0.5 mm/day	26.9	67561	77.7	68275
Increase 2°C – 0.5 mm/day	0.9	2312	30.1	26449
GISS model	0.0	0	68.7	60367
GFDL	100.0	251181	0.0	0

(Source: Williams and Liebhold, 1995)

A review of the evidence for the enhanced greenhouse effect has established that global temperature records have increased with the warmest seven years of this century occurring since 1970. Measurements have also confirmed that the concentration of some greenhouse gases, namely carbon dioxide, methane, nitrous oxides and halocarbons have increased during the last 200 years. Most people now accept that human activities like deforestation and the burning of fossil fuels have contributed to these increases, but the role of natural factors cannot be dismissed as negligible. There is much less certainty and more disagreement on the impact these changes will have on the earth's systems. The impact of another global problem, ozone depletion, is also a controversial topic.

3.6.5 Ozone layer and ozone depletion

Most ozone is held between 15 and 40km altitude in the stratosphere, forming a concentrated, thin layer around 25km, more commonly referred to as the ozone layer. Stratospheric ozone is important because it shields people, animals and plants from harmful UV-B radiation. Without the ozone layer neo-Malthusians like Ehrlich and Ehrlich have argued (1990, p. 124) that 'life on land would be transformed into something like life under an ultraviolet sterilizer ... it would essentially be impossible'.

Moreover incidence of cancer from radiation would also increase and ecosystems would be damaged because UV-B radiation damages genetic material in all organisms. Neo-Malthusians argue that human actions are upsetting the natural processes that form and destroy ozone, resulting in lower levels of ozone in the stratosphere. Ehrlich and Ehrlich express concern that ozone depletion could adversely affect crops and fisheries and, in part, limit future food production. The possible impacts of ozone depletion are described in Box 3.8. Cornucopians such as Michaels *et al.* (1994), Singer (1995), Bailey (1994), Budiansky (1996) and Simon (1995) cannot see what all the fuss is about. They argue that reductions in ozone values are slight and the anticipated affects on human health have been over-exaggerated.

Box 3.8 Impact of ozone depletion

Evidence from field and laboratory experiments suggests that ozone depletion will increase levels of UV-B radiation reaching the earth's surface and adversely affect human health, terrestrial plants and animals (Blaustein and Wake, 1995; Caldwell *et al.*, 1995), terrestrial and aquatic biogeochemical cycles (Zepp *et al.*, 1995), aquatic ecosystems (Häder *et al.*, 1995) and tropospheric air composition and quality (Tang and Madronich, 1995).

Destruction of the stratospheric ozone layer may result in an increase in the incidence of skin cancer, eye and infectious diseases (Longstreth *et al.*, 1995). A thinning of stratospheric ozone will allow more ultraviolet radiation, especially UV-B, to penetrate through the atmosphere, which can be harmful to those organs exposed to sunlight such as the eyes and skin. Excessive exposure to ultraviolet radiation can damage DNA contained within human skin cells, causing the cells to mutate. Mutated cells (known as *p*53) function differently to normal cells and when damaged they do not self-destruct. Thus, if normal cells surrounding a *p*53 cell are damaged by over-exposure to sunlight the mutated cell may replace them with its own progeny. The growth of mutated cells in this fashion results in the formation of a tumour, commonly known as skin cancer (Leffell and Brash, 1996).

High UV-B radiation can also damage human eyes and cause conjunctivitis, cataract formation and cornea degeneration. It is, however, difficult to estimate the magnitude of increases in ocular problems. Ozone depletion may also affect the ability of human immunological system's, especially those in the skin, to work efficiently. Research indicates that UV-B radiation will suppress the immune system's response to certain ailments such as skin cancer, allergies and forms of autoimmunity. This effect may encourage viral infections as the human body will be less able to fight viruses such as leishmaniasis, malaria, herpes and trichinosis (Longstreth *et al.*, 1995).

Enhanced levels of UV-B radiation will have both direct and indirect effects on plants. Inhibition of photosynthesis, reduced plant growth, DNA damage, phenological changes (such as reduced flowering) and biomass accumulation all occur in response to higher UV-B radiation. Changes in plant growth in response to lower stratospheric ozone levels may have profound implications for agricultural production. A six-year experiment on sensitive soyabean cultivars to contrasting levels of UV-B radiation showed a reduction in crop yield in four years by 19–25 per cent. Tolerant soyabean species respond better to higher radiation levels with yields increasing by 4 per cent to 22 per cent. This suggests that new radiation-resistant breeds of crops may be needed to maintain levels of global food production (Caldwell *et al.*, 1995).

The world's oceans provide over 30 per cent of animal protein for human consumption. Ocean food webs commence with phytoplankton that live in the upper euphotic zone of the water column. Not only are phytoplankton a major food source, they are also an important carbon dioxide sink. Research has shown that over-exposure to increased UV-B radiation can adversely affect survival rates of phytoplankton by changing their orientation mechanisms and mobility. Changes in phytoplankton populations have already been recorded in Antarctic waters and solar UV-B radiation is thought to be responsible for damaging the early development cycles of other marine life including fish, shrimps, crabs and amphibians (Häder *et al.*, 1995; Blaustein and Wake, 1995).

Because of the limited nature of the research it is not possible to quantitively predict the anticipated effects of ozone depletion on terrestrial and aquatic ecosystems. Furthermore the interaction of ozone depletion and predicted climate change could significantly alter the response of terrestrial and aquatic ecosystems. An experiment to measure the combined effect of increased atmospheric carbon dioxide and higher radiation levels on crops of rice, wheat and soyabean reveals that increased carbon dioxide promoted growth and seed yield in rice but this was cancelled out by higher radiation levels and caused a reduction in wheat, but had minimal effect on soyabean (Teramura *et al.*, 1990). Thus, the interaction of factors must be considered if societies are going to make realistic assessments of environmental change.

However, a number of people have questioned the impact of ozone depletion on human health. Singer (1995) argues that claims of upward trends in UV-B radiation caused by ozone loss are based on statistically faulty data and the links between skin cancer, ozone loss and UV-B radiation are also uncertain. Simon (1995) argues that there has been no increase in skin cancers from ozone, even if the ozone layer does pose a threat to societies, people have 'large modern capacities' to alter conditions in the atmosphere or to protect themselves.

The formation of the ozone layer is a natural process, a combination of unique climatic conditions and chemical reactions. Ozone is a gas consisting of three oxygen atoms, created in large amounts close to the equator before atmospheric winds transport it to the poles, making the tropics ozone poor compared with higher latitudes. However, ozone is a dynamic molecule in the stratosphere because it is constantly being created and destroyed by a natural, photochemical reaction (see Box 3.9; Chapter 1).

Box 3.9 Processes of ozone formation and destruction

Ozone is naturally created and destroyed in a self-perpetuating cycle in the stratosphere. Ultraviolet radiation splits oxygen molecules into atoms. These oxygen atoms can then combine with an oxygen molecule to reform ozone. Ozone can then combine with another oxygen atom to form two oxygen molecules or be split by ultraviolet radiation into an oxygen atom and molecule. This process is summarised in reaction one.

▶

Reaction 1: photochemical reactions of ozone

$$O_2 = Uv \rightarrow O + O$$

$$O + O_2 \rightarrow O_3$$

$$O_3 + O \rightarrow O_2 + O_2$$

$$O_3 + Uv \rightarrow O + O_2$$

The following chemical reactions show the various ways ozone is thought to be destroyed at the poles in the stratosphere. These chemical reactions are thought to take place on the surface of polar stratospheric clouds (PSCs) in the presence of ultraviolet radiation.

(i) Ozone destruction by chlorine nitrate and hydrochloric acid

Chlorine nitrate reacts with hydrochloric acid to produce nitric acid and molecular chlorine. Molecular chlorine is split into atoms by ultraviolet light in the stratosphere. The chlorine atoms then react with ozone, producing oxygen molecules and chlorine monoxide (equation a).

Equation a: $\overbrace{ClONO_2 + HCl}^{PSC} \rightarrow Cl_2 \rightarrow HNO_3$

$$Cl_2 \rightarrow U_V \rightarrow Cl + Cl;\ Cl + O_3 \rightarrow ClO + O_2$$

Chlorine monoxide then forms a dimer (Cl_2O_2), which is broken down by ultraviolet light into chlorine atoms and oxygen molecules (equation b) or it can combine with nitrogen dioxide to form chlorine nitrate (equation c). The chlorine then reacts with ozone.

Equation b: $ClO + ClO \rightarrow Cl_2O_2 \rightarrow Cl$ and O_2

Equation c: $ClO + NO_2 + M \rightarrow ClONO_2 + M$ (M = an air molecule)

The role of CFCs: producing a source of chlorine nitrate

The role of CFCs as stratospheric ozone depleters was publicised in the scientific community during the early 1970s (Molina and Rowland, 1974; Stolarski and Cicerone, 1974). By the 1990s the origin of stratospheric chlorine had been firmly linked to the breakdown of CFCs. Once released, CFCs are very inert in the troposphere and remain intact until they are eventually carried into the stratosphere. Here, CFCs can be broken down by ultraviolet light to create free chlorine atoms that are available to form chlorine nitrate. Chlorine then reacts with ozone to produce chlorine monoxide. Chlorine monoxide can combine with other gases in the atmosphere like nitrogen dioxide (NO_2) to form chlorine nitrate as shown in equation. Chlorine can also react with methane to form hydrochloric acid (equation e).

Equation d: $CFC + UV \rightarrow Cl + Cl, Cl + O_3 \rightarrow ClO, ClO + NO_2 \rightarrow ClONO_2$

Equation e: $Cl + CH_4 \rightarrow HCl$

Chlorine nitrate and hydrochloric acid act as reservoirs for inert chlorine until polar stratospheric clouds form.

(ii) Ozone destruction by dinitrogen pentoxide and hydrochloric acid

Dinitrogen pentoxide (N_2O_5) also acts as a reservoir for nitrogen in the stratosphere. It can react with hydrochloric acid to produce nitric acid and nitryl chloride (ClONO) as shown in equation f.

Equation f: $N_2O_5 + HCl \rightarrow HNO_3 + ClONO$

and/or $N_2O_5 + H_2O \rightarrow 2HNO_3$

The nitric acid is used to form PSCs and nitryl chloride is unstable in the presence of ultraviolet light and breaks down into chlorine.

(iii) Ozone destruction by bromine

Bromine in the form of radicals Br and BrO are responsible for ozone destruction. Fan and Jacob (1992) describe how bromine is present in a reactive form to destroy ozone. Once in the stratosphere, both Br and BrO are rapidly converted into non-reactive forms HBr, HOBr and $BrNO_3$ (equation g). However, the presence of polar stratospheric clouds and sulphate aerosols in the Arctic stratosphere provide a reactive surface to produce Br_2 from these non-reactive forms;

$HOBr\ (aq) + Br^- + H^+ \rightarrow Br_2\ (aq) + H_2O$

$BrNO_3$ can be converted to HOBr as it is scavenged by the aerosol and hydrolysed:

Equation g: $BrNO_3(g) \xrightarrow{H_2O} HOBr(aq) + HNO_3(aq)$

Br_2 can be photolysed to Br, which destroys ozone.

Ozone depletion can be enhanced by denitrification of the stratosphere. Nitric acid produced during the reaction of chlorine nitrate and hydrochloric acid can be removed from the stratosphere by precipitation. This prevents chlorine atoms from forming chlorine nitrate so they remain active.

Sources: Kemp, 1990; Toon and Turco, 1991; Fan and Jacob, 1992; SORG, 1996.

It has been suggested that industrial and agricultural processes are upsetting the balance of natural ozone formation and destruction by adding more chlorine and bromine into the atmosphere, which break down stratospheric ozone molecules by a series of chemical reactions as outlined in Box 3.9.

Brune (1996) and Russell *et al.* (1996) suggest that approximately 80 per cent of the halogens contained in the stratosphere are almost exclusively anthropogenic in origin. Aircraft, for example, release nitrous oxides, soot, water and sulphur dioxide into the troposphere and the stratosphere (SORG, 1996; Pearce, 1997a), and human-made halocarbons containing chlorine, fluorine or bromine also destroy stratospheric ozone. These include CFCs that are used as coolants in refrigerators, in air-conditioning, aerosol propellants, plastic forms, foam insulation and in some solvents (Haas, 1991). Originally believed to be non-toxic, non-carcinogenic and non-flammable, CFC use peaked in 1974 with production concentrated in the industrialised First World. For example, in 1986 western Europe and the USA produced 36 per cent and 35 per cent respectively, but demand for CFCs is now increasing more rapidly in underdeveloped countries (Haas, 1991). Anderson *et al.* (1991) and Russell *et al.* (1996) suggest that stratospheric chlorine derived from CFCs has been measured over Antarctica and its concentration now exceeds natural background emissions by more than five times.

Halons are used in fire extinguishers and are being used as transitional replacements for CFCs. The atmospheric concentration of halons (Halon-1301 (CF_3Br), Halon-1211 (CF_2BrCl), Halon-1202 (CF_2Br_2) and Halon-2402 ($CF_2Br\text{-}CF_2Br$)) is also increasing and has the potential to carry approximately 20–40 per cent of the total amount of bromine into the stratosphere (Butler *et al.*, 1992).

Pesticide use can contribute to ozone depletion by adding ozone-destroying chemicals into the atmosphere. Methyl bromide (MeBr) is a colourless, odourless gas used to fumigate the soil with approximately 35Gg released each year (20–60 Gg/yr) (Buffin, 1992). Methyl bromide can lead to ozone depletion by providing a bromine source (see Box 3.9). Atmospheric concentrations of around 10pptv (parts per trillion (10^{12}) by volume) have been measured in the northern hemisphere and over Antarctica (Longstreth *et al.*, 1995; SORG, 1996; Fan and Jacob, 1992) and current models suggest that methyl bromide may account for between one-tenth and one-twentieth of observed annual ozone loss (Buffin, 1992; Fan and Jacob, 1992).

SORG (1996) argues that total, vertical and polar ozone measurements confirm that the addition of human sources of ozone-destroying chemicals have had an adverse impact on stratospheric ozone levels. They identify the following pattern of decreasing total and vertical stratospheric ozone levels, while ozone lost at the poles is described in Box 3.10.

Total ozone

(1) Mid-latitude levels have declined in the northern hemisphere, with the largest reductions occurring in winter and spring, approaching 7 per cent per decade.
(2) Over southern mid-latitudes total ozone levels have decreased by 3 6 per cent per decade.
(3) In the tropics ozone trends are slightly negative but are assumed to be insignificant.
(4) Ozone loss in Antarctica and the Arctic occurs annually every spring and levels of ozone decline are increasing.
(5) Accumulated loss of ozone annually and globally averages 5 per cent.

Vertical ozone

(1) The bulk of ozone loss over northern mid latitudes occurs between 15 and 25km. Measurements suggest the amount lost varies between –7 per cent ± 3 per cent and –20 per cent ± 8 per cent per decade.

(2) Ozone loss of around 10 per cent per decade has been measured between 35 and 45km, but ozone levels are small at this altitude.

(3) Since the early 1990s all ozone is destroyed between 15 and 20km in Antarctica in September.

Box 3.10 Stratospheric ozone loss at the poles

Stratospheric ozone values naturally decline over Antarctica during late August and early September, and reach a minimum level in October before increasing in November (Stolarski, 1988). However, since the mid-1970s measurements suggest that stratospheric ozone loss during the southern spring has become more pronounced and has continued into January and February over Antarctica. For example, springtime ozone values decreased by 40 per cent between 1977 and 1984

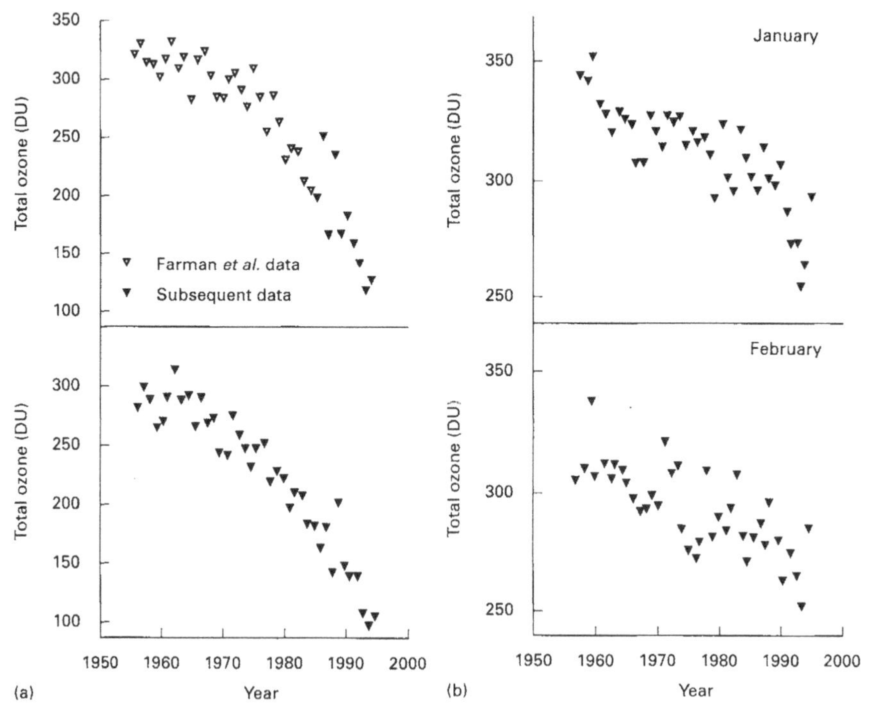

Figure B3.10 Recent ozone fluctuations over Antarctica. (a) Declining ozone values between 1950 and 1994 during October. (b) Declining ozone values between 1957 and 1994 in January and February (Source: Jones and Shanklin, 1995)

(Stolarski, 1988) and have continued to fall, creating an area of extremely low ozone values in the stratosphere that is approximately the size of the Antarctic continent. Jones and Shanklin (1995) and SORG (1996) suggest this is the result of human activity (see text) and changing climatic conditions in recent years (Figure B3.10).

Meteorological conditions contribute to the natural depletion of the ozone in the stratosphere. A weather system known as the polar vortex forms in June as the Antarctic winter commences and remains in place until November, stopping ozone-rich air travelling from the equator reaching the South Pole until it dissipates in late spring (Toon and Turco, 1991). The chemical reactions that break down ozone, outlined in Box 3.9, take place inside the polar vortex on the surface of polar stratospheric clouds, which form when temperatures fall to –80°C (see Toon and Turco, 1991 and Santee *et al.*, 1995).

Ozone loss over the Arctic is less apparent because the stratosphere is warmer and this restricts polar stratospheric cloud formation in the Arctic polar vortex from a few days to two months (Manney *et al.*, 1994; SORG, 1996). Because the Arctic polar vortex breaks down in late winter only a limited amount of sunlight reaches the near pole regions and this limits the amount of ozone loss as sunlight is required to destroy ozone (Santee *et al.*, 1995).

First reports of the presence of chlorine dioxide in the Arctic occurred in 1988 (Kerr, 1988). Since then depletion of stratospheric ozone by around 20 per cent has been observed over the Arctic since January 1989 (Hofmann *et al.*, 1989; Manney *et al.*, 1994; Profitt *et al.*, 1990). Cornucopians do not doubt that chemical processes are causing ozone depletion at the poles; however, they question whether human action is reponsible and argue that the decrease below an average mean ozone value is minimal (up to 3.2 per cent between 1957 and 1987) (Simon, 1995).

Singer (1995) argues that decreases in global and polar ozone levels caused by human action and the resultant impact may be insignificant once natural factors are taken into account. Ozone levels can be influenced by a variety of natural factors including volcanic eruptions, seaspray, sunspot cycles, the quasi-biennial occillation (QBO) and the junge layer (see Drake, 1995). For example, when volcanoes erupt they eject sulphur dioxide and chlorine in the form of hydrogen chloride that is readily available to react with ozone if it is transferred into the stratosphere. Low levels of ozone were measured following the eruptions of Mount Pinatubo in the Philippines and Mount Hudson in Chile that released an estimated 20 million metric tonnes of SO_2 and 4.5Mt of HCl into the atmosphere, although the quantity that is thought to have reached the stratosphere is much smaller, around 0.5 to 5Mt (Tabazadeh and Turco, 1993). Brasseur (1992) reports that ozone concentration between 24 and 25km above Natal in Brazil and the Ascension Islands were 15–20 per cent lower for the first three to six months after Mount Pinatubo erupted. Hofmann *et al.* (1992) recorded unexpectedly low levels of ozone during the spring at altitudes between 11 and 13 and from 25 to 30km over Antarctica in

1991. They suggest that aerosol particles released by Mounts Hudson and Pinatubo entered the polar vortex and enhanced ozone-destroying chemical reactions by encouraging polar stratospheric cloud formation. The release of sulphur by volcanoes has also perturbed ozone concentrations in the tropics. Bekki *et al.* (1993) describe how sulphur dioxide catalyses mid-stratospheric ozone production while SO_2 clouds absorb solar radiation and reduce ozone production. However Brune (1996) argues that volcanoes and seaspray only contribute about 5 per cent of stratospheric hydrogen chloride and Russell *et al.* (1996) state that only a few per cent of observed stratospheric chlorine is derived from natural sources.

Section summary

Ozone depletion provides another example of societies influencing natures on a global scale, but the extent of that influence and its resultant impact is hotly contested. The cause of ozone depletion, chemical reactions involving chlorine and bromine, has been established and it is apparent that industrial and agricultural processes emit these chemicals into the atmosphere. Studies have also shown that increased UV-B radiation could have an adverse effect on vegetation, animals and humans, but the role of ozone loss in this process has been questioned. However, agreements are now in place to lower the concentration of ozone-depleting chemicals from human sources (see Chapter 4).

3.7 Regional and local environmental costs of industrialised societies

There is also concern about the cleanliness of Planet Earth. Neo-Malthusians have argued that human action has created widespread pollution and waste (see Box 3.11) and led to a deterioration in environmental quality. Ehrlich and Ehrlich (1990), for example, argue that industry, motor vehicles and energy generation have created acid rain, and suggest that this problem is becoming 'truly global' (p. 123) as more industrial regions outside the developed world are affected by acidic atmospheric pollutants and the incidences of ocean pollution are rising.

Box 3.11 Definitions of pollution and waste

Pollution is defined by Holdgate (1979, p. 17) as 'the introduction ... into the environment of substances or energy liable to cause hazards to human health and harm to living resources and ecological systems, damage to structures or amenity, or interference with legitimate uses of the environment'.

Contamination is 'the presence of elevated concentrations of substances in water, sediments and organisms'.

Waste is defined by the OECD (1988) as 'materials ... intended for disposal for specified reasons' or 'natural substances produced in quantities which do not impair those ecosystems'.

Waste becomes problematic when harmful to humans or the environment. Such waste is often referred to as either 'hazardous', 'toxic' or 'radioactive'. Hazardous waste can be defined as 'waste that has physical, chemical or biological characteristics that cause or contribute to threats to human health, or adversely affect the environment when improperly managed' (Cutter, 1993, p. 114; Blowers, 1996, p. 155). Toxic waste has toxic properties (BMA, 1991) and radioactive waste is 'any material that contains or is contaminated with radionuclides at concentrations or radioactivity levels greater than the "exempt quantities" established by the competent regulatory authorities and for which no use is foreseen' (IAEA, 1992). Blowers (1996) provides a comprehensive review of the classification of wastes.

Ehrlich and Ehrlich (1990) argue that most of the blame for creating excessive amounts of waste and pollution lies with industrialised countries. Chapter 2 outlined the origins of the industrial societies and increases in pollution levels coincide with the beginning of the industrial revolution. Since then Nriagu (1996) has argued that levels of metals in the environment have increased (Figure 3.12) as societies are releasing large amounts of metals into the environment each year.

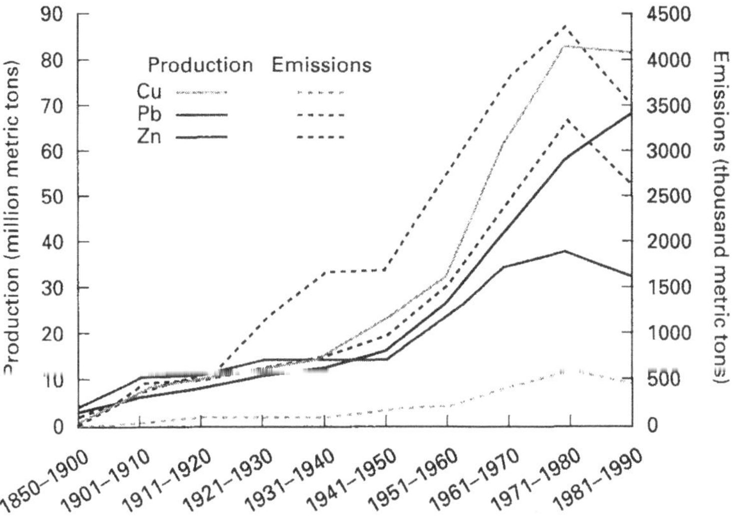

Figure 3.12 Recent historical changes in mine production and anthropogenic emissions of trace metals to the atmosphere (source: Nriagu, 1996)

For example, cadmium production increased sharply between 1900 and 1970 and again in the late 1970s. By 1996 it was estimated that 7000 metric tonnes are released into the atmosphere annually through mining and smelting (Elinder and

Järup, 1996) and gold mining operations annually release between 400 and 500 tonnes of mercury (Green, 1993).

Pollution now occurs in virtually every country, irrespective of its stage of social and economic development, and it is becoming an ever-increasing global problem simply because the incidence of major accidents involving toxic chemicals is increasing. Between 1984 and 1992, 106 major accidents occurred, 15 of which exceeded the Bhopal disaster in both quantity and toxicity. The combined effects of industrial and agricultural pollution are also noticeable in developed countries. One example is the Great Lakes region, where 40 million people are exposed to more toxic chemicals than anywhere else in the USA according to the US National Research Council. Nearly 300 chemical compounds, many of which are dangerous to humans, have been found in the Great Lakes (Cobb jun., 1987), but the extent of the pollution varies between each lake (see Baumol and Oates, 1995). New Zealand suffers from toxic waste generated by forestry, agriculture and industry (Szabo, 1993) and Poland has been ravished by decades of unabated heavy industry (Fischhoff, 1991), while incidents of pollution are common in underdeveloped countries, for example, the chemical explosion at Bhopal (Morehouse, 1994).

All facets of the industrial process and energy generation cause pollution: the extraction of resources from the earth's crust (mining), the processing and working of the resource into a product (eg smelting or burning fossil fuels in power stations to produce electricity) and the transportation of the resource in its raw form or as a product. These processes generally create pollution on a regional and local scale across the world – examples include acid rain, radioactive leaks and heavy metal contamination. The main sources of pollutants released by urban and industrial processes are summarised in Box 3.12.

Box 3.12 Urban and industrial sources of pollution

Pacyna (1986) divides production-related anthropogenic emissions of trace elements into six groups:

(1) Stationary fuel combustion includes elements released from coal and oil combustion. Arsenic (As), cadmium (Cd), chromium (Cr), copper (Cu), nickel (Ni), tin (Sb), and zinc (Zn) are the main elements emitted into the atmosphere. The actual value of individual elements varies according to the type of oil and coal and the efficiency of the power plants. Some arsenic, copper, cadmium, nickel and lead can be released from wood combustion.

(2) Internal combustion engines contribute a significant amount of tetraethyl lead and other petrol additives that contain a variety of trace elements. Proliferation of motor car use from the 1970s made gasoline lead additives the single most important source of atmospheric lead pollution. Approximately 95 per cent of lead within the biosphere is anthropogenically produced but the phasing out of leaded petrol has seen a dramatic fall in lead values. However, overall global lead consumption rates have not declined, suggesting that lead exposure will continue to pose a health risk (Smith and Flegal, 1995).

▶

(3) Non-ferrous metal manufacturing releases the major non-ferrous metals (Al, Cu, Zn, Pb and Ni). Lead zinc ores also contain high amounts of Cd and As. The bulk of emissions of these elements occurs during the smelting process (Brown *et al.*, 1990).
(4) Iron, steel and ferroalloy plants and foundries release Zn, Pb and Mn from blast furnaces during iron production, whereas steelworks are an anthropogenic source for Cd, Cu, Cr, Mn, Ni, Pb and Zn.
(5) Concentration of trace elements in refuse incinerators can be as high as 2400 μg/g. Pb reaches 8.9 per cent and has an emission factor of 17.6 g/t of refuse. Sewage sludge incineration also contributes trace elements to the atmosphere. Concentrations of chromium and lead have reached 30,000 and 26,000 μg/g of dry sludge respectively. World-wide, municipal sewage releases 2300 tonnes of silver, 3000 tonnes of cadmium, 15,000 tonnes of lead, 17,000 tonnes of nickel, 42,000 tonnes of copper and 100,000 tonnes of zinc annually into the environment (Lottermoser, 1995). Emission rates are dependent upon the composition of mineral, metal ore or sewage sludge, combustion temperature, volatility of the metal and efficiency of pollution controls (Pacyna, 1986).
(6) Cement production releases mainly cadmium and lead.

Use, wear and tear of consumer and commercial products also releases trace elements (Ag, As, Cr, Cd, Hg, Pb and Zn) into the environment. Sources include the weathering of paints and pigments, incineration of discarded pharmaceuticals, batteries, electronic tubes, photographic film, electroplated surfaces and the combustion of treated wood (Brown *et al.*, 1990; Tarr and Ayers, 1990).

3.7.1 Pollution and energy generation

As noted earlier energy generation also produces pollution. Two forms of energy generation will be considered here. First, acid rain, which has been linked with fossil fuel combustion, the dominant form of fuel since industrialisation (see Chapter 2), and then pollution caused by nuclear power generation, a fuel source developed since the first controlled nuclear fission experiment was conducted in 1942 (Simmons, 1997) and often regarded as the energy source for the twenty-first century.

Acid rain

Virtually all rain is naturally acidic, however the term 'acid rain' refers to the additional amount of acids that are created in the atmosphere resulting from the emission of carbon dioxide, sulphur (SO and SO_2 which are referred to as sulphurous oxides or SO_x) and nitrogen compounds (NO and NO_2 which are collectively termed nitrous oxides or NO_x) primarily from coal-burning power stations and vehicles (Kulp, 1995).

Fulkerson *et al.* (1990) argue that the combustion of fossil fuels accounts for about 80 per cent of the SO_2 and most of the NO_x released into the atmosphere by

Box 3.13 Formation of acids in the atmosphere

Nitric acid

Once nitric oxide is released into the atmosphere it can be converted to nitrogen dioxide by the following reaction:

$$2NO + O_2 \rightarrow 2NO_2$$

Nitrogen dioxide can then react with ozone to form NO_3 or nitrogen pentoxide (N_2O_5). Nitrogen pentoxide or NO_2 can combine with water to form nitric acid. These reactions are summarised in the following equations:

$$N_2O_5 + H_2O \rightarrow 2HNO_3$$

$$3NO_2 + H_2O \rightarrow 2HNO_3 + NO$$

$$2NO_2 + H_2O \rightarrow HNO_3 + HNO_2$$

Sulphuric acid

Sulphuric acid is produced in the atmosphere by the oxidation of sulphur dioxide, as shown in the following equation:

$$SO_2 + \tfrac{1}{2}O_2 + H_2O \rightarrow H_2SO_4$$

human activity. Acid rain has largely been restricted to industrialised regions of the developed world, where emissions of these compounds has been greatest. For example, coal-fired power plants account for approximately 70 per cent of anthropogenic sources of SO_2, 30 per cent of NO_x and 35 per cent of CO_2 in the USA and for approximately 80 per cent of SO_2 emissions and 40 per cent of NO_x in the European Community in the early 1980s (McCormick, 1998). However, industrialising regions of the underdeveloped world are beginning to experience problems of acid rain. For example, East Asia now contributes one-third of the world's output of SO_2 and China is set to become the world's biggest air polluter, affecting an estimated 40 per cent of its land and spreading further afield to Japan (Hadfield, 1997).

Other industrial sources of nitrogen compounds include stationary furnaces and internal combustion engines and Kennedy (1992) estimates that approximately 50 million tonnes of nitric oxide and nitrogen dioxide are released into the atmosphere each year. Human activities, especially coal burning and the smelting of pyrite ores, contribute approximately 100 million tonnes of sulphur annually to the atmosphere, which is rapidly oxidised to sulphur dioxide. Once sulphur dioxide and nitrogen oxides are released into the atmosphere they are converted into acids. This process and the principal ways in which acids are transported are outlined in Box 3.13 and Figure 3.13.

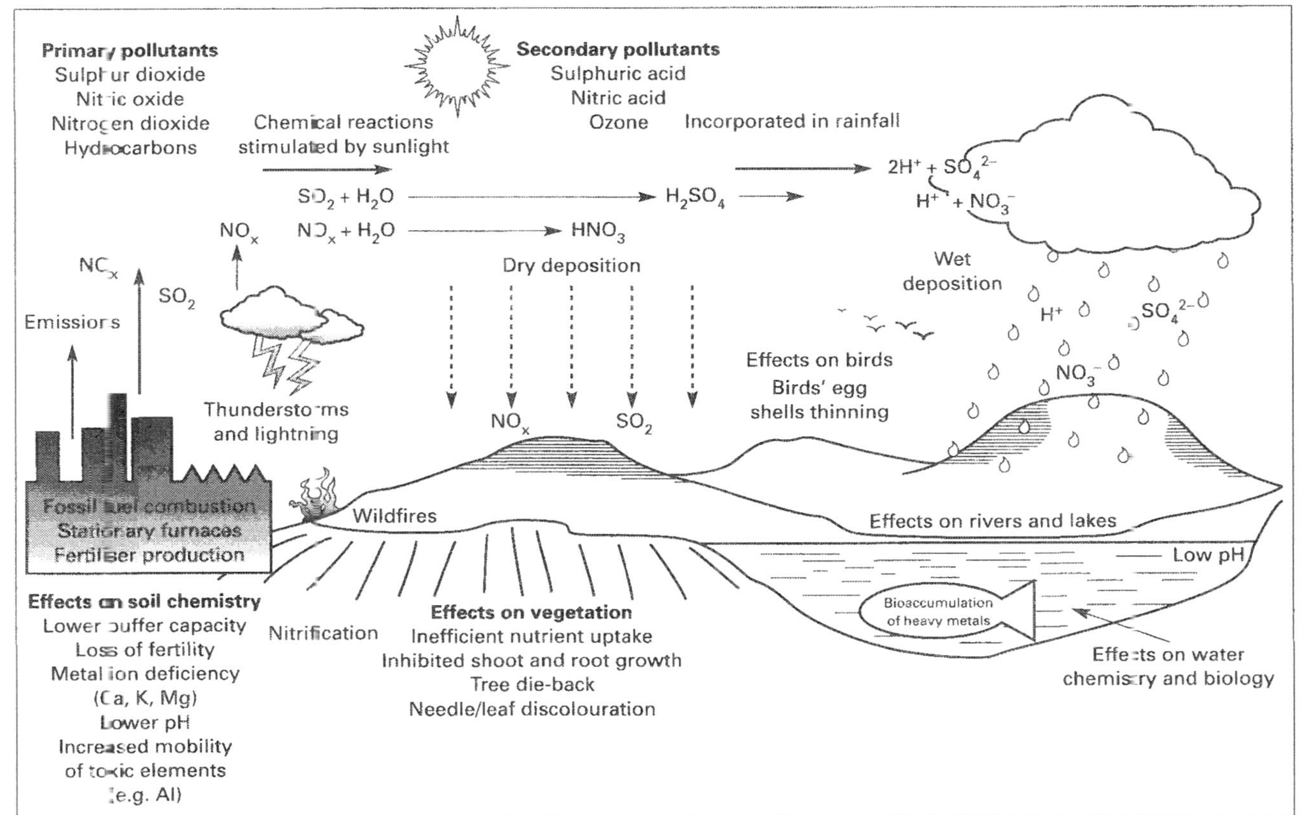

Figure 3.13 Impact of acid rain on terrestrial and hydrological ecosystems (after ApSimon and Warren, 1996)

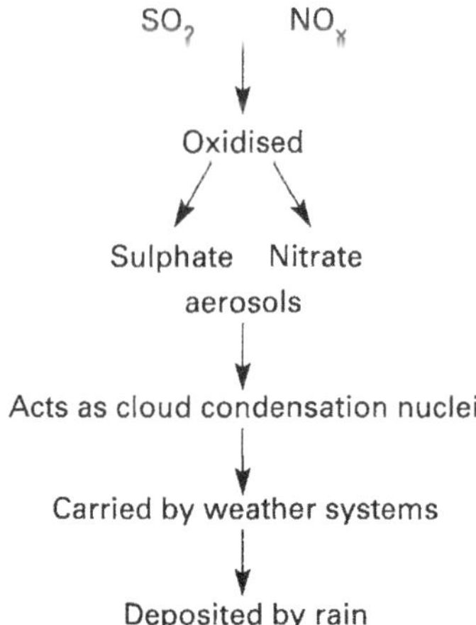

Figure 3.14 Dispersal of atmospheric acid pollutants

Once in the atmosphere acids are carried by weather systems and can be distributed over large areas. Acid atmospheric deposition occurs via three processes: wet, dry and occult deposition (Fowler *et al.*, 1985). Wet deposition carries and deposits acids in rain, hail or snow, a process that is thought to carry acids over long distances. Figure 3.14 summarises how acids are carried in clouds; approximately 70 per cent of wet deposition is sulphuric acid and 30 per cent nitric acid.

Dry deposition occurs when SO_2, NO_2 and nitric acid (HNO_3) are directly absorbed through the stomata of leaves or deposited directly on to the soil surface. SO_2 and NO_2 are oxidised into SO_4 and NO_3 during dry deposition (Fowler *et al.*, 1985). Occult deposition can occur from the impaction of mist, fog or cloud droplets on to vegetation (Kennedy, 1992; ApSimon and Warren, 1996).

Data from the EMEP-W model provides a spatial estimate of sulphur deposition across Europe (Figure 3.15). Calculation of critical loads of sulphur and nitrogen (nitrate and ammonium) can crudely indicate areas sensitive to acidification. A critical load for acidity is the equilibrium value for deposition of acidic species balanced by the capacity of a soil to supply acid neutralising basic cations, eg Ca^{2+} (cations are positively charged ions while bases are substances that react with hydrogen ions or release hydroxyl ions). ApSimon and Warren (1996) compare critical loads with maps of acid deposition. Areas where deposition exceeds critical load establishes areas of exceedance or those areas particularly vulnerable to acidification and the map shows that a large proportion of Europe is affected by acid deposition. Critical load models provided data to determine emission reductions of sulphur and nitrogen compounds for the Second Sulphur Protocol, a replacement for the 30 per cent club. For a discussion of the use of critical loads to define areas affected by acidic atmospheric deposition see ApSimon and Warren (1996), Battarbee *et al.*, (1996) and Løkke *et al.* (1996).

Soil acidification is, however, a process that would occur without human pressure (Reus *et al.*, 1987). The natural processes of soil acidification are outlined in Box 3.14, but human action appears to have accentuated the outcome of these

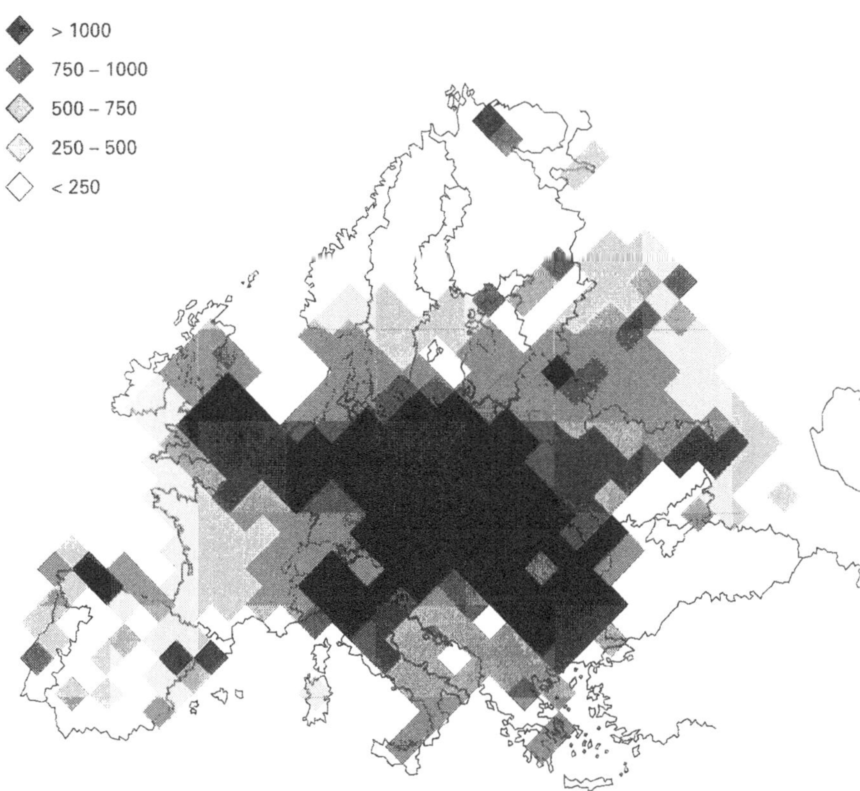

Figure 3.15 An estimation of sulphur deposition across Europe in 1990 (source: ApSimon and Warren, 1996)

processes, especially in areas that cannot fully neutralise the acids. In a study of acidification of Scottish lochs, Battarbee *et al.* (1988), for example, reveal that lakes were more sensitive to acidification in moorland catchments under acidic bedrocks (eg granites), but lakes in areas underlain by more calcium-rich bedrock experienced slight acidification.

Even though catchment characteristics can influence soil and water acidification Flower and Battarbee (1983), Battarbee (1989) and Battarbee *et al.* (1988) argue that anthropogenic sources of acids are primarily responsible for the problems of acid rain, as most lakes began to become more acidic during the industrial revolution (see Chapter 2) and a strong correlation exists between the most acidified lakes and areas of high sulphur deposition. Moreover, observations by Henriksen *et al.* (1988) suggest that natural organic compounds derived from humic acids removed from soils and bedrocks with low buffering capacity could not account for lake acidification.

A combination of natural and anthropogenic processes can also produce bases in the atmosphere. Atmospheric dust particles formed by basic materials, such as calcium carbonate and magnesium carbonate, are released into the atmosphere by fossil fuel burning, cement manufacturing, mining, forest fires and wind erosion. When in the atmosphere or on the ground, basic dust particles can neutralise acid rain. Atmospheric dust particles provide an important source of basic cations to the soil, which maintains its buffering capacity (Hedin and Likens, 1996). However, in recent decades levels of atmospheric dust have fallen over Europe and North America by as much as 49 to 74 per cent. The decline in dust may have exacerbated the impact of acid rain by depleting the input of calcium and magnesium into lakes and forests. Ironically, legislation to improve air quality is partly to blame for the drop in dust levels.

Box 3.14 Natural processes of soil acidification

Acids and bases are measured using the pH scale. Solutions that have a pH from 1 to 7 are acidic; those between 7 and 13 alkalic or basic; those with pH 7 are neutral (Hedin and Likens, 1996). Natural sources of acids are produced by various sources (see Catt, 1985; Rowell and Wild, 1985). Volcanic eruptions produce high amounts of hydrogen sulphide and sulphur dioxide that can be converted into sulphuric acid, thunderstorms can produce nitrogen oxides and wildfires produce acids volatilised from the combustion of reduced sulphur compounds, nitrates and ammonia held in organic materials (Kennedy, 1992).

Acidic soils form when carbonic acid is produced by water and carbon dioxide in the atmosphere or in the soil. If the soil pH is above 5.6 carbonic acid dissociates, releasing hydrogen ions, the reaction being:

$$H_2CO_3 \rightarrow H^+ + HCO_3^-$$

Malmer (1976) suggests that increased soil acidity intensifies the degree of leaching, especially of basic cations like calcium (Ca), magnesium (Mg) and potassium (K). At low pH a soil becomes dominated by hydrogen ions or aluminium and releases basic cations that can be leached. Decomposition of organic matter and microbial respiration can also encourage leaching (see Bache, 1985; Ellis and Mellor, 1995 for more details).

Nitrification – the conversion of ammonium to nitrate – is another acid-forming reaction. Ammonium is first converted into nitrite, releasing hydrogen (Kennedy, 1992):

$$NH_4^+ + \tfrac{3}{2}O_2 \rightarrow NO_2^- + 2H^+ + H_2O$$

Finally, needleleaf afforestation can also promote soil acidification, because the leaf litter produced by coniferous woodland is acidic. Aluminium and sulphate concentrations are higher under these types of vegetation compared with deciduous woodland, and coniferous trees produce more hydrogen ions in mor humus and uptake basic cations, both of which promote soil acidification (Hornung, 1985; Ellis and Mellor, 1995).

Acid atmospheric deposition has caused damage to hydrological, ecological and urban environments as well as human health. For example, the increased deposition of acidic pollutants has contributed to the lowering of pH in lakes across north-west Europe (in particular southern Scandinavia), Ontario in Canada, in the north-eastern part of the United States (Henriksen *et al.*, 1988; Woodin and Skiba, 1990) and in Sweden, where 14,000 of 83,000 lakes are damaged or dead as they have become acidified (Bishop and Hultberg, 1995). Acidic water mobilises toxic elements such as aluminium, which has been linked with problems with fishing including poor fluid regulation, the loss of salts like sodium, lower egg production and mucus formation on gills causing the fish to suffocate. In particular acid surges – a peak in acid levels in water following a snowmelt or heavy rain – carry large amounts of acids into freshwater lakes and rivers releasing aluminium (Morris and Thomas, 1987). Fish kills as a result of acid surges have been described in the Lake District (UK), Scotland (Dudley, 1986) Scandinavia (Wright and Snekvik, 1978), and Jensen and Snekvik (1972) correlate the wipe-out of salmon and brown trout populations from lakes and rivers in southern Norway with low pH levels.

Other animals are also being adversely affected by acid rain. Smith (1997) reports that birds such as the great tit and the black tern are failing to lay eggs because of calcium deficiency. Birds that nest in areas of calcium-poor soils cannot obtain sufficient calcium in their diets to form eggshells. Studies have shown that black terns living on calcium-rich soils hatch four times as many eggs compared with birds nesting in calcium-poor areas. Mortality of chicks in acid-rich areas is being caused by poorly developed skeletons and broken bones.

Atmospheric acidic deposition has affected terrestrial ecosystems. Soils suffer from metal deficiency, increased toxicity and become unsuitable for plant growth. Root and shoot growth and branching can be inhibited by soil acidity (see Kennedy, 1992). Forests such as the Black Forest in Germany, others in central Europe and the Adirondack Mountains in upstate New York, USA appear to have been affected by acid rain (Kennedy, 1992; Schulze, 1989; Rehfuess, 1985). Hauhs *et al.* (1988) link forest dieback (Figure 3.16) with soil acidification processes caused by high sulphur and nitrogen emissions from human sources. For example, Gregory *et al.* (1996) and Ardö *et al.* (1996) suggest that there is a good correlation between forest damage and high levels of acidic pollutants in Czechoslovakia and Poland.

Commercial forest plantations, trout fisheries, nature reserves and national parks can also be adversely affected by acid rain. Pearce (1995) reports that across Europe around 70 per cent of nature and biosphere reserves are located in areas where acid critical loads are exceeded. However, acid rain and soils are only partly to blame for the damage to forest ecosystems. Poor silviculture, drought, tropospheric ozone and pest attacks can also have a detrimental impact (Gregory *et al.*, 1996).

Despite the difficulty in demonstrating a cause and effect relationship between air pollution and human health (Figure 3.17), there is a growing belief in the medical world that acid deposition has direct and indirect effects on human health (Maugh II, 1984). Amongst the complaints linked to acid rain are chest tightness, exacerbating asthma, chronic bronchitis, cystic fibrosis, emphysema and lung damage (Maugh II, 1984; Luoma, 1988). Increased atmospheric sulphur dioxide is also thought to cause asthma attacks. Studies in the USA have shown that chest

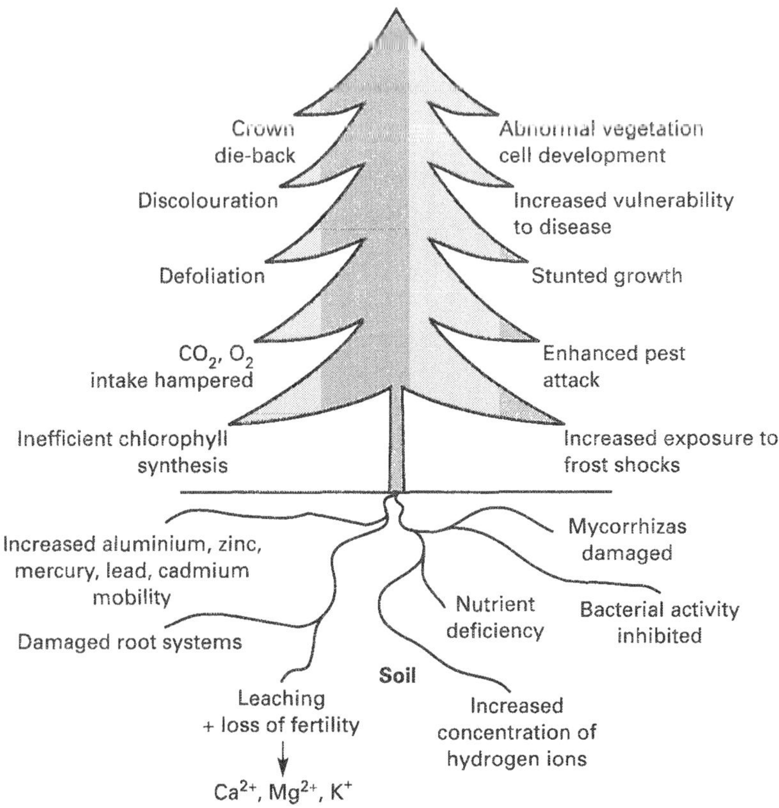

Figure 3.16 The effect of acid rain on soils and vegetation (sources of information: Kemp, 1990; Rehfuess, 1985; Kennedy, 1992)

tightness and wheezing symptoms are more common in mild asthmatics when sulphur dioxide levels reach 100 parts per billion. Another study indicated that child bronchitis and chronic coughs were more common in US cities with higher levels of air pollution and, in 1984, it was estimated that acid rain-linked pollution accounted for approximately 50,000 premature deaths in North America (Luoma, 1988). Indirect effects of acid rain include higher levels of lead and aluminium in drinking water, which has been linked with Alzheimer's disease. Lead in water exceeding EPA standards has been measured in the Adirondack Mountains and acid rain can also mobilise mercury, cadmium, nickel, zinc and selenium that can then bioaccumulate in vegetables such as spring wheat, potatoes and carrots, and in seafood such as shellfish (Oskarsson *et al.* 1996).

Radioactive leaks and nuclear power

All forms of natural life are exposed to naturally occurring forms of radiation: X-rays, gamma radiation, alpha and beta particles, neurons, protons and cosmic rays. The discovery of radioactivity provided the industrial societies with another

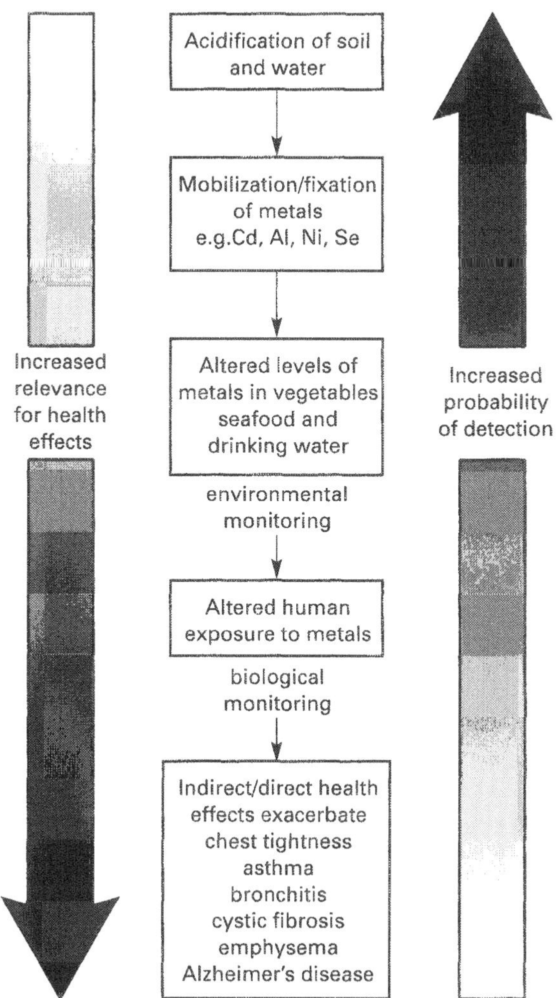

Figure 3.17 A scheme for the processes involved in human health effects caused by acidification of soil and water (after Oskarsson *et al.*, 1996)

method of energy generation, Simmons' (1991) so-called 'nuclear age', but has also increased the risk of exposure to harmful levels of radiation. The nuclear power industry increases the movement of radioactive material through the earth's major systems. Uranium must be mined and refined, nuclear fuel must be fabricated and waste either recycled or disposed of. Eisenbud (1990) estimates that mill tailings contain 30,000 Ci of Ra-226 and emanate radon.

Exposure to radiation from nuclear reactors working under normal operating conditions is miniscule. Dangerous levels of exposure will only result from leaks from the main reactor core, when levels could increase above 100 mrem – the average dosage a person naturally receives in one year. Cohen (1995a) claims that routine releases of radiation during nuclear power plant operations and reprocessing will cause an extra 0.25 cancer deaths over the next 500 years and technology should continue to reduce releases. While normal working activities of nuclear power plants might represent a minor threat the possibility of a major accident is more alarming; Chernobyl, for example, does prove it could happen. Respent reactor fuel contains I-129, Kr-85 and tritium, which pose a potential threat if they enter biogeochemical cycles. Strontium-90 and plutonium-239 are also extremely dangerous. Transportation of radioactive material containing an estimated nine million curies of radiation takes place each year and could be released into the environment if an accident occurred. So far some radioactive waste containers have failed but Eisenbud (1990) suggests the consequences have been relatively minor and Cohen (1995a) argues the probability of the safety procedures failing are low.

Along with the development of nuclear weapons, nuclear power has increased the risk of exposure to harmful levels of radiation. Stochastic effects of exposure include cancer and genetic damage, and non-stochastic effects include suppression of bone marrow function, damage to the gastrointestinal tract, skin burns, cataracts and temporary sterility in males. Small accidents can be lethal. In the USA there have been 34 known cases of 'accidental criticality' – 'the inadvertent accumulation of radiative material to cause a spontaneous shower of deadly radiation' – killing seven people (Zorpette, 1996a, p. 24). Similar incidents have occurred in the former Soviet Union and in the UK in nuclear processing and weapons plants.

The effects of larger-scale nuclear incidents such as Chernobyl are difficult to quantify and there has been some debate about the magnitude of health problems relating to the Chernobyl disaster, especially discerning between the direct and indirect effects of radiation exposure. Shcherbak (1996) estimates a minimum value of 90 million curies was released when the nuclear reactor at Chernobyl exploded, which could affect human health for a long time. Likhtarev *et al.* (1995) found increased concentrations of iodine-131 in the thyroid gland in 150,000 Ukrainians, suggesting that a rise in cancer victims is a direct consequence of the Chernobyl nuclear explosion. Hohenemser (1996) suggests that thyroid cancer in the Chernobyl region has increased dramatically and increased in Belarus, Ukraine and Russian Federation from less than 1 per million to 36, 22 and 3.1 respectively, but Greenhalgh (1995a and 1995b) suggest that mortalities in Ukraine may have resulted from psychological stress and the fear of radiation ('radiophobia') following Chernobyl rather than from the effects of radiation exposure itself. Cases of leukaemia have been attributed to an accident at the nuclear plant at Mayak in the former Soviet Union over a timescale of 5 to 20 years after exposure to radiation (Segerstahl, 1996). Other long-term effects, such as cardiovascular disease, nervous and respiratory disorders, have been attributed to stress, and radiophobia may be pathological (Perera, 1995). Other knock-on effects include the contamination of meat and milk in parts of Wales, Poland, Germany, Austria and Sweden (Warner, 1996; Shcherbak, 1996).

To date deaths from exposure to large doses of radiation are low (31 at Chernobyl) but the long-term consequences of such accidents are difficult to predict. The explosion at Chernobyl contaminated an area of 2590 km^2 and winds carried radioactive material over much of western Europe (Stiling, 1992). Worldwide the impact of Chernobyl is also difficult to predict. Hohenemser (1996) suggests the radiation released in the explosion is the equivalent of 28,000 extra cancer fatalities, with cancer mortality rates exceeding background levels by 1 per cent in the Chernobyl region, 0.007 per cent in western Europe and 0.00005 per cent in the USA. The threat to human health is disputed. Jaworowski (1996) claims that the deaths caused by Chernobyl are much lower, with 48 people dying from radiation and/or injuries. Despite Chernobyl, the hazards of nuclear power may be less than other energy sources – Cohen (1995a) compared the eventual deaths caused by a standard-sized nuclear power plant with conventional fuel sources and suggests that coal pollution is more harmful over a 500-year period.

Large-scale nuclear war remains a remote possibility and could result in millions of deaths: the bombing of Nagasaki and Hiroshima killed some 200,000 people (Eisenbud, 1990).

Regional and local problems of extracting, transporting and manufacturing resources

Extraction of resources has also created widespread environmental damage on a local scale (also see Chapter 2). For example, mining and smelting, construction and demolition produce waste and metalliferous materials that release heavy metals into the environment from sources such as opencast mines, spoil heaps and landfills. The impact of mining can be extensive. For example, the north-east of the Brazilian Amazon is the location of the Greater Carajas Project that aims to open up 900,000 km^2 of forest for mining and industrialised agriculture. Opencast iron ore mining and bauxite mining produce 2.5 million tonnes of pig iron and 800,000 tonnes of bauxite annually. As many as 610,000 hectares of forests are cleared each year to produce charcoal for pig iron furnaces (Treece, 1989). Enrichment of heavy metals in soils and hydrological systems occurs in all major metal mining areas (Hildyard, 1989); examples include copper and gold mining in the Amazon (Filho and Maddock, 1997; Lacerda and Marins, 1997; Pestana *et al.*, 1997; Melamed *et al.*, 1997), contamination of molybdenum in the Knabeåna-Kvina drainage basin in Norway (Langedal, 1997a; 1997b) and lead, copper, zinc and cadmium in the Harz Mountains in Germany (Gäbler, 1997).

Soil acidification and contamination is linked to industrial processes such as mining and smelting. Metallurgical industries, such as the nickel-copper processing plants in the cities of Nikel, Zapolyarniy and Monchegorsk, located on the Kola Peninsula, are responsible for elevated levels of copper, nickel, lead and zinc in top soils (Boyd *et al.*, 1997). Oil and gas drilling produce drilling fluids and petroleum products that contaminate soils. A study in Padre Island National Seashore in Texas, USA showed elevated levels of heavy metals (barium, chromium, lead and zinc), sodium, salinity, pH and petroleum hydrocarbons, but the long-term effects on ecosystems are unknown (Carls *et al.*, 1995).

The release of toxic chemicals and elements into the environment from industrial processes poses threats to plants, animals and humans. For example cyanide, leached from mine spoils, has had disastrous consequences on fauna. In one incident over ten million gallons spilled into a South Carolina river killing approximately 10,000 fish, and around 10,000 animals were poisoned by cyanide in Nevada as a result of heap leach operations between 1986 and 1991 (Greer, 1993). This has led to higher levels in plant and animal tissues from where it can be transferred to humans through food chains. For example, cadmium in the soil can be readily taken up by foodstuffs such as rice, grain and vegetables, and absorbed into the human body. Rising levels of cadmium have been recorded in food during the twentieth century and increased exposure to cadmium has been linked with renal damage, bone disorders, lung cancer and cardiovascular disease (Elinder and Järup, 1996). Serious risks to human health can occur indirectly through the food chain. For example, mercury released from gold mines has entered the food chain, contaminating fish in marine seafood taken from the Gulf of Davao off the Philippines (Greer, 1993), and Amazonian villagers have high levels of mercury in their bloodstream from eating fish poisoned by mercury leached from mining waste (Aks *et al.*, 1995; Hacon *et al.*, 1997). Poisoning can be fatal, for example 43 deaths occurred in Minamata, Japan between 1953 and 1983 (Ellis, 1989; Barrow, 1995), and acute mercury poisoning can cause skin irritation, diarrhoea, nausea, breathing difficulties, loss of memory, impaired sense of smell, vision and speech, severe tremors, kidney, brain and placenta damage (Greer, 1993).

As noted in Chapter 2, the emergence of industrialisation has developed hand in hand with the emergence of transportation that, amongst other things, has allowed the redistribution of energy and waste. Both are often environmentally damaging. Oil and radioactive waste, for example, are commonly transported between countries for use or disposal. The unevenness of reserves means that oil must be transported to meet global demand. Pipelines are commonly used to carry gas and oil across land, and both their construction and leaks can pollute terrestrial and hydrological systems. In the Oriente region of Ecuador, for example, an estimated 16 million gallons of petroleum was leaked from the Trans-Ecuadorean Pipeline. However, oil tankers are the most popular form of transportation and accidents regularly occur due to poor weather conditions and, in some cases, poor judgement by shipping crews. Major oil spills have occurred in recent years, including the Exxon Valdez off the coast of Alaska and the Sea Empress off the coast of South Wales, and they have disrupted coastal marine ecosystems. For example, in 1989 the Exxon Valdez oil tanker crashed into Bligh Reef in Prince William Sound off the coast of Alaska spilling 37,000 tons of oil, which spread over approximately 1750 kilometres of shoreline (Holloway, 1996).

Oil spills upset the balance of coastal marine ecosystems, adversely affecting all parts of the food chain. Birds have suffered after other major oil spills. For example, following the Exxon Valdez disaster it is estimated that up to 400,000 birds died, including around 150 bald eagles, while guillemots have not bred since the oil spillage and it may take around 70 years for their numbers to recover (Hodgson, 1990). Numbers of eider ducks, black guillemots and great northern divers around Sullom Voe, Shetland, close to where the Braer spilt 84,000 metric tonnes of crude oil in January 1993, have declined dramatically (Pearce, 1993). Attempts to rehabilitate birds affected by oil are costly and there is some evidence to suggest that the birds do not survive (Schmidt, 1997).

Animals that occupy the coastal zone have also died as a result of oil pollution – oil can cause blindness, nose bleeds, emphysema, impair liver and kidneys and lead to loss of insulation. Approximately 3500 to 5500 sea otters are thought to have died after the Exxon Valdez disaster, and many others carry abnormally high hydrocarbons in their tissue, which disrupts reproduction. Toxic by-products of the oil have also moved through the food chain – brown bears and bald eagles ingest oil by feeding on the dead carcasses of seabirds and fish. The long-term effects oil contamination may have on breeding patterns is unknown (Pearce, 1993; Hodgson, 1990).

Manufactured products also can be harmful when released into the atmosphere, hydrosphere and biosphere. The release of organic chemicals frequently occurs during product-related processes. Polychlorinated biphenyls and hexachlorobenzene, for example, have been used as transformer and capacitor oils, additives for paints, sealants and hydraulic fluids, and they appear to have contributed to the decline of European otters, Baltic grey and harbour seals between 1950 and 1980 (Olsson *et al.*, 1992; Vörösmarty *et al.*, 1997) as well as contaminating Arctic and Antarctic seals. On incineration or during metal reclamation these compounds produce other undesirable chemicals such as polychlorinated dibenzo-p-dioxins, dibenzofurans and coplanar biphenyls that have been redistributed world-wide, and they have been detected in fur seal blubber in the Arctic and Antarctic (Oehme *et al.*, 1995). The Beluga whale population has shrunk from an estimated 16,200 in 1866 to 500 after hunting, pollution by DDT, PCBs and Mirex, and HEP has degraded its habitat on the St Lawrence River, USA. Pollution is now the most dangerous threat to the whales as it is transferred from mother to calf through milk, causing cancerous tumours and damage to internal organs (Beland, 1990). Production of organic chemicals does centre on industrialised regions such as the USA, Japan, West Germany and the former Soviet Union, but newly industrialised countries, such as Brazil, Mexico and China, are emerging as new centres of production of compounds like ethylene and benzene (see Brown *et al.*, 1990).

Use of agrochemicals

Agrochemicals, namely fertilisers, pesticides, fungicides and herbicides, have been used excessively over the last 40 years to improve yields, maintain soil fertility and to combat against crop pests and diseases. The increase in crop yields during this period does reflect the success of these products in sustaining global food production, but the environmental cost of agrochemicals has been high.

Nitrogen is an extremely important nutrient. Without it humans and animals could not survive because it is a vital ingredient in DNA and RNA, the molecules that store genetic information. Humans and animals are dependent on plants and nitrogen-fixing organisms, such as *Rhizobium* bacteria and cyanobacteria, to convert nitrogen into an available form for plants. Once in the food chain both humans and animals digest nitrogen through the consumption of leguminous plants (Smil, 1997).

Reliance on nitrogen fixers to provide nitrogen has created many practical difficulties. High rates of population growth have placed enormous stress on agricultural systems to produce sufficient food. To increase food production additional nitrogen has been needed to prevent soils from suffering from nitrogen depletion. Ammonia production using the Haber-Bosch synthesis process fulfilled this role. Synthetic fertilisers provide about 40 per cent of nitrogen contained in crops and approximately one-third of the protein in human diets, and over 70 megatonnes of nitrogen fertiliser is used each year (Smil, 1997). As such fertiliser is considered to be the primary source of nitrogen pollution in industrial countries (Bleken and Bakken, 1997). Nitrogen compounds are released by livestock farming (which produces 78 million tonnes of manure annually) and leakages from factories.

Fertiliser and manure increase nitrate and phosphate levels in water sources, promote cultural eutrophication (Figure 3.18), lower biochemical oxygen demands and create unpleasant odours (Skinner *et al.*, 1997). Rivers, lakes and coastal areas are now contaminated with excessive levels of nitrogen and phosphorus. Examples include the coastal lagoon of the Great Barrier Reef, which receives approximately 15 million tonnes of sediment, 77,000 tonnes of nitrogen and 11,000 tonnes of phosphorus each year from agricultural land (Bell and Elmetri, 1995) and the water of Chesapeake Bay, a 64,000 km^2 estuary located off the coast of Maryland and Virginia, USA, which experiences zero oxygen levels, enhanced metal solubility and algal blooms caused by high levels of nitrogen and phosphorus levels derived from farmland, fertiliser and sewage (D'Elia, 1987; Horton, 1993; Owens and Cornwell, 1995). The ecological damage caused by cultural eutrophication is discussed in Box 3.15.

Box 3.15 Cultural eutrophication

Water oxygen deficiency, development of sulphur bacterial mats, algal blooms and the reduction of benthic fauna and demersal fish all characterise the eutrophication of hydrological systems (Gunnarsson *et al.*, 1995). The North Sea, the Baltic, Arabian and Black Seas, Chesapeake Bay, the Bay of Bengal, the Gulf of Mexico and the Pacific coasts of North and South America suffer dead zones due to oxygen deficiency caused by algae (Abramovitz, 1997). Part of the lagoon that separates the coast of Queensland, Australia and the Great Barrier Reef has been adversely affected by high nutrient levels and increased rates of sedimentation. Hard corals have been replaced by other benthos, sea grasses and soft corals. Algal blooms and phytoplankton growth has reduced the diversity of reef builders through competition for space and light and the change in water quality. Blooms of one nitrogen-fixing phytoplankton, *Trichodesmium*, have increased in concentration due to 20,000 tonnes of nitrogen being added into the lagoon ecosystem each year. The occurrence of the Crown of Thorns starfish, pennate diatoms and small flagellates are typical indicators of eutrophic conditions. High nutrient levels increase the survival rate of Crown of Thorns starfish larvae (Bell and Elmetri, 1995). In Chesapeake Bay fishing harvests dramatically declined from the early 1960s to the late 1970s as a result of eutrophication of the Bay's water. Harvests of American shad decreased by 35 per cent, herring by 95 per cent and fishing for striped bass was banned. White perch, yellow perch, weakfish and other species numbers also fell, while oyster colonies were wiped out (Horton, 1993). Eutrophication has also affected major rivers, such as the Rhine and the Mississippi, San Francisco Bay, New York's Long Island Sound (Smil, 1997), the Baltic Sea (Gunnarsson *et al.*, 1995) and lakes (Havens *et al.*, 1996).

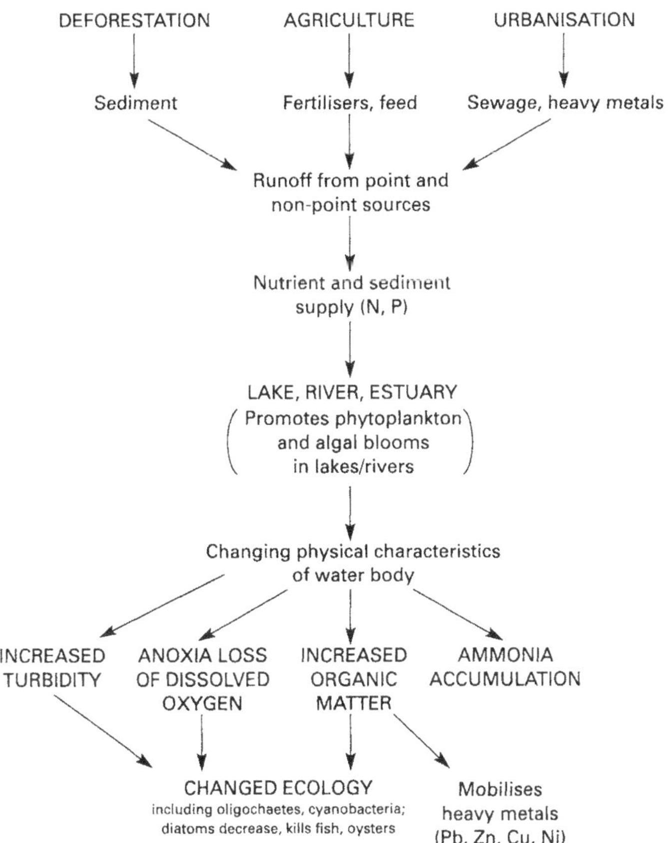

Figure 3.18 Schematic diagram for cultural eutrophication (sources: Havens *et al.*, 1996; D'Elia, 1987)

Excessive amounts of nitrate in ground and surface water, released from agricultural sources (Burt and Haycock, 1991; Croll, 1991; Croll and Hayes, 1988; McCracken *et al.*, 1994), can affect human health and cause methemoglobinemia (blue baby disease). Nitrate is converted to nitrite in the stomach where it combines with red blood cells and prevents oxygen circulating around the body, causing the skin to turn blue. Gastric cancer has been linked to higher nitrite levels. Nitrite reacts with secondary amine to form N-nitroso compounds that can modify DNA and stimulate cancer (Forman *et al.*, 1985a,b; Smil, 1997). However, the link is tenuous, as Addiscott *et al.* (1991) note that incidence of gastric cancer has declined while nitrate concentrations in water have risen, but efforts are being made to restrict nitrate releases from agricultural land (eg Archer, 1994).

Pesticides (including herbicides, insecticides and fungicides) such as pentachlorophenol (PCP) and DDT have been deliberately released into the environment to kill fungi and insects. For example, some pesticides like DDT have been extensively used world-wide and have a high persistence in the environment, particularly by bioaccumulation (see, for example, Rowell, 1991). Production of DDT peaked in the late 1950s and early 1960s when it was used as a delousing agent, to protect people against water-borne diseases (*eg* malaria) and in agriculture (Brown *et al.*, 1990). As soon as it was considered to pose a serious environmental threat (a view championed by the publication in 1962 of Rachel Carson's *Silent spring*) production in North America and Europe ceased and by 1984 Asia and Africa accounted for 74 per cent of DDT consumption.

Pesticides can adversely affect human health. Unintentional occupational poisoning by pesticides has affected several million people globally (Skinner *et al.*, 1997). For example, exposure to methyl bromide may cause skin burns, eye damage and central nervous system disorders if inhaled. Sales of the fumigant reached 67,000 tonnes in 1990 globally, until use of the gas was banned under protocols protecting stratospheric ozone (Buffin, 1992). In the UK a report by the Ministry of Agriculture claims that nearly one-third of common fruit and vegetables contain pesticide residues, the long-term effect of which remains unknown despite links with some human ailments (Nuki, 1996). Organophosphate sheep dips have damaged the health of farmworkers, producing symptoms such as acute poisoning, neurological effects, psychological and behaviourial changes and respiratory problems. Other pesticides such as atrazine have been linked with potential breast carcinogens (Skinner *et al.*, 1997).

There is no doubt that pollution has increased on a regional and local scale and it can be argued that the exploitative relationship has become more intense since prehistory. During prehistoric times pollution caused by industrial activities was very localised (see Chapter 2); since then, especially from the industrial revolution onwards, incidences of pollution have multiplied as different regions of the world become more industrialised. Today the problems of pollution have worsened to the extent that industrial regions appear to be affected by pollution, in particular acid rain and, after Chernobyl, radioactive fallout. The so-called underdeveloped and developed world is littered with examples of local pollution caused by industrial processes to the point where it can be suggested that pollution is a world-wide phenomenon. As stated in Chapter 1 there is no part of the earth untouched by the impact of societies. If this process continues unabated the environmental cost could be enormous. Poland, a country devastated by decades of heavy industry and no environmental protection, is a clear warning to other countries. Pollution adversely affects the physical environment, often rendering it useless for human use; it affects human health and, ironically, it has threatened other forms of economic activity. The cost of cleaning up the effects of industrial pollution is also enormous. If our industrial societies are to have a sustainable relationship with natures, they must address the problems of pollution. These attempts are discussed in Chapter 4.

However, the views expressed in the previous paragraph are essentially neo-Malthusian, and they are not shared by everyone. Simon (1995), for example, has argued that pollution is a short-term problem that will be overcome by scientific and technological developments. He and Baumol and Oates (1995, p. 445) argue that the world has, in many respects, become cleaner, and the view that 'environmental deterioration has been universal and an accelerating process whose source is modern industrialisation and population growth' is naive. They argue, for example, that pollution levels, such as heavy metals and DDT in the Great Lakes, have fallen since the the mid-1970s and that trends in water quality and particulates show improvement. Indeed societies have developed technology to control pollution, and these methods are discussed in Chapter 4.

3.8 Summary

Modern forms of exploiting natures are still expanding world-wide and have resulted in an ever-increasing demand for energy and raw materials. World trade has grown tremendously since mid-century, with exports of primary commodities and manufactured products increased elevenfold. Trade can be good or bad. The modern technologies of industrial production have been responsible for a physical impact on nature and are a consequence largely of western social institutions that drive world trade (Giddens, 1997). However, the impact of this on the natural world is a subject of controversial debate.

Neo-Malthusians identify the spread of industrialisation and population growth as the main causal factors behind contemporary environmental problems. They argue that pollution, the production of discarded waste or the expulsion of gases into the atmosphere and the depletion of some resources (fish, land by erosion, oil) is destroying our resource base and that if these trends continue unabated it seems they will all reach global proportions. The examples looked at in detail in this chapter, namely the enhanced greenhouse effect, ozone depletion and acid rain, show that modern industry, technology and science are not always beneficial in their consequences, and a review of the extent of the damage makes it easy for people like Giddens (1997, p. 530) to argue that 'human onslaught on the natural environment is so intense that there are few natural processes uninfluenced by human activity'. The evidence presented in this chapter suggests that renewable resources are more vulnerable to depletion. Current trends suggest that world fish stocks, rates of soil erosion and tropical rainforests are under threat. Societies must be careful that the short-term blips in resources like ocean fish catches and grain production do not turn into long-term trends. However, supplies of non-renewable resources, especially minerals, are abundant and they will be readily available in the foreseeable future, and even the life expectancy of oil and natural gas appears to be secure, especially if alternative supplies can be successfully extracted.

Cornucopians provide very contrasting views on the depletion of resources, waste production and pollution, displaying a total belief in human ingenuity – mainly through technology and science – to overcome the problems of environmental degradation. However, less extreme cornucopians do accept that some environmental problems need addressing, namely the protection of biodiversity (see Sedjo and Clawson, 1995), environmental quality (Baumol and Oates, 1995) and water (Adelman, 1995). Cornucopians also claim that the quality of human life has improved dramatically over the last century. The statistics bear this out, but neo-Malthusians counter that millions of people live in poverty and that the environmental cost of attaining those measures of human welfare in this context has been high. It is possible to improve environmental quality as shown by Beisner and Simon (1995), but societies must show a willingness to develop the methods to do so.

Mitigating against environmental problems will not be easy. The review of the environmental problems associated with industry and agriculture shows that they are complicated by inter-linkages with the earth's major systems, such as the use of nitrogen in world food production. To expand food production between 1950 and 1980 required the massive introduction of usable nitrogen into natural systems, directly or indirectly. This contributed to several environmental problems including eutrophication, soil acidity, human health problems, ozone depletion and the enhanced greenhouse effect. Because nitrogen is so important for producing food it is not easy to control its use, which complicates the management of human activity on the environment. Rice production, for example, produces high quantities of methane, a greenhouse gas. A huge proportion of the underdeveloped world (especially in tropical Asia and China) depends upon rice as a staple food source. Lowering methane levels by abandoning rice production is not a viable option. Futhermore attempts to curb sulphur dioxide emissions may exacerbate global warming as sulphate aerosols cool the troposphere; so reducing one problem, acid rain, may be at the expense of another (Wigley, 1991; Michaels, 1995).

Overall, it is difficult to argue that societies will reach their limits to growth in the foreseeable future. A review of the evidence does not fully support the predicted neo-Malthusian disaster. It does, however, show that there are environmental problems that require some form of mitigation, and people must meet the challenge that environmental problems have given societies to avoid serious ecological and environmental damage at local, regional and possibly global scales. Chapter 4 looks at the possible solutions.

Chapter 4

Managing Problems in the Environment

4.1 Introduction

Chapter 3 argued that the relationship between society and the physical environment is still, in many ways, exploitative and that, if this type of relationship persists, the consequences for humans could be disastrous. Concern over the exploitative relationship has led neo-Malthusians to argue that there are natural limits for economic and population growth (embodied in the 'limits to growth' argument) where there is a limited carrying capacity for population, and productive capacity (for all types of resources); while people such as Beck (1992; 1995) and Giddens (1990) argue that society has now entered a new era of 'risk' or 'reflexive modernity', where new kinds of risk have been produced by technology and science through industrialisation. Giddens (1990), for example, argues that industry and capitalism have transformed the world of nature in ways that were unimaginable to earlier generations, leading to the formation of a 'created environment' – one that is physical but not natural. This transformation is seen by Giddens (1990, p.77) to substantially alter the pre-existing relationship between society and the environment to create what he describes as 'one world', 'a world in which there are actual or potential ecological changes of a harmful sort that affect everyone on the planet.'

Dobson (1990) suggests that these views have helped to instil the belief in society that environmental management is necessary. However, various forms of environmental management have arisen from different perceptions and views about humans' relationship with nature, which were outlined in Chapter 1. Because human nature and its relationship with nature varies it is not surprising that there are different views on how the environment and human-made environmental problems and hazards should be managed (Pepper, 1996; Wilson and Bryant, 1997). Management and control are a feature of modern human society and the concept of management has been a major philosophy in human–nature relationships. This is reflected by those people who argue for a managerial approach to environmental problems. This approach encompasses both the 'accommodators' and some ecocentric perspectives outlined in Chapter 1, which see possible solutions to environmental problems without making fundamental changes to society, its consumption and production patterns. These forms of management try to work with nature, or at least accommodate it, and include many of the technological and regulative methods of management outlined in this chapter, such as recycling, waste minimisation, organic farming, renewable energy, energy conservation and efficiency.

Other, more extreme views on managing the environment also exist. Some cornucopians believe that resources and the physical environment do not need managing because there are no environmental problems that threaten humanity. They believe in a 'free market' or 'no-management' approach that advocates that private ownership will enable people to exploit the earth's resources, and that it is in their own interests to develop less environmentally damaging forms of exploitation or to use technology and science to overcome any environmental problems when they arise. Free-market environmentalists generally oppose state intervention in environmental management, favouring market-driven trade in environmental 'goods' and 'bads' that can be bought and sold. Various forms of free market principles are now evident in environmental management and include commercial trading of animal products, permits to pollute and ecotourism. Methods of using the economic free-market to limit and/or regulate over-exploitation are also described in this chapter. While this view is essentially an extreme technocentric view, some ecocentrics also argue that natural laws will succeed in the long run and deal with human-created problems (and possibly humans!) if people do not adapt and live within the laws and limits of nature. An example of this approach is the Gaia theory that, as discussed in Chapter 1, argues that the earth is a self-regulatory system that will continue to function and manage itself even if people do exploit the earth's resources to the point of collapse. They argue that the earth's systems will not collapse, but society's use of the earth will.

This chapter will investigate how people have tried to preserve, protect and manage the agents of natures, and discuss whether the methods and approaches adopted by them are effective or not. First we will introduce some of the general issues involved with management, and this will be followed by a discussion of six common forms of managerial methods that people have used to manage and/or control environmental problems. Each form of management will be illustrated using a series of examples that demonstrate that they can be effective.

4.2 Managerial methods

4.2.1 General issues

The dominant approach used to ascertain the extent of environmental problems and hazards, and the need for management, is to measure them in terms of risk and safety. The concept of making something safe by reducing its risk to society has always been promoted as a good thing, and this philosophy provides the core beliefs behind many forms of environmental management. Risk is a working concept that can be applied to natural hazards such as volcanoes and earthquakes, and human-made environmental problems such as pollution from local to global scales (Ellis, 1989). According to Gerrard (1995, p.301) risk 'is the likelihood, or probability, that a particular set of circumstances will occur, resulting in a particular consequence, over a particular time period'. To resolve environmental problems the practices of risk evaluation or assessment and management have emerged, which aim to achieve high levels of safety and low levels of risk with the development of environmental impact assessments (EIAs) and environmental risk assessments (ERAs) to monitor and manage the effect of hazards and environmental problems (see Box 1.3 and Mitchell (1989)).

Risk management 'means reducing threats to life, property and the environment posed by hazards while simultaneously maximising any associated benefits' (Smith, K., 1993, p.46). To manage a risk or hazard it must be identified, usually by some form of measurement and quantification, and then assessed to estimate the probability of that risk posing an actual threat to society, before a strategy to minimise it can be formulated (see Ellis, 1989; Smith, K., 1993). Most risks that are physical, chemical or biological in nature can be identified and directly measured to determine the magnitude and frequency of occurrence and the social consequences (see, for example, Vesely, 1984; Smith, K., 1993). Measurements, however, need to be accurate: ie to provide an answer that is close to the real one and to be precise, ie the measurements must be consistent. A range of tests must be conducted in order to cover the majority of possible sequences. If the collected data proves to be both inaccurate and imprecise it will invalidate the risk assessment procedure. Too often risk assessments have to be made when insufficient data are available or are too subjective in nature to make a sensible decision and rely on the intuition of 'experts'. Some environmental problems can persist over long timescales and therefore they need to be monitored regularly to review whether the management process is working and is still valid – a process known as environmental auditing (Ellis, 1989; Gerrard, 1995).

Another consideration of risk assessment is the cost of environmental policies against the benefits to society. Indeed, cost–benefit analysis has become a fundamental managerial approach (see Box 1.3) used by decision makers to determine the cost of implementing a policy in relation to the benefits and to avoid rigid forms of environmental protection that end up regulating activities that pose little threat to the environment (Reisenweber, 1995). For example, Clifford (1996, p.194) criticises the US Clean Air and Clean Water Acts because they force society to 'regulate nonexistent risks too much and ignore larger, documented risks'. He argues that the US government is spending billions of dollars cleaning up the last 10 per cent of pollution of outdoor air yet pays no attention to improving the quality of air indoors, although people such as Ott and Roberts (1998) argue that indoor pollution may pose a significant health risk. Ritter (1995) and Reisenweber (1995) argue, therefore, that capital is being wasted that could be spent more productively on preventing pollution. Clifford (1996) argues that the US Clean Air Act of 1990 has forced businesses to spend an estimated $20 billion to meet air pollution regulations while only saving $12 billion in health costs. In other cases, where absolute safety cannot be determined, a notion of acceptable level of risk for a given activity or situation is used (Smith, K., 1993).

Once a risk or hazard is identified and assessed the decision maker must decide what action is needed, who takes it and how it should be implemented (Russell, 1995). A fundamental issue for any environmental manager is whether to be 'proactive' and prevent environmental problems before they have a detrimental impact or to be 'reactive' and cure or ameliorate the impact of problems when they happen.

The precautionary principle approach and the best environmental practice approach are the two main forms of preventive environmental management. They work on the premise that preventive action should be taken before a definite causal link has been established between an activity and specific environmental damage, by using scientific data to ascertain thresholds beyond which permanent environmental damage occurs (Collins, 1995). Advocates of this approach argue that it is better to act as soon as possible to avoid or tackle a problem, despite any uncertainty, rather than risk facing more disastrous consequences in the future (Dovers and Handmer, 1995). Examples of the precautionary approach include conserving biodiversity. Myers (1993a; 1993b) argues that the majority of fauna and flora living in the rainforest remain unknown to humans and it is estimated 30,000 species become extinct each year as the rainforests are cleared. If their preservation was assured before they are discovered these species could be valuable to society in the future, supplying medicines, genetic material or food sources.

Thus, the precautionary principle responds to uncertainty by using management to set limits to exploitation well within environmental thresholds, and utilises concepts such as sustained yield and carrying capacity to determine maximum yields of resource exploitation and use without degrading the environment (see Fairlie, 1995; Xu *et al.*, 1995; Wilson and Bryant, 1997 and Box 3.2 for more details and examples of these concepts). However, the precautionary principle approach has its critics, because it could mean that governments spend vast sums of money to tackle a problem that may not materialise or be threatening to society.

Ideally it would be nice if people could pre-empt environmental problems. However, accidents do occur and societies can only respond once they have happened. As discussed in Chapter 3, deliberate or accidental introduction of waste or a pollutant into the environment can occur due to a variety of factors, including poor weather conditions, poor judgement or even through negligence or ignorance – as is often the case with major oil spills. These kinds of environmental problems are very difficult to foresee and control. Once the pollutant is released the only option is to clean it up or leave nature to deal with the problem.

A series of approaches exists to managing resources and reducing environmental degradation. These can be divided into:

(1) Technological methods that use advances in technology to improve the efficiency of industrial processes and to reduce the emission of pollutants.
(2) Regulative methods that place restrictions on the emission of pollutants or the extraction of resources, such as imposing taxes or fines.
(3) Efficiency measures that use resources more efficiently.
(4) Resource substitution, which encourages the use of alternatives or substitutes that are more environmentally friendly.
(5) Managing with nature, which encourages people to work with nature to manage a problem.
(6) Using the free market, which removes any state intervention in environmental and resource management.

The next section outlines examples of each management method to demonstrate that in particular circumstances they can all effectively ameliorate an environmental problem.

4.2.2 Technological methods

Technology has played a complex role in the interrelationship between human societies and natures. The role that technology can play in creating environmental problems is evident in Chapter 3, but technology can also provide a way of avoiding or ameliorating environmental problems. For example, structural technologies such as flood levees and earthquake-proof buildings, and non-structural technologies, like satellites to provide data to forecast weather, can both provide a means of defence against environmental hazards (Jones, 1990). Technology can also play a crucial role in understanding how the earth's systems respond to human activities and provide policy makers with management options. While some ecocentrics and technocentrics have argued that technology alone cannot solve environmental problems (see, for example, Hardin (1968)), societies have also developed technologies to mitigate against environmental problems. The examples presented below show how technology can lower pollution and use resources more efficiently in the developed and underdeveloped world.

It was argued in Chapter 3 that pollution was a major form of over-exploitation. Technology can reduce the amount of pollution released by people's activities; for example, technological developments in the motor industry are able to reduce emissions of oxidised nitrogen compounds that cause atmospheric acidic deposition. Lean-burn engines increase fuel consumption efficiency and emit 70 per cent fewer pollutants. Catalytic converters have been developed that reduce emissions by up to 90 per cent. One advantage of catalytic converters is that they do not substantially increase the cost of the car and hence the price can be passed on to the consumer, which makes this is an example of an economically competitive piece of environmentally friendly technology (Dietz *et al.*, 1991; Van der Straaten, 1996). However, while technologies such as catalytic converters may reduce pollution emissions from individual cars they may not lead to a drop in total pollution levels if the number of cars continues to increase. The development of new ways of propelling vehicles is another option to control pollution. Electric vehicles can reduce greenhouse gas emissions and pollution because they are up to 90 per cent more efficient at converting electric into kinetic energy; at present they are not competitively priced, although Sperling (1996) suggests that the price will fall, making electric-powered cars affordable during the first decade of the next century.

Technology has also been employed to curb other forms of pollution. Corcoran (1991) and Ridley (1993) describe several clean coal technologies that limit sulphur dioxide and nitrogen dioxide emissions from fossil fuel power stations (Figure 4.1). For example, flue gas desulphurisation units have been installed at fossil fuel power stations that remove up to 90 per cent of sulphur dioxide (see also Hadfield, 1997).

Such technologies, however, are not without problems, in that they consume a large amount of electricity, increase carbon dioxide emissions and generate large quantities of waste. Other options do, however, exist, including coal cleaning (which separates pyritic sulphur from coal using a flotation method – pyritic sulphur has a higher specific gravity compared with coal) and fuel switching (Corcoran, 1991).

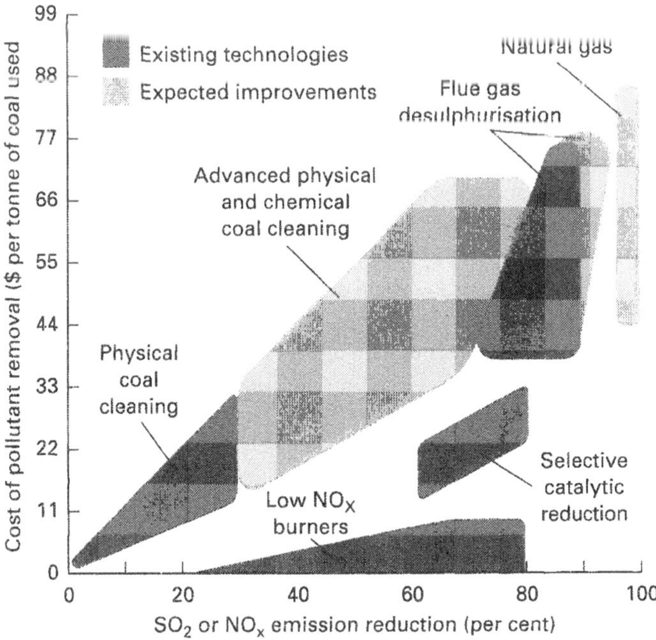

Figure 4.1 Clean coal technologies (source: Ridley, 1993)

Technological advances could help to control the amount of pollution released into the environment and, advantageously, produce no environmentally harmful by-products. For example, Hutchings *et al.* (1996) describe a treatment process for volatile organic chemicals released from industrial stack emissions using uranium-oxide-based catalysts that can destroy a range of pollutants such as toluene and benzene at moderate temperatures. Cooper and Holloway (1996) argue that developments such as these should continue to prevent pollution and provide clean technology that will not stifle economic growth.

As well as preventing pollution at source, chemical technology can be used to clean up or dispose of wastes and unwanted substances after harmful substances are deliberately or accidentally released into the environment. Chemicals can contain, absorb, chemically alter, immobilise or neutralise a harmful substance (see Bridges, 1991). For example, soil contaminants can be rendered immobile by using binding agents such as lime silicates or organic substances that are injected into contaminated soils or applied after the soil is removed. Point sources of reactive metals like mercury can be immobilised by encouraging them to react with other adsorbent compounds like activated charcoal, organic substances such as wool, peat, hair and even chicken feathers, which can remove up to 90 per cent of mercury compounds within 24 hours of contact. Water-based mercury can also be removed using chemical techniques, such as hydrous ferric oxides and sodium sulphide, which chemically react to precipitate mercury out of a solution (Veiga *et al.*, 1995).

The addition of lime into lakes can combat against acidification by immobilisation or neutralisation. Liming can also reduce the release of elements like mercury, but it is more commonly associated as a method of ameliorating lake acidifcation. The addition of basic cations to lake waters neutralises acidic pollutants and prevents the lake from acidifying. Lime is added to lakes by dumping powered limestone from a lorry, boat or helicopter; the amount of lime needed is dependent upon the size of the lake, especially the volume of water, its pH and the rate at which the water is turned over in the lake (Woodin and Skiba, 1990).

The practicality of liming has been questioned, not least because it needs to be replenished regularly to prevent acidification. Since 1982 Sweden has been adding lime to approximately 5500 lakes at an annual cost of SKR 100 million every year (Van der Straaten, 1996). Many Scottish lochs have a short turnover period, which means additional liming is also necessary to prevent the water reverting back to its acidic status. Woodin and Skiba (1990) report that only four out of 39 lakes treated with lime remained above pH 6 for longer than a year and 18 sites did not respond to liming at all. Individual liming treatments have been effective, however. In Ontario, Canada, reduced smelter emissions and liming have reversed the acidification of lakes in the area (Woodin and Skiba, 1990). The use of such techniques is, however, limited because they are expensive, produce indifferent results and only offer a short-term solution.

As well as dealing with environmental impacts technology can be used to alleviate other environmental concerns. For example, biotechnological developments have pushed agriculture towards higher productivity, with crop yields per hectare increasing dramatically this century (Brown, 1998). The so-called 'green revolution' of the 1960s and 1970s demonstrated how food production could be increased using biotechnology (see Box 3.3). New biotechnological methods could minimise the impact of insects, disease, drought, pollution and acidity on crop production and promote sustainable and high-yield agriculture. Biotechnology has developed hybrid, high-yield varieties and virus-resistant plants. For example, the resistance of the plant to pests can be improved by isolating the gene that produces a protein that is toxic to leaf-eating caterpillars from a rod-shaped bacterium, *Bacillus thuringiensis,* and transplanting it into the plant (Plucknett and Winkelmann, 1995). The potential to genetically engineer food is really still in its formative stages but the technique offers tremendous possibilities to produce more food and fight disease and pests. However, with advances in cloning and concern over the safety of genetically modified food, there are moral and ethical issues that need to be addressed (see Box 1.2).

However, for technology to provide the maximum social benefits, incentives are needed to convince people to adopt new technology. There is no point providing a subsistence farmer with a tractor if the fuel to run it or replacement parts are too expensive (a fault of the 'green revolution'), but providing appropriate technology in a form that is user-friendly and cheap can be effective. In the underdeveloped world the introduction of 'appropriate technology' has been one option (also see Chapters 1 and 5). For example, efficient cookstove designs can help to combat deforestation, soil erosion and greenhouse gas emissions, especially when it is remembered that nearly six billion people cook food and heat their homes with traditional fuels, especially wood and charcoal. One example is the ceramic jiko, a metal stove with a ceramic interior that is 25–40 per cent fuel efficient. Approximately one million are now used in countries such as Kenya, replacing old metal stoves that were comparatively fuel inefficient at 10–20 per cent. The increased efficiency of the new jiko stove can save an average household 1300 pounds of fuel or $65, the equivalent of one-fifth of an average annual salary for urban dwellers (Kammen, 1995).

Appropriate technology can also help to improve food production. There are millions of people who practise unsustainable agriculture, and future problems may be most acute in developing countries. Examples include slash and burn agriculture in the tropical rainforests of Brazil, Malaysia, Indonesia and Latin America (see Sponsel *et al.,* 1996). To counter the negative impacts of certain forms of agriculture on biodiversity and soils, Brookfield and Padoch (1994) argue that more attention should be paid to the diversity and dynamism of traditional farming. For example, results from 106 'sustainable programmes' under way in 26 countries indicate that food production and rural economies can be improved by promoting appropriate, environmentally friendly and sustainable forms of agriculture. Reij *et al.* (1996) and Pretty (1995) argue that after the failure of mechanised forms of agriculture to sustain food supplies and the soil, sustainable agriculture programmes, which use a combination of indigenous, traditional agricultural practices and soil and water conservation techniques, may be more effective. Pretty *et al.* (1995), Syers *et al.* (1996) and Toulmin (1997) argue that the problems of land degradation and food production can be addressed, and recommend various strategies to promote sustainable land management and agriculture using case studies. Mabogunje (1995) also describes how farmers have invested in their land and protected it from environmental degradation in several regions of Africa with a relatively dense human population.

Examples of sustainable agricultural programmes include the New Forests Project, which is helping subsistence-level farmers utilise their land without creating soil erosion and destroying the forests. Simple changes to the methods of farming, including contour farming and planting of alley crops between rows of trees, allow the land to be farmed continuously and generate higher yields. Yields of beans and corn tripled in four years with the introduction of soil conservation techniques east of Gaimaca in Central Honduras (Pye-Smith, 1997). Confinement rearing prevents livestock from damaging the land by using forage trees in restricted areas. Family woodlots containing fast-growing tree species produce fire wood, and watershed improvement strategies help maintain the longevity of springs (New Forests Project, 1991). Soil and water conservation methods employed in the district of Machakos in Kenya demonstrate that smallholders can adapt their agricultural practices successfully in response to socio-economic conditions such as high population growth. A fivefold increase in population has been matched by improved land management in an area of unreliable rainfall: 95 per cent of arable land has been terraced to minimise soil loss and, combined with stream diversions, to improve water conservation. Areas of land previously abandoned due to environmental deterioration are now productive in a district that supports 1.5 million people (Brookfield and Padoch, 1994).

Brookfield and Padoch (1994) and Thompson and Hinchcliffe (1997) argue that if policies are introduced to promote successful forms of sustainable agriculture in developing countries, using appropriate technology, the problems that impede food production and promote environmental degradation encountered by farmers can be alleviated, and farmers can then concentrate on conserving their resources. Political decisions, education, land reform, diversification of farm economies and local scale investment are needed to revitalise food production in underdeveloped countries. If efforts such as those briefly summarised here work and are widely adopted, Brookfield and Padoch (1994) argue that food production using appropriate, traditional, subsistence level agriculture could keep pace with population growth in developing countries.

4.2.3 Regulative methods

Regulative methods are a popular way of managing the environment and resources. Regulation of environmentally poor practice occurs in numerous ways, including the adoption of hard laws, soft laws, voluntary practices, taxes, bans, fines and litigation.

As will be discussed in Chapter 5, notions of environmental issues rose to prominence in the international political arena in the early 1970s. Globalisation of managing environmental problems evolved with the realisation that some environmental problems were too big to be solely managed under the auspices of individual countries. Prior to the 1960s environmental problems such as water pollution were visible and localised, therefore they were amenable to clean up and did not necessitate international co-operation. Since then the view has shifted – it has become more widely recognised that environmental problems are widespread and cut across national boundaries. French (1994) claims that between the 1960s and the 1990s over 170 international treaties have been drafted in order to protect the environment. There have been several success stories as countries abide by hard laws, including the cleaning up of the River Rhine (see Malle, 1996) and attempts to protect the ozone layer, but there have also been some notable failures, such as the moratorium on whale hunting and efforts to curb greenhouse gas emissions.

Hard laws

O'Riordan (1995a) describes hard laws as a basic set of principles, obligations and rules that bind behaviour. More often than not hard laws are international and are used to protect the interests of all nations. Hard laws are usually created by means of protocols, treaties or conventions that impose mandatory obligations on countries.

For a hard law or agreement to be successful Mitchell (1995) proposes that successful treaties need to be formulated to give enforcers and users the political and economic incentives, the practical ability and the legal authority to monitor and adhere to any agreement. To do this, Mitchell (1995) argues that treaties must target those parts of any industry that are most susceptible to alter their behaviour. For example, the international oil pollution treaty 'MARPOL' was effective at curbing oil discharges from tankers by introducing equipment standards, such as the installation of segregated ballast tanks, that help prevent oil pollution from tankers. Imposing discharge limits proved less successful because it is not easy to regulate discharges from every tanker, especially if they are at sea. Mitchell (1995) argues the former treaty was most effective because tanker constructors had the practical ability to install equipment and it is required for the correct classification and insurance papers needed to trade, which forces tanker owners into action; whereas it is much more difficult to monitor or provide incentives to stop tanker captains discharging waste oil at sea. Mitchell (1995) also argues that policies that work target the part of industry most likely to alter its behaviour (owners rather than operators), because it is most susceptible to monitoring and enforcement. For example, more effective enforcement of oil tankers was possible when a treaty legalised daily reporting that produced an accessible database that can identify ships that consistently violate the treaties. If tanker operators fail to adhere to the treaty it could lead to bad publicity, economic boycotts or the tanker could be detained, and this threat provides sufficient incentives for good behaviour and compliance of the treaty (Mitchell, 1995).

French (1994) argues that trade incentives can secure compliance to a hard law. For example, the Rhine Action Programme, designed to clean up the river, worked because the countries benefit from supplies of drinking water, uncontaminated sediments and a pollution-free North Sea. Lessons learnt from the Rhine are now being used to reclaim other European rivers, including the Elbe, the Volga and the Danube (Malle, 1996).

Agreements work when there is political, scientific and public acceptability. One example is the 1987 Montreal Protocol, signed by some 140 nations that committed themselves to lowering their countries' emissions of CFCs following the rising scientific expressions of concern about ozone depletion (see Chapters 1 and 3). Amendments to the Protocol were agreed in London in 1992, Copenhagen in 1994 and Vienna in 1995 and, by the end of 1994, 148 countries had ratified the Montreal Protocol, 100 the London amendments and 37 the Copenhagen amendments (Haas, 1991; Parson and Greene, 1995). The obligations drawn up in Copenhagen are summarised in Table 4.1.

Table 4.1 Timetable for phasing out ozone-depleting chemicals

Substance	Year	Montreal Protocol
Halons	1994	100% phase-out of production
CFCs, CCl_4, CH_3CCl_3	1996	100% phase-out (*phase-out of CFCs and CCl_4 by 1995 in EC*)
HBFCs	1996	100% phase-out
HCFCs	1996	Freeze on calculated consumption at 2.8% of CFC consumption in 1989 plus total HCFC consumption in 1989 (*calculated at 2.6% of CFC consumption in EC*) 2020 Phase-out, with a 0.5% tail until 2030 to service existing equipment (*phase-out by 2015 in the EC*)
CH_3Br	1995	Freeze on production and consumption at 1991 levels
	2001	25% reduction from the above (*25% reduction by 1998 in the EC*)
	2005	50% reduction
	2010	Phase-out, with possibile exemption for critical agricultural uses

Source: SORG, 1996)

According to Haas (1991, p.234) the Montreal Protocol offers a blueprint for future agreements and a new paradigm for international co-operation because it combines the views of government, industry and science to create a realistic framework for the phasing out of ozone-depleting chemicals. Haas (1991) argues that the success of the Montreal Protocol was driven by the interplay of three forces. First, the power of knowledge, in that a global network of scientists agreed on the causes of ozone depletion, the role of CFCs and the policies needed to solve the problem. Such seemingly almost universal causal beliefs within the scientific community were crucial for political acceptance of the dangers of stratospheric ozone loss and led to a spate of legislation to curb CFC emissions in the USA and Europe. The second key criterion is political support to protect the ozone layer. Finally, there is the technology available to market CFC alternatives. Du Pont, a leading manufacturer of CFCs, came to support CFC controls as it believed it could gain a competitive edge over its rivals by developing alternatives (but see Chapter 5). Thus there was industrial agreement that CFCs should be phased out. This meant that the Protocol does not require countries to substantially alter existing industrial practices or necessitate economic and political changes. Limited global opposition, therefore, greatly facilitated the relative ease with which an agreement to protect the ozone layer was reached.

The Montreal Protocol is also flexible in accepting that CFC use is needed in essential areas such as medical applications; in addition, some underdeveloped countries were given a ten-year exemption for compliance. Otherwise, loopholes were avoided and the Protocol adopted strict rules. Innovative voting rules meant that recalcitrant nations were forced into abiding by majority voting and, in certain cases, dissatisfied nations could not withdraw from the agreement (French, 1994).

However, political and economic obstacles may still delay the phasing out of ozone-destroying chemicals. These include, first, the discovery of new anthropogenically derived ozone depleters that will require the Protocol to be amended. For example, in 1994 the production of halons was halted and a decision was made to phase out the use of methyl bromide by 2010. Second, the financial costs of implementing the protocols may deter countries from fulfilling their obligations, especially if more ozone depleting chemicals are discovered. A third issue is funding, in that although $750 million has been pledged to implement the Protocol and help underdeveloped countries phase out CFCs only $226 million had been collected by 1991, and countries such as France and the USA insist on paying their contributions using promissory notes, which complicate administrative procedures. Underdeveloped countries are expected to phase out CFCs, carbon tetrachloride and methyl chloroform by 2015, but they will need financial assistance to do so (see Kerr, 1996a; Parson and Greene, 1995; French, 1994; SORG, 1996).

So, a protocol must be realistic to prevent non-compliance. Victor and Salt (1995) suggest that five key procedures should be followed:

(1) ensure flexibility over targets and timetables;
(2) collect the necessary data and allow sufficient time for analysis and political consideration;
(3) increase use of national reports to allow data comparison;
(4) develop a secretariat to review the progress of all countries;
(5) evolve a consultation process to analyse the effectiveness of policies.

Mutual coercion through public accountability may also make agreements work. Reports, explaining how the aims of an agreement are fulfilled, are sometimes used to try to enforce a treaty. Some nations have been caught out. For example, the United Kingdom was alleged to have violated the spirit of the 1972 Oslo Convention on Ocean Dumping by discharging coal ash into the North Sea after information was made available to public pressure groups like Greenpeace (French, 1994), and public 'outings' of poor environmental behaviour may force law breakers to abide by regulations.

There is evidence that hard laws can be effective. For example, first evidence that the restriction on the emission of chlorine-based chemicals to protect the ozone layer was beginning to work was reported in 1990, when the rise in atmospheric chlorine began to decline. Butler *et al.* (1992) and Russell *et al.* (1996) have argued that the growth rates of CFC-11 and CFC-12 have declined significantly since their phase out. Growth rates of other halocarbons (eg H-1301, H-1211, CH_3CCl_3) are also falling, and Montzka *et al.* (1996) suggest that anthropogenically derived tropospheric chlorine peaked in early 1994 and was decreasing by 25±5 parts per trillion by the middle of 1995. However, the recovery of the ozone layer will take much longer, as ozone-destroying catalysts remain resident in the troposphere before being transported to the stratosphere. Computer modelling of the fluctuation of atmospheric chlorine concentration suggests that if the agreement holds ozone should recover to the 1979 level by around 2050 (Kerr, 1996).

Although attempts to control acid rain pollution have been made with varying degrees of enthusiasm by fossil fuel burning countries like the UK (see Chapter 1; Dudley, 1986), efforts to lower emissions have been relatively successful. Since the introduction of a series of protocols sulphur dioxide emissions have been reduced by 50 per cent in Europe and by one-third in North America (see Wright and Hauhs, 1991; Bishop and Hultberg, 1995; ApSimon and Warren, 1996; McCormick, 1998; and Table 4.2). Lower emissions appear to have resulted in lower sulphate levels in precipitation and pH reversals in lakes in acid rain affected regions (Dillon *et al.*, 1986, 1987; Allott *et al.*, 1992) but the recovery of terrestrial ecosystems is much slower (Wright and Hauhs, 1991; Bishop and Hultberg, 1995; Likens *et al.*, 1996). However, Van der Straaten (1996) argues that acid rain policies have been less effective in the Netherlands where the Dutch government is under pressure from industry to relax emission controls and he argues changing industrial practices or economic recession is responsible for lower emissions in Europe.

Soft laws

O'Riordan (1995a, d) describes soft laws as agreements that are normally created by custom and framework conventions that are based on more ambiguous wording when compared with hard laws. O'Riordan (1995d) argues that soft laws can be interpreted in a very flexible way and that, despite their ambiguity, they contain an element of calculated persuasion that some lawyers refer to as 'legal custom': countries support broad agreements even though they are not legally binding. He goes on to argue that soft laws are of great importance in managing areas of the global commons: (i) in areas where there is scientific doubt and a need to regularly update obligations; (ii) because they allow freedom of action whereby an individual nation

Table 4.2 Changes in sulphur dioxide and nitrogen oxide emissions between 1980 and 1993 for selected countries

Country	SO_2 emissions			NO_x emissions		
	1980	1993	% change	1980	1993	% change
Canada	4614	3042	-34	1959	1952	0
USA	23779	20621	-13	18672	18217	-2
Austria	397	71	-82	246	182	-26
Sweden	507	103	-80	424	391	-8
Germany	7486	3896	-48	3440	29904	-16
UK	4898	3069	-37	2392	2752	+15
Spain	3319	2200	-34	950	1257	+32
Greece	400	510	+28	306	306	0
Czech Republic	2257	1419	-37	937	574	-39
Poland	4100	2725	-34	1500	1140	-24
Russia (European)	7161	3456	-52	369	443	+20
Norway	142	37	-74	186	225	+21
Croatia	150	180	+20	60	83	+31
Totals						
North America	28393	23663	-17	20631	20169	-2
European Union	27009	13795	-49	12885	13188	+2

(Source: McCormick, 1998)

can decide how to meet an agreed target; (iii) because they can aid 'social learning' and help countries reinterpret and adapt from an earlier negotiating position; and (iv) because the ambiguous wording of many soft laws can be used to 'save face' and get a broad consensus. Examples are 'Agenda 21' and the 'UN Framework Convention on Climate Change' that both emerged from the so-called Earth Summit in 1992 (see Chapter 5) urging nations to adopt sustainable development and stabilise greenhouse gas emissions at all levels of organised society, from global treaties down to individual action. O'Riordan (1995d, p. 351) argues that the Climate Change framework was accepted by some countries like the USA only because of its ambiguous wording and the flexibility it gives countries to respond to their obligations. For example, the Convention aims to stabilise greenhouse gas emissions to levels that 'would prevent dangerous anthropogenic interference with the climate system' and 'this should be adopted in a time frame to allow ecosystems to adapt naturally'. To overcome the ability of individual nations to reduce emissions the framework states that 'action by each nation should be carried out in an equitable manner, according to historic responsibilitiy, state of development and capacity to respond'. This allows individual nations to mitigate against greenhouse gas emissions 'using methods that are most appropriate to their country', such as incorporating rice straw into paddy fields to reduce methane, carbon sequestration, fuel switches and increased efficiency (see, for example, Dixon *et al.*, 1996; Bai and Wei, 1996; Dabas and Bhatia, 1996; Schneider, 1996; Houghton, 1996; Reid and Goldemberg, 1998; Yong-Kwang Shin *et al.*, 1996). Although this could be deemed to be fair, O'Riordan (1995d) argues that it is often seen as a 'cop out' by some nations, especially the USA, which has accused underdeveloped countries of not fulfilling their obligations. The United States government has consistently argued that underdeveloped countries should be made to do more to reduce their greenhouse gas emissions because they already account for 60 per cent of global sources of methane and 40 per cent of carbon dioxide (Houghton *et al.*, 1992) and these figures are expected to increase dramatically. The USA has used this argument to stall on fulfilling its own obligations, and this limits the effectiveness of soft laws as countries negotiate rather ineffectual emission targets. Fossil fuel burning countries also cite the uncertainty of global warming predictions and the fact that the contribution of anthropogenic greenhouse gas emissions to global warming is unknown (see Anon, 1992; 1995a; 1995b). Because of the uncertainty surrounding future climate change it is difficult to convince the public and politicians that it is in their own interests to take action (Anon, 1997b).

Voluntary codes of environmental practice

Voluntary codes of environmental practice are a popular form of management. Businesses adopt non-regulatory environmental standards and codes of practice to arrest environmental degradation, using sustainable manufacturing and production methods, and thus avoid the need for governments to place a heavy financial burden and restrictive working practices upon industry through national legislation. Examples include the Coalition for Environmentally Responsible Economies' Principles (CERES) (see Box 4.1), the Responsible Care Code in the United States (Nash and Ehrenfeld, 1996b), the Eco-auditing and Management Scheme (EMAS) (O'Riordan, 1995), eco-labelling in the European Community (Collins, 1995) and the creation by the American Petroleum Institute (1992) of Strategies for Today's Environmental Partnership (STEP) with the goal of improving environmental, health and safety performance in the petroleum industry while portraying a positive image as an energy provider.

Nash and Ehrenfeld (1996) outline the main advantages of these green codes:

(1) Shift costs from the public to the private sector.
(2) Avoid using public resources that can be better employed elsewhere.
(3) Encourage self-regulation. For example, EMAS in Europe is soon to become mandatory (O'Riordan, 1995b), but if companies demonstrate that they can act responsibly while the scheme is voluntary the European Community may reconsider.
(4) Designed to be flexible. Firms can introduce environmentally safe working methods without over-compromising their business interests. A firm can tailor environmental management strategies to suit its individual needs.
(5) Build up public confidence. If industry is seen to be cleaning up its act this will improve relations with the public and environmental organisations. In the long term this will result in less distrust of industry and less legal action against companies for poor environmental performance.

Well-known companies like Volvo, Volkswagen, ICI and Shell have started to pay more attention to their environmental image (see Chapter 5) and have chosen to adopt self-regulatory practices such as releasing annual environmental reports, formulating environmental policies and creating advisory councils (O'Riordan, 1995b).

Box 4.1 CERES – a green code of practice at work

CERES is a voluntary environmental code of practice that has attracted support from industry. It was developed by non-industry groups that designed a series of environmentally safe working practices that companies agree to adhere to or keep within the spirit of. They encourage industry to monitor their environmental performance and disclose their achievements (or lack of them) to the public. The wording of the CERES code of practice is persuasive in nature rather than forcing companies to comply with regulations.

Codes of practice, including CERES, generally ask firms to address particular issues such as those listed below:

(1) **To assess their environmental problems**
CERES does not require firms to assess hazards but encourages them to disclose data on chemical use, waste generation and resource consumption.

(2) **To establish environmental goals and targets**
CERES asks firms to submit information on how they will reduce greenhouse gas emissions, ozone-destroying chemicals, hazardous wastes and chemicals.

(3) **To measure key activities systems**
Companies measure their key activities that have an environmental impact on a regular basis. CERES does not require this to be done.

(4) **To provide training**
Some green codes encourage industry to make sure that employees receive the appropriate training. CERES asks firms to document worker training programmes.

(5) **To self-audit**
Companies that adopt the CERES principles must explain how they implement them.

(6) **To recognise worker performance**
Companies must explain how they recognise and reward outstanding environmental behaviour.

(7) **Verification**
Companies must provide an annual report describing how they have complied with the CERES principles and key issues. A public report introduces an element of accountability and the findings can be challenged.

Moreover, the CERES code of practice encourages companies to eliminate harmful products and provides consumer advice to prevent unsafe use of their products. There is also a clear statement on sustainable working practices, insisting that 'corporations must not compromise the ability of future generations to sustain themselves'. Companies are also advised to use renewable resources and energy systems where possible and to conserve non-renewable resources. By doing so companies are seen to be able to promote an environmentally friendly image, remove the distrust of the general public and environmental groups over the motives behind corporate environmentalism. Several major companies have ▶

backed the CERES code of practice, including the Sun Company, General Motors and Polaroid. Some of the supposed advantages of green codes over regulations are listed in Table B4.1

Table B4.1 The advantages of green codes compared with other regulations

Regulation	Private Codes
Instituted by government	Instituted by private sector
Enforced by government	Enforced by firms themselves with some third-party verification
Compliance mandatory, with direct sanctions	Compliance voluntary, with indirect sanctions such as peer pressure
Largely medium specific (air, water etc)	Integrative, focusing on life-cycle impacts 'beyond the fence line'
Places emphasis on product and process standards	Places emphasis on management systems
Defines standards for emissions or technology	Lets each firm define own performance, with requirement for continuous improvement
Provides public access to information on compliance	Provides public access to information only in select cases

Source: Nash and Ehrenfeld, 1996, *Environment*, 38(1).

Ecotaxation

Ecotaxation is a form of taxation imposed on activities that are resource depleting and environmentally destructive (O'Riordan, 1997a). Three types of ecotaxation exist according to S. Smith (1997, p.23):

(1) Measured emission taxes, which are taxes paid as a result of metered or measured quantities of a pollutant.
(2) A tax similar to value-added taxation, which adds an additional cost to the price of a product that may be environmentally damaging.
(3) Non-incentive taxes, which are used to generate revenue that can be used for environmental monitoring and abatement schemes.

Turner (1995) argues that ecotaxes are possibly the most efficient way of achieving some predetermined level of of environmental quality, and argues that taxes have several advantages when compared with other types of regulation:

(1) Taxes are a more cost-effective method.
(2) They incorporate a dynamic incentive for polluters to use resources more efficiently, lower consumption rates and reduce the emission of pollutants.
(3) The money raised can be used to replace existing stocks of resources or repair any damage caused by pollution.

(4) Ecotaxes can be introduced to control a wide range of activities, including waste minimisation and recycling through a landfill tax (Powell and Craighill, 1997), efficient use of water (Herrington, 1997), energy generation and forest use (Schneider, 1996), and can be used in combination with other schemes. For example, ecotaxes have been called for to reduce food surpluses in the European Community alongside set-aside schemes, quotas and reducing pricing support for arms production. Clunies-Ross (1993) argues that ecotaxes could also lessen the use of mineral fertilisers and pesticides to lower food production and protect the environment from cultural eutrophication and pesticides, but Pan and Hodge (1994) argue in favour of land use permits.

Taxation is one strategy that could be used to meet greenhouse gas emission targets agreed under the United Nations Framework Convention on Climate Change in 1995. As 75 per cent of current emissions are produced by fossil fuel comsumption, economists propose that the carbon content of fossil fuels should be taxed (Barker, 1995; Muller, 1996), and Loske (1991) argues that a carbon tax will encourage more efficient energy use and stimulate the development of zero emission technology.

Barker (1995; 1997) and Muller (1996) argue that fiscal, economic, political, social and environmental considerations may render taxation less effective. For example, politicans must set a reasonable price to pay. As O'Riordan (1995c, p.5) argues, if the price is too low the penalty will not have the desired effect and could result in further destruction because the implementation of the tax would appear to legitimise the environmentally damaging behaviour, and if the price is too high it might result 'in a real social cost in terms of excessively diverted investment towards environmental clean up where the net improvements would not be justified'.

Muller (1996) argues that it is unlikely that an ecotax will ever achieve the economically optimal reduction of greenhouse gas emissions because lobbying can weaken the political will to implement a tax that will induce the desired environmental benefits. For example, such action has already undermined energy taxes in the USA, where President Clinton proposed to introduce a broad-based energy tax – the US Btu tax – in 1993, based on the energy content of fuels, with higher rates on petroleum. However, only an amended version of the bill was passed by the House of Representatives in June 1993 after the tax was revised to provide special exemptions for oil used in refineries and coke in steel production. Despite the exemptions, the oil industry and the National Association of Manufacturing still mounted a vigorous publicity campaign warning of the threat of the tax on industry and employment, without any forceful counter-measures from environmentalists or the government. Finally, the bill was defeated in the Senate, which chose to put a small increase on the price of gasoline (Muller, 1996).

Fines and litigation

Fines and litigation can act as a deterrent to curb environmentally damaging practices. For example, in the United States the owners of five boats were fined $8.5 million after they were caught illegally fishing for scallops, cod and other fish in 1995 (Safina, 1995), and Canada sent out a clear message to other countries' fishermen who violate its fishing rights by arresting the crew of the Estai, a Spanish boat accused of over-exploiting halibut stocks, even though the boat was in international waters. Local inhabitants of Newfoundland are concerned about fish stocks, which has seriously reduced their own fishing fleet, while other nations, in this instance Spain and Portugal, continue to fish the area to the point of exhaustion (Usbourne, 1995; Cathcart, 1995).

Litigation against parties responsible for damaging the environment is now commonplace. Legal action has been used to attribute responsibility and to determine compensation after incidents of pollution such as the Bhopal chemical plant explosion (Morehouse, 1994) and the Exxon Valdez oil spill (Shaw and Bader, 1996). However, the results of legal action have created controversy, especially when there are no legal precedents. One major difficulty resulting from foreign investments has been apportioning blame. A case in point is the disaster at Bhopal, in Madhya Pradesh, India, where it was unclear which legal system should deal with any litigation. The choice was between the country that provided the technology (USA) and the host country (India). As a result of the legal wrangling and confusion the processing of compensation claims to people poisoned by the chemical fumes was extremely slow. In 1994 Union Carbide still refused to accept legal responsibility for the explosion and under 3000 compensation claims out of a total of 15000 had been dealt with by 1993 (Morehouse, 1994; Barrow, 1995).

The Exxon Valdez disaster also highlighted the complexities involved with legal action. Statutory legal requirements to clean up the oil and to calculate the extent of the damage have proved unsatisfactory. Shaw and Bader (1996), for example, claim it is extremely difficult to demonstrate 'resource injury' to the ecosystems of Prince William Sound because legally the investigators must compare the change to the environment with a natural environmental baseline. To do this it would be necessary to have studied the area before and after the oil spill. The lack of detailed scientific studies prior to the Exxon Valdez oil spill meant that the scientists had no environmental baseline to begin with, and once the oil pollutes a coastline it is impossible to reconstruct the health of the pre-spill ecosystem. To evaluate a dynamic environment such as a coastal ecosystem in this way is virtually impossible, because coastal environments undergo large variations in their natural state over a range of time-scales. Thus it is difficult to decide what constitutes an environmental baseline. Furthermore, the scientific studies were designed to meet litigation goals rather than to discover how the ecosystem changed *per se.* Conflicts then emerged between conducting a proper, independent scientific study and the collection of data used by the opposing legal parties, leading to independent observers questioning the validity of the findings and the legal process.

4.2.4 Efficiency

Chapter 3 discussed concerns of meeting our future energy needs and the problems of resource depletion. Over-production and consumption of resources and the need for large quantities of energy have been blamed for contributing to an environmental crisis. One way to avoid the limits to growth scenario is to lower rates of production and consumption by using resources more efficiently. Efficiency is a measure of how much useful energy is derived from the input of fuel, stated as a proportion of the energy input (Edge and Tovey, 1995, p.320). Hirst (1991) argues that the benefits of improving efficiency during the manufacture of a product, either by reducing the number of ingredients or by using less energy, include increased economic productivity and competitiveness, lower consumer prices, less reliance on imports, waste reduction and lower environmental impact and/or resource consumption rates. Examples include lowering production of wastes and pollutants by waste minimisation, reduction and recycling, and energy efficiency.

Waste reduction, minimisation and recycling

By the end of the 1980s governments were still slow to recognise the benefits of waste reduction and minimisation. Out of $16 billion allocated to environmental protection by the US government in 1986 only 25 per cent was used on waste reduction schemes (Oldenburg and Hirschhorn, 1987). Threats to human health, increased generation of hazardous waste, the cost and the lack of suitable sites for land disposal have gradually forced a change in attitude towards waste reduction and minimisation. Waste reduction is a term given to describe 'the practices that reduce, avoid or eliminate the generation of hazardous wastes or pollutants', whereas waste minimisation involves efforts to recycle or treat wastes once they have been produced and, therefore, combines the ethic of prevention with control (Oldenburg and Hirschhorn, 1987, p.17).

Oldenburg and Hirschhorn (1987) and Caincross (1994) argue that industry is developing ways of reducing the amount of waste produced during the manufacture of a product because such developments reduce costs, avoid the need to comply with regulations, lower risks to public health and increase their market competitiveness. For example, Exxon Chemicals and Polaroid have replaced organic solvents with water-based ones in their manufacturing process, while 3M saved $482 million over a 15-year period by altering its copper sheet cleaning process by using pumice scrubbers rather than a concoction of chemicals to reduce waste and lower the cost of disposing of hazardous chemicals. Waste minimisation schemes have also resulted in the reduction of by-products. For example, Duke Power's McGuire power station in North Carolina reduced waste by 80 per cent using techniques such as daily input reductions, recycling of liquid waste, inventory controls and product substitutions (Poteat, 1995; Devarakonda and Hickox, 1996).

Another way of reducing waste is through recycling or returning materials to their raw material components and then using these again to supplement or replace new materials in the manufacture of a new product (Carless, 1994, p.172). Recycling of items like aluminium cans, glass and paper converts rubbish into resources. Advantages of recycling are numerous: it eliminates waste for land disposal or incineration, it conserves resources and it can lead to savings in capital and energy consumption. For example Alcoa, an aluminium company, recycles aluminium cans using a plant build at one-tenth of the cost required for mining and refining bauxite. Production of recycled paper apparently uses between 23 and 74 per cent less energy, creates 74 per cent lower air pollution and 35 per cent water pollution compared with the production of virgin paper. According to Carless (1994) potentially 80 per cent of household waste could be recycled in the future and offer a serious alternative solution to waste disposal. Bleken and Bakken (1997) argue that recycling agricultural waste could reduce the amount of excess nitrogen in the agricultural system and improve the efficiency of transferring nitrogen through the various trophic levels of food production by approximately 30 Gg N/yr, while reducing atmospheric levels of nitrous oxide, a greenhouse gas in the troposphere and an ozone depleter in the stratosphere.

However, Van Voorst (1994), Rothbard and Rucker (1994) and Boerner and Chilton (1994) argue that recycling domestic waste has now become a victim of its own success. The collection of recyclable goods has outpaced the ability of local governments or industry to process them into resaleable products and this has led to stockpiles of paper, plastics and glass in the USA and Germany. A glut in recyclable products has lowered their market price and cuts profit, and scrap paper price fell from $120 per ton to $30 between 1988 and 1994 in the US. Finally, recycling schemes can be expensive to run.

However, despite the growing use of waste minimisation and recycling schemes, waste production levels are still rising in many countries; for example municipal waste generation in the US rose 105 per cent between 1960 and 1988 (Baumol and Oates, 1995). The effectiveness of recycling schemes has been questioned by deep green critics such as Porritt (1986) and Dobson (1990), who argue that they still encourage consumption and production, that recycling uses resources and energy and is an industrial activity, which therefore will not reverse the 'limits to growth' scenario (see also Chapter 5). Boerner and Chilton (1994) argue that state recycling schemes in the USA and Germany that mandate that the manufacturer of a product is responsible for recycling will also prove to be counter-effective because they do not address the true economics of recycling. They argue that a free market approach must be employed if societies are going to benefit from recycling because markets work most efficiently without state interference. Boerner and Chilton also argue that policy makers should look at all waste management schemes in terms of their efficiency and cost effectiveness, as alternatives may be a better option. For example, there is plenty of space to develop landfills in the USA and the environmental concerns connected with landfill are misguided. On the other hand, Beisner and Simon (1995) note that recycling and reuse of waste plays a major part in diminishing the environmental impact of ever-increasing levels of waste generation and the contribution of recycling increased by 96 per cent in the US between 1960 and 1988. They go on to point out that trends in pollution levels in the Great Lakes and drinking water quality have improved in the US and waste reduction methods, by implication, can make a difference.

Energy efficiency

The idea of efficient use of resources has been touted as a major management tool to help ensure the long-term availability of energy supplies. If McGowan (1991) is correct, and renewable energy sources will contribute only a small proportion of future energy demands (possibly no more than 6 per cent), then as Flavin and Durning (1991) argue energy efficiency measures may be the most effective solution to meeting future energy demand. However, Boyle (1989) argues that lowering energy use does not have to have a detrimental impact on industry. Since the early 1970s many countries have seen their GDP grow considerably without major increases in energy consumption. For example, Japan used 6 per cent less energy in 1989 compared with 1973 but its GDP grew by 46 per cent. Hirst (1991) argues that energy efficiency also could be the most effective way to reduce greenhouse gas emissions and acidic atmospheric pollutants.

Most predictions, however, suggest that energy demand will increase. Hirst (1991) argues that if current trends continue energy consumption in the USA would increase by 81 quadrillion British thermal units (Btu) to 102 quadrillion Btu by 2010, but if cost-effective energy efficiency improvements were adopted by residential, commercial, industrial and transportational sectors, the 2010 figure could be lowered to 88Btu. This represents a significant reduction in the predicted figures for future energy consumption and demonstrates that efficiency measures could work.

Edge and Tovey (1995) argue that measures to improve energy efficiency are many and can be applied at power stations both to improve the conversion of fossil fuel burning into electricity and in its end use. For example, replacing old energy consuming products with efficient ones, such as heating or air-conditioning units, fridges, washing and drying machines, can lower energy demand, and lighting and heating systems can be controlled by timers and thermostats that regulate temperature and light more efficiently (Boyle, 1989; Hirst, 1991).

Hirst (1991) argues that although many energy efficient products are more expensive they prove cheaper over the longer term. For example, Bevington and Rosenfeld (1990) have shown that good window design can lower heating requirements; although energy efficient windows cost 20–50 per cent more than single pane glass windows, that is recouped through energy savings within four years. Designing buildings with insulating materials can also reduce energy consumption. Smart materials (polymers that respond to external stimuli such as temperature by changing their characteristics) are being developed, and buildings constructed of these materials should be more energy efficient as they will respond to changing environmental conditions and regulate heating (Pickering and Owen, 1994). The development of low-emissivity windows, electronic ballasts and high efficiency supermarket refrigeration systems lead to primary energy savings of about 250 trillion Btu per year (or $1.5 billion in 1995) and net saving to consumers is estimated to be $10 billion over the lifetime of the technology (Geller and McGaraghan, 1998).

Boyle (1989) argues that energy efficient technology can lower consumption. For example, in 1989 the European Commission argued that member countries could save energy by a further 20 per cent, and if Britain switched to energy efficient appliances, lighting and motor vehicles it could lower its electricity demand by up to 70 per cent. Energy saving schemes and technology are not just confined to the developed world – underdeveloped countries can remove inefficiencies while producing energy. The Pakistan National Energy Conservation Centre, for example, estimates that the country could reduce its consumption of electricity by up to 30 per cent from 1989 to 2004 (Boyle, 1989).

However, energy efficiency opportunities are not being introduced for a variety of political, monetary and social reasons (see Grubb, 1998; Edge and Tovey, 1995) and environmental managers tackling the problems of energy generation (see

Chapter 3) have often focused more upon controlling pollution from conventional energy generating methods or by mitigating their effects through other forms of regulation such as energy taxes, rather than looking to energy efficiency improvements. Edge and Tovey (1995) and Flavin and Durning (1991), however, argue that encouraging efficiency may be the best long-term option.

Food production

Being efficient can be defined as being productive with the minimum of waste or effort. Making food production more efficient is one method to maintain food production (see Chapter 3). Advances in agricultural systems or the adoption of traditional ones may also improve food production and lower rates of environmental degradation. More efficient use of resources like water will help sustain agricultural production. In Israel, for example, highly efficient water irrigation systems have reduced the amount of water applied to crops by 36 per cent and tripled the area under irrigation. Creating conditions that provide incentives for farmers to adopt sustainable, efficient agriculture is also necessary. Rosegrant and Livernash (1996) suggest that agricultural production is influenced by macroeconomic fluctuations in trade, exchange rates, credit, prices and marketing. Improving the productivity of the rural poor by providing basic needs and increasing access to resources (credit, education, training) has been seen as a means of encouraging development of less environmentally damaging and sustainable agricultural systems. Policy reforms can work. The creation of water markets in Chile, for example, apparently improved water efficiency by 4 per cent between 1976 and 1992, which was enough to irrigate 264,000 hectares and enabled crop diversification. Policies make farmers use their resources efficiently and lead to soil and water conservation. Bender (1997) argues that food consumption and demand is a product of people's physiological requirements and dietary patterns. He suggests that public policies that encourage people to follow healthier diets would reduce pressure on food production and more efficient food delivery systems from the farm to the consumer would eliminate wastage. It is estimated that losses from end-use inefficiency equal 30–70 per cent of the actual amount of food consumed in some countries.

4.2.5 Substitution as a management strategy

Cornucopians argue that resources will never run out because alternatives and substitutes will be found to replace them. An example of substituting one product with another to alleviate resource scarcity is the use of biofuels instead of petroleum. Brazil increased production of biofuel from 900 million to 4.08 billion litres between 1973 and 1981, following the oil crisis of the early 1970s, saved over 200,000 barrels of gasoline per day and the industry created over 700,000 jobs (Reddy and Goldemberg, 1990). The use of alternative vehicle and power station fuels has other advantages. Gray and Alson (1989) argue that the widespread use of methanol would dramatically reduce the quantity of harmful chemicals being released into the atmosphere. For example, ground-level ozone could be reduced by up to 90 per cent and sulphur-rich coal could be used for methanol production rather than being burnt in fossil-fuel power stations – this would lower sulphur emissions, a major cause of acid rain, while natural gas can be converted into methanol and reduce methane emissions. Methanol is also cheaper when compared to other fuels such as ethanol and gasoline/petrol, and its use would also rule out the need to legislate against vehicle emissions, a solution that appears doomed to failure unless there is a significant shift away from the private automobile. The main disadvantage of using methanol as a vehicle fuel is that many of the processes that convert gas and coal into methanol are still in the early stages of development and therefore the costs of a fuel switch in the short term will be uneconomic while petroleum prices remain low. Several adjustments in car design and/or improvements in engine efficiency are also necessary because burning methanol produces half the energy compared to gasoline.

Renewable energy

The use of substitutes has also been seen as a method to limit environmental damage and resource scarcity over longer time-scales. One example that has received a lot of attention (as discussed in Chapter 3) is the concern expressed over the ability of society to meet future energy requirements (see for example Edge and Tovey, 1995; Flavin and Durning, 1991). These people argue that the availability of oil and gas may fall over the short to medium term and that there are environmental problems associated with coal and nuclear power that may restrict their use. A dilemma exists about meeting future energy demands because people such as Hoagland (1995) have argued that world-wide demand for fuel is expected to increase by 30 per cent and demand for electricity by 265 per cent by 2025. It is not surprising, therefore, that interest in developing alternative energy sources has increased. However, there is some concern as to whether alternative energy sources can produce sufficient amounts of energy at a viable economic rate and at what environmental cost.

At present, the contribution of renewable energy only accounts for a minor percentage of the total amount of energy produced, but recent evidence suggests that certain types of renewable energy (wind power, geothermal energy, hydroelectric power (HEP)) are on the verge of making a larger contribution to the world energy market because technical innovations have improved their efficiency over the last 20 years. For example, Sims (1991) and Mackay and Probert (1996) argue that hydropower is already the most successful renewable energy source, accounting for 20 per cent of the world's electricity consumption, and it could provide much more energy – especially in underdeveloped countries that utilise only 8 per cent of potential capacity. Humans are prodigious dam builders, with 36,000 dams over 15m high in operation world-wide (Vörösmarty *et al.*, 1997), and substantial development of HEP has occurred in western Europe and North America, which exploit 98 and 83 per cent of their potential capacity respectively.

Coles and Taylor (1993) argue that wind power will provide a significant amount of energy in the future. At present, it provides only 1 per cent of the world's total and its use is concentrated within certain countries, especially the United States and Denmark, but more countries, including the UK, China, India, Egypt, Germany and The Netherlands, are beginning to realise its potential, and the EC expects wind power to supply 10 per cent of Europe's electricity by the year 2030. Trainer (1995), however, argues that many areas are unsuitable for wind power (wind speeds are too low), including many parts of Asia and Africa, so wind power, like many of its renewable counterparts, is likely to contribute energy on only a regional scale unless cheap, reliable methods of storing and transporting energy over long distances can be developed.

DiPippo (1991) and Dickson and Fanelli (1994) argue that countries located in areas with regions of above normal geothermal gradients could derive a substantial proportion of their energy from geothermal energy sources, but only less than 1 per cent of possible production in 1988 was utilised. For example, hydropower and geothermal energy now provide approximately 13 per cent of the Philippines and 12.8 per cent of El Salvador's total electrical capacity, and 72 per cent of Iceland's gross energy consumption, which has reduced fossil fuel imports (Mackay and Probert, 1996). McGowan (1991) argues that if other countries, including the UK, Mexico and Hungary, exploit geothermal energy sources the total amount of global installed geothermoelectric capacity could nearly double by the year 2000.

Sampson *et al.* (1993), Hall *et al.* (1993) and Hall and House (1995) argue that biomass burning, especially wood for fuel, is already a major energy source in underdeveloped countries, accounting for approximately 14–15 per cent of world energy use. Biomass refers to all forms of plant-based material including wood, sugar cane, crop and forestry residues and dung (Sampson *et al.*, 1993; Hall and House, 1995). They argue that burning biomass is a viable option for energy generation in developed countries as agriculture, industry and domestic rubbish all provide burnable materials. For example, Denmark generates energy from 75 per cent of its available wastes and Oliver *et al.* (1991) argue that biomass burning could reduce UK fossil fuel demand by 10 per cent. Moreover, Hall and House (1995) argue that 10 per cent of usable land in Europe could be given over to biomass production, especially ex-agricultural land, and if this potential is realised biomass energy in OECD Europe could account for 17–30 per cent of the total amount of energy produced. Pimentel *et al.* (1994) estimate that approximately one-third of the 32 billion tonnes of biomass growth per year in the USA could be converted into energy, supplying about 27 per cent of the country's total energy use. However, Trainer (1995) doubts the potential of biomass generating sufficient energy to meet the demands of developed countries and argues that some estimates are highly optimistic. Shea (1988) agrees with this assessment, arguing that energy derived from biomass taken from forest, crop and animal wastes would provide less than 5 per cent of energy for western countries such as Germany. Therefore exploitation of such biomass sources would not be a long-term solution to meet future energy needs.

On the basis of the statistics presented here renewable energy could, albeit regionally, make a significant contribution to world energy supply, but several forms of renewable energy are still in the early stages of development and do not offer a short-term solution. McGowan (1991) argues that renewables will only provide a minor proportion of people's energy requirements because its development is blocked by a range of barriers that are discussed in the next section.

Barriers to renewable energy

Expansion of renewable energy is taking place because obstacles, namely government policies, technology, cost, environmental impacts and planning, are slowly being overcome. All renewable energy sources are, however, constrained by their ability to integrate into existing energy systems on a viable commercial basis.

Research and development

McGowan (1991) argues that the relatively low level of funding for renewable energy has restricted its development as governments and power generating companies have opted for the cheapest, most reliable method of energy production, which limits opportunities for investment in alternatives. Not surprisingly the success of renewable energy has been largely dependent upon the approach adopted by individual countries. Street and Miles (1996) argue that the uptake of renewable energy is dependent upon favourable government attitudes towards alternative energy. For example, from 1978 the United States has provided tax incentives to power generating utilities to develop wind power, which resulted in the installation of 1800MW within six years. Cumulative investment in wind power now exceeds $2000 million and the energy produced by wind turbines saw a capital return of approximately $100 million/year in 1988 (Grubb, 1988). The Danish government also adopted an interventionist energy policy to promote wind power, by giving a capital subsidy for each newly constructed wind turbine and introducing a levy on conventional energy supplies (Danielsen, 1994; Street and Miles, 1996). Governments that promote renewable energy policies are beginning to see a return on their investment whereas other countries, such as Britain, that have not been as enthusiastic are lagging behind, and this is reflected by the relatively constant proportion of research and development money allocated to develop alternative energy sources: roughly 7–13 per cent of the total International Energy Agency (IEA) budget.

Cross (1993) argues that national energy policies are rigged in favour of conventional fuel sources for a variety of reasons including safeguarding national security, protecting employment, preventing the abuse by monopolies and revolts against governments. Such protectionism, especially subsidies (see Roodman, 1997; Edwards, 1997), further limits opportunities for renewable energy. If subsidies are removed and external costs are added to conventional fuel sources renewable forms of energy may then become the cheapest source of energy (Kelly, 1997). Arguably, without a free energy market it is difficult to judge which energy sources are most profitable and efficient. To achieve economic competitiveness has proved difficult because construction and operating costs of renewables are too high, especially when the technology is in its formative stages. Another problem is converting renewable energy into electricity efficiently. Hydropower has a 90 per cent efficiency conversion rate but solar power and geothermal energy are low at only 5–15 per cent (Lund, 1990; Trainer, 1995).

Costs

Costs have played a major role in curbing the development of renewable energy. According to Simon (1991b) tidal power offers potential but it is considered to be too expensive and restricted to a limited number of suitable locations (Baker, 1991;

Oliver *et al.*, 1991). For example, a tidal barrage across the River Severn could generate 8000 megawatts of electricity but would cost an estimated £8200 million (Ross, 1990). Others now claim certain forms of renewable energy are economically competitive. For example Grubb (1988) and Coles and Taylor (1993) claim wind power is already economically superior to nuclear power, while Charters (1991) believes that electricity generated from solar power could be sold at competitive prices (8–10 cents/kWh) in semi-arid and arid areas and Hall and House (1995) argue that electricity produced by biomass burning compares well with fossil fuel combustion. However, if measures such as carbon taxes are introduced to curb pollution the cost of electricity generation using fossil fuels will probably increase, strengthening the position of renewable energy in the global energy market. Kelly (1997) reports that the Department of Trade and Industry (UK) believes wind power will cost 0.04per kWh by 2005 compared to current prices of 0.03 and 0.08–0.11per kWh for fossil fuels and nuclear power respectively, indicating how quickly renewables are becoming economically competitive.

Environmental benefits or problems?

Renewable energy sources have environmental and economic advantages over their conventional counterparts. Limiting greenhouse gas emissions is one of the main reasons for the current popularity of renewable energy. For example, McGowan (1991) argues that renewable energy could save in excess of 900 tonnes of carbon for every GWh of coal-fired power displaced; geothermal energy use in Iceland is expected to lower carbon dioxide emissions by between 1 and 2 per cent when compared with conventional fuel sources (Mackay and Probert, 1996); while Kelly (1997) argues that wind power in the UK already saves 400,000 tonnes of CO_2 per annum. Sampson *et al.* (1993) argue that biomass use could also reduce carbon dioxide emissions when substituted for coal at similar conversion efficiencies and lower the amount of methane produced by decomposing rubbish disposed in landfill sites. Hall and House (1995) claim carbon dioxide emissions will decrease (by up to 90–120 Mt compared to coal, 72–96 Mt to oil and 50–67 Mt to gas) if biomass energy is fully exploited in western Europe.

Table 4.3 Carbon dioxide and sulphur emissions from different energy sources

	Emission levels	
Energy source	CO_2 (kg/MWh)	S (kg/MWh)
Coal	1000	11
Oil	850	11
Gas	550	0.005
Geothermal	96–11	<6
Hydropower	0	0
Nuclear	<1	0
Solar	140–0	0

(Sources: Dickson and Fanelli, 1994; Mackay and Probert, 1996)

However, renewables are not entirely free from environmental problems. Oliver *et al.* (1991) and DiPippo (1991), for example, argue that biomass burning and geothermal plants still release air pollutants such as carbon dioxide, methane and hydrogen sulphide, but emission levels are much lower when compared with conventional fuel (Table 4.3). Rosa and Schaeffer (1995) have argued that an expansion of hydroelectric power could also result in an increase in greenhouse gas emissions. Using a global warming potential index model (see Lashof and Ahuja, 1990), they calculated that the release of greenhouse gases from decomposing vegetation submerged by HEP, especially methane, could be the same as fossil-fuelled power station emissions depending upon the size of the area flooded, its biomass and the lifetime of a dam.

Walker (1995b) and Di Pippo (1991) also argue geothermal energy plants release chemicals like arsenic, mercury and boron into surface groundwaters that could adversely affect aquatic fauna. Biomass production for energy generation is also criticised by Pimental *et al.* (1994) and Hall and House (1995), who argue that continued harvesting from biomass could severely deplete the soil of nutrients and create high rates of soil erosion, and that fertiliser application would be needed to sustain yields, which would be costly in both monetary and environmental terms as excessive fertiliser use causes eutrophication. Furthermore biomass monocultures and plantations might have an adverse effect on ecology and increase the chances of exotic species and pests disrupting native ecosystems. Degens *et al.* (1991) and Hall and House (1995) argue that the use of agrochemicals to increase biomass production could promote eutrophication in local water bodies, while dam construction can affect water quality and aquatic ecology. For example, Humborg *et al.* (1997) argue that dam construction on the Danube has altered the biochemistry of surface waters of the Black Sea by decreasing the dissolved silica concentration by 60 per cent. These changes could have an adverse impact on the ecology of rivers and waters entering coastal areas worldwide, as approximately 36,000 dams are in operation. For example, plant diversity dropped by 15 per cent on riverbanks and by 50 per cent near large storage reservoirs in Sweden (Anon, 1997a).

Walker (1995b) argues that HEP and tidal schemes can disrupt the hydrological cycle by altering river regimes. For example, tidal barriers can affect river flow and sedimentation patterns, disrupt wildlife (especially migratory birds) and accumulate pollutants behind the barrage (Baker, 1991), but HEP projects on the Kemijoki river in Finland have avoided major disruption to the river regime by carrying out extensive environmental impact studies prior to dam construction. After an initial decline in oxygen content river quality has actually improved, allowing the successful introduction of fish such as the peled whitefish.

Geothermal energy plants and wind turbines also cause noise pollution, although this is seen as a problem only when renewable energy power plants are located close to people. To reduce disturbance from noise in California wind farms are located well away from urban areas, while in Denmark the pro-wind energy attitude of people, combined with the use of smaller turbines that create less noise, has limited complaints (Grubb, 1988; Thayer and Hansen, 1988; Walker, 1995b; Munksgaard and Larsen, 1998). Legislation setting limits can negate the problem of noise but electromagnetic interference can occur if turbines scatter or block electromagnetic signals, interfering with TV and radio reception (Street and Miles, 1996).

In general renewable forms of energy also require larger amounts of land per unit of electricity. For example, Walker (1995a) argues that every megawatt of electricity produced by nuclear power occupies 630 m^2 of land, compared to solar power at 100,000 m^2, HEP at 265,000 m^2 and wind power at 1,700,000 m^2. However, DiPippo (1991) argues that geothermal energy plants utilise similar amounts of land per megawatt when compared to other energy sources, and Charters (1991) states that solar thermal and solar photovoltaic methods occupy a similar amount of land to a coal mine over a 30-year period. Trainer (1995) disagrees, and argues that solar power requires a large amount of land. Conflicts over land use for renewable energy and food production look increasingly likely if a growth of large-scale biomass energy projects continues and more land is required for biomass production. However, with food production peaking in the early 1990s and with global population rising, it is likely that extra land will be converted for agricultural purposes (see Manshard, 1985). Dam construction also involves submerging large areas of land. For example, the Narmada project in India will submerge over 550,000 hectares of land, and the Xingu Dams planned in the Amazon would flood a minimum of 26,000 km^2 but the total area could reach a staggering 250,000 km^2 (Hildyard, 1989)!

This brief summary shows that there are environmental problems connected with all renewable energies. However, it can be argued that environmental impacts are mainly small-scale in comparison with conventional fossil fuel sources.

Social impacts

Renewable energy also has a history of social impacts that make its development controversial, especially in underdeveloped countries (eg Alvares and Billorey, 1987; Roy, 1987; Pearce, 1991b). HEP schemes cause social problems by displacing hundreds of thousands of people, conflicting with local agriculture and other local needs and promoting water-borne diseases, especially in tropical areas (Walker, 1995b). For example, the Selingue Dam in Mali displaced about 12,000 people, and social problems emerged in the villages to which people were relocated. However, this is gradually being resolved, aided by the construction of new schools, and the local economy has benefited from new fishing grounds (Sims, 1991). McCully (1996) argues that many of the perceived benefits of dams never come to fruition. For example, the majority of the 4000 people displaced by the construction of the Hirakud multipurpose dam project in 1946 still await resettlement land. The construction of the Three Gorges Dam in China represents possibly the greatest social upheaval caused by dam developments (see Box 4.2).

Sims (1991) also argues that hydropower development in tropical countries can increase the incidence of diseases such as malaria, bilharzia, schistosomiasis and river blindness, which are prevalent and require public health measures to control. For example, schistosomiasis increased between 25 and 82 per cent in communities close to the Diama dam and irrigation schemes on the lower Senegal river in Senegal, West Africa, and outbreaks of malaria in Africa and Brazil occurred after dams and irrigation schemes were completed (McCully, 1996).

Box 4.2 Three Gorges Dam project

Scheduled for completion in 2009, the Three Gorges Dam project will be the largest hydroelectric generating capacity in the world. Authorised by the Chinese government the dam will create a reservoir 370 km long at a cost of between $17 and $75 billion. The aim of this venture is to enhance the economic prosperity of an area historically renowned for its poverty. However the project, which involves the construction of a dam across the Yangtze river, will cause massive social and environmental disruption and conflict. Between 1 and 1.9 million people will have to be resettled, as 13 cities and 1400 rural towns and villages will be submerged (Kwai-Cheong, 1995; Zich, 1997). Claims of widespread corruption are rife as money is made available for resettlement programmes. Kwai-Cheong (1995) warns that past economic neglect, illegal immigrants, rising population, fuel and water problems in new urban areas will place great pressure on money allocated for resettlement, and the perceived advantages may be illusionary. Previous attempts by the Chinese to resettle people have been seen to have failed and critics argue that sedimentation, pollution from sewage, the loss of agricultural land and archaeological sites and threats to endangered animals outweigh any benefits. Approximately 1208 archaeological sites, extending back 50,000 years, will be submerged in water, together with some 240,000 acres of cropland. The soils of the Three Gorges catchment are particularly vulnerable to erosion, especially as large areas of forest have been cleared for agriculture, and the effect on river regimes and water quality could adversely affect species like the Chinese river dolphin and the paddlefish that are close to extinction.

Others argue that the project will bring economic prosperity and reduce the use of fossil fuels for electricity production: 18,200 megawatts of electricity, the equivalent of 18 nuclear power plants, will be produced (Zich, 1997). Kwai-Cheong (1995) reports that spread over a 20-year period the resettlement should not pose a serious financial and logistical burden. It is also believed that areas outside the dam's waters provide ample opportunities for agriculture, fishing, mineral mining and tourism that will help people adjust. Trial resettlement schemes appear to have been successful, with incomes increasing from $40 to $130 per year. The success of the dam will determine whether this is the final chapter in the history of 'mega economic projects' turning into social and environmental disasters or a blueprint for future pathways to sustainable development for millions of people.

Public opinion towards renewable energy is generally favourable, but opposition to schemes can be vociferous amongst local communities, making public acceptability an important factor in any decision-making process. For example, attempts to develop geothermal energy in Hawaii met with public disapproval and demonstrations, civil disobedience and law suits disrupted the construction of a geothermal plant with a 500MW electrical capacity. Walker (1995b) suggests that careful management, small-scale, community-based schemes, compensation, more effective public involvement in the planning process, and education to rise the awareness of renewable energy could placate a sceptical public.

Based on current trends, it is evident that renewable energy could play a major role in suppling energy in the twenty-first century. Several forms of alternative energy are now commercially viable and, with improvements in technology, the cost of generating electricity using renewables should continue to fall. However, it is unlikely that one form of renewable energy will ever replace 'the carbon economy' that dominates the world energy market. Alternative energy sources could contribute a sizeable proportion of energy on a regional scale, with the type of source dependent upon the individual region's suitability and energy requirements. This pattern of energy development is already in progress, with the expansion of HEP where water resources are adequate, solar power in areas receiving high amounts of solar radiation, geothermal energy is exploited in tectonically active areas, and wind power in areas where wind speeds generate economically competitive electricity.

The long-term success of alternative energy remains unknown and certain projects, such as the Californian wind farms, have not been immune to failure, but if technology continues to improve the efficiency of renewable energy production many of the short-term problems should be resolved. Political resolve to expand energy provision will also influence the role of renewables. Initiatives like the Non-Fossil Fuel Obligation (NFFO) will increase the capacity of renewable energy. Imposed on power companies in the UK under the 1989 electricity act, the NFFO requires 8 gigawatts of electricity to be generated by non-fossil fuel sources per year (Kelly, 1997).

Success of renewable energy will also be controlled by the future availability of oil, gas and coal. Proven reserves for fossil fuels have increased as new oil and gasfields are discovered and technology makes former non-economically retrievable stocks more viable. The key, therefore, possibly lies in the methods of abating environmental problems associated with fossil fuels. If methods such as scrubbing pollutants, using sulphur-free coal and energy efficiency measures reduce the environmental impact of conventional fuel sources, the expansion of alternative energy may be restricted (see Fulkerson *et al.*, 1990; Cross, 1993). Trainer (1995) states that the success of renewable energy will be based on economic costs and the ability of each source to supply electricity and/or liquid fuel to richer countries after accounting for losses through conversion, storage and transportation.

Ultimately the exploitation of renewable energy will also depend on future demand, and estimates vary. For example, the IEA suggests a 40 per cent increase in primary energy, whereas Johansson *et al.* (1993) predict a decrease by 20 per cent (Hall and House, 1995). Another possibility to fill an 'energy gap' is nuclear fusion. Improvement in technology has increased the amount of energy released by fusion experiments over the last 25 years. Optimistically, the construction of nuclear fusion power plants may occur early in the next century (Furth, 1995) and other forms of renewable energy may be commercially viable in the twenty-first century. Ocean wave energy, located offshore or in the coastal zone, could meet current global energy demand in theory. In practice, like several of its counterparts, practical and financial factors have curtailed its development (Falnes and Lovseth, 1991).

4.2.6 Managing with or against nature

Managing the problems caused by human intervention in the physical environment have revolved around solutions such as the introduction of technology.

Ideologically, the deep green movement seeks to shape society so that it interferes with the natural world as little as possible. They advocate that interference for anthropocentric reasons is part of the problem rather than a solution. Greens such as Porritt (1986) and Dobson (1990) argue that society should adopt a 'hands-off' approach, especially if people do not know the outcome of intervention and it is potentially dangerous. Thus working with nature may be more successful than an interventionist form of management, and evidence from attempts to manage oil spills, river channels and coastlines supports this idea.

One reason society's attempts to clean up oil spills have been interventionist or technologically oriented is because public pressure demands that governments remove environmental problems as quickly as possible. To be seen to be doing nothing gives the impression of an uncaring, unresponsive political attitude, so in this situation quick but inappropriate measures are often used.

However, recent research suggests that leaving nature to deal with the oil might be the most appropriate form of management, because inappropriate clean-up techniques can cause as much damage as a pollutant. Containment methods, like mechanical booms that prevent oil from dispersing and make it easier to remove oil from the water because oil floats (see Kirby, 1993), are widely used, but they are expensive, and in some cases it is seen to be preferable to disperse the pollutant. For example, hot water jet sprays and chemical dispersants have been used to disperse oil, but studies have shown that hot water jets create more damage because they sterilise everything and kill off those organisms that have survived the spill. Research by the National Oceanic and Atmospheric Adminstration (NOAA) revealed that unclean beaches were healthier than their cleaned, sterilised counterparts, and marine life survived until the oil was removed naturally. Combined with the manual removal of oil, hot water jets also disturb areas of the shoreline that might not have suffered serious damage by working oil into the sediments. For example, studies have shown that hydrocarbon values in the intertidal and subtidal zones were ten times higher following jetwashing further upshore. Chemical dispersants also force oil down the water column and harm organisms living on the seafloor. For example, the dispersants used after the Torrey Canyon sank off the Cornish coast in 1967 were toxic to algal, mussel and barnacle populations. Less damage would be incurred if the ecosystem was left to recover naturally (Holloway, 1991).

MacDonald's (1998) analysis of natural oil spills in the Gulf of Mexico supports the idea that nature is best left to deal with oil. Oil and gas naturally ooze from the seafloor and research suggests that the oil provides a source of chemical energy, via symbiotic bacterial relationships, for some mussel and clam species. The lesson to emerge from this work is that some oil spills are damaging while others are benign, and society must learn when to intervene and when to leave the oil alone.

Management methods that work with nature, such as natural biodegradation or bioremediation, may be an efficient method of dealing with oil spills and may have potential in cleaning up other types of pollution. Although bioremediation is an expensive option both Holloway (1991) and Wheelright (1996) argue that nature is most efficient at cleaning up oil. Studies following the Amoco Cadiz oil spill of 1978 showed that the Brittany coast recovered without any attempt to clean up the oil because it was broken down into its constituent parts by sunlight, micro-organisms and bacteria. A similar approach was used in Prince William Sound, when a fertiliser such as Inipol EAP22 was used to increase micro organism populations and the rate of biodegradation (Hodgson, 1990). Exxon claimed such an approach speeded up oil breakdown tenfold, although other researchers suggest that the figure was closer to threefold.

Interventionist methods of managing natural hazards have also proved unsuccessful. River channel modification and coastal defences are two examples. River channels are modified by humans to protect them from flooding. Modification and/or restoration of a river channel normally involves a process known as channelisation (Keller, 1975; 1976) and river courses and banks are modified and concreted to suit human needs. However, the short- and long-term consequences of direct intervention has created problems upstream and downstream of the channelised section of the river. Channelisation of a section of the Blackwater river in Johnson County, Missouri, for example, doubled the gradient of the river channel and increased the rate of river erosion. Bridges had to be rebuilt or strengthened in areas threatened by river bank erosion, and one bridge actually collapsed. Downstream increased movement of sediment has had detrimental effects on wildlife and fish populations and farmers have lost land. The costs of channelisation appear to have outweighed any benefits (Emerson, 1971). Construction of coastal defences to protect coastlines from erosion often does not work. Younger (1990) assessed the effectiveness of sea defences in north-east Norfolk, where attempts to fix the problem by engineering methods have failed for over 150 years. He suggests that one option is to stop constructing coastal defences and allow the sea to erode the coastline, compensate people for damage to property and forbid construction work in vulnerable areas.

4.2.7 Using the market

According to Pepper (1996) one of the problems of managing global resources is that people fail to internalise the environmental costs of their actions because they do not care or do not accurately perceive these costs. Whereas many ecocentrics would argue for a radical and democratic solution whereby resources would be 'owned' communally by small-scale societies with little use for money, free market environmentalists argue that one way of providing an incentive to curb environmentally damaging behaviour is to create a market-regulated private property system where pollutants or resources are owned and can be traded (see Chapter 5). Jeffreys, K. (1994) argues that the free market represents societies' most tangible asset, which enables humans to harness and utilise natural resources efficiently with less environmental impact compared with other options. t' Sas-Rolfes (1994) and Fisher, M. (1994) argue that the free market would promote conservation because individuals will not overuse their resources for fear of devaluing them. This approach can also be used to generate revenue and protect resources. Technocentric, free market environmentalists or interventionists (see Chapter 1), such as Jeffreys, K. and Fisher, are critical of state controls that limit commercial ventures. Jeffreys, K. argues that state laws that forbid humans access to natural resources are another way of separating humans from nature, and that capitalism is the best way to reconcile humans and nature.

One free market approach to lower rates of exploitation and unsustainable environmental practices is to place a monetary value on nature's 'free' services. This idea is known as 'environmental accounting'. Natural resources such as those listed in Table 4.4. are often thought of as nature's free services (see Chapter 5).

Table 4.4 Natural resources that are often thought of as nature's free services

Raw materials production (food, fisheries, timber and building materials, non-timber forest products, fodder, genetic resources, medicines, dyes)
Pollination
Biological control of pests and diseases
Habitat and refuge
Water supply and regulation
Waste recycling and pollution control
Nutrient cycling
Soil building and maintenance
Disturbance regulation
Climate regulation
Atmospheric regulation
Recreation
Cultural
Educational/scientific

(Source: Abramovitz, 1997)

Abramovitz (1997) argues that such resources are undervalued, and this is typified in the way countries produce their national accounts. Repetto (1992) argues that the United Nations System of National Accounts (SNA) fails to recognise the value of natural resources and the potential economic loss as a result of environmental degradation, so the economic health of a nation is misrepresented. Another, more traditional way to measure the strength of a national economy is to use the Gross National Product (GNP), but Dobson (1990) argues that this also does not measure spending on pollution or waste or account for the degradation to natural ecosystems. An alternative method of measuring economic wealth is the GPI (Genuine Progress Indicator) which considers the depletion of natural habitats, pollution costs and crime. In the USA between 1950 and 1994 the value of GPIs decreased, whereas the more commonly used GDP increased. The loss of natural resources could lead to a significant decrease in future economic production, lower income and reduced employment, especially in underdeveloped countries that rely heavily on natural resources to generate wealth. The problem of the current system is summed up by Repetto using a simple example:

> Should a farmer cut and sell the timber in his woods to raise money for a new barn, his private accounts would reflect the acquisition of a new income-producing asset, the barn, and the loss of an old one, the woodlot. In the national accounts income and investment increase as the barn is built and the wood is cut but the loss of the woodlot is not measured (Repetto, 1992, p.66).

A failure to properly account for loss of natural resources can hide the true status of a country's natural assets and give a misleading impression of its economic wealth – Costa Rica is a prime example. To generate economic capital it has exploited its natural resources: 30 per cent of Costa Rican forests have been cut for timber and converted to unproductive pasture and hill farms, causing high rates of soil erosion. Erosion rates average 300 tons per hectare from arable land and 50 tons per hectare from pasture. This destruction will eventually have an impact on Costa Rica's economy, as 17 per cent of its national income is generated from forests, farming and mining. Erosion, for example, washed away 17 per cent of the value of annual crops and 14 per cent of livestock products. Such losses are not accounted for in Costa Rica's economic accounts (Repetto, 1992). If these trends continue unabated Costa Rica will exhaust the natural resources that generate its economic wealth. Until policy makers recognise that the loss of natural resources has important economic implications they cannot implement policies that manage the resources in an economically and environmentally sustainable way.

The assumption that the greatest value that can be derived from an ecosystem is to maximise the production of a single commodity may also be incorrect. A study in India showed it would be cheaper to keep mangrove forests in Bintuni Bay and exploit fish and other non-forest products, as well as control erosion, rather than market the timber. By not cutting the forests $25 million per year could be made by the fishing industry, and other local uses of the area could produce an income of $10 million. Abramovitz (1997) argues that these examples show that understanding and valuing nature's services in non-monetary and monetary terms could ensure the resources are used equitably and within non-destructive limits. This kind of approach could preserve the tropical rainforests and provide an economic income at local (indigenous peoples) and national level (see Box 4.3).

Box 4.3 Valuing nature's free services: an example from the tropical rain forests

Harvesting non-wood products grown in the rain forest could generate higher economic returns and prevent the destruction of rain forests. Rain forests not only provide a rich diversity of trees but also fruits, nuts, latex, medicines, rattan, edible oils, gums, dyes, rubber and resins that are currently traded in local and, in some cases, global markets (Peters, 1991) and could represent up to 90 per cent of the potential market value of the rain forests (Tyler, 1990).

Robinson (1985) describes several small-scale projects that exploit alternative rain forest products. These include developing aspects of 'garden hunting': the procurement of meat through natural animals of the rain forest such as green iguanas and pacas. Various forms of aquaculture could be developed to raise fish and turtles or even floating plants like the Azolla fern that fixes atmospheric nitrogen and therefore could be a source of fertiliser. Forest gardening on small plots may also allow surrounding areas of rain forest to be maintained and plant products to

▶

be exploited. A study conducted by Peters *et al.* (1989) determined the potential value of non-wood products in one hectare of Peruvian rain forest. The results showed that the net potential value (NPV) of fruits and latex trees is about $6330 compared to approximately $1000 of net revenue of marketable timber liquidated in one felling at a sawmill. In order to increase the profit of the hectare sustainable logging would produce an NPV of $420, based on the removal of 30 cubic metres of wood per hectare every 20 years. However, to ensure that sufficient capital returns could be maintained Peters (1991) recognises that an expansion of existing and the creation of new local and international markets, secure land rights and the productivity of the non-wood products must be maintained by educating collectors of the importance of non-destructive harvesting practices.

Concern about the ability of harvesting alternative rain forest products to ensure the long-term survival of indigenous peoples has been voiced. Corry (1993) argues that rain forest products are at the mercy of market forces and profit-driven dealers. Little of the capital generated by such schemes filters down to indigenous people. So far efforts by the US-based 'Cultural Survival' and the UK-based Body Shop to finance projects for Amazon Indians have been problematic, as only a small section of their communities have benefited and one of the most important issues, secure land rights, has been largely ignored.

It is estimated that over half the world's ten million species of plants and animals live in tropical rain forest (Bird, 1991). Because the rain forests contain such a diverse range of plant life the amount of genetic material available to the pharmaceutical industry to create medicines and to breed pest- and disease-resistant crops is vast. Kleiner (1995) reports that the tropical rain forest contains undiscovered drugs worth up to $147 billion and so far 47 major drugs have been derived from rain forest plants. Rice *et al.* (1997) estimate that 35,000–70,000 plant species provide traditional remedies world-wide. Drugs such as quinine used to combat malaria, curare as a muscle relaxant and vincristine in the treatment of leukaemia originate from tropical plant species (Myers, 1989; 1992). All plants that contain secondary metabolites and/or alkaloids like vincristine may have some use in human medicine. Now organisations like the Worldwide Fund for Nature are working alongside indigenous tribes to learn about their traditional forms of medicine to develop new medicines. Researchers are working in West Africa with the Cameroons Centre to discover medicinal plants in the Korup rain forest. They are looking at plants related to *Zeulania quidonia* that contain a number of secondary metabolites to treat cancer. So far 70 per cent of plants that are active against cancer cells derive from the rain forest, and the potential market in drugs of this kind is enormous: globally $50 billion was spent on drug prescriptions containing active plant ingredients in 1991 (Bird, 1991). In the US 25 per cent of all prescriptions and 60 per cent of non-prescription drugs dispensed between 1986 and 1990 were extracted from natural products, primarily plants (Abramovitz, 1997). Finally, tropical rain forests also play a major role in absorbing carbon dioxide from the atmosphere (Bunyard, 1987).

Another market-driven scheme to use resources more efficiently is certification, a scheme that make companies promote an environmentally friendly image by being allowed to put a certificate of good environmental practice on their products (Schneider, 1996). For example, this approach has been used to protect the earth's dwindling rainforests. Under the banner of the Rainforest Alliance's Smart Wood Program, timber firms are formally approved if their logging operations meet certain conditions and big retailers, such as IKEA and Home Depot, are allowed to label their products as green. How effective efforts will be to slow rates of deforestation using certification is unclear. Schneider (1996) argues that certification needs to be widely practised to be effective but many timber operations in Europe and the USA do not participate in the scheme, which makes it difficult to persuade underdeveloped countries to take part, and a more wide-ranging approach is needed to tackle all the causes of deforestation.

Certification has also been used to conserve fish stocks. A newly formed Marine Stewardship Council, formed by the World Wide Fund for Nature and Unilever, will certify fish caught using sustainable working practices. However, MacKenzie (1996) argues that the Council may run into political and economic difficulties as governments cannot agree on what constitutes a sustainable method of fishing and, furthermore, have ignored appeals to cut fish catches. However, the certificate might be effective if consumers favour products carrying the label and persuade fishermen to adopt sustainable fishing technology. Other free market approaches include pollution trade permits (see Pepper, 1996), trade in endangered plants and animals by breeding them on commercial farms (see Fisher, 1994; t' Sas-Rolfes, 1994).

However, while interventionist technocentrics advocate free market mechanisms to deal with environmental problems, other people have identified problems with them. Daly (1993) and Brooks (1992), for example, have argued that free trade agreements have tended to work against the introduction of environmental legislation largely because these agreements employ the lowest common denominator principle. For example, widespread pollution along the Mexico–USA border has been linked with US-owned companies relocating to Mexico following a free trade agreement. This movement occurred because it is cheaper to pollute in Mexico than to dispose of waste and pollutants safely in the USA. Over 100 types of industrial waste have been identified in the New River, which is so toxic signs warn people to go nowhere near it, while up to 12 million gallons of toxic chemicals and raw sewage are discharged into the Tijuana daily. US-owned firms such as Rimir continue to release untreated toxic chemicals like the solvent Xylene into public waterways, and levels of Xylene now exceed the legal limit of Mexico by at least 6300 times. Disproportionate cases of anencephaly (where parts of the brain do not form) have been reported from people living on both sides of the border and have been linked with Xylene released from the Rimir factory in Mexico (Brooks, 1992).

Benton (1996) and Bhagwati (1993), however, view free trade agreements more optimistically. Bhagwati (1993), for example, argues that environmentalists are wrong to fear the effects of free trade, because efficient policies and trade will generate capital to fund pollution abatement and other measures to protect nature. Bhagwati claims that economic growth does not always increase pollution, and he argues that in US cities an increase in per capita income coincided with a reduction in sulphur dioxide levels. Benton argues that the North American Free Trade Agreement (NAFTA) effectively highlights that the environment has become an important factor in international trade, as this agreement included environmental protection by resolving to promote sustainable development and the environmental community, albeit divided, participated in its formulation. Benton (1996) suggests that NAFTA is the greenest free trade agreement ever negotiated. Agreements like NAFTA may increase the mobility of capital through countries like Mexico, eliminate tariffs and provide more revenue to invest in cleaner industries. The environmental success of NAFTA and other trade agreements appears to be dependent on the strength of all participants to adhere to environmental legislation.

4.3 Problems of management (1): the nature of the physical environment

4.3.1 Spatial and temporal aspects of physical processes

Pollution is an international management problem because it has no national boundaries and occurs in virtually every country, irrespective of its stage of social and economic development, therefore countries must co-operate to resolve the problems of pollution. Mukerjee (1995b) argues that pollution control is of great importance because the potential toxicity of many pollutants is unknown. As a consequence, the potential risks to human health and to parts of nature are also unknown, which places environmental managers in a difficult position because they have to decide whether to manage, at considerable expense, a substance that may be harmless, or else risk leaving a substance that is potentially dangerous.

The potential threat of a pollutant is variable because their residence time and spatial scale (either concentrated or diffuse) varies in the environment (Figure 4.2). For example, plutonium-239 has a half-life of 24,000 years, CFCs persist from 300 to 20 years, oil can persist in water-based sediments from months to decades, and volatile organic compounds for days to hours (see Brown *et al.*, 1990). The form and composition of a pollutant or waste also controls its toxicity to all life forms. For example, a small quantity of plutonium-239 is highly toxic if allowed to accumulate in the lungs and in bone marrow (Blowers, 1996) but

> it has been said that it would be fairly safe to sit on a lump of plutonium wearing only a stout pair of jeans. On the other hand, it could be fatal to inhale even a very small particle of it (HMSO, 1986, p.xvi).

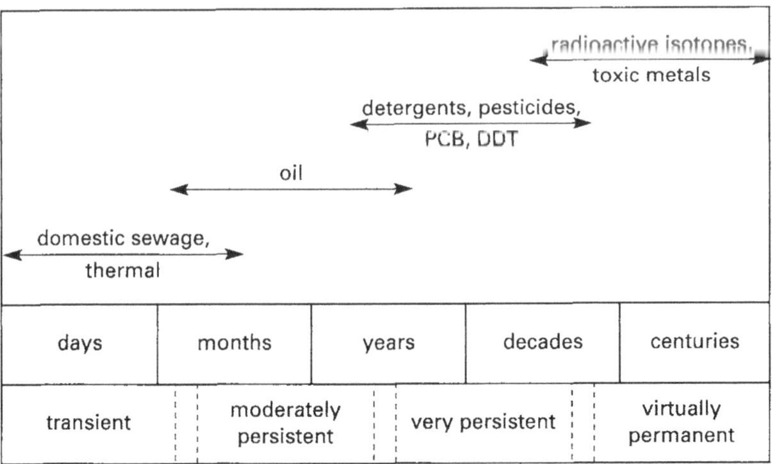

Figure 4.2 Subdivision of pullutants and their persistence by time (source: Smith and Greene, 1991)

Dealing with waste over time-scales of thousands of years is a challenge. The problems of disposing of nuclear waste are discussed in Box 4.4.

Box 4.4 Problems of persistence: nuclear waste disposal

The management methods that have been outlined so far in the text deal with pollution and waste that can be neutralised. However, radioactive wastes cannot be treated using conventional methods because it is too dangerous. In fact no successful method of dealing with radioactive waste has yet been developed.

Disposing of radioactive waste, from either military or energy use, has become one of the most controversial environmental problems of the post-war period. At present, the only option open to the nuclear industry is to store the waste until an appropriate method of disposing of it safely is developed. Because of the life expectancy of radionuclides, disposal in permanent subterranean storage repositories is considered to be the most cost effective option for nuclear waste. In the USA there are proposals to store nuclear waste in an underground repository at Yucca Mountain, approximately 160 km north-west of Las Vegas. The repository will store 70,000 metric tons of waste in containers buried approximately 300 metres below the land surface and between 240 and 370 metres above the water table (Whipple, 1996). However, simply storing radioactive waste is not that easy. Life expectancy of waste, the design of storage containers, potential leakage and contamination via earth movements and earthquakes must be considered.

Research in Sweden on waste container design suggests that a suitable container can be made, but storage containers and tanks used for storing spent fuel rods have corroded and leaked at sites in the USA (Whipple, 1996). It is possible that leaked radioactive waste could sink down to the water table from the canisters and, although the pathways by which radionuclides move through soils and into water supplies are poorly understood, it is clear that leakages could contaminate irrigation systems and human water supplies unless the leaking canisters are retrieved and repaired and/or the industry avoids siting repositories close to human settlements in order to reduce the chance of contamination.

The amount of radioactive waste is expected to increase in the future. For example, electricity generated by nuclear power plants in the USA still accounts for about 38 per cent of the total, despite a scale down of the nuclear industry following the incident at Three Mile Island, a rise in operating costs due to more stringent safety requirements and a fall in electricity consumption between 1970 and 1980 (Brown, G.E., 1987). Over the last 50 years the USA has accumulated approximately 420,000 metric tons of high-level radioactive waste and spent fuel rods from nuclear reactors. Over 15,100 cubic metres of spent fuel had been generated by 1988 and this figure is expected to rise to 24,000 cubic metres by 2005 (Pasztor, 1991; Anderson, 1995; Whipple, 1996).

The scaling down of nuclear arsenals and the safe management of nuclear waste generated from weaponary also needs to be addressed. Approximately 100 metric tons of weapons-grade plutonium will be placed in storage by 2005 until it can be disposed of safely (Devolpi, 1995). This figure may increase globally if the dearming of nuclear warheads continues. Thus, in the short to medium term the amount of waste produced by the nuclear industry will increase, and it has been argued that the storage capacity within the USA is insufficient to meet the country's future waste production and that a second repository will be needed (Whipple, 1996).

Furthermore, it is difficult for scientists to predict what will happen during the lifetime of a repository because it takes several hundred thousand years for radioactive material to decay and it could take longer than the US Environmental Protection Agency's suggested limit of 10,000 years (Pasztor, 1991; Whipple, 1996). For example, earth movements caused by earthquakes and volcanism could result in the leakage of waste. A recent study suggests that crustal deformation at Yucca Mountain may have been underestimated by a factor of ten, which therefore increases the possibility that there may be an earthquake or volcanic eruption during the lifetime of a repository (Kerr, 1996; 1998). Other storage options include burying radioactive waste under the sea-bed. Vast mudflats forming part of abyssal plains beneath the oceans located in the middle of major tectonic plates have been suggested as a suitable, permanent resting place for high-level nuclear waste. Subsea-bed disposal would utilise sediments that have been geologically stable for over 50 million years, in areas well away from seismic activity and volcanoes and that are composed of clay that is suited for containing any leaks with low permeability to water, high adsorption capacity and plasticity. However, at least $250 million is estimated to be needed to finance research into the feasibility of sea-bed disposal, and international opposition might be expected and would need to be placated by revising the international ban on ocean dumping (Hollister and Nadis, 1998).

Advances in methods to decommission and decontaminate nuclear facilities may allow the safe clean up of contaminated sites. Methods include carbon dioxide blasting for dismantling structures, descouring methods such as soda blasting concrete and steel surfaces, and the use of robots to reduce risk of radiation exposure to workers. Methods are also being developed to remove or immobilise radioactive isotopes from waste with, for example, titanium coated zeolite being used to remove plutonium ions from highly alkaline wastes (Devarakonda and Hickox, 1996). At least five other storage and disposal options for nuclear waste exist. Nuclear transmutation involves transmuting radioactive material by neutron bombardment into other substances that would remain radioactive for shorter periods of time. Other less popular suggestions for nuclear waste disposal include: (i) shooting nuclear waste into space or the sun, although this option is considered too expensive; (ii) storing waste under the polar icecaps, but high-level waste generates excessive heat; (iii) dissolving waste uniformily over the world's oceans (Anon, 1995d); and (iv) containing waste in glass, a process known as vitrification. Surplus weapons plutonium is mixed with radioactive waste and moulded into a special type of glass or ceramic that is then buried. This process immobilises radioactive atoms and makes it extremely difficult for subsequent deliberate extraction, although the glass itself would not provide a shield against radiation (Hollister and Nadis, 1998).

The cost of dealing with nuclear waste is enormous. For example, it is estimated that the US government will have to spend well over $375 billion to clean up four nuclear weapons sites such as Hanford; $2 billion has already been spent on the evaluation of Yucca Mountain as a nuclear waste repository and another $1–2 billion will probably be spent before construction commences (Hollister and Nadis, 1998; Zorpette, 1996b). Hitherto the US government has allocated relatively little money to finance the clean-up costs. In 1996 the budget was $6 billion for all nuclear weapon complexes. Compared to the $28 billion the government will spend on intelligence it calls into question the commitment to tackling this problem. Political opposition to the location of a repository has also delayed the construction of a waste facility. In 1982 the US Department of Energy (DOE) pledged to start building a national waste repository at Yucca Mountain by 1998, but the state of Nevada is strongly opposed to the project and it is unlikely that any waste will be stored before 2015. Once the DOE officially designates the site as a repository a licence must be obtained from the Nuclear Regulatory Commission before it becomes operational. If a licence is granted and the state of Nevada objects, a decision must be made by the US Congress. Even if the proposal gets that far the financial commitment to the site still has to be resolved (Brown, 1987; Whipple, 1996).

Once released, pollutants can disperse through a number of different pathways, as shown in Figure 4.3. Pollution is generated from traceable 'point sources' such as power stations or sewage works, and untraceable 'non-point sources' such as runoff from agricultural land, which account for 80 per cent of degradation of US waters, mainly because it is unregulated, insidious and therefore more difficult to manage (Mitchell, 1996). Runoff from urban areas and agriculture cause the greatest amount of non-source pollution. Heavy metals, oil and fertiliser from lawns contribute to urban pollution, while organic matter, sediment, fertiliser, herbicides and bacteria run off agricultural land; all affect human health and wildlife at source, along pathways and at recipient sites. Once in the natural environment the processes involved with dispersal are complex. Miller (1997), for example, describes how fluvial geomorphic processes play a role in the dispersal of heavy metals from mining sites and Melamed *et al.* (1997) describe how mercury can occur in a variety of different mobile forms once it is released into a river system. Mercury is, for example, leached from mine tailings, directly dumped into rivers when placer ores are dredged and processed on river barges and washed out of the atmosphere in rain showers where it can be mobilised, dispersed and taken up by plants and animals dependent upon physio-chemical conditions (eg redox conditions, pH, solid-metal species interaction) of the water.

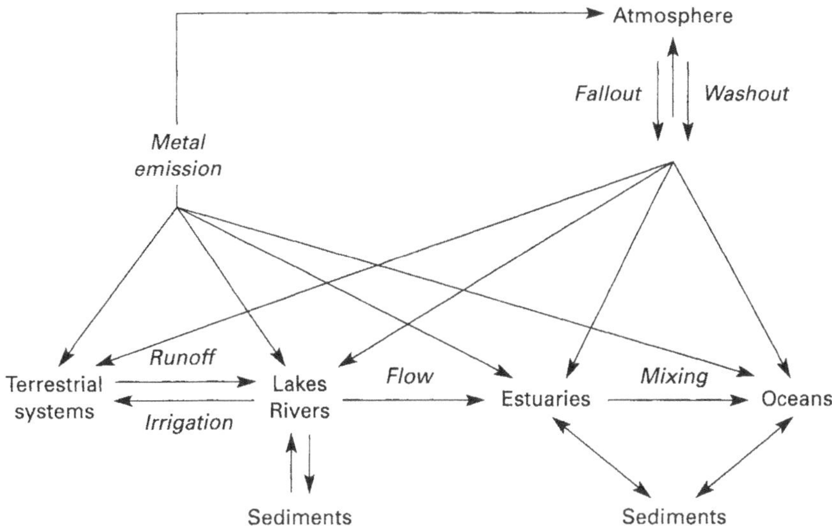

Figure 4.3 Relationship between industrial activities and environmental releases (source: Brown *et al.*, 1990)

Because many pollutants occur in a variety of forms, are distributed over a range of spatial scales and pathways, and can persist in the environment for a long time, different types of responses are needed to contain, remove, convert or transfer them into a safe form or place, from either their source or their place of deposition. Sometimes there is no choice but to manage a pollutant at 'source' or 'end point'. For example, ozone-depleting CFCs must be controlled at source because once they are released into the atmosphere it is impossible to prevent them from reaching the stratosphere. However acid rain pollution, for example, can be tackled at source by filters or at end point by liming (Woodin and Skiba, 1990).

Furthermore, people do not always understand the long-term impact of pollutants. Research so far has shown that dealing with oil spills is dependent upon the chemical make up of oil, its toxicity, weather conditions, and commercial, practical and political factors. Yet Hodgson (1990) argues that although the major impact of oil spillages disappears after three years their long-term consequences are still poorly understood, and therefore it is unclear over what time-scale the clean up and management of pollution affected areas should be.

Environmental management must, therefore, address all problems and hazards at different scales: local, regional and global (Klinger, 1994) and over different time-scales, but global environmental problems cannot be easily compartmentalised to one scale because problems on one scale may accentuate problems at another. For example, Wigley (1991) argues that reducing acid rain pollutants (a regional problem) can induce greater global warming (a global problem) (see Chapter 3). Irvine and Ponton (1991) similarly argue that a change in one element can change another. For example, although nuclear power may alleviate the problems caused by acid rain it still contributes to global warming. These interlinkages between regional and global environmental problems mean that they should not be managed in isolation, as the consequences of any action may prove more detrimental in the future.

4.3.2 Scientific uncertainty and the use of science to manage environmental problems

Chapter 3 discussed the nature of many contemporary environmental problems and used scientific knowledge as evidence that these problems exist and pose, in some cases, a global threat to society. Looking at science in this context provides it with a positive image. However, a number of people have challenged this view of science.

As discussed in Chapter 1, people have begun to question the information provided by scientists and so-called 'environmental experts'. Wynne (1989) and Pepper (1996), for example, argue that people mistrust scientists over the information they provided about the effects of pesticides and herbicides in the 1950s and 1960s, the impact of radioactive fall-out from the Chernobyl nuclear explosion, and more recently the BSE crisis (see Chapter 5, especially Box 5.2). Pepper (1996) and Giddens (1990) have suggested that this mistrust has undermined the credibility of science and the general public has become more sceptical about scientific findings. Pepper, for instance, uses the example of nuclear power, that:

> is so complex and dangerous that it has to be surrounded by secrecy and private police. It cannot be controlled and understood by ordinary people; only by the hierarchy of experts and managers backed by massive resources at the command of the state and big corporations (Pepper, 1996, p.93).

O'Riordan (1995c) and Pepper (1996) argue that scientific research is often tied to the wants of dominant political and economic groups so that the established ways of conducting science now reflect and reproduce elitist and exploitive aspects characteristic of all instruments of power (see also Chapter 1). Pepper (1996) cites the arguments made by Albury and Schwartz (1982), who suggest that scientific and technological advances have served the particular needs and ideologies of the capitalists who funded them. They cite as examples the development of telecommunications and the micro-electronics industry, which revolutionised the ability of people to expand trade, and the conscious decision of the Rockefeller Institute to develop the 'green revolution' to suit plantation-type agriculture used by multinational corporations.

Other people have criticised science and technology for creating, as opposed to solving, environmental problems: examples here might include nuclear power, pollution and pesticides. It has also been argued that scientists often use, and even create, ideas of crisis to persuade people to fund their research. Blowers and Lowry (1987), for example, claim that UK government officials, who were pro-nuclear power, tried to create an agenda whereby scientific experts would create scientific data to justify or legitimise their proposed plan of dumping nuclear waste at sea. Pepper (1996) notes, however, that there have been clear instances where scientists have been used by both sides of the environmental debate to legitimise claims. Such examples demonstrate that scientific research and scientists are not isolated from the socio-economic and political consequences of environmental problems and can be 'bought' to secure public acceptance. One example is the scientific investigation to determine the ecological cost of the Exxon Valdez oil spill. In this case, different people or organisations use scientific data to serve their own interests because scientific data can be interpreted in different ways and this creates different views on the need for management (see Box 4.5).

Box 4.5 The Exxon Valdez oil spill

The results of research undertaken by Exxon, the Department of Fish and Game and the Department of Environmental Conservation following the oil spill were controversial because of the opposing conclusions about the severity of the ecological damage in Prince William Sound (Pain, 1993). Holloway (1996) argues that the methodological approach of the post-oil spill research by these organisations seriously undermined the reliability of the findings because most of the research teams worked in isolation and the research was not subject to peer review. Each study developed its own methodology to assess the impact of the oil spill. One study concentrated on the intertidal zone, another investigated the effect of oil on all organisms living at different elevations in the coastal zone and comparisons were also sought with data collected from unspoiled areas. The way the impact was measured also varied. Some of the studies looked at species composition and diversity while others concentrated on particular species. The complexity of studying coastal ecology also helped to create opposing opinions on the recovery of contaminated areas. One study suggested 91 per cent of the area had recovered by 1990 while another study showed that the recovery was misleading as species presence and abundance fluctuated dramatically over a longer time-scale, making it very difficult to state categorically that a recovery had taken place. The ecology of the coastal zone is so variable that it is virtually impossible to establish a benchmark or set of criteria to quantify its health. Nobody really knows what the pre-spill ecology of Prince William Sound was like, so it is difficult to assert that the area has returned to its pre-spill status. Furthermore different parts of the ecosystem recover in different ways and at different rates. For example, pink salmon numbers have improved but herring have not. There has been little attempt to view the coastal ecosystem across a variety of time-scales or ecological zones and to consider how each component interacts.

Another problem was the politically sensitive nature of the findings. Exxon wanted to portray a successful, efficient clean-up operation and to scale down the magnitude of the disaster to negotiate favourable legal settlements and reduce their financial commitments. Exxon quickly responded to studies that concluded the damage caused by the oil spill was shortlived. By May 1993 Exxon claimed the oil spill would have little long-term effect on Prince William Sound (Pain, 1993). In contrast, local residents and industries directly affected by the oil spill would perhaps perceive the disaster on a much larger scale to receive maximum compensation. Thus the search for the truth may be overshadowed by the conflicting interests of each party that financed the research. In hindsight, the whole episode drew attention to the problems associated with using science to solve environmental problems and the need to try and assess environmental damage *per se*, without external influences, in order to remove what Holloway describes as the 'scientific uncertainty principle'. Holloway argues that long-term monitoring of ecosystems and the collection of scientific data are needed to improve our understanding of how natural systems function, change and evolve through time. This will enable scientists to recognise the consequences of any changes, especially those stimulated by human action, and implement management strategies (Jennings and Polunin, 1996).

Rather more general criticisms of the role of science and scientists are made by people such as Giddens (1990) and Beck (1992; 1995). Beck, for example, argues that society has now reached a point where people are no longer able to perceive environmental risks through their normal senses but rather have to rely on the scientists and experts to identify the causalities of risk (see also Yearley, 1991; MacNaghten and Urry, 1998; and Chapter 1). Thus society finds itself in a situation where environmental knowledge is 'scientised' because people depend upon science to identify and measure risks for policy purposes, but science is confronted by greater uncertainties that it is unable to neutralise (Eden, 1998).

Beck similarly argues that the environmental problems created by science, such as pesticide ingestion and fall-out from nuclear explosions or nuclear war (see Turco *et al.,* 1983; Turco *et al.,* 1984; Perry, 1985), are without a scientific solution and suggests that economic and scientific progress is being overshadowed by these new risks or hazards.

Beck and Giddens also argue that people rely on science to overcome environmental problems, and they express concern about the ability of so-called experts to manage these risks because the experts do not always fully understand the physical, chemical or biological systems that operate on earth. There are many natural or human-made environmental problems that are very complex, including hazards such as earthquakes, volcanic eruptions, tsunamis, tornadoes, wildfires and floods. One school of thought argues that people will never fully understand certain natural systems and hazards because they operate in mysterious ways or appear to be completely unpredictable, and experts cannot predict or forecast their occurrence or magnitude with any degree of confidence even though they represent a serious threat to society. For example, natural hazards associated with weather are destructive: since the beginning of 1997 over 16,000 people have been killed and $50 billion worth of damage has been caused world-wide with floods, winter storms, forest fires and hurricanes, attributed to the unpredictable effects of El Niño (Miller, K. 1998 and see Box 3.5). In November 1998 Hurricane Mitch passed over Central America killing over 15,000 people in Honduras and Nicaragua, and forecasters expect more extreme weather as atmospheric temperatures rise in response to greenhouse gas emissions. If these natural hazards and environmental problems are inherently unpredictable and do behave in a disordered and random fashion (ideas that characterise chaos theory, see Prigogine and Stangers, 1985) then, as Wilson and Bryant (1997) argue, the traditional positivist approach of being able to measure, monitor and rectify environmental problems using scientific knowledge are ineffective.

Taylor and Buttel (1992) claim science-centred environmentalism is vulnerable to deconstruction and open to alternative explanations that also cast doubt on the certainty of predictions. In particular they suggest that the underlying assumptions made by scientists are open to question. Examples include the 'limits to growth' model (see Chapter 3), predicting El Niño events (Betsill *et al.*, 1997) and the construction of global circulation models to predict future climate change. Predictions of how climate may change over the next century, for example, have been simulated using general circulation models on global and regional/local scales (see Chapter 3), and these predictions are made even though scientists do not fully understand how and why climate changes, and they cannot measure accurately all the processes and parameters that influence climate (see Bergman *et al.*, 1981; Stone, 1992; Schwartz and Andreae, 1996). Widely recognised problems in these models include:

(1) That some climatic parameters such as ocean eddies, storms and cloud activity operate on small scales that are difficult to measure and hence their influence on weather and climate must be estimated.
(2) That scientists cannot accurately predict how human activities may change through time. For example, if deforestation rates continue at their current level more carbon dioxide will enter the atmosphere. However, there are numerous attempts to stop deforestation and so the amount of CO_2 released into the atmosphere could be far less than the figures entered into current models. As a consequence different scenarios are often run in the models (see Houghton *et al.*, 1994; 1996).
(3) Different GCMs (General Circulation Models) have also produced different results (see Chapter 3) which adds further uncertainty to any decision making process.

Despite all these problems with the scientific data used in the models the results have been widely accepted and used to formulate policies to combat climate change. Not only must policy makers and politicans balance the positive and negative impacts of global warming before they can implement policies, they need also to be reasonably confident as to the magnitude of any impact. Without an appreciation of all the possible scenarios they may end up spending substantial amounts of money on unnecessary and wasteful efforts to reduce greenhouse gas emissions.

Taylor and Buttel (1992) also challenge the notion that scientific problems always need a scientific solution. They argue that science monopolises people's understanding over environmental problems, stating that 'science documents the existing situation and ever tightens its predictions of future changes' (p. 403). Taylor and Buttel (1992) and Redclift (1992) argue that scientists pay too much attention to mitigating environmental problems using science at the expense of looking at its causes in human behaviour and physical processes. Thus they argue that the dominance of physical climate research may not be the most productive avenue to follow because global warming is a social problem: 'since it is through industrial production, transport and electrical generations and tropical deforestation that societies generate greenhouse gases ... yet it is physical change that is invoked to promote policy and social change'.

They suggest that the hierarchical physical–natural–social arrangement of investigation is not always needed in the construction of environmental problems and 'social environmental problems require social diagnosis and response' (see also Box 4.6). Beck (1992; 1995) and Giddens (1990) argue that science should not dictate environmental risk assessments and policies through a one-way process but science should be exposed to greater scrutiny by society: an approach Beck (1995) describes as 'reflexive scientization'. To remedy this situation many social scientists are calling for uncertainty and risks to be managed in an interdisciplinary way involving scientists, the public and authorities where the dissemination, legitimisation and contestation of science is debated (see Eden, 1998).

Box 4.6 Scientific uncertainty and earthquake prediction

Another way in which society can mitigate against many environmental problems and natural hazards is to predict or forecast the timing, magnitude and frequency of their occurrence. If natural hazards, such as earthquakes, tsunamis, volcanic eruptions and floods, can be predicted then humans can introduce emergency plans to avoid loss of life and limit damage. Because millions of humans live in tectonically active areas, earthquakes pose a threat. It is not surprising therefore that since the turn of the century approximately 20,000 lives have been lost each year due to earthquakes (King, 1988).

In order to protect people from earthquakes scientists have tried to predict them since ancient times, but with little luck. Numerous prediction methods have been tried and tested, including the presence of mysterious rainbows, fog and light, abnormal animal behaviour (Tributsch, 1982), geologic methods such as fault mapping (Fischmann, 1992; Vittori *et al.*, 1991; Wuethrich, 1995), seismic methods like the seismic gap theory (Savage *et al.*, 1986; Mestel, 1995), migration of crustal activity (Kasahara, 1981), measuring dilatancy (Scholz *et al.*, 1973), geodetic measurements (Massonnet *et al.*, 1993, Bock *et al.*, 1993), geochemical and hydrologic monitoring (Davidson, 1994; Arieh *et al.*, 1974; Igarashi *et al.*, 1995; Tsunogai and Wakita, 1995). Despite such an array of methods society still cannot accurately predict earthquakes because we still do not fully understand the mechanisms of the earth and no method has proved reliable (Lomnitz, 1995).

▶

Some earthquakes have been predicted but without preventing widespread damage and loss of life because it has not been possible to predict accurately (on long, intermediate or short time-scales) the exact time of the quake or its magnitude. An earthquake along the Cocos plate in Mexico was predicted, but because its epicentre or the exact time could not be deduced 9000 people were killed when it occurred on 19 September 1985, registering 8.1 on the Richter scale (Boraiko, 1986). Seismologists predicted that an earthquake of a magnitude between 6 and 7 would take place in the Yunnan region of China in 1996 but, because a 'total prediction' (exact time and place) was impossible, it killed 240 people and injured 4000 on 3 February (Pringle, 1996). Despite intense efforts to master earthquake prediction, nobody anticipated the Kobe earthquake in Japan in 1995 which killed 3000 people.

The dilemma for policy makers and governments is that earthquakes are incredibly destructive if humans live in earthquake-prone areas. Earthquake prediction programmes cost the Japanese government $14 million in 1991 (Geller, 1991) yet scientists cannot predict when the next quake will strike. Assuming that it is unrealistic to move people away from these areas it could be argued that policy makers should abandon this expensive scientific research. Lomnitz (1995, p. ix) suggests: 'even if we could predict earthquakes to the nearest split second we might not necessarily save more lives or more dollars'; and perhaps money should be invested in designing buildings that are resistant to quakes and preventing damage. After all it is the destruction of buildings and roads during an earthquake that causes loss of life.

Unless scientists can reduce the uncertainties natural hazards and global environmental problems will remain a potential threat to society and make decision making an uncertain process. Miller (1998), however, reports that there is progress with predicting certain aspects of the weather. For example, weathermen predict 82 per cent of extreme weather events and average tornado warning times have doubled to ten minutes because radar can discern subtle shifts in a storm's behaviour, but even when predictions and forecasts are possible, for example for hurricanes, it does not always solve the problem. Hurricane Andrew hit the Florida coast in 1992 and caused $30 billion worth of damage.

Gerrard (1995) argues that traditionally in the western world risk management has been under the jurisdiction of those in power, who consult specialist groups like scientists to analyse a problem and develop guidelines to resolve it. However, it has been argued that this can create conflict and it may be unsatisfactory to exclude non-scientists from the debate, especially when the scientific community cannot achieve a common consensus and the scientific process might be flawed because the wider implications of management might be ignored. Gerrard (1995) suggests that the decision-making process must encompass the free exchange of ideas between all social groups, including the public, policy makers, scientists and risk predictors.

Scientific uncertainty is also used by governments as an excuse to do nothing. By 1995, after the euphoria and expectation that followed the so-called Rio Earth Summit, little progress had actually been made on tackling future climate change. Only a few countries will hit targets for carbon dioxide emissions, with the USA and Europe exceeding them by approximately 6 per cent. Representatives of the fossil fuel burning countries often cite the uncertainty of global warming predictions as a reason for non-compliance, highlighting, for example, that scientists have not agreed upon how much anthropogenic greenhouse gas emissions will enhance radiative forcing (see Anon, 1992; 1995a). Because of the uncertainty surrounding future climate change it is difficult to convince the public and politicians that it is in their own self-interest to take action (Anon, 1997b).

4.4 Problems of management (2): the nature of people

4.4.1 Humans as property owners

In section 4.2.7, 'Using the market', some of the arguments for free market environmentalism were outlined. One of the principle arguments was that placing resources under private ownership would, in the words of Pepper (1996, p.76), 'make private owners care for their environment because they know they would reap the benefits of doing so'. Free market environmentalists argue that communal property rights fail to protect resources because no individual feels a specific interest in looking after them and it is more difficult to identify the culprits of poor environmental practice. Garrett Hardin explores this issue in more detail in an essay titled the 'Tragedy of the Commons' (Hardin, 1968) and he argues that common resources, such as the oceans and the atmosphere, are treated as free resources and people who use them often do so in an environmentally damaging way.

Tragedy of the commons

Hardin (1968) argues that people have been locked into systems that 'foul our own nest' because individual and group actions are strongly influenced by a desire to be competitive and economically successful, and these desires control our behaviour towards the environment. Hardin suggests that if humans have free access to common areas or resources these will be ruined because, more often than not, individuals will use the commons to maximise their own benefit (any profit is 'internal' to the individual user) without considering any potential damage that may result from misuse (which is 'externalised' or shared by all the users). Hardin argues that most people will place their own short-term needs before the long-term preservation of that resource or the interests of other people, and that most users will exploit the resource as much as possible because they think if they do not somebody else will. Hardin argues that, if this course of action dominates everybody's thinking, eventually the carrying capacity of the commons will be exceeded and the resource ruined.

Hardin proposes that this situation can be rectified by 'mutual coercion' (action that is mutually agreed upon by all users of a resource) to stop environmentally damaging behaviour, rather than state control or unrestricted privatisation. Hardin suggests that people should not be allowed the freedom to behave irresponsibly and that freedoms could be limited by mutually agreed forms of management such as taxes, introducing prohibitive legislation and long-term education to breed environmental responsibility as part of normal social behaviour. Although aspects of Hardin's essay have been criticised, particularly the assumption that all people behave in the same way (see Stillman, 1983; McEvoy, 1987), elements of Hardin's tragedy have been recognised in some recent environmental problems (Kurien, 1993; Butler *et al.*, 1993). Pepper (1996) argues that many mainstream scientists and economists have drawn from Hardin's arguments and generally favour as much privatisation as is practicable.

A review of the fishing industry suggests that free access can lead to resource depletion and scarcity. To protect fish supply, unfettered access to commercial fishing areas has gradually been stopped and fishing areas have come under more 'private' or, more accurately, state ownership.

First, the *International Council for North Atlantic Fisheries* was established by a convention in 1949, with the aim of providing statistical information and co-ordinating regulatory measures such as total allowable catches, net mesh size limits and gear and vessel restrictions in the commercial fishing industry. However, this initiative appears to have had little impact as, for example, redfish, halibut, cod, mackerel and herring were still intensively caught in North America and off the coast of north-west Europe, resulting in a serious decline in numbers by the late 1960s despite a series of measures to regulate the industry, including controls on mesh size, catch limits and quotas, closed and open seasons, and gear and vessel restrictions (Coull, 1984; Mitchell and King, 1984). Further moves to counter over-fishing culminated with the introduction of economic exclusion zones (EEZs) imposed on 1 January 1977, which effectively placed 99 per cent of the world's fishing grounds under state control or ownership. EEZs placed a 200-mile fishing limit around a country's coastal waters to prevent foreign fishing fleets from exhausting another country's fish stocks (Coull, 1984), and these new laws evolved into the *United Nations Law of the Sea Convention* in 1982. In each economic exclusion zone coastal states determine the allowable catch and allocate quotas to local and, sometimes, foreign fishermen. Finally, in December 1995, a treaty designed by the United Nations to protect global fish stocks by international law was signed by 28 nations. The creation of a series of property zones would, according to Hardin's arguments, lead to greater concern to preserve fish stocks because the impact of their decline would be internalised rather than externalised.

However, restricting free access to the resource has not solved the problem as 70 per cent of the world's marine fish stocks are considered by the United Nations Food and Agriculture Organisation to be heavily exploited, over-exploited, depleted or slowly recovering (Russell, 1996). It could be argued from a free market perspective that the problem is that there has not been sufficient privatisation of property rights. Hence while the global commons of the seas has, for fishing purposes, been broken down into a series of national properties, these are still in a sense 'commons' in that individual fishermen can maximise their profits by externalising the negative impacts of over-fishing. Canada, for example, introduced a licensing and catch quota system, regulated by a system of surveillance, inspection and record-keeping, to monitor fish catches (Coull, 1984). These measures regulated foreign fishermen but increased the proportion of fish caught by Canadians, and the expansion of the Canadian fleet and fish processing industry led to serious structural problems: excess harvesting and processing capacity, and fluctuations in market prices. A review by the Canadian government in the early 1980s changed fish resource allocations and restructured the processing industry, but the industry still faces problems of fish exhaustion in the 1990s (Myers *et al.*, 1996).

Other interpretations of over-fishing see it more as a measurement problem rather than one that is intrinsic to property rights. So, for example, people such as Garcia *et al.* (1996) have highlighted how it is hard to calculate fish stocks and set quotas even though new methods of tracking fish using electronic tags may help us to understand migratory patterns and define biologically realistic quotas (Metcalfe and Arnold, 1997). It is also very difficult to monitor and enforce any agreement; it is an arduous task to check whether fish quotas are kept to and whether illegal technology, such as nets with a small mesh size, is being used or whether illegal catches are being landed.

Besides monitoring problems, economic and employment conflicts have been at the forefront of political decisions over the management of the fishing industry (Mitchell and King, 1984). Whitmarsh and Young (1985), for instance, argue that governmental weakness or unwillingness to implement regulations at the expense of economics and employment is a key factor in the continued failure of fishing regulations and state ownership. The fishing industry is a major employer. In Canada, for example, the number of people employed in catching fish in the mid-1980s exceeded 40,000, with a further 20,000 being employed in fish processing plants, and the Atlantic coast fishing industry accounted for nearly 10 per cent of the GDP in commodity-producing industry in Atlantic Canada (Coull, 1984). Given this importance it is not surprising that governments are reluctant to take measures that might adversely affect the industry. So, for example, when the mackerel fishing fleet off the south-west coast of England suffered a dramatic decline in catches by the late 1970s the government was anxious to resist calls to cut quotas both to maintain employment and to gain a favourable outcome from the EEC Common Fisheries Policy. This policy calculated quotas on the basis of the historical performance of individual countries' fishing fleets, which meant that there was less of an incentive to cut fishing quotas because this would lead to further cuts in the future. The evolution of the Common Fisheries Policy for the European Community, which allocates quotas to member states based on estimated fish stocks, has not improved the situation and an EC Commissioner even threatened to dismantle the whole of the EC's fishing policy (Gwyer, 1991)! However, regulations have little impact if governments dilute their effectiveness due to economic and political factors. Indecisive political action, for example, did not help the Canadian fishermen in the long run as the eventual depletion of fish at the Grand Banks resulted in 40,000 lost jobs when a moratorium was introduced to stop fishing completely so that fish stocks could recover.

Measures taken to impose limits on fishing have also created further problems. One example is the 'bycatch problem' that has occurred in Europe after the imposition of a quota system. The bycatch is a term used to describe those fish that are trapped, especially in large-scale drift nets and longline hooks, when fishing for a commercial species. Usually they are unwanted or it is illegal for fishermen to land these fish so they are thrown back into the sea. Globally the bycatch represents approximately a third of the regular commercial catch, or 27 million tons of fish annually (Safina, 1995) (Figure 4.4).

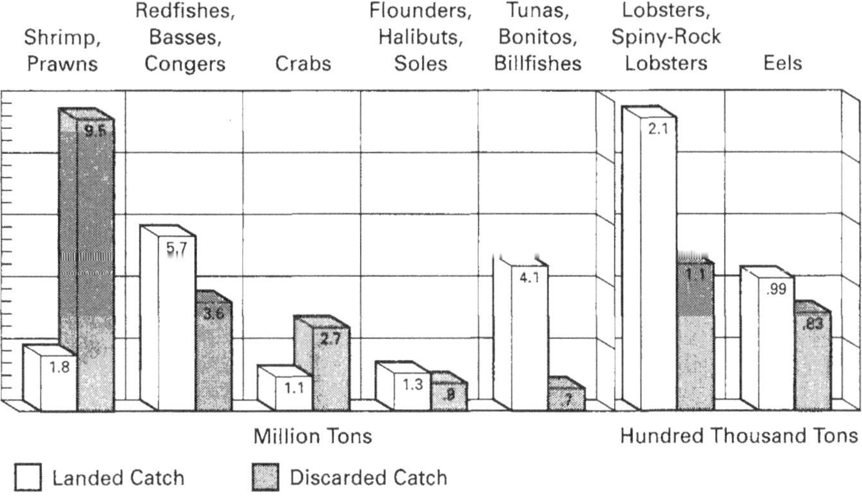

Figure 4.4 The global bycatch problem (source: Safina, 1995)

Despite the collapse of some fish stocks the industry has been slow to recognise the long-term detrimental impact of over-fishing. This may be because it has been easier politically to lay the blame elsewhere. The collapse of cod stocks off the east coast of Canada has been attributed to climate change, seal predation and ecosystem changes as well as over-fishing (Myers *et al.*, 1996). So far ownership and legislation have not been the panacea needed to ensure a long-term future for the Canadian fishing industry, and this pattern of events parallels most of the world's major fishing areas. At present commercial fishing, with or without regulation, results in over-exploitation.

Ownership can, however, be effective when the policies of the owner are abided by and all parties are prepared to compromise and recognise the need for management. For example, Watson (1997) argues that closing small areas to fishing can lead to increased catches in the remaining area. In Shimoni, Kenya fish catches fell during the 1980s so a coral-marine reserve was excluded from fishing; this resulted in increased numbers of fish being caught elsewhere because no-take areas allow fish to grow to their maximum size and this can enhance reproduction. For example a 61 centimetre snapper produces 212 times more eggs compared with one 42 centimetres long, so protecting areas increases the opportunity for fish stocks to recover.

It has been suggested that ownership at a national level often fails to protect the environment because of weak government. When, however, strong government legislation is introduced it is not popular with everyone. For example, in the USA the Federal Environmental Protection Agency (EPA) has come under intense scrutiny. The EPA set up programmes to implement the *Clean Air Act*, the *Clean Water Act* and the *Endangered Species Act* to protect the environment. Opinions are clearly divided regarding the success of the EPA (see Reilly, 1994; Environmental Protection Agency, 1994; Claussen, 1994; Fisher, 1994; Knopman and Smith, 1993; Sanjour, 1994; Prosser, 1994; Orme, 1994 and Bean, 1994). An analysis of two of the pieces of legislation demonstrates the difficulties involved in implementing environmental laws and regulations, especially resolving the conflicts that emerge between social, political and economic factors and the environment, and how problems can persist with state control.

4.4.2 Humans as law and policy avoiders

In section 4.2.3, 'Regulative methods', the role of regulations such as hard and soft laws was discussed as a managerial approach to environmental problems. The discussion suggested that the successful implementation of a law or agreement required the correct incentives, the practical ability to meet the specified obligations and an effective method of enforcing and monitoring any agreement or law.

To ensure the success of a policy it must be correctly structured and implemented. Agreements are especially difficult to structure, and take time to evolve before individuals or countries adopt their recommendations. For example, Boehmer-Christiansen (1986) reviews the formulation of a global agreement on ocean dumping of radioactive waste. Stumbling blocks during the 1970s and early 1980s include defining waste, defining what constitutes dumping at sea, agreeing on the level of waste to be dumped, where to dump it, issues of safety and how to monitor the whole process. This saga demonstrates that it is extremely difficult to construct a universally acceptable policy. Until an agreement is reached individuals and/or countries follow their own policies, which generally reflect their own vested interests. For example, those nations accumulating radioactive waste, such as the UK and USA, favour ocean dumping. If this becomes common practice it may result in the global community sharing the risk based on the least regulated and most malicious country.

Many agreements fail to work because countries will not abide by them. A declaration to meet a target is not always translated into action. For example, Victor and Salt (1995) argue that Australia and Austria have done little to meet targets on greenhouse gas emissions agreed at a conference held in Toronto in 1988 that called for a 20 per cent reduction in carbon dioxide by 2005. French (1994) summarises the problems with agreements. First, the final draft of most treaties is written so as to satisfy the views of the most reluctant party, or else the treaty will not be signed. One example is the Declaration on Climate Change produced at the so-called Rio Earth Summit in 1992. Negotiators from the USA refused to sign the paper until the requirement to reduce or stabilise carbon dioxide emissions to 1990 levels by the year 2000 was made to be not binding, seemingly worried about the effect the treaty would have on the US economy at a time close to the presidential elections (see Chapter 5; Beardsley, 1995).

Countries that consistently refuse to abide by agreements render these agreements useless. One example is the International Whaling Commission (IWC), which has been polarised by non-abiding countries. The purpose of the IWC is 'to provide for the proper conservation of whale stocks and thus make possible the orderly development of the whaling industry' (Motluk, 1996, p.12). Created in 1946 the IWC, representing 40 member countries, introduced a moratorium on commercial hunting in 1986 due to concern that whale stocks were seriously depleted (Cherfas, 1986; Motluk, 1996). Norway, Japan and the former USSR, however, regard whale hunting as part of their culture and heritage and have consistently objected to the moratorium and continue to hunt whales (Taylor, 1992).

One of the major problems for the IWC has been to estimate the number of whales in the world's oceans accurately. Since 1986 the IWC has tried to devise a scheme known as the 'revised management procedure' to estimate catch quotas, but the estimates vary, and this has been a point of conflict between the IWC and commercial hunters (see Holt, 1985; Taylor, 1992). For example, the number of minke whales is now thought to approach 60,000, not 86,700 as previously estimated. The reluctance of the Norwegian government to accept the revised figure means that whalers resumed hunting of 425 minkes in May 1996. Norway has continually refused to adhere to the IWC moratorium on hunting minkes, and insists that a catch of approximately 400–500 whales per season will not threaten the survival of the species (Masood, 1997d; Motluk, 1996).

The inability of the IWC to estimate whale stocks has antagonised Japan, Norway and the former USSR, countries that wish to resume whaling. The IWC cannot provide a common definition of what a whale actually is! Some nations include bottlenose whales, killer whales and white whales under the regulation of the IWC while others, notably Denmark and Japan, argue they are not whales but 'small cetaceans' (Holt, 1985). Nations cannot agree on the number, the function and authority of observers, which technology to use to control whaling, and what is a reliable method of ensuring hunters keep to their quotas. Designing an accurate counting procedure, especially for stocks spread out in the world's oceans, is virtually impossible. Losses through natural mortality, reproduction rates and age structure of whale populations (whales commence breeding at a certain age) complicate the issue and only a crude maximum sustainable yield can be calculated (Gulland, 1988).

Furthermore, many nations now object to hunting on the grounds that it is cruel, and have no intention of ending the moratorium. Greenpeace and other environmental non-governmental organisations claim that whaling should be banned to preserve the balance of marine ecosystems, irrespective of whether a species is endangered. This has led to accusations that anti-whaling countries are deliberately ignoring the scientific evidence that could lead to the resumption of hunting (Skåre, 1994). The increasing popularity of whale watching also strengthens the argument that there is no social or economic need for whale hunting. For example, in Japan during 1992 nearly 30,000 people went whale watching, spending over £500,000, and whale watching generates £130 million annually in the USA (Hoyt, 1993). Unless an agreement is reached soon it may be increasingly difficult for the IWC to exert any control over whale hunting countries, as they will simply leave the IWC and resume uncontrolled hunting (Motluk, 1995; 1996).

Other socio-economic issues are testing the ability of the IWC (see Skåre, 1994). After the removal of the grey whale from the endangered species list in the USA, a small Indian tribe known as the Makah asked the US government to support their request to hunt five whales per year for subsistence, ceremonial and trade purposes. Whale hunting has been an important part of the Makah culture for 2000 years. Granting the request should not have a significant impact on grey whale numbers, but it is going to intensify the friction between the IWC, the USA, large-scale commercial whaling countries and environmental groups (De Alessi, 1996).

The role of science, politics, cultural heritage and values, conservation and economics all intertwine to make an equitable formula for the management and future protection of whales a distant reality. The lack of power given to the IWC means that it is virtually impossible for it to enforce its decisions, further undermining its influence. Legislation or agreements without powers of enforcement can be ineffectual and appear to be reliant solely on goodwill. One novel idea to embody the IWC, or a similar organisation, with more power is to give them sole ownership of whales by buying out existing whaling operations. They could then sell rights for controlled commercial hunting to the highest bidder. Those opposed to whaling could purchase all or part of the quota and prevent commercial whaling (Gulland, 1988), although one might note that state ownership has so far not stopped problems of over-fishing (see above).

On a global scale, French (1994) suggests that problems of using laws and treaties to protect the environment might be resolved if individual nations were willing to recognise one international authority with sole responsibility for ensuring laws and treaties were implemented. Such organisations may help resolve the polarisation between underdeveloped and developed countries that marred the 1992 so-called Rio Earth Summit (see Chapter 5). However, as the problems facing the IWC demonstrate, any regulatory body must have the power to force reluctant parties to abide by its regulations; and if individual countries are unwilling to relinquish power to make their own decisions strong international legal frameworks to regulate resources are needed if they are going to be successful (Rogers, 1993; Barrow, 1995).

Enforcement is a big problem and laws and regulations must be enforced if they are to be successful but this can be difficult. For example, it is impractical to locate every non-point source of pollution or monitor every fishing trawler and whale hunter. For example, D'Elia (1987) argues that the ban on the release of phosphates to reverse the effects of enhanced nitrogen and phosphorus in Chesapeake Bay by the EPA and the states of Maryland and Virginia will take an expensive monitoring programme to isolate all the sources of these nutrients. Cohen (1996) also argues that the ban on mahogany and virola harvesting for two years by the Brazilian government will be also difficult to enforce. Even when regulations are made and regulatory agencies exist this does not prevent poor environmental practice. For example, in 1989 a Federal raid at the Rocky Flats nuclear complex near Denver exposed the violation of regulations and the dangerous working conditions. The main contractor at the plant, Rockwell International, subsequently paid $18.5 million to avoid prosecution (Zorpette, 1996b).

4.4.3 Humans as short-term or long-term decision makers

An analysis of any environmental issue reveals that it is extremely difficult to get a consensus of opinion on the management of a resource and the environment. Opposing viewpoints can be found on virtually every environmental problem, exemplified in the collection of papers published by Polesetsky (1991), Cozic (1994) and Sadler (1996). Many conflicts arise because any policy regarding the environment has short-term political, economic and social implications.

Most conflicts of interest arise when environmental management limits commercial opportunity. Recent examples include the ban on logging in hardwood forests to protect the spotted owl (Wilcove, 1994; Rice, 1996) and the restrictions on oil exploration in Alaska (*Independent on Sunday*, 1996), and it is cases like these that create political and economic conflicts that have, for example, threatened to dilute the US Endangered Species Act (see Box 4.7).

Box 4.7 Loss of biodiversity and the US Endangered Species Act

Species extinction appears to be occurring at an unrivalled rate during the early 1990s. Reports by the World Conservation Monitoring Centre and IUCN (World Conservation Unit) suggest that 25 per cent of mammalian species are threatened with extinction. Out of a total of 4327 mammal species 1096 are regarded as 'at risk', while 169 are classified as 'critically endangered'. Loss of species diversity has become a global problem and unintentional and intentional human modification of habitats is the main threat to species (see World Conservation Monitoring Centre, 1994; IUCN, 1996; Doyle, 1996c, 1997b, 1997c). Barnes (1996) suggests that if current trends are not reversed up to 50 per cent of the world's species will be extinct in the next 50 years, with the annual estimated rate of extinction between 10,000 and 20,000 species.

Abramovitz (1997) argues that human activities are creating a biodiversity deficit by destroying species and habitats faster than replacing them with new ones by 100 to 1000 times. All areas of human activity appear to be threatening biodiversity. Commercial exploitation, illegal hunting and poaching, land use changes and pollution all have direct or indirect effects on biodiversity (see for example Jolly, 1988; Chadwick, 1991; Hornocker, 1997; Line, 1993; Repetto, 1987). Abramovitz (1997) argues that whole ecosystems are under threat, and reducing their size and integrity inhibits nature's capacity to evolve and create new life. For example, Jennings and Polunin (1996) argue that tropical reef ecosystems are diminishing as a result of habitat-destructive fishing techniques and mining that adversely affect all trophic levels, and wetlands and mangrove cover are dwindling as they are reclaimed for agricultural and residential use, submerged by dam/barrage schemes, used as rubbish dumps and for aquaculture (Gujja and Finger-Stich, 1996; Huggett, 1998).

The 1973 Endangered Species Act was introduced by the US government to try to save plants and animals from extinction, but it has resulted in political turmoil and economic conflict. Since the 1500s at least 500 species and subspecies have become extinct as a result of human activities in the USA (Chadwick, 1995). The 1973 Endangered Species Act works on the very ecocentric principle that all life forms may prove to be valuable sometime in the future and therefore each is entitled to exist for its own sake. Thus, the Act contains powers to protect any species that is considered, based on the results of scientific research, to be close to extinction, and 109 species were deemed to be either endangered or threatened in 1973; the number has since risen to over 900, with a further 3700 officially recognised ▶

candidates (Chadwick, 1995; Mann and Plummer, 1995). Two agencies, the Fish and Wildlife Service and the National Marine Service, are responsible for assessing the status of species. Once a species is listed a government department cannot proceed on a project without prior approval from one of these agencies.

The Endangered Species Act has been successful in preventing the extinction of some wildlife. Californian grey whale numbers have increased from a few thousand to over 24,000, the number of breeding pairs of the bald eagle has risen from 400 in the 1960s to approximately 4000 in 1995 and other birds, like the peregrine falcon and brown pelican that were being poisoned by chemicals such as DDT, have also benefited. Only six listed species have become extinct, but the rate of recovery of species is slow, with only seven successes to date (Chadwick, 1995). Since 1973 22 species have been taken off the list for a variety of reasons, including incorrect listing, extinction, discovery of additional stock or recovery (Mann and Plummer, 1995). Critics argue that the Act does not, therefore, aid the recovery of endangered species, while supporters claim that rejuvenation of plants and animals will become more noticeable with time. Other species, placed on the list after their numbers fell below a viable level, are unlikely to recover.

Despite its success, the 1973 Act has received criticism for its stringent rules in favour of nature rather than economics. Critics argue that the Act restricts economic development, limits growth, creates unemployment and contravenes property rights. Opponents of the Act suggest that it encourages the destruction of wildlife, especially if legislation impedes economic development. Populations of the San Diego mesa mint and the St Thomas Island prickly-ash, two species that were close to being listed, were destroyed by development companies. Thus, for the success rate of the Endangered Species Act to improve, it is argued that incentives to landowners to practise conservation are needed, rather than more stringent regulations. Further expenditure also incites bad feelings amongst politicians (Mann and Plummer, 1995). Recently, anti-green Republican politicians, such as the former leader of the US Congress Newt Gingrich, proposed to dilute the Act, save money and weaken its powers, arguing that economic growth and jobs are more important. In March 1995 Republican politicians planned to introduce a re-authorisation bill to compensate land owners if their property lost value as a result of the Endangered Species Act and to halt new species listings or designation of critical habitats (Reichhardt, 1995).

Incidents such as the hardwood forest logging restrictions to protect the northern spotted owl (Wilcove, 1994; Rice, 1996) and the Telico dam project have created fierce controversy. Dam construction was halted when the snail darter was judged to be endangered. Subsequent research revealed that shoals of the snail darter occurred elsewhere in the USA and it has now been officially reassigned to the lesser 'threatened' list. Congress responded by setting up a committee 'to permit the extinction of a species if it caused undue social and economic hardship' (Chadwick, 1995, p.26) and special measures ensured the dam was completed.

In reality the conflict between economics and nature is not as polarised as some politicians wish to believe. Consultation procedures between the Fish and

Wildlife Service, the National Marine Service and developers are held to protect both the endangered species and the proposed development project. Out of approximately 98,000 consultations between 1987 and 1992 only 55 projects were abandoned (Chadwick, 1995). This has not stopped the House of Representatives and Congress suspending environmental regulations to allow accelerated logging, opening up sensitive wetlands for development and freezing all new listings of endangered species (Lean, 1995b).

The inability of countries to combat global climate change is confounded by short-term political and economic pressures. The Framework Convention on Climate Change at the UNCED Earth Summit, ratified by 147 countries, aims to see a

> stabilisation of greenhouse gas concentrations in the atmosphere at a level that would prevent dangerous anthropogenic interference with the climate system. Such a level should be achieved within a time-frame sufficient to allow ecosystems to adapt naturally to climate change, to ensure that food production is not threatened, and to enable economic development to proceed in a sustainable manner. (Muller, 1996, p.13).

To accomplish this Kelly (1998) argues that carbon dioxide cuts of 60 per cent are needed. Developed nations, primarily responsible for high emissions of greenhouse gases, are committed to reducing emissions but many look likely to fail to meet the year 2000 target. For example, Japan's emissions of carbon dioxide increased by 0.5 per cent during 1995 and total emissions increased by 8.3 per cent between 1990 and 1995 (Anon, 1997b) while emissions in the USA have increased by 7 per cent since 1990 and are expected to rise by 25 per cent above 1990 levels in 2010 (Macllwain, 1997; Masood, 1997a, 1997b, 1997c). Short-term political decisions account for the failure to meet targets. Kelly (1998), for example, argues that although the US government agreed that action is necessary to control emissions, the US Senate will only approve any measures to limit emissions if employment, the economy and their political appeal are not adversely affected, which means that short-term economic and political pressures may override concerns for the long-term impact of climate change.

In order to manage resources and the environment over long time scales the concept of sustainable development has been widely advocated by politicans, policy makers and academics.

Sustainable development: a long-term management strategy?

According to Soussan (1992) the ideas that underpin the concept of sustainable development have evolved as people have re appraised their concerns over the relationship between economic and social development, patterns of resource availability and use, and the resultant environmental impact. This is encapsulated in the concept of sustainable development defined in the 1987 Brundtland Report titled *'Our Common Future'* as 'development that meets the needs of the present without compromising the ability of future generations to meet their own needs'

(World Commission on Environment and Development, 1987, p.43). Since that conference the idea of sustainable development has become synonymous with the long-term management and/or development of society and the world's resources, despite people formulating different views on the concept (see Box 4.8).

Box 4.8 Sustainable development: definitions and theoretical considerations

Since ideas of sustainability and sustainable development have entered the political and academic arena both concepts, as Redclift (1987; 1992) and O'Riordan (1997b) amongst others have noted, have come to mean different things to different people. Two broad approaches have emerged: developmental and radical.

(1) Developmental

The Brundtland Report emphasised that the role of sustainable development is to meet human needs. It identified two key concepts: first that the basic needs of all people must be provided – food, water, security, employment, etc.; and second that there are no limits to development but that development is a function of existing technology, the socio-economic organisation of society and its impact on the environment.

From an economic viewpoint sustainable development looks to improve material wealth, especially for the poor, and provide lasting and secure livelihoods. Material wealth is defined by Redclift (1992) using indices suggested by Barbier (1989): food, real income, educational services, health care, sanitation, water supply and emergency aid. Thus the emphasis of this approach is to meet social and economic needs. Following this route sustainable development is used to maintain future levels of production and consumption. Good science and ecological principles are analysed, using concepts such as carrying capacity, cost–benefit analysis, economic efficiency, natural capital and valuation to achieve such aims (O'Riordan, 1997b). Economics attempts to place a value on the environment. Conservation is defined as the need to protect the resources necessary to feed international markets or, alternatively, to encourage society to make fewer demands on resources. Meeting such criteria is difficult. Redclift (1992) suggests that per capita levels of resource consumption, renewable energy issues, use of clean technology and environmental accounting in production all need to be resolved. Pearce *et al.* (1989) discuss this approach in more detail.

A departure from the Brundtland view leads to the possibility that sustainable development 'may be defined by people themselves to realise an ongoing process of self-realisation and empowerment' (Redclift, 1992, p.397). Essentially expressing still an anthropocentric view, Redclift (1992) argues that conditions must be met that allow people to practise sustainable development. Most poor people behave unsustainably out of a need to survive (see Chapter 5). To overcome this some social, economic and environmental trade-offs are necessary to allow the poor to conserve their environment and resources.

▶

(2) **Radical** The focus of sustainable development is to conserve the earth's natural resource base. Sustainable development in this sense takes a deep green approach, where ecological stability is the key issue. This approach addresses environmental degradation, loss of soil and water quality and air pollution, and determines sustainable yields of renewable resources. Ecocentric ideas of stewardship of the earth by humans, rather than for humans, feature strongly in this view of sustainable development. Embodied in this approach is the idea of ecojustice: 'treating life-support systems of the Earth with a degree of care and respect that can only emerge if society also treats itself and its offspring similarly' (O'Riordan, 1997b, p.11).

Because the concept of sustainable development is so broad it can accommodate many different perspectives, which creates potential for misunderstanding. It has been criticised as a slogan without substance, as ambiguous, impracticable and even unnecessary (Karp, 1996)! O'Riordan (1997b) suggests sustainable development has, if nothing else, provided a platform for dialogue between environmentalists, economists and politicians, because everybody perceives it to be a good thing and worth achieving. Thus it acts as a 'social energising force'.

The concept of sustainable development is still evolving. Articles by Tisdell (1988), Dixon and Fallon (1989), Lele (1991), Redclift (1992a), Dobson (1996) and O'Riordan (1997b) discuss the theoretical evolution of sustainable development and ask as many, if not more, questions as they provide answers.

Examples where the concept of using resources sustainably have been put into practice include tourism and coastal management (Briguglio *et al.*, 1996; Knight *et al.*, 1997), land use and agriculture (Wallace, 1994; Eger *et al.*, 1996; Syers, *et al.*, 1996), urban development (Wennan, 1996; Youjing and Jiadong, 1996; Molebatsi, 1996) water resources (Arntzen, 1995) and waste management (Yaliang, 1996). However, there are some major barriers to introducing sustainable development as a long-term management strategy at global, regional and local levels.

At a global level the WCED argues that seven key strategic objectives must be addressed for sustainable development to become a reality:

(1) Revive growth.
(2) Change the quality of growth.
(3) Meet essential human needs.
(4) Ensure a sustainable level of population.
(5) Conserve and enhance the resource base.
(6) Reorientate technology and manage risks.
(7) Merge environment and economics in decision making.

Further, to achieve these goals the following must take place:

(1) a political system that allows effective citizen participation in decision-making processes;

(2) an economic system that is able to generate surpluses and technical knowledge on a self reliant and sustained basis;
(3) a social system that provides solutions to the tensions that arise from present forms of unequal development;
(4) a production system that respects the obligation to preserve the ecological base for development;
(5) a technological system that can search continuously for new solutions;
(6) an international system that fosters sustainable patterns of trade and finance;
(7) an administrative system that is flexible and has the capacity for self-correction.

The Brundtland Report (World Commission on Environment and Development, 1987) and the Brandt Commission's Report of 1980 go so far as to identify development and environmental priority issues and to suggest courses of action to resolve them. However, Soussan (1992) argues that this agenda only sets out a series of well-intended statements, without offering a strategy to confront many of the issues that need to be resolved to meet the 14 points listed above. To accomplish these objectives society must undergo massive restructuring to eliminate the causes of unsustainability that can be identified on a global level, including resource consumption, exploitation of raw materials, international trade, poverty, population growth and energy consumption. Practically, the recommendations are fraught with difficulty and there is no common consensus on how best to implement them, not least because there is a need to confront a diverse range of social, political, economic and environmental factors to make the transition to a more sustainable society. Soussan (1992, p.21) argues that 'human institutions responsible for managing the planet are a long way from turning the goals of sustainable development into a clear programme of achievable actions'.

A similar situation exists at regional and local levels, where complex economic, social and political issues confront the transition to sustainable forms of development. Dixon and Fallon (1989) argue that compromises and choices will have to be made, and there still may be winners and losers. For example, the development of the rubber industry in Malaysia can now be maintained indefinitely after the conversion of forested uplands into large commercial rubber plantations. However, to do so it was necessary to replace a 'natural' ecosystem of forest with an 'artificial' ecosystem. The conversion process changed the type of fauna and flora resident and simplified biodiversity. Thus Dixon and Fallon (1989) argue that a trade off occurred with the replacement of one ecosystem with another to create conditions that are conducive for sustainable rubber production. Cater (1995) and Goodall (1995) also argue that efforts to create environmentally sustainable tourism also involve trade offs and that environmental damage may still occur through secondary resource consumption, such as through the provision and use of accommodation and transportation (see Chapter 2).

As noted in Box 4.8 sustainable development means one thing to one person and a different thing to another. Dixon and Fallon (1989) argue it is not always possible to develop different forms of sustainable development at the same time in the same place, and therefore it cannot always serve the interests of everybody. For example, the Bacuit Bay area on the island of Palawan in the Philippines has three main industries, fishing, tourism and logging, which appear to be incompatible, operating alongside each other. This incompatibility arises because logging is environmentally disruptive, causing high rates of soil erosion, and is visually unaesthetic. This deters tourism and increases problems of sedimentation and water quality in the fishing grounds (Hodgson and Dixon, 1989). Thus, the Philippine government had to make a choice between a logging industry that generated foreign exchange and exports and the protection of tourism and fishing, both of which support the local subsistence economy. Eventually, to slow down the rate of forest clearance, a logging ban was introduced in 1989. Doing so, Dixon and Fallon (1989) argue, allows sustainable resource use at a local level, but the costs of losing capital through the logging industry will effect fiscal policies at a national level and possibly prevent development elsewhere. Not only does this example illustrate the complexity of making decisions but it also shows that sustainable development projects can lead to a conflict of interest.

To use sustainable development as a long-term management strategy is fraught with difficulty. Other important decisions must be made when dealing with the exploitation of resources. These include deciding how much of a limited stock of resources to leave for future generations and to what time-scales societies should be operating upon. These issues are complicated by our limited ability to predict future needs, values and technical developments. Soussan (1992) argues that resource values can change through time; for example, before the development of nuclear power uranium had no real value to society. Dixon and Fallon (1989, p.83) summarise many of the problems with the notion of sustainable development when they state: 'while the concept of sustainability and its extension, sustainable development, engender superficial agreement, on a substantive level no consensus has been reached with regard to the meaning and implications of the concept.'

4.4.4 Humans as ignorant beings

Nations, companies or individuals often simply refuse either to accept responsibility for their actions or commit themselves to managing an environmental problem, especially if these are legally binding. Refusal to recognise a problem when it occurs or to sign a treaty means that an individual or country can continue to exploit the environment and delay any recovery. Such tactics are commonplace. For example, countries have objected or threatened to withdraw from agreements on ocean dumping (Boehmer-Christiansen, 1986), acid rain (Dudley, 1986) protection of endangered species (Chadwick, 1991) and greenhouse gas emissions (Reid and Goldemberg, 1998).

Environmentally damaging actions can also come about through lack of knowledge or ignorance: people have simply been unaware of the environmental consequences of their actions. Zorpette (1996a) argues that the problems of neutralising nuclear waste have also arisen through either ignorance or neglect. Knowledge of how to dispose of nuclear waste is way behind the science of generating energy from nuclear fuels (see Box 4.4). In the early stages of nuclear power generation it was widely assumed that waste would be reprocessed and only a minor proportion of unusable waste would need to be disposed of. It was recognised that nobody had developed an acceptable way of disposing of nuclear waste but it was felt that this would soon happen: an illustration of interventionist faith in human abilities. Waste was therefore simply stored, awaiting the development of new technologies. However, as the debate over how to dispose of nuclear waste has continued the amount of non-processed waste has grown significantly, leading in some cases to serious environmental problems. At the Hanford nuclear complex in the USA, for example, nuclear waste has been intentionally or accidentally dumped from the 1950s onwards by the Department of the Environment, with over 1.3 billion cubic metres of contaminated waste being released into the soil and the atmosphere. Sixty-seven out of 149 storage containers containing high-level waste, which were designed with a 25-year life expectancy, are now leaking after being used for over 40 years. Despite this, irrationally, nuclear waste was still being stored using these containers in the 1980s, and radioactive tritium has irretrievably leached into the Colombia river. It is estimated that billions of cubic metres of soil, groundwater and surface water have been contaminated in the USA. Because scientists still do not know the best method of cleaning up and storing the waste generated at sites like Hanford the US government is forced to spend up to $600 million annually to safely maintain the site. Even when the dangers of industrial processes are known it does not always prevent a problem from materialising: until the 1982 Nuclear Waste Policy Act (NWPA) there was no coherent strategy for disposal in the USA (Brown, 1987). The example clearly illustrates the problems society could face through ignorance, unwillingness or inability to accept responsibility for its actions.

Another example of the dangers of lack of knowledge or ignorance is the introduction of chemicals into the environment without proper knowledge of their environmental impact. For example, Mukerjee (1995b) suggests that the potential toxicity of 75 per cent of the 70,000 substances used to create approximately five million products world-wide is unknown. Irvine and Ponton (1988:34) suggest that 'over half a million chemicals are in common use and approximately one thousand are added each year' and yet 'we know little of their interaction and combined environmental effects'. One example is bis (4-chlorophenyl) sulphate (BCPS), used to produce high temperature polymers and in reactive dyes. Traces of BCPS are reported to have contaminated fish, seals and sea eagles in Baltic and Swedish coastal areas. Human ignorance of the potential environmental dangers of the chemicals used in industrial processes means that they are being released into the environment with the possibility that they could pose a serious threat to organisms at different trophic levels (Olsson and Bergmann, 1995).

Ignorance should not, however, be an excuse to exploit. Dovers and Handmer (1995) suggest that schemes such as 'ignorance auditing' should become integrated into policy decision making. Constanza and Cornwell (1992) suggest that an assurance bond system, where money is paid to cover the best estimate of future adverse environmental impacts, would be an effective way of insuring exploiters minimised environmental damage or loss. These bonds would only be repaid once the exploiter proves no damage or loss has occurred.

As we have shown in this chapter management is often seen to be a technical, regulative and scientific issue. However, if these managerial approaches are going to be effective then people must behave in particular ways. If they do not, then management strategies fail. This then raises the question of how to respond. One common approach is to seek to alter people's behaviour or to manage people. Educating people to behave in a particular way is one approach to overcome managerial failures. This can be seen as a form of social disciplining (see Chapter 2) but it is also often described as 'educating' people to adopt behaviour that is in their best interests, although they may not always realise this. Education is hence commonly seen to represent a viable, long-term option to control environmental problems. For example, exposure to toxic substances and pollutants may be reduced through educating technical and non-technical personnel at risk of poisoning from industrial processes, which should lead to safer working practices. Such an initiative is under way in Brazil, where informal gold mining (see, for example, Rodrigues *et al.*, 1997 and Chapter 2) known as 'Garimpo activity', exposes miners to mercury. It is estimated that 2000 mining sites are operational in the legal Amazon region, directly and indirectly involving approximately 4.5 million people producing at least 100 tonnes of gold per year (Veiga *et al.*, 1995). Mercury is used to separate gold from its ore, volatilising about 70 tonnes of mercury each year. Soils and fish in the gold mining region contain excessive amounts of mercury that have entered the human food chain; miners also accumulate mercury in their lungs through inhalation, which then diffuses into the bloodstream and is transferred into vital organs such as the brain, kidneys and liver (Veiga *et al.*, 1995).

To warn miners of the dangers of mercury and to convince the *garimpeiros* to adopt safer working methods educational measures have been introduced. Both preventive and remedial measures are part of the educational process. Measures include visual communication media, material in booklets, instruction manuals, meetings and courses to explain how to handle mercury and to stress its toxic effects. Inexpensive devices that curtail gaseous mercury emissions have also been given to the miners. Finally, computer programs outlining symptoms of mercury pollution and how to deal with them are being disseminated throughout the mining industry to rapidly diagnose the pollution potential (Veiga *et al.*, 1995). Once the measures are accepted by miners their knowledge will be passed down to future generations of miners, both legal and illegal, and provide a long-term solution to mercury pollution. Educational methods have also been employed to persuade people to stop dumping waste and pollutants, and to introduce better soil management practices (Mitchell, 1996).

Non-formal education about the environment through nature recreation is another way of changing public atitudes towards conservation. Popularising biodiversity issues in places like schools, zoos, herbariums and botanical gardens can raise public awareness, which will ultimately help enforce laws that protect biodiversity (Bagarinao, 1998).

4.5 Summary

This chapter has outlined six major approaches to managing natures on global, regional and local scales. Examples have shown that all of them can be effective in alleviating environmental problems.

The globalisation of environmental problems, such as global warming and ozone depletion, and of societies towards the end of the twentieth century has required people to find solutions that are global in nature, requiring international co-operation and a global scientific and political consensus. This has seen the formulation of international hard and soft laws. Regional conflicts over acid rain, water resources and pollution have also arisen between neighbouring states. To resolve these types of conflicts countries have sometimes drawn up agreements to work alongside each other; the evidence presented in this chapter suggests that these forms of management can be effective, especially when they provide incentives to limit environmentally damaging actions and they are flexible, easy to implement and avoid excessive bureaucracy.

People have generally measured the success and failure of these managerial approaches by the overall benefits and costs to societies. In other words, people are often more willing to participate in those managerial approaches that are deemed to be socially beneficial and cost effective. They also favour policies that do not substantially alter their behaviour. There may well, however, be problems with both these perspectives. For example, notions of societal/public benefit may involve the neglect of important minority views and interests. Similarly, policies that do not impact on behaviour may not alleviate problems if the cause of the problem is the way people act.

This chapter has also reviewed the problems of managing environmental problems. While some of the benefits of science and technology, such as the green revolution, biotechnology and renewable energy were highlighted, the chapter also focused on the problems caused by the complexity of natural and human-made environmental problems such as the impact of global warming. A lack of scientific knowledge has led to scientific uncertainty amongst politicans and academics, which means it is difficult to choose and implement the most effective policies.

The success of many forms of new technology is dependent upon government funding. Many governments appear to be either funding research inadequately, providing funding in the wrong area or providing none at all. More often than not funding is tied politically and therefore decisions are made for political rather than environmental reasons. Continuing research and development to obtain alternative and environmentally friendly technology and to understand how natures respond to human impacts is also crucial if society is going to make further inroads into solving environmental problems. Data collection and availability also emerge as important aspects of policy decision making, in the formulation of laws and regulations and the use of technological methods to solve local environmental problems such as oil spillages. Many environmental decisions are hampered by inadequate data to evaluate a problem. Wider access to information would help measure the performance and economic competitiveness of environmentally friendly technology and products.

The discussion of the merits of free market environmentalism and private ownership as managerial approaches to managing the earth's resources also highlighted the importance of economics in the long-term success of environmental solutions. Chapter 3 showed that both democratic, market driven economies and undemocratic, centrally planned economies have created environmental problems. Incorporating economics into environmentalism or vice versa could curtail the environmental damage that has resulted from unchecked economic development. New technology such as renewable energy, energy efficiency measures and biotechnology need a fair economic playing field to make an impact. Clearly, a stronger commitment is needed by government, industry and the public to realise the economic and environmental potential of new and more efficient technology (in terms of energy or resource consumption) against the costs of using conventional technology. However, problems still exist over the role of the state in environmental management. Economic, political and social pressures have resulted in some noticeable failures, including commercial fishing, and may strengthen the argument that free market environmentalism may prove to be more effective in some instances.

An additional dimension is changing the way in which humans behave. Behaviourial aspects of environmental problems like pollution cannot be solved by technological approaches. Non-targeted factors, especially human attitude towards the environment, must be considered when developing management strategies. Perhaps some of the main reasons for the failure of some of the environmental solutions that have been reviewed in this chapter is the reluctance of individuals or nations to adhere to laws and policies, a failure to implement the correct policies or the design of ineffectual policies that do not tackle the root causes of poor environmental practice, and also through ignorance of the true cause and effects of human actions, for example the release of pollutants into the natural world. There has been some success in changing people's attitude towards nature. The introduction of voluntary codes of practice, recycling and waste minimisation schemes and using nature to tackle environmental problems is a step forward. The prominence of sustainable development on the international political agenda provides people with the opportunity to consider the long-term consequences of their actions.

Chapter 5

Environmental Problems in Social Context

5.1 Introduction

The preceding chapter looked at how environmental problems may, or may not, be successfully managed. The notion of 'managing' nature can be seen as an essentially 'technocentric approach' and more specifically linked into O'Riordan's notion of 'accommodators'. As we will discuss in the next chapter such approaches may be challenged by people adopting a more 'ecocentric' perspective because they feel technocentrics fail to appreciate and value nature and the natural sufficiently. In the present chapter, however, we will examine a rather different line of critique that has been levelled at managerial approaches, namely that they fail to consider sufficiently both the social causes of environmental change and environmental management/mismanagement.

As has been demonstrated in the previous chapter, there is a clear tendency for advocates of environmental management to see the causes of environmental degradation and hazards as lying essentially in the failure to create or adopt the right sort of environmental knowledge. In this chapter we want to challenge this assertion by suggesting, amongst other things, that there is a need to recognise that in many social situations, or 'social contexts', there may be very good or 'rational' reasons for undertaking environmentally damaging actions, and that therefore the prevention of these actions requires social as much as intellectual or technological change. We will argue that in the contemporary world a key social context of human–environment interaction is capitalism, and that this may have inherent tendencies that will encourage people to undertake environmentally damaging actions. We will also suggest that the consequences of environmental degradation within capitalist societies, and indeed many other societies, are unequally distributed.

This chapter is divided into five main parts. First, we will briefly illustrate how social context may be seen as crucial in understanding the application and non-application of environmental good practice. Second, we will highlight different ways in which the responsibility and consequences of environmental degradation are commonly assigned to particular social groups. Third, we will outline how quite seemingly disparate social groups and environments may be seen to be situated within, and affected by, capitalism. Fourth, we will consider claims that capitalism may not only produce environmental problems, but may also be made solve them. Finally, we will explore attempts to develop global sustainable economies. In each section we will develop arguments using a narrow set of environmental 'problems' – namely deforestation, famine and industrial pollution – although the arguments are also applicable across the full range of issues covered in the previous two chapters.

5.2 Environmental degradation in the age of environmental awareness

5.2.1 The implementation gap and its culprit

There is a tendency within discussions of environmental issues to see them as a new concern. However, as outlined in Chapter 2, human use and degradation of environments has a long history and indeed many contemporary environmental 'problems' or 'crises' have actually been long recognised. This point has been well made by Blaikie and Brookfield (1987b), who have noted how the impacts of people upon land quality have been widely recognised since the late 1930s. They also add that in spite of this, 'whole United Nations agencies and a worldwide environmentalist movement' (Blaikie and Brookfield, 1987b, p. xvii) have been unable to make more than a marginal impact upon levels of soil erosion. They suggest that this is largely because the 'land managers' – that is the various people and agents who are responsible for controlling the use of areas of land, the 'peasants, pastoralists, commercial farmers, state forest departments and so on' (Blaikie and Brookfield, 1987b, p. xvii) – are 'unwilling or unable' to stop undertaking environmentally damaging actions. There therefore appears to be something of an '*implementation gap*', whereby seemingly logical policies or courses of action are not being implemented in people's practices (see also Chapter 4 for examples in respect of other environmental issues). As Cloke and Little (1990, p. 96) highlight, there are two common reactions to such a gap. First, there is heightened emphasis placed on the design of policy, with issues of implementation becoming 'relegated to the technical backroom'. Second, and relatedly, implementation problems become something of a 'scapegoat' for any inadequacies in policy and policy making.

These two reactions are clearly evident in discussions of environmental management. Blaikie and Brookfield (1987b), and also Blaikie (1985), have argued, for instance, that issues relating to the implementation, or non-implementation, of soil erosion policies have often been neglected, being seen as something of an unopened 'black box'. They argue that this feature in part stems from what they see as the unhelpful division between the physical and social sciences, with the former playing a considerable role in environmental policy design but seeing issues of implementation as being beyond their remit and best left to social scientists. Social scientific analysis of environmental policy making, on the other hand, has been limited both by being marginalised from discussions of policy design by physical scientists and managerial professionals and because social scientists themselves have felt that issues of nature and environment are not within the disciplinary focus of the social sciences.

Blaikie and Brookfield also illustrate that discussions of environmental policy often engage in scapegoating the process of implementation and, more particularly, in scapegoating the implementors (or more precisely the non-implementors) of environmental policy. Blaikie and Brookfield (1987b, p. 34), for example, comment that there has long been a practice in discussion of environmental policy to 'decry the "stupidity" or the "conservatism" or the "uncaring idleness" … or the "ignorance"' of farmers and other land managers and to suggest that these people are to blame for the non-implementation of the knowledge created by 'governments and international experts' (p. 35), which could help prevent, reduce or alleviate environmental degradation.

Both these reactions to the implementation can be criticised from a number of different directions. One line of argument is that knowledge of environmental problems and strategies for their resolution go back even further in history than the 1930s and are to be found in many more sites than the offices of government officials and international experts. In Chapter 2, for instance, it was remarked how Amazonian Indians such as the Kayapo had established complex sets of practices to use the tropical rain forest, practices that were superior in terms of their long-term sustainability than were the agricultural practices developed along the lines of modern European environmental knowledge. In this instance one not only has environmental awareness not being implemented, but it being erased through the processes of economic and cultural imperialism. Another example of this is described by Watts (1983) in a discussion of famine in the Nigerian Hausaland. Watts argues that in this area of northern Nigeria there was, prior to British colonialism, a highly sophisticated 'subsistence ethic' or way of life that enabled the peasant farmers of the area to deal quite effectively with any periods of drought (see Box 5.1). However, with the onset of British colonialism, this way of life was eroded and replaced with agricultural practices that meant that periods of drought were increasingly associated with periods of famine (the distinctions and associations between drought and famine were discussed in Box 3.4).

Box 5.1 The pre-colonial subsistence ethic in the Nigerian Hausaland

Watts (1983), drawing on the work of Scott (1976), suggests that in the period prior to British colonial rule in 1903 there existed in northern Nigeria a 'subsistence ethic' that operated in three principal ways. First, peasant farmers adopted 'risk aversion' or 'safety first' strategies towards agricultural production. This, involved, for instance, the planting of drought-resistant rather than maximum-yield crops. There was also the intercropping of drought-resistant crops with other less drought-tolerant crops so that there would be some crop yield in both cases of rainfall or drought. A second element of the subsistence ethic involved the establishment of mutual support mechanisms to store and redistribute foodstuffs. An important element of this was extended kinship and clan grouping, which meant that people could turn to a range of other people in times of hardship. Other constituents were religious and civic beliefs and ceremonies that involved the redistribution of foodstuffs from elite groups to the poor. For example, a *sarka noma* or 'king of farming' was elected on the basis of being able to produce in excess of 1000 bundles of grain, which became a store from which others could borrow interest free until harvest time. The third element of the subsistence ethic was the existence of a 'feudal' system of 'state' support for those in difficulty. In pre-colonial Nigeria there was a hierarchical system of political organisation and obligations stemming from Emirs through district fiefdoms and down to village chiefs. At each level there was appropriation of foodstuffs, often through taxation in kind (see text), which was then made available in times of famine. Redistribution would generally start at the village level, but if these stores were exhausted then redistribution would move up to the level of the fiefs and then, in extreme cases, to the level of the Emir.

Cases such as the Kayapo Indians and the farmers of the Nigerian Hausaland highlight three problems with the still widespread understanding of environmental problems, which argues that they largely stem from an 'implementation gap' caused by people, or some groups of people, being unwilling or unable, through ignorance, to implement the knowledge of 'environmental experts'. First, they devalue other forms of knowledge, such as the 'local knowledges' of the Kayapo and pre-colonialists in the Nigerian Hausaland. Second, they can be seen to overvalue the knowledge of contemporary 'environmental experts', seeing them as being almost by definition better than other knowledges and able to exert considerable control over nature. Third, they can be seen create to categorical differences between people.

As discussed in Chapter 1, the second issue in particular has been a focus of attention by ecocentrics who are sceptical about claims to control nature. While many people might not articulate a clear ecocentric philosophy there is certainly evidence of a quite widespread scepticism about the claims of environmental experts. As Blaikie and Brookfield (1987a, p. 35) record, studies have suggested that there is extensive hostility to governmental environment advisers within peasant societies, while MacNaghten and Urry (1998) argue, on the basis of focus groups with people in Lancashire, Britain, that there is a widespread feeling that state institutions are 'part of "the system" which generates environmental problems ... rather than benign agents committed to solving them' (MacNaghten and Urry, 1998, p. 231). People were also seen to be very sceptical about the environmental information they were being given, and instead preferred to rely on their own sensory experiences of the world around them. This reluctance to trust the knowledge of experts may be clearly warranted, as was starkly demonstrated in the BSE crisis in Britain (see Box 5.2).

Box 5.2 BSE and the risky business of eating British beef

According to Beck (1996), the handling of bovine spongiform encephalopathy (or BSE or 'Mad cow disease') in Britain is indicative of the formation of what he had earlier described as a 'risk society' (Beck, 1992). As MacNaghten and Urry (1998, p. 254) note, such a society can be seen to be characterised by an 'intense public awareness of the riskiness of … daily life … [and] of the manifest uncertainty of those risks'. In other words people know that their pursuit of every-day life carries endemic risks but they do not feel that there are any secure assessments of these risks and, in a sense therefore, no rational strategies to avoid them. This can be seen to have clear resonances with the BSE crisis, in that what was until then an 'every-day and taken-for-granted social practice' (MacNaghten and Urry, 1998, p. 262) suddenly became a practice with both potentially deadly consequences and affecting an incalculable number of people. It was also a risk that was not open to calculation through the senses – one cannot see or smell or feel whether beef carries BSE. As a result people are highly reliant on 'expert systems' (Giddens, 1990; 1991), that is on sources of risk assessment provided by supposed experts, such as scientists and government officials and ministers. However, the BSE crisis showed clearly that experts can get things wrong. The disease was first detected in 1985, and appears to have emerged as a result of a government decision to change the process by which cattle feed using animals' offal and other animal by-products was manufactured. As a consequence of this decision it appears that sheep offal infected with the disease scrapie came to be fed to cattle. Many environmentalists argued that feeding a herbivore such as cattle with feed made from other animals was unnatural and therefore potentially dangerous. The government quoted scientific research to claim that there were no dangers, that scrapie from sheep meat could not infect cattle and that there was no 'conceivable hazard to human health'. This remained their stated position from 1985 through to 1996, although a number of scientists and others began to express reservations about this view. Suddenly, on 2 March 1996, the government announced that it, or more precisely its advising scientists, now considered that there might be a connection between eating meat from cattle infected with BSE and a variant of the fatal Creutzfeldt-Jakob Disease (CJD). The impacts of this announcement were dramatic. Sales of beef plummeted and beef farmers and abattoirs were faced with an almost complete loss of income, made more severe by a European Union imposed ban on the export of British beef and beef products. Furthermore, as MacNaghten and Urry note, public faith in the 'expert systems' of government and science was shattered. This faith in expert knowledge was misplaced and in the end was clearly shown to be wanting. As MacNaghten and Urry conclude:

> British people were placed in a laboratory in which cattle were turned into carnivores and on occasions into cannibals, but where there was no knowledge of when the outcomes of the scientific experiment would be known … it is clear that no one should have been exposed to such dangers in the laboratory of the English dinner table (MacNaghten and Urry, 1998, p. 263).

The third problematic feature of 'implementation gap' interpretations of environmental degradation is that they often reconstruct categorical differences between people. In other words environmental problems are seen as being created by a particular group with a particular set of attributes. Moreover these attributes are seen not to be shared by the environmental experts: while non-implementing 'land managers' are ascribed as being lazy, ignorant and conservative, these attributes are not seen as applicable in any sense to environmental experts. In large part this is because there is very little attention paid to the lives and social circumstances of the 'experts': it is their knowledge that is seen as being of importance and often very little consideration is paid to how this knowledge was created and to the lives and characteristics of these knowledge producers. As Escobar (1996, p. 50) has argued, there is a need to pay more attention to 'who is this "we" who know what is best for the world', because too often it has been an uncritical acceptance of the figures, knowledges and patterns of action of 'the (white male) western scientist-turned-manager' and an uncritical rejection of other people, knowledges and practices. This difference in focus of attention and ascription of characteristics can be seen as a particular instance of the process of 'othering' (see Box 5.3).

Box 5.3 The concept of 'othering'

Recent work in social and cultural studies has often employed the notion of 'othering'. This involves the establishment of differences between people *and* the evaluation of some forms of difference, and hence some people, above others. It has been suggested that this form of evaluation often centres around people identifying a particular feature of themselves or people like themselves that becomes a benchmark against which other people are evaluated and distanced. Hence the process of 'othering' involves a process of constructing a 'Same' (people like me) as well as a construction of 'Others' (people not like me). A clear discussion of such constructions can be found in Crang (1998).

Examples of this process can be found in much of the environmental literature, including many of the central texts, such as Thomas Malthus's *Essay on the principle of population* (Malthus, 1798). Although, as outlined in Chapter 3, this text has often been viewed as providing a general model of how society may be limited by nature, it should be noted that the original text placed the responsibility of over-population squarely on the working class, claiming that 'the poor are arbiters of their own destiny, and what others can do for them is like dust in the balance compared to what they can do for themselves' (Malthus, 1800, p. 25). Over-population was effectively caused, so Malthus maintained, by the lack of moral constraint amongst the working classes.

This argument was empirically highly dubious. There are many historical studies that have revealed that the working classes in many ways had a stricter moral code than many members of the middle and upper classes. The argument, however, found much favour amongst the ruling classes in Britain because it effectively absolved the rich and powerful from doing anything to alleviate the situation. It is also an argument that has many contemporary re-enactments, such as the ascription of global warming to deforestation created by the 'unknowledgeable' actions of peasant farmers (see Figure 5.1).

Figure 5.1 Constructions of the environmental culprit

Deforestation is certainly a highly significant practice, although rates of deforestation vary between different regions and forests still cover 30 per cent of the terrestrial land surface, although once again there are considerable spatial variations. As Table 5.1 demostrates, much of the largest decline in forest cover, in relative and/or absolute terms, is currently occurring in tropical areas, where according to Doyle (1996c) the amount of 'natural forest' has declined by one-fifth between 1960 and 1990. It has been estimated that the current level of deforestation is about 1.5 acres per second, equivalent to totals of approximately 100,000 km^2 per year (Sioli, 1985), or an area the size of the US state of Ohio, or England and Wales (New Forests Project, 1991). Approximately 2 per cent of rain forest was destroyed world-wide in 1991, with 14 areas accounting for 43 per cent of that figure (Myers, 1993b). In some areas the levels of deforestation has meant that relatively little forest now remains. Africa was once a leading exporter of tropical hardwoods but many of its tropical forests have been overcut and few remain productive. In the Gabon over 60 per cent of tropical forests have been lost and in Ghana and the Ivory Coast 80 per cent of original rain forest areas have disappeared (Lewis, D., 1991). A similar pattern of destruction exists in south-east Asia, where Thailand has lost 80 per cent of its rain forest cover and the Philippines over 90 per cent (Lewis, D., 1991). Given that forests are, as Soussan and Millington (1992, p. 80) observe, of considerable economic and social significance in that they form both 'the basis of a range of industries' and are also 'home to millions of people', then their loss is of considerable local significance. It is also, as Figure 5.1 highlights, seen to be a global issue, not least because forests act to absorb CO_2 and release oxygen into the atmosphere. Conversely, however, the burning of wood, either in the process of deforestation or subsequently, acts to release CO_2 that, as discussed in Chapter 3, may contribute to an enhanced atmospheric greenhouse effect and thereby, arguably, global warming (see Bunyard, 1987; Cartwright, 1989; Sage, 1996). Furthermore, as Soussan and Millington (1992) also note, forests may perform a variety of other 'vital environmental services', such as reducing both the impact of rain on earth surfaces and overland and channel flow of water, building and binding soils, and in providing habitats for a large proportion of the earth's flora and fauna. Tropical forests are said to contain at least 50 per cent of all the earth's species (see Cartwright, 1989; Myers, 1989; Bird, 1991; Sage, 1996) and constitute a largely untapped source of genetic material of potential use for medicines and pharmaceutical drugs (see Box 4.3), but each year up to 10,000 species may become extinct, which Sponsel *et al.* (1996) argue is the highest rate of extinction in 'geological history'. Farrar (1998) suggests that 1 in 12 tree species is threatened with extinction and 7000 species are at immediate risk, although such high extinction rates have been disputed by Simon and Wildavsky (1995).

Table 5.1 Rates of annual deforestation 1980–90, for selected countries

Country	Amount deforested (,000 km^2)	Percentage change	Total area of forest (1990) (,000 km^2)
Japan	0.0	0.0	238
Madagascar	1.3	0.8	158
Ecuador	2.4	1.8	120
Philippines	3.2	3.4	78
USA	3.2	0.1	29,660
India	3.4	0.6	517
Thailand	5.2	3.5	127
China	8.8	0.7	1,247
Indonesia	12.1	1.1	1,095
Russian Federation	15.5	0.2	7,681
Brazil	36.7	0.6	5,611

(Source: World Bank, 1997)

While levels of deforestation are clearly highly variable and their precise levels open to dispute, these issues will not form the primary focus of attention here (but see Soussan and Millington (1992) for a review of the significance of such issues). Rather we wish to focus upon the social causes of deforestation, and in particular the validity of explaining deforestation as the outcome of the actions of unknowledgeable peasant farmers. We would argue that such an explanation, encapsulated by Figure 5.1, is misleading, not least because a significant proportion of the deforestation in many tropical rain forest areas is not done by people living and farming in the area, but is rather undertaken by commercial logging concerns for timber, for large cattle barons or for the demands of state governments. For example, official estimates in Brazil between 1966 and 1975 suggest that peasant farmers cleared only 17.6 per cent of the total area deforested, while large-scale cattle ranchers and highway constructors accounted for over 60 per cent (Plumwood and Routley, 1982). Even these figures probably overstate the amount of forest cleared by small peasant farmers, particularly in the more recent periods when peasant colonisation programmes have been wound down and the activity of large corporate ranchers and logging companies has been stepped up. According to Arnold (1987) shifting cultivation was responsible for only 35 per cent of deforestation in Latin America, although this rose to 50 and 70 per cent in the Asian and Africa continents. Sponsel *et al.* (1996) estimate that approximately 85 per cent of Amazonian deforestation in Brazil is caused by some 5000 cattle ranchers. In 20 years of road development in Amazonia cattle numbers have increased from virtually nothing to five million. By 1983 it is estimated that 15,611 square kilometres of land had been converted to pasture in the Brazilian state of Rondonia alone (Moran, 1996).

Part of the problem is the extensive manner in which commercial logging is generally carried out in tropical forests (see Uhl *et al.*, 1994; Rice *et al.*, 1997). Only a relatively small number of tree species are desired for timber, with mahogany being amongst the most prized. It has been calculated that the number of trees extracted generally lies between 2.9 and 9.3 per hectare. However, because of the means by which the timber is extracted many more than the desired trees are felled or damaged. Bulldozers, for example, are generally used to open up the forest to allow for the mechanical extraction of the felled trees. This process is said to decrease canopy cover by about 35 per cent and damages approximately 150 trees per hectare. Furthermore some 25 per cent of the trees felled for timber are not retrieved because the collectors never find them.

The significance of the logging companies and the cattle barons in comparison with the small-scale peasant farmers is further emphasised in many countries, particularly in Latin America, if one looks at the sequence of land occupation. Small-scale farmers often practise shifting cultivation whereby an area is cleared, farmed for a few years and then abandoned when the fertility of the soil is exhausted, and a new area of forest cleared. However, shifting cultivators in tropical rain forest, generally clear secondary growth of vegetation on land already cleared by the logging companies and cattle ranchers. Indeed, as Blaikie and Brookfield (1987c) note, small-scale colonist farming is often a relatively minor element in the general sequence of deforestation and land use in tropical rain forest, which begins with logging companies clearing the forest. The colonists then move on to the land to practise a form of shifting cultivation until crop yields decline substantially (see Sioli, 1985). At this point the colonists seed the land with grass for cattle and shortly afterwards commercial cattle producers then buy the land for ranches at a cheap price. However, land degradation continues apace, because with only grass cover nutrient loss and soil erosion occurs at an extremely high rate and many areas are soon abandoned by the cattle ranchers as being unfit for grazing. The end result of all these land uses and transfers is what has been termed a 'red desert' (Goodland and Irwin, 1975) in that what is left is a poorly vegetated and easily erodable reddy-coloured podzolic soil.

A similar picture emerges when one looks at deforestation in south-east Asia. An estimate of deforestation in Indonesia by the United Nations' *Food and Agriculture Organization* (*FAO*), for example, suggests that some 2000 hectares are felled annually by shifting cultivators, while some 800,000 acres are cleared by commercial logging companies (Plumwood and Routley, 1982). Again, much of this land cleared for shifting cultivation has already been cleared for timber and roads. Furthermore, many of the farmers clearing forest for shifting cultivation are people who have themselves been displaced from another area of the forest by the logging companies (Plumwood and Routley, 1982). Indeed, in many tropical rain forests population increase is not the result of high birth rates (and, if one applied the arguments of Malthus, thereby reflecting lax moral behaviour amongst peasant population), but is the result of the in-migration of people into the area. This in-migration has been actively encouraged by many of the governments of these areas, with population resettlement into rain forest areas occurring in many countries including Brazil, Colombia, Peru and Indonesia. This again raises the question as to who really might be said to be the cause of environmental degradation: is it the peasant who moves into a rain forest to start sedentary agriculture? Or is it the governments that encourage – and force – people to move into these areas? Or is the issue even more complex than this?

5.2.2 Universalising environmental problems

We will look again at the issue of who, or what, might be said to be the cause of environmental degradation associated with practices such as deforestation or agricultural production in a later section. For now, we want to highlight another widespread but problematic construction of contemporary environmental problems, namely a tendency to 'universalise' or 'globalise' them: that is to see them as being caused by everyone and/or affecting everyone.

Environmental problems such as global warming or depletion of the ozone layer are often presented as being the outcome of myriad individual actions that more or less everyone is seen to perform. Hence, for example, by the act of breathing everyone is contributing to the production of CO_2 and every-day acts such as burning fossil fuels to heat homes, provide light and to travel to school, work or the shops are seen to contribute even more. There is a construction of universal guilt, an 'ecumenical we' as Bookchin (1979) has put it.

Such interpretations are widespread, particularly within ecocentric interpretations of contemporary society–nature relations, but they have been subject to some criticism. Hilyard (1993, pp. 30–1), for example, has argued that while atmospheric ozone depletion is often blamed on a more or less universal consumption of aerosols and refrigerators, it should be blamed on the producers of CFCs such as *Du Pont* that have, in his words, used 'their global reach to globalise sales of CFCs and other-ozone depleting chemicals regardless of the known environmental impact'.

We will discuss the case of CFC production and Du Pont later in this chapter, but Hilyard's basic argument is that placing the blame on myriad consumers seems to absolve producers from any blame, many of whom have many more resources to determine environmental costs than do the consumers, and who should therefore share a proportion and, indeed, the bulk of any blame. Dickens (1996, p. 159) makes another, rather more general criticism of universalistic constructions, when he suggests that holistic constructions of environmental issues, while in some ways addressing the form of contemporary society–nature relations, are problematic in that, 'the kind of "whole" to which they refer 'bears precious little relationship to the "whole" of a complex modern economic, social and political system'. Buttel and Taylor make a similar point, claiming that:

> within both science and politics, the 'globalization' of the environment has served to steer attention to common interests in environmental conservation, and away from analysing the difficult politics that result from the different social groups and nations having highly variegated – if not conflicting – interests in contributing to and alleviating environmental problems (Buttel and Taylor, 1994, pp. 228–9).

Environmental degradation is also universalised and globalised in that it is often portrayed as something that affects everyone equally. There are many expressions of concern in the media and by some politicians about the 'global impacts' of environmental changes such as global warming and ozone depletion. In Chapter 1 it was noted how notions of the earth as a finite resource emerged in the 1970s alongside imagery of 'Spaceship earth' and 'Only one earth'. Another metaphorical expression that was also used at this time was the earth as a 'lifeboat', and it was frequently suggested that everybody, wherever they lived on the earth, was 'in the same boat'. As MacNaghten and Urry (1998, p. 214) remark, the concept of the 'same boat' implied that everybody in the world shared common and finite 'planetary resources', and that unless care was taken with these resources 'we risk common catastrophe'. They argue that this concept was most strikingly applied to the notion of a 'nuclear winter', 'but was then used in relationship to other environmental risks such as resource use, acid rain, ozone depletion and greenhouse warming' (MacNaghten and Urry, 1998, p. 214).

As we shall discuss in more detail later in this chapter, the notion of common, global fates was often connected to notions of common, global causes. However, as Figure 5.1 also demonstrates, it was frequently also connected to the construction of specific, 'othered', culprits. As the cartoon highlights, the Amazonian peasant farmer is seen as not only degrading his local environment, but also threatening to produce a global calamity.

While Figure 5.1 can be seen to be using humour to rightly question the notion that the major culprit in deforestation and global warming is the small peasant farmer, it is also important to recognise that the construction of deforestation as 'a global environmental problem' can be also be seen as problematic. In particular, it can be argued that the construction of global environmental problems often acts to displace attention away from localised problems. In the case of deforestation, for example, it is important to recognise that while it may *at some point in the future* have consequences that will be felt by people quite distant from the sites of deforestation, there are many people living in the areas undergoing deforestation who are *already* experiencing considerable local impacts, including considerable environmental degradation.

Tropical forests hold most of their nutrients in the 'biomass', that is in the living and decaying plant matter, rather than in the soils, as is the case in temperate ecosystems. As a result of the structure of the tropical ecosystem, the removal of the biomass through deforestation destroys or removes the very environmental elements that make the tropical rain forests productive. As a result, any attempt to farm land through felling leads to environmental degradation. Subsistence techniques of shifting cultivation minimised the impacts of this degradation. However, as land is increasingly being consumed by ranching, logging and construction interests, and as state governments encourage the migration of people into such areas, there is in many areas a serious problem of land degradation. The majority of the logging companies view such land degradation as simply not being their problem. It is a more serious problem for cattle ranching, which depends upon the growth of grasses to feed cattle, but as discussed earlier the response of many cattle ranchers is not to attempt to repair, or even to prevent or slow down soil erosion, but simply to clear a new area of forest every few years and move their cattle on to it.

The logging companies and the cattle barons have the power to ignore or escape the environmental consequences of their actions. This is not often the case with the small farmer households, who frequently have to live with the consequences of their own and other people's environmental degradation – consequences that include declining soil fertility and crop yields. In the Amazon many of the colonists have struggled to produce enough food even to feed their own households, and many have effectively merely swapped their landless poverty in areas such as north-east Brazil for being extremely poor landowners.

Deforestation is not merely undertaken to clear land for cultivation: timber is also extremely important as a fuel supply. Indeed timber is the principal fuel source for most of the population in the so-called least developed countries. However, continued felling of trees for timber can create serious fuel shortages and also has further knock-on effects in terms of agricultural production, in that the poor in a degraded environment increasingly have to spend time gathering timber from further and further away, and therefore spend less time in agricultural activities. Redclift (1984), for example, notes that in The Gambia fuel is so scarce that some 360 women days a year per family is spent on gathering it. The term 'women day' here is highly significant, in that in underdeveloped countries the task of collecting wood, and indeed undertaking most of the heavy manual tasks, is generally undertaken by women.

The gender inequality in the burden of work is not just a moral issue about inequality: it also has a direct causal relationship with the prospect of economic and social development. One study of women's health in Kenya, for example, has concluded that the daily burden of collecting water (which typically involved women, many of whom were pregnant or weaning their children, and meant carrying 20–25 kilogram loads for 3.5 kilometres once or twice a day over very rough terrain in temperatures of over 40°C) has two major negative impacts. First, it takes up much time and effort that would otherwise have been put into direct farm work. Second, it has been argued that the cumulative effect on women is considerable:

> high nutritional demands go unmet and presupposes women to a range of debilitating diseases. In many cases prolonged exposure to such stress leads to chronic disablement and the burden on the more healthy women is concomitantly increased (Fergusen, 1985, p. 13).

Environmental degradation can therefore be seen as a crucial initiator of a variety of downward economic and social cycles amongst those people who are already amongst the most deprived. Environmental degradation is, hence, clearly not in the interests of such groups, and yet many still continue to undertake actions that will create such downward economic and social cycles, or resist attempts by 'environmental' agencies to counteract the impacts of environmental degradation. This is the crux of the implementation issue and it is to explaining such apparent contradictions that attention needs to be directed.

5.2.3 Explaining the implementation gap: some initial direction markers

In the preceding two sections we have discussed two ways in which the presence of environmental problems in an age of environmental awareness is frequently understood. First we argued that environmental problems are often effectively 'localised', 'individualised' and 'othered', in that they are seen to be caused by particular groups of people who are unwilling or unable by dint of particular personal characteristics to implement the required course of actions to prevent detrimental environmental change. This construction often leads to strident calls for people to change their ways of acting, often supported by various behavioural 'incentives' and 'sanctions'. At other times, however, it can lead to indifference and exclusion, in that the people who inhabit degraded environments are viewed as have brought this upon themselves by failing to act in the correct manner.

It was also argued that, converse to this construction of 'categorical social difference', discussions of environment issues have often adopted a 'universalising/globalising' perspective. In these cases environmental problems are seen as being caused by and/or affecting everyone on the earth. It was argued that this view fails to recognise the presence of clear differences in the values, interests and power held by various people and organisations.

If there are problems with both these interpretations, then what is the alternative? We would suggest that there is a need to develop interpretations of environmental actions and society–nature relations that steer a course that both recognises social and geographical differences and, at the same time, pays attention to social and geographical inter-connection and commonality. We would argue, to use the words of Buttel and Taylor, that there is a need to recognise:

> the multi-faceted reality of global change, which invokes very complex ecological relations (between locally functioning and globally functioning ecological systems) and complex socio-environmental relations (within and between local, meso, and global levels) (Buttel and Taylor, 1994, pp. 228–9).

Even more specifically we will argue through the course of the rest of this chapter that there is a need to:

(1) Situate environmental actions in both historical and contemporary 'social relational' or 'supra-individual' (bigger than the individual) contexts.

(2) Recognise that any one person or organisation will be in a range of different social relational contexts and that these will each be constituted at a range of difference spatial scales.
(3) Consider how people are differentially positioned within social relational contexts and how their positioning will affect their ability to perform particular environmental actions.
(4) Explore the extent to which social relations contexts can be modified by individual or localised action
(5) Explore the extent to which one social relational context may be modified by features external to it.

We will begin by exploring whether one particular social relational context – capitalism – can be seen to be the cause of unsustainable environmental actions.

5.3 Capitalism as the originator of unsustainability?

5.3.1 Situating action in social context: a case study of deforestation

According to people like Blaikie (1985), Blaikie and Brookfield (1987a) and Redclift (1984), who all adopt a variant of what is termed a '*political economy perspective*', the answer to the seemingly paradoxical situation that people who suffer some of the major consequences of environmental degradation continue, even in an age of environmental awareness, to perform actions that are, at least in part, the causes of this environmental degradation, may be seen to lie in the social context in which these environmental actors live. Blaikie and Brookfield (1987b, p. 3), for example, argue that those they term 'land managers' often have to change their use of land in response to 'changes in their social, political and economic circumstances quite independently of changes in the intrinsic properties of the land they employ'. They claim that small peasant farmers are frequently placed in situations where they are compelled by external forces and social groups into growing crops or undertaking land management practices that have adverse impacts upon the environment, and in the long term upon the sustainability of their livelihoods. O'Riordan (1988) has produced a useful summary of some of these forces, both diagramatically and textually (see Box 5.4), and it is possible to illustrate some of O'Riordan's arguments with reference to deforestation in Nepal.

Box 5.4 O'Riordan on the forces of unsustainability

According to O'Riordan the notion of 'sustainable development' can be considered as 'the refuge of the environmentally perplexed' and as a 'contradiction in terms for a modern capitalist culture' (see also O'Riordan, 1988; O'Riordan, 1989, p. 93). Less rhetorically he identifies a series of forces that act against the creation of sustainability. Drawing from a range of studies of what he terms 'three great environmental dilemmas facing the Third World, namely soil erosion, desertification and tropical forest depletion' he identifies the following as general explanatory features of unsustainable activity:

(1) Most of the non sustainable action is taken by the accumulation of small decisions made at the household level by people who are trapped into undermining their own livelihood.

(2) Such actions are essentially uncontrollable unless the structural conditions that induce poverty and desperation are altered.

(3) Middlemen who take advantage of the desperation of the poverty stricken and the landless exploit any propensity to accumulate surplus by expropriating capital through extortion and debt creation.

(4) Militarism, and especially civil war, which is now commonplace in poor Third World countries, strikes against any successful approach to sustainable development by drawing away capital into arms, removing able-bodied labour into warfare, and physically destroying the vital infrastructure of rural development. It is unlikely that any long-term agricultural programme built on sustainability can remain unscathed.

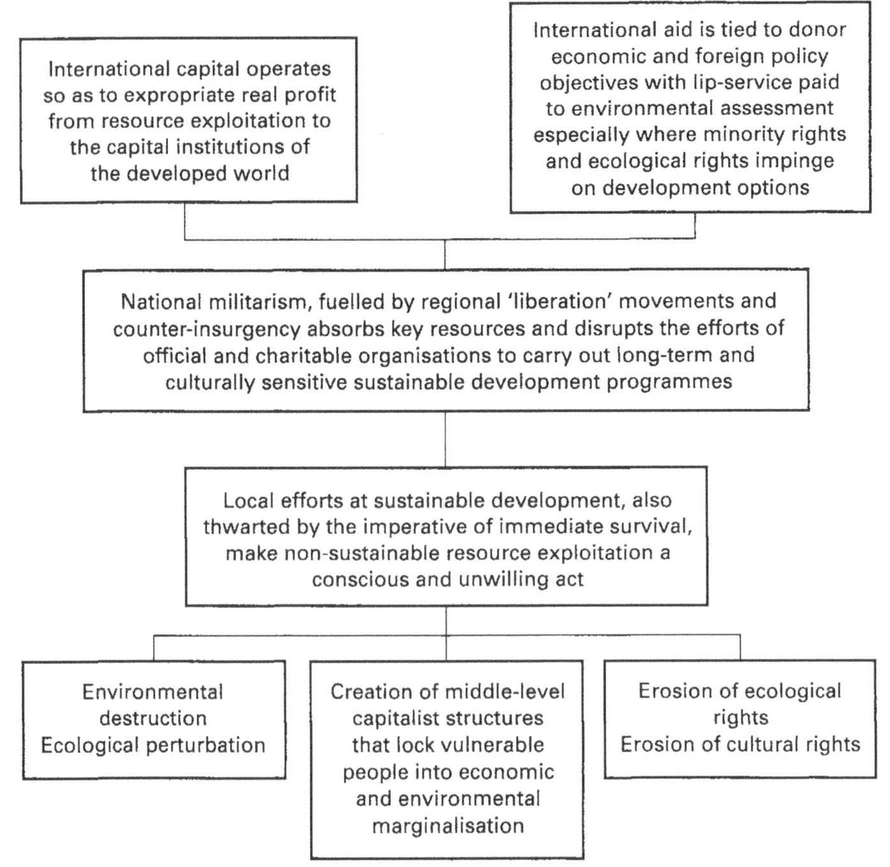

Figure B5.4 Non-sustainable resource draw and international capitalism (source: O'Riordan, 1988)

(5) International aid is not geared to sustainable development at the micro-scale. Aid is linked to established political structures and to a degree is dependent on recipient government support. Recent studies of World Bank aid, even those programmes that allegedly have a specific environmental component, indicate outcomes that are socially divisive and environmentally destructive (O'Riordan, 1988, p. 42).

O'Riordan also expresses these arguments diagrammatically, presenting an illustration of the pressures for what he terms 'non-sustainable resource draw', or 'over-exploitation', which stem from beyond the immediate 'land manager', and that he suggests are aspects of 'international capitalism'.

Nepal is an immensely varied environment: its boundaries include parts of the Himalayan mountains, a high altitude desert and part of the Gangetic plain. It also used to have considerable areas of forest, although today much of this forest has been removed, with the result that there is extensive soil erosion (see Chapter 3). As both Blaikie and Brookfield (1987b, pp. 37–41) and Mannion (1991b, pp. 250–4) have documented, the precise amount of deforestation and erosion has been a matter of some dispute, but it is widely recognised as being a major problem both for Nepal and for the states of India and Bangladesh through which the Ganges and Brahmaputra rivers flow. It has also been widely argued that it is Nepalese peasant farmers who are to blame for this deforestation and erosion, although once more there is some debate over details. Some people, such as Myers (1986) argue that a major cause of the deforestation and associated environmental problems is peasant deforestation for fuel wood. As mentioned in Chapter 2 this view, on the one hand, has led to the imposition of restrictions on the use of timber for heating homes and, on the other hand, has been criticised for ignoring the impact of deforestation related to tourism. However, just as peasant householders were being exempted from being major contributors to this form of deforestation, an alternative reason to find them culprits emerged. Studies by the Asian Development Bank (1982), Echolm (1976), Mahat *et al.* (1986a; 1986b; 1987a; 1987b) and many others argued that deforestation was a consequence of a growth in the peasant population and a resultant over-extension in their area of cultivation to include highly inclined slopes. These slopes were formally wooded, and this tree cover kept soil erosion to acceptable levels. With the growth in their population numbers peasant farmers cleared higher and higher up these slopes, building complex series of agricultural terraces on which to grow crops. The result, so the argument went, was the establishment of a cycle of environmental degradation and poverty that accounted for the current parlous state of Nepal's economy, which is currently amongst the world's poorest countries.

Blaikie and Brookfield (1987b) suggest, however, that this explanation of Nepal's current economic and environmental situation needs to be moved away from its starting point of poor environmental management by peasants in two directions. First, it needs to be 'pushed back' historically to consider the legacy of past actions on the present conditions; and second, and relatedly, explanation needs to move 'upwards' from the peasant households to consider the economic, social and political structures operating in Nepal. Specifically, Blaikie and Brookfield claim that the peasant householders do recognise the environmental impacts of their action, but they, and their predecessors, have had virtually no choice but to continue these practices. In particular they are constrained by the amount of money and labour power they have. The lack of money makes it impossible within an increasingly capitalist-oriented economy to purchase alternative fuels for heating, and also means that they are able to purchase only the poorer quality lands. This illustrates O'Riordan's (1988) first argument (see Box 5.4) that non-sustainable actions are created by 'the structural conditions that induce poverty and desperation'. This lack of money is related to the poor prices paid by merchants and by the high levels of taxation levelled by the Nepalese state, a point that illustrates O'Riordan's third argument that 'middlemen' exploit any propensity to create surplus. The Nepalese state has, in the past, also been responsible for diminishing the amount of labour available to peasant households, in that it required its citizens to do a period of work or service, often military service, for the state. The significance of military service was in turn related to the political context in which Nepal has found itself: first as a British colony and more recently as a satellite state of India. Furthermore, the actions of Nepal have been further constrained by the state of its finances. The country is now heavily dependent upon external finance and aid and is caught in a vicious poverty trap, with no money to invest in the development necessary to improve its economic situation.

Nepal is not the only country in this situation, and many of the current practices leading to environmental degradation can be directly related to the capitalist structure of the world economy. This structure affects not only the small peasant farmer but also influences the operations of that other agent of environmental degradation: the multinational logging company. In Guyana, for example, the government permitted a series of multinational logging companies to purchase large tracts of its forests. This sale of logging rights was part of an *International Monetary Fund* (IMF) 'economic recovery programme' for Guyana, designed principally to pay off an estimated $1.7 billion debt that the country had accrued with western countries (Lewis, D., 1991).

This discussion of deforestation in Nepal and Guyana illustrates the need to situate environmental degrading actions in both historical and social contexts. Attention has been paid in this case study to a range of contexts, from militarism and inter-state relations through to flows of money and labour. These influences on the land manager have been seen to stem from both national and more global origins. In the next case study we will explore a global context in more detail, and in particular consider the extent to which the principal contextual determinant of unsustainable actions is the '*capitalist world economy*'. We will be examining this argument with respect to the creation of famine.

5.3.2 The capitalist world economy and the perpetuation of famine

It is one of the most tragic contradictions of the contemporary world that there are currently more people on earth who can be said to be experiencing hunger and malnutrition than ever before, and yet food production per capita has been steadily rising (see Chapter 3). In the period 1980–90 food production per capita in the so-called 'developing countries' increased by an estimated 2 per cent, and yet in these countries an estimated 800 million suffer from hunger, of whom some 500 million are '*chronically malnourished*' (United Nations Development Programme, 1996, see Box 3.4). Furthermore, in many of the so-called 'developed countries' of the world land is actually being taken out of agricultural production and food put in storage because of a so-called food surplus. In the European Union, for instance, there have been so-called butter and grain 'mountains' and wine and milk 'lakes', while it has been recorded that in 1972, a year in which there was widespread famine in the Sahel, the US government paid its farmers $3 billion to take 50 million hectares of land out of production, land that grew sufficient wheat to have averted the famine in Africa (Bradley and Carter, 1989, p. 112). Here one has once more the paradoxical situation where a crisis appears to be avoidable and yet is still allowed to occur.

As discussed in Chapter 3, a variety of explanations of famines have been proposed, including many that link them to a failure to implement a correct or adequate form of environmental knowledge. In this section we want to suggest, however, that the perpetuation of famine and hunger in this age of environmental awareness is very much a manifestation of the inherent structures of the contemporary capitalist world economy, although we would add that the capitalist world economy should not be seen as a total explanation of the perpetuation of famine.

Capitalism involves the production of goods for exchange in a market, and with the rise of a capitalist world economy people increasingly have to obtain the goods and services that they need to live by participating in a market. As Bradley and Carter (1989, p. 109) say, this means, with reference to food, that people can obtain it in two ways: 'They can either grow it or buy it'. They add that, 'it is with the factors that control their ability to do one or other that the real causes of hunger lie' (Bradley and Carter, 1989, p. 109). In other words, the causes of famine lie in the ability of people to either produce and/or purchase food.

The significance of the first issue is generally recognised in discussions of famine: as discussed in the last chapter, famine is often seen to be the result of a breakdown of agricultural production. Watts, for example, has observed how in the famines of the Sahel and elsewhere it is generally those people who live and work on the land who perish for lack of food: 'those who died were those who produced' (Watts, 1983, p. 26). Watts has worked particularly on the causes of famine in the 'Nigerian Hausaland' and he has argued, in a manner very similar to the arguments of Blaikie and Brookfield discussed in the preceding section, that many explanations of famine in this area have: (a) ignored the historical dimensions of famine, and (b) failed to link individual behaviour of peasant producers to 'the wider socio-economic context' in which this behaviour is situated. With reference to the first point, Watts notes that drought and famine are not new phenomena to the Sahel, but rather have been a 'recursive', or reappearing, feature of the region. What is new, Watts argues, is the ability, or more precisely the inability, of the peasant farmers and nomads to deal with the situation. He argues that the ability of peasant farmers in areas such as northern Nigeria has been seriously undermined by colonialism and the associated integration of these farmers into a world economic system.

As noted earlier (see Box 5.1), prior to British colonial expansion into the area peasant farmers adopted what Watts describes as a 'subsistence ethic', which enabled them to deal quite effectively with any periods of drought. However, with the onset of colonialism, the three elements of the subsistence ethic that helped support people during periods of drought came under attack. In particular colonial taxation, export commodity production and commercialisation meant that the poorer peasant producers became less capable of responding to, and coping with, drought and food shortages. This occurred because the colonial taxation system took no account of fluctuations in the economic fortune of the peasants, unlike the pre-colonial taxation system that, although roughly the same in its size, was altered to take account of poor harvests. The problems created by the colonial taxation system were further heightened by the time at which the tax had to be paid: taxes were levied prior to the cotton harvest, 'leaving the rural cultivator little choice but the sale of grain when prices were lowest or, alternatively, the money lender' (Watts, 1983, p. 30).

Mention of money lenders and the sale of crops links into the other two detrimental features of colonialism for rural producers: export crop production and commercialisation. Britain encouraged the export of two crops from northern Nigeria: cotton and groundnuts. Neither of these are food crops, which meant that if a producer transferred to producing such crops there would be a reduction in production of foodstuffs available in the region. It might be objected that the decision to change from food to cash crops was still in the hands of the peasant producer. However, their ability to choose was effectively determined by the increasing commercialisation of peasant society. Money was increasingly becoming the medium through which economic transactions were being carried out in northern Nigeria. One of the clearest examples of this, and one of the changes that can be most clearly traced back to the establishment of colonialism in the area, was the change in collecting tax from '*goods in kind*', whereby, for instance, a grain producer would pay his or her tax in grain while a nomadic cattle herder would pay in cattle, towards collecting tax in cash. This change meant that peasant farmers had to sell at least one crop to a market, and in many cases the most financially lucrative 'cash crop' was a non-food one.

In a number of overlapping ways, therefore, the establishment of colonial rule in northern Nigeria undermined the ways of life associated with the subsistence ethic. Monocultural production of cash crops replaced intercropping: high yields to bring high monetary incomes became more and more important to peasant farmers, thereby decreasing the possibility of growing lower-yielding but more drought-resistant crops; taxation and commercial pressures undermined mechanisms of mutual support and the colonial state was essentially unwilling to support those in difficulty, even in terms of simply deferring payment of taxes. The result was, Watts argues, that peasant households became more vulnerable during periods of drought. Their cash crops were more likely to suffer during periods of drought, which meant that they would fail to generate much monetary income, which further increased the burden of monetary taxation on them, which in turn might lead them to taking out loans with local money lenders at high rates of interest, which could lead them into even further trouble in later years. It could, for instance, lead to them having insufficient money to purchase seed crops for the next season.

Again, such action might seem 'irrational', but one has to consider what options peasant households would have when in debt. With the demise in the subsistence ethic there would be few if any mechanisms of mutual support and, with their conversion to cultivation of cash crops, they would themselves need to purchase foodstuffs on the market simply in order to survive. This need to purchase foodstuffs would be greatest in periods of drought, just when food prices were at their highest and when the returns from their cash crops would be at their lowest. The stark choice for many peasant households is therefore to spend all their money on foodstuffs, or perish. Even this degree of choice is likely to be reduced by the next season – if the household does not have enough money to purchase the seed for its next year's crop then there will be no money in the future to purchase foodstuffs. In this way the penetration of capitalist social relations can be seen to provide an account of why it is the agricultural producers of the Sahelian region who have died in periods of drought and famine.

It might be argued that such problems are problems of the past: we are now after all in a post-colonial era. However, as discussed in Chapter 2, while political control of territories may have ceased there are still highly damaging economic linkages between countries. Bradley and Carter (1989), for example, identify at least five ways in which economic control of peasant livelihoods is still being removed from the peasant household and placed in the hands of agents connected to the world capitalist economy. These economic linkages of domination are:

(1) The control of credit, input resources and marketing by transnational corporations, and, one should add, more indirectly by international finance agencies such a the World Bank and the International Monetary Fund (IMF).
(2) An 'urban bias' in pricing policy, in which the political support of the rapidly growing urban population in the underdeveloped world is literally bought by keeping food prices at a minimum (see Lipton, 1977). This has led, first, to food production being increasingly unprofitable in many of these countries and, second, to an increase in food imports, in part to make up the shortfall in domestic production, but also because imported food is often cheaper than domestically produced foodstuffs.
(3) The private appropriation of common land and the commercialisation of land that increasingly becomes exchanged through the medium of money.
(4) The increasing dependence on wage labour, with many members of peasant households finding it necessary to sell their labour power, either to another farmer or within urban labour markets.
(5) The 'modernisation' of cultural attitudes and the widespread adoption of western attitudes has lead to the 'wholesale adoption of western clothes, corrugated iron roofs and a whole panoply of consumer items all available only through cash purchase' (Bradley and Carter, 1989, p. 115). Modernisation has been pushed both by western 'development experts' and by multinational corporations through advertising and marketing.

Bradley and Carter argue that in these and other ways peasant households are increasingly being pushed into the world economy. They further suggest that the nature of this world economy is such that it will not feed those people who need food the most – the starving. As Bradley and Carter explain:

> As food is now a commodity, the forces which govern its production are those relating to buying and selling at a profit. If the poor of the Third World cannot purchase it, then production and distribution will fail to adjust to their real needs, even though there may be widespread malnutrition and starvation. Thus we observe a paradox whereby a global production system has the capacity to feed the world, but does not do so because people are too poor. *In extremis*, such is the logic of capitalism that the very fact that the poor of the Third World are starving is the reason why they cannot be fed (Bradley and Carter, 1989, pp. 120–1).

The conclusion is, therefore, that once incorporated into the world capitalist economy a person's susceptibility to hunger and starvation is determined largely by the amount of money or wealth they have. This means that starvation and malnutrition can, and do, occur even in the context of widespread localised wealth. On the global scale famine, as opposed to drought (which is merely a shortfall in expected rainfall), is clearly associated with levels of economic prosperity, in that it is the poorest countries of the world that generally experience famine.

One implication of such a statement is that the way to prevent famine is to overcome poverty. Poverty is however a self-reinforcing phenomenon. If you have no money you are less likely to be able to generate money than someone who has money. A key issue is, therefore, whether people and countries can obtain money for the economic 'development' necessary to escape poverty and, in some countries, famine. Alternatively it may be that famine can be avoided if mechanisms for distributing food and/or developmental resources can be found that do not rely on the possession of money. Let us look briefly at these two strategies for dealing with famine.

5.3.3 Money for development: trade, debt and aid

Trade, borrowing and aid can be seen as three ways of trying to escape the trap of poverty within capitalism. Each can be seen as a mechanism for distributing the money necessary to sustain life within a capitalist economy. In this sense they can be seen as mechanisms for development. However, it will be argued here that, rather than serve to overcome or ameliorate the constraints on development, these mechanisms of distribution frequently serve to heighten dependence upon the capitalist world economy and while there are some clear instances where a particular person, region or country has benefited from employing these development strategies there are as many, if not more, instances where these strategies have not worked. Furthermore, it can be argued that the success and failure of particular development strategies is influenced more by the general constitution of the world capitalist economy than it is by the particular policy decisions made by governments, organisations and peoples.

Beginning with trade, it has frequently been suggested by development strategists that poverty problems can be solved by expanding the amount of goods that are sold on world markets There are, however, at least four factors that such advocates tend to ignore. First, they ignore the extent to which the pattern of trade in many underdeveloped countries is still heavily influenced by these countries' experience of colonialism: indeed that is why we use the term underdeveloped rather than such alternatives as 'Third World' or 'developing countries' (see Box 2.8). As mentioned in Chapter 2, many of the colonies tended to be used either as markets for industrial products of the imperial core countries or as sources of cheap raw materials. Trade for many countries still exhibits the dual features of colonial trade: high import dependency on manufactured goods from the core and over-concentration in one or two raw material exports. A second problem is that the 'terms of trade', or the relative price of goods exported and imported, are often structured to operate in a way that makes it hard to expand trade. This, together with the structure of trade, means that the former underdeveloped colonies tend to be exporting low-value goods and importing high-cost goods. This makes it hard to increase the wealth of the underdeveloped countries through trade: money tends instead to move from periphery to core. Third, countries in the core are largely unwilling to alter their terms of trade and indeed frequently act through various protectionist strategies (such as tariffs, quotas and subsidies) to ensure that they remain in their favour. Fourth, the success of any developments in trade are in large part determined by earlier success in trading, given that current returns from trade generally form the principal source of future investment. This problem is particularly acute in Africa, where the level of exports by 'low-' and 'middle-' income countries appears actually to have declined in the 1980s and remains low in the 1990s (see Table 5.2). The continent also has a large proportion of the world's poorest countries, many of whose exports are predominately primary products (see Figures 5.2 and 5.3). It can therefore be argued that African countries, with some exceptions (including most notably South Africa), rather than gaining access to money for development are at best staying where they are economically, or in some cases appear to be in an increasingly worsening position, with declining trade leading to declining investment leading to declining trade and so on.

Table 5.2 Annual growth rates in exports in low and middle income countries

Regions	Annual growth in exports		
	1965–80	1980–9	1990–5
Overall	3.3	5.4	7.2
Sub-Saharan Africa	6.1	–0.6	0.9
East Asia	10.0	10.0	17.0
South Asia	2.2	6.2	8.6
Europe, Middle East and North Africa	3.7	5.8	1.1
Latin America and Caribbean	–1.0	3.6	6.6

Notes: Growth rates calculated as a percentage of export level at start of each period. Figures for 1990–5 may not necessarily be directly comparable with earlier figures due to changes in methods of calculation and geopolitical changes in territorial states

(Sources: World Bank, 1991; 1997)

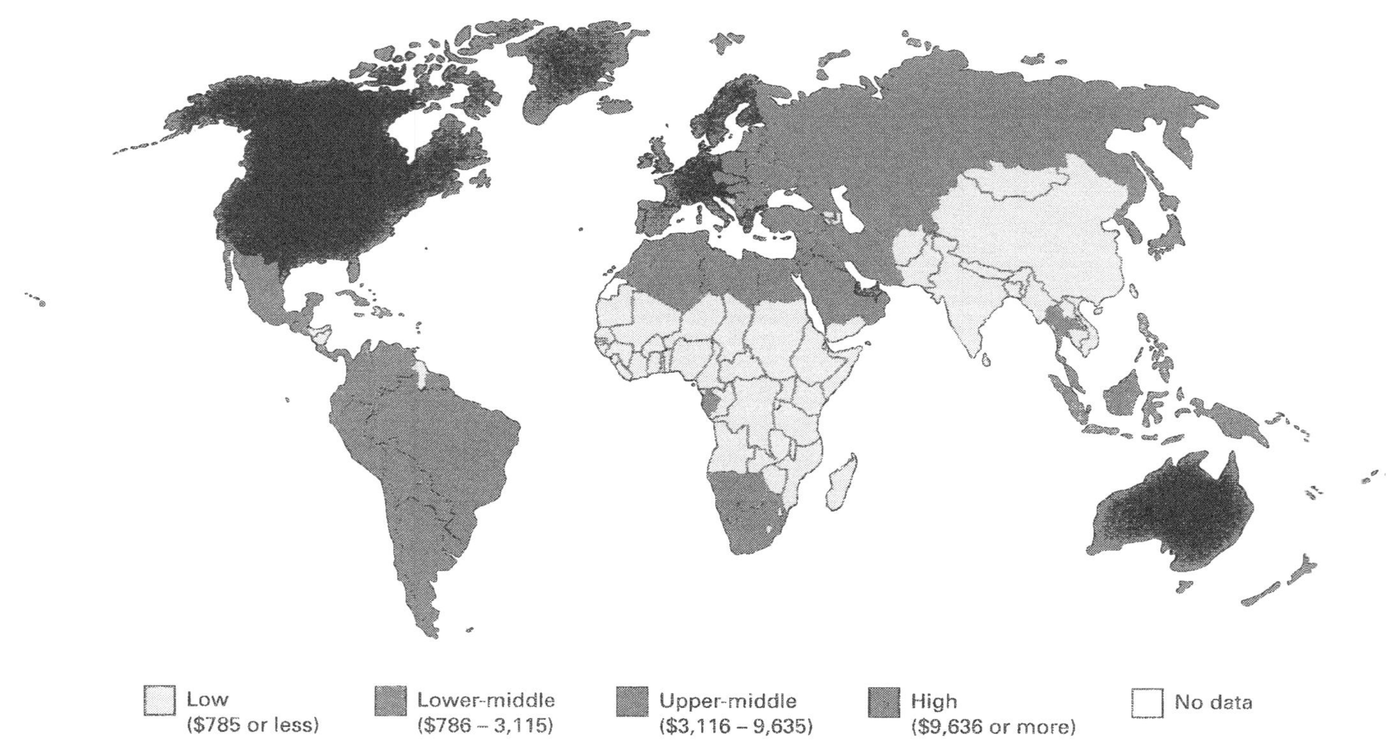

Figure 5.2 Per capita income, 1996 (source: IBRD, 1998, *World Bank Atlas* p. 38)

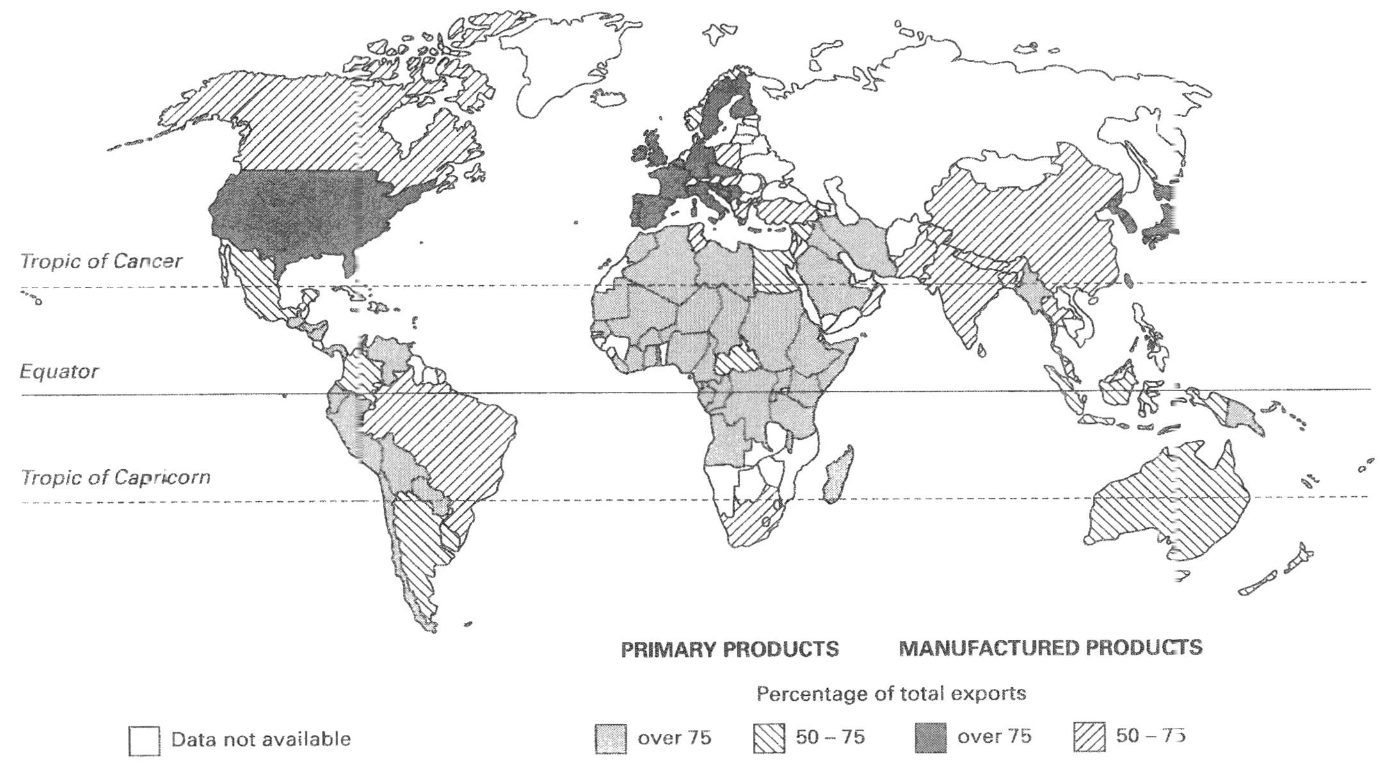

Figure 5.3 Structure of exports, 1990. Source: Dickenson et al., 1996.

One way out of such a cycle may be to borrow in order to invest. Indeed African countries are amongst the highest borrowers in the world (see Table 5.3). However, borrowing for development works only if the investment yields positive returns. Much of the growth of the so-called *newly industrialised countries* (NICs), and indeed Europe after the Second World War, has been fuelled by borrowing. Unfortunately not all borrowing results in successful investment, and furthermore much of the current debt run up by the poorest underdeveloped countries is not really borrowing for development. As George (1988) highlights in her book, *A fate worse than debt*, much of the borrowing is undertaken to finance war rather than development, and in some instances simply to finance extravagant lifestyles amongst the ruling elite. Such use of potential investment money has been rightly criticised, but of even more general significance has been the borrowing of money to pay for increased import bills, and in particular to pay for oil after large price rises in 1973–4 and 1978–9. Faced with a rapid rise in their import bills many low-income countries turned to commercial lending institutions that at this time were relatively awash with money, related to, first, having received a large input of investment from oil-producing countries, whose price rises had generated large surplus revenues that they wanted to invest, and second, a decrease in commercial lending in the 'developed countries' that were experiencing economic recession, again related to the rise in oil prices. The recourse to commercial banks was seen at the time as merely a temporary measure to overcome the immediate impacts of the oil price rises. However, by the late 1970s the economies of the 'developed world' started to adjust to the oil price rises and the surplus of investment capital started to decline. As a result interest rates were increased and many low-income countries found it increasing difficult to 'service' their debt, that is to pay the interest payments on their loans. Governments started to borrow more money from commercial banks and other lenders simply to repay the interest on existing debts, while some, such as Mexico and Brazil, became involved in lengthy rescheduling negotiations that altered the size of repayments due by spreading loans over longer periods. The size of the debt of some countries far exceeds their level of domestic production. In 1995, for example, the figures for debt as a percentage of gross domestic product for Congo, Mozambique and Nicaragua were 365, 444 and 590 per cent respectively (World Bank, 1997). The problem of paying interests on debts, let alone repaying the original loans, remains a severe problem for many countries and has come to absorb a large proportion of any income generated by exports: in Zambia, for example, debt servicing in 1995 was equivalent to 174.4 per cent of the country's exports (see also Table 5.4).

Table 5.3 The size of national debt in low and middle income countries

Regions	Size of debt as percentage of Gross National Product (GNP) 1980	1995
Overall	21.0	39.6
Sub-Saharan Africa	30.6	81.3
East Asia and Pacific	17.3	32.9
South Asia	17.4	30.5
Europe and Central Asia	9.9	39.9
Middle East and North Africa	18.3	37.3
Latin America and Caribbean	36.0	41.0

(Source: World Bank, 1997)

Table 5.4 Total debt service as percentage of exports in low and middle income countries

Regions	Size of service payments as percentage of value of exports	
	1980	1995
Overall	13.0	17.0
Sub-Saharan Africa	9.8	14.5
East Asia and Pacific	11.5	12.8
South Asia	11.7	24.6
Europe and Central Asia	7.4	13.8
Middle East and North Africa	5.7	14.9
Latin America and Caribbean	36.3	26.1

(Source: World Bank, 1997)

The problems created by debt have led some people to suggest that financial aid should be given to the poorest countries to stimulate development investment. Such aid would involve no or minimum interest repayments. Many countries, including many that suffer from drought and famine, have received considerable *Official Development Assistance* or *ODA* as it is commonly referred to. However, there are a number of problems with aid, including:

(1) The giving of aid is rarely as innocent as it may at first appear. Britain, for example, now receives more income from aid-related contracts than it gives away as aid.
(2) The level of aid given frequently does not correspond with the need of recipient countries but instead reflects the economic and political situation in the core, donor, countries. In the 1980s, for instance, ODA to African countries fell at just the time when drought and famine were at high levels in the continent.
(3) Aid that is targeted at improving the economic performance of particular groups or organisations can undermine other producers who cannot compete with the producers who have received investment aid. The result can be that overall production levels decline, even though the recipients of the aid may have expanded their production.
(4) Aid may heighten dependency relationships, particularly when the donation of aid is, as is often the case, linked to the receipt purchasing technology from the donor country.

5.3.3 Commodities for development: transfers of technology and food

Heightening dependency relationships between periphery and core in the capitalist world economy has been a subject of considerable concern. E.F. Schumacher, whose work was discussed in Chapter 1, has long argued that part of the problem of the development project as espoused by academics, governments and international development agencies was that it increased reliance on western technology (see Schumacher, 1973). Schumacher argued that much of this technology was 'inappropriate', both environmentally and economically. Western technology, particularly that applied to agriculture, has been designed for use in temperate climate and, as was noted in Chapter 3, has often been found to be unsuitable for use in many other parts of the world. As discussed in Chapter 4, a series of attempts have been made to develop more 'appropriate technology'. Of particular significance in the context of famine has been the development of high-yielding varieties of staple foodstuffs such as wheat and rice. This is commonly known as the 'green revolution'.

The success of these attempts to harness the power of western science to develop 'appropriate technology' has been the subject of considerable debate (for example see King, 1973; Farmer, 1977; Griffin, 1979; Bayliss-Smith and Wanmali, 1984; Harris and Harris, 1989). There have, for instance, been serious questions raised about whether these high-yielding crop varieties, which require large inputs of oil-based fertilisers and pesticides, are really appropriate for countries already heavily dependent on imported energy supplies. Furthermore, as Harris and Harris (1989) outline clearly, the green revolution has led to significant social change in rural areas, which broadly can be seen to have involved the polarisation of a 'differentiated peasantry' – that is one where there are clear gradational differences between, say, a rich, middle, poor and very poor peasantry – into the two relational classes of capitalists or bourgeoisie and a proletariat or working class. Figure 5.4 seeks to illustrate how the green revolution can contribute to this process of social polarisation and class transformation by initiating cycles of *accumulation* and *pauperisation* that can lead what were initially only slightly differentiated peasant households into quite different courses of action and into quite different livelihood situations. In other words, the green revolution can be seen to have initiated, at least in some areas, a movement towards capitalist social relations of production and to the creation of both a relatively wealthy agrarian capitalist class and a very poor agrarian, or non-agrarian, working class.

One final mechanism for achieving 'development' has been the sending of food aid. This has largely been undertaken in 'crisis' situations and suffers from many of the problems associated with financial development aid: namely it is often influenced by political considerations, reflects the economic and political situation in the core rather than the needs of the periphery, it can undermine existing producers and can heighten dependency relationships. It is arguably the form of action most removed from the dictates of capitalism: the search for continued profits. It does, however, have repercussions that may serve to heighten dependence on the world economy. Baker (1987 p. 160), for instance, notes that, 'even during some of the hardest periods of domestic food shortage, many African countries pursued policies of using agricultural land to earn foreign revenue at a time when they were on international food relief.' Baker cites the example of Mali, where cotton production has increased over precisely the period when food aid was being given to the country. Once more one has a situation where land is being taken out of food production at precisely the point where there is most acute hunger. Food aid was, at least in part, enabling this seemingly irrational change, but as Baker notes:

> It is too easy to blame the policy makers and their foreign advisers for this [situation]. These countries have little option but to raise the foreign exchange as the impact of debt renunciation on the international monetary and banking system would be catastrophic (Baker, 1987, p. 161).

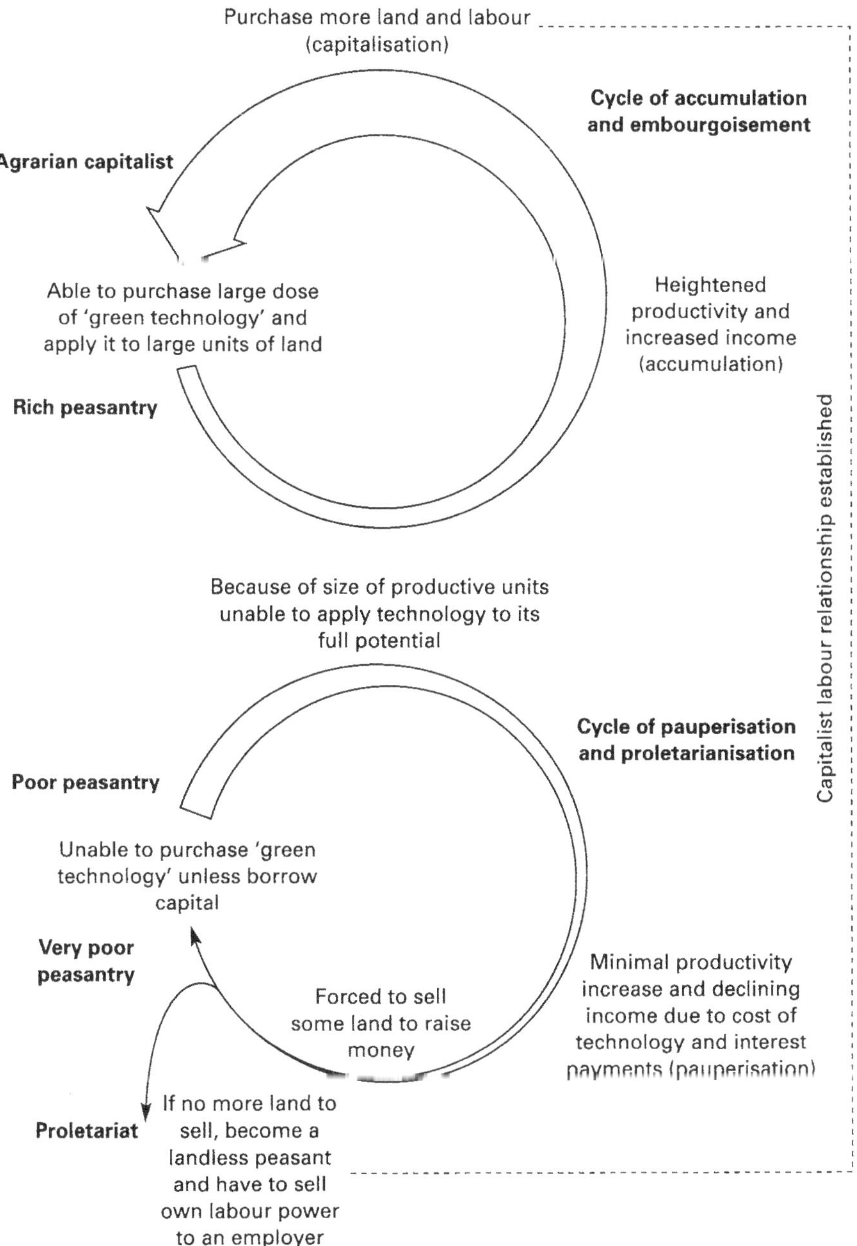

Figure 5.4 The green revolution, social polarisation and the class transformation of a differentiated peasantry

Baker does not ask catastrophic for whom, but it is pertinent to note that some people have claimed that the only development option for some underdeveloped countries is to default on their loans. Such action would go against all established ways of acting within a capitalist world economy but if, as Bradley and Carter (1989) suggest, and as we have also sought to demonstrate, this economy seemingly will never solve the problem of famine, then perhaps such action may appear rational.

5.4 Aside from capitalism?

In the preceding two sections we have been moving towards an explanation of environmental degradation that locates the origin of seemingly quite different environmental problems occurring in different places on the same phenomenon, namely the capitalist world economy. In this section we will explore some counterarguments to this claim, and in particular explore whether or not there are other 'social relational contexts' apart from the capitalist world economy to which environmentally unsustainable actions can be attributed. We will do this by discussing the creation of environmental change in two quite distinct places: Amazonia, and Eastern Europe and the former Soviet Union. Our argument will be that there are 'social relational contexts' apart from the capitalist world economy, but that these generally act in association with the capitalist world economy.

5.4.1 Amazonian deforestation

In Chapter 2 we mentioned how European colonists often constructed the Amazon as being an uninhabited wilderness, a place purely of nature. We noted then, and have further demonstrated here, how misleading this interpretation is. However, it is still a widespread image of the Amazon, reproduced in countless photographs of densely packed vegetation through which intrepid explorers have macheted their way. However, other images of the Amazon are also widespread today, such as its being a massive 'unexploited resource', the 'lungs of the earth', a 'global commons', a store of 'gene plasma' and a 'place under threat of total destruction'.

Earlier in this chapter we presented a critique of those amongst this last set of interpretations that sought to place the blame for the destruction of Amazonia on environmentally ignorant peasant farmers. By contrast we highlighted the agency of the logging companies and cattle barons in deforestation. Following on from the subsequent discussion of deforestation in Nepal and famine in Africa, it is also probably already apparent that connections can be drawn from the actions of these two agents, and also from the peasant farmers, towards capitalist social relations, including those constituted within a capitalist world economy. In this section we will briefly outline some of the contours of such an interpretation, before considering critiques of it developed from what, following on from Chapter 1, might be described as ecofeminists. We will end by noting some other influences on deforestaton in Amazonia.

Multinational logging companies and large cattle ranchers can both be quite directly connected to the operation of a capitalist world economy: they after all are producing products for sale rather than direct consumption, and many of these products are exported to other countries, including those in the developed world. Exponents of green consumerism have indeed often targeted this trade, organising consumer boycotts of tropical hardwoods and 'hamburgers' produced from cattle grazing on deforested land. While significant these direct linkages are, however, often overplayed, particularly in the context of Amazonia. As Hecht and Cockburn (1990, p. 98) note, most of the timber extracted from the Amazon goes to fuel the domestic requirements of Brazil's rapidly expanding economy, and even in Central America 'where the "hamburger connection" has been most elaborately argued, exports only [account for] about 15 per cent of its beef'. Hecht and Cockburn go on to criticise interpretations of Amazonian environmental degradation that focus solely on the connections to the world economy, arguing that such interpretations effectively neglect the role of 'national elites' and local sources of capital. They suggest that there is a need to develop more 'penetrating analysis' that focuses on the composition of a 'triple alliance' of international, domestic and state companies and that recognises the complex 'history and political economy of both the region and of Brazil' (Hecht and Cockburn, 1990, p. 100).

One illustration of the value of such an argument can be found by considering the diversity of one of the three agents of deforestation identified at the beginning of this section, the peasant. We have used this term widely in discussions so far and, although we have noted suggestions that there may well be important differentiations within the peasantry, we have not really explored these in great detail or commented on their significance. It is, however, clear in the context of Amazonia that quite divergent social groups are lumped together under the label of peasant and that there is a need to disaggregate them to get a reasonable level of understanding of the changes taking place in Amazonia.

One of the most immediate distinctions that can usefully be made is between indigenous cultivators, the Amazonian Indians, and colonist cultivators. Both groups can be seen to be peasants in the sense that they may cultivate land to produce for consumption rather than exchange, although the former may be seen to have a more tribal form of organisation than do household-focused colonist cultivators. The two groups may also exhibit different attitudes to the forest environments. Hecht and Cockburn, for example, note both how use of the forest is understood by tribes such as the Kayapo through Shaman religion and also how it can be interpreted as being good environmental practice within the terms of modern environmental science:

> The Kayapo stimulate forest succession in their fallows by making sure that their agricultural sites incorporate the necessary elements to recuperate forests ... This includes the creation of suitable environmental conditions and the manipulation of successional processes themselves (Hecht and Cockburn, 1990, p. 39).

By contrast the colonists, Hecht and Cockburn argue, have very different forms of knowledge, many of which are much less environmentally successful:

> Colonists within the Amazon are more likely to use fire judiciously and to turn their agricultural plots to something in the nature of a succession orchard, in the pattern of indigenous forest reconstruction. Colonists who turn their plots to monocultures such as rice and pasture areas are more likely to engage in destructive burning (Hecht and Cockburn, 1990).

This last quote raises another line of difference, namely between colonists from within Amazonia and colonists from without. Some of the discussion on deforestation in Amazonia implies that it is a very recent phenomenon, or else a very recent and a very ancient phenomenon. However, as Hecht and Cockburn demonstrate extremely clearly, there has been a long, complex and highly contested history of environmental exploitation within Amazonia. As well as deforestation by indigenous shifting cultivation and deforestation by such modern agents as logging companies, cattle ranchers and recent agrarian settlers, there are what Hecht and Cockburn term the *caboclos* or 'backwoods folk', formed, as they put it, 'out of the long history of detribalization, miscegenation, and extraction, from each immigrant wave that left people behind in the region' (Hecht and Cockburn, 1990, p. 147). Ever since its colonial contact with the Portuguese Amazonia has been the site of in-migration by people seeking to exploit some aspects of its nature, be this its mineral reserves such as gold; the produce of plants such as the rubber tree or the Brazil-nut tree, its timber; its land for grazing or settlement, its waters for hydroelectric power. Although most phases of in-migration were accompanied by hopes of sustained exploitation, in most cases these hopes were not realised, and many of those involved in the particular phase of exploitation abandoned Amazonia or else moved on to a new form of exploitation. However, as Hecht and Cockburn note, some people have remained from each phase of in-migration and have sought to continue making a living from their own particular form of nature exploitation. As a result within Amazonia there is a diverse and often disparate group of settlers engaged in making their living from mining, rubber tapping, agriculture, fishing and hunting. These groups are a marginal population, generally finding it difficult to secure their living, often being overlooked by commentators, politicians and policy makers, and often having tenuous property rights to the natural resource they are seeking to exploit. Many of them are hence described as 'squatters' or illegal settlers, and many more have to rent land and rights to use land from other people. Hecht and Cockburn (1990), however, suggest that these groups may have learnt very efficient ways of utilising the Amazonian environment. This group constitutes what Hecht and Cockburn call the 'Amazonian colonists'.

More recently these long-established colonists have been joined by new sets of colonists related to the various governmental settlement schemes (see Box 5.5). As Hecht and Cockburn note, this group appear to be particularly prone to clear forest for essentially monocrop cultivation. As discussed earlier this practice is frequently linked to one of shifting cultivation, in which land is farmed for a few years until soil fertility is lost, at which point a new area of forest is cleared for agricultural production. It was noted that this practice is often the focus of considerable criticism, although its overall significance is not as great as is commonly thought, in part because it is often one element in a sequence of land occupation involving loggers, settlers and cattle ranchers.

This sequence of land occupation is often given an environmental logic: it is the decline in soil fertility that leads first the colonists and then the cattle ranchers to abandon the land. Social processes are, however, of equal if not greater significance in the sequence, not least because many of the settlers do not give up farming an area because it has lost fertility but because they have been more or less forcibly ejected from the land. Plumwood and Routley (1982) argue that the cattle ranches are frequently run by notorious cattle barons who take over land fraudulently or by violent means. Indeed, in Brazil cattle have become a way of establishing property claims, following government legislation in 1964 that allowed people to claim property rights in Amazonia if they could demonstrate effective use of an area for a year and a day. As Hecht and Cockburn observe, one consequence of this law was that people started clearing large areas of forest and grazing cattle on it so as to be able to claim effective use of the area and establish ownership rights to large areas of land. This strategy was employed by both wealthy cattle ranchers and individual colonists and can be seen to explain why the Amazon rain forest came to be subject to what Hecht and Cockburn have described as 'one of the most rapid and large scale enclosure movements in history' (Hecht and Cockburn, 1990, p. 107).

At times this enclosure movement – that is the taking of land into private ownership – appears to have quite comic results. Hecht and Cockburn (1990, p. 148), for example, have recorded how in some areas of Amazonia ownership claims exceeded the total amount of land. However, while the granting of property rights to over 160 per cent of one's administrative territory might appear comic it points to the presence of a situation that in many instances has had quite tragic consequences. Basically the figures reveal the extent to which there are often multiple claims to the same piece of land, claims that are often resolved through the use of violence and other forms of intimidation. Often it is the cattle barons who wield the power and who thereby gain the final rights to property, but as Hecht and Cockburn highlight it is possible to discern complex series of private appropriations of property with, for example, 'Amazonian colonists' such as tree tappers displacing indigenous Indian tribes, only to be displaced themselves by more recent settlers, who again lose out to cattle ranchers and other powerful agents such as multinational logging companies or state enterprises. As Hecht and Cockburn note, this complex and contested process of constructing private ownership rights in the Amazon 'exactly coincides with the explosion of deforestation in Amazonia'. They add that while followers of Hardin's '*Tragedy of the commons*' rush to condemn communal forms of ownership they ignore the fact that virtually all deforestation in the Amazon has occurred on privately held land, or is used to assist in the passage of public land into private hands (Hecht and Cockburn, 1990, p. 107).

For Hecht and Cockburn, therefore, deforestation should be viewed much more in the context of a long, complex and contested history relating to emerging private ownership of rights to land and resources, rather than as a direct consequence of a global capitalist economy. This is an important point to make but it should be noted that, as Hecht and Cockburn themselves demonstrate, much of this history was constituted by waves of nature exploitation and settlement that had at least some of their roots in Portuguese colonialism or First World consumption demands. This is not to say these are the only important factors. Hecht and Cockburn, and also Allen (1992), for example, highlight the importance of military concerns within government resettlement programmes (see Box 5.5). As Plumwood and Routley note, such militarism was frequently associated with a barely disguised racism: a member of the Brazilian ruling junta, for example, is recorded as stating: 'When we are certain that every corner of the Amazon is inhabited by genuine Brazilians and not by Indians, only then will we be able to say the Amazon is ours' (quoted in Plumwood and Routley, 1982).

Box 5.5 Settlement programmes in Amazonia

Programmes to settle people in areas of rain forest have been in operation in a range of countries, including many that have boundaries that encompass parts of the Amazonian rain forest. As Plumwood and Routley (1982, p. 9) remark, the justification for these resettlement programmes is often one of social justice and benefit, such as 'that of providing for the landless poor'. They argue, however, that the real motivation for such schemes is rather more to do with 'providing a political alternative to the redistribution of existing cropland along more egalitarian lines'. A clear illustration of this can be seen in programmes of colonisation in the Brazilian Amazon, many of which have been promoted as a way of alleviating problems of poverty in north-east Brazil. The latter area has long been one of widespread poverty and landlessness, related to a highly unequal distribution of land in which a relatively small number of people own large plantations, or *lattifundia*, while a much larger number of people live on very small plots of land, or *minifundia*. In the early 1960s the area had been a centre of anti-government revolt and in 1964 a military government seized power, drawing on support from major landowners. This government wished to ameliorate the problems of north-east Brazil but did not wish to restructure land ownership. The policy arrived at was to look to relocate the landless people of the area by 'developing' Amazonia. In 1966 *Operation Amazonia* was begun with the aim of opening up Amazonia for both external investment, to develop large resource extraction projects, like the recovery of the Carajas iron ore deposits, and for settlement by people from outside the area, particularly from problem areas such as the north-east but also the highly crowded cities of the south-eastern parts of Brazil. A development agency, the *Superintendency for Development of Amazon* (SUDAM) and a development bank, the Bank of Amazonia, were both established to help facilitate the movement of people and capital into the area, and a programme of road-building was initiated for the same purpose. So, for example, the Belem-Brasilia Highway that was begun in 1960 was completed, and the BR-364 was extended, first through Mato Grosso and Rondonia in the 1960s, and then on into the state of Acre in the mid-1980s. The largest project of all, the Trans-Amazonia Highway that ran east to west across Amazonia, was initiated as part of a 'national integration plan' (PIN) and was built between 1972 and 1977 (see Fearnside, 1987, 1989; Allen, 1992; Shankland, 1993; Reid and Bowles, 1997). Feeder roads were constructed from these main roads to create a system of access points into the rain forest and agrarian colonists were encouraged to settle and farm the areas alongside these roads. By 1984 some 160,000 colonists had settled in Rondonia, although in some of the more central areas the level of settlement was often markedly lower.

A subsequent series of government plans adopted various other strategies to continue the dual policy strands of economic growth and population relocation. In the *POLAMAZONIA* scheme, for example, some 15 centres were designated as

▶

'growth centres' for primary industry and large-scale, capital-intensive agriculture, while in *PROTERRA* finance was made available for the purchase of large land holdings for redistribution to small farmers. These attempts to develop Amazonia have continued since the ending of military rule in 1985. Examples include the *Calha Norte* project and the *2010 Plan*, which both aim to increase economic development within northern Amazonia. Projects include dam construction, mining and capital-intensive agriculture (Hildyard, 1989). Also, even though these projects have been initiated by civilian governments, they have quite clear military motives. Indeed the Calha Norte project was initiated secretly by the military in 1985 (Allen, 1992), the same year as a civilian government took power. It was argued that there was a need to integrate the area, which lies on the north-west margins of Amazonia, economically and politically into the rest of Brazil. This integration was said to be necessary in order both to deal with internal social dissent and disorder and to reinforce Brazil's national security and sovereignty in the area. Much of the area had been transferred to Brazil's sovereignty from the neighbouring countries of Venezuela and Colombia during the late nineteenth and early twentieth centuries (see Allen, 1992), and there were some fears that future governments of these countries might try to reinstate their claims for sovereignty in the area. Furthermore much of the area was populated by indigenous peoples, which was seen to pose a further challenge to national sovereignty (see text). The 'integration' plan involved the designation of 'Indian colonies' for the protection of Indian property rights, but these were much smaller than the areas of indigenous usage prior to the Calha Norte project and therefore actually represented a loss of property rights. The project also involved the establishment of army and air force bases along a 150-kilometre corridor of the Colombia and Venezuela border, HEP projects, road construction (including 'a border-hugging Transfrontier Highway') and settlement schemes (Allen, 1992, p. 29). It is therefore possible to see the Calha Norte, and many of the other Amazonia projects, as an exercise of geopolitics rather than a disinterested promotion of economic development.

Furthermore these projects have had a series of material impacts on nature. As well as contributing to deforestation every year the resettlement of people into Amazonia was accompanied by the movement of disease. Indeed while the settlement programmes can be seen as a form of internal economic and political imperialism, with modern Brazil colonising and exploiting indigenous Brazil and its natures, one can also see them as another instance of ecological imperialism (see Chapter 2). It has, for example, been argued that in some areas of Brazilian Amazonia about 85 per cent of the indigenous population has died from diseases (such as malaria and measles) introduced by settlers, or else through violence as they have sought to resist the loss of their land (Shankland, 1993). Damage to forest reserves and the illegal conversion of forest to pasture is further exacerbated by non-enforcement of laws and under-resourcing of development agencies and programmes for the country's indigenous peoples (see Fearnside, 1984).

As discussed in Chapter 1, a relatively recent development in the study of society–nature relations has been the emergence of ecofeminist perspectives that posit lines of connection between gender relations and society–nature relationships. A series of studies has demonstrated the relevance of such arguments to the study of deforestation. Hecht and Cockburn's (1990) book, *The fate of the forest*, which we have been discussing extensively here, while making no explicit mention of gender issues, does include some suggestive findings. For example, when discussing the presence of mining in Amazonia by small-scale, 'petty-capitalist' producers they remark on how the majority of these were 'young men between the ages of fifteen and twenty five' (Hecht and Cockburn, 1990, p. 146). This raises a question as to the gender composition of the other agents of deforestation discussed by Hecht and Cockburn and whether, in seeking to differentiate them on the basis of a complex history and political economy, they omit to reflect on a striking element of similarity, namely how predominantly male many of these agents of deforestation are. The heads of big multinationals and more local businesses are, for example, predominantly men, as are the politicians and the military in Brazil. The majority of the people employed to cut down the trees and work the deforested areas for the male-dominated businesses are also men: consider, for example, the gender composition of loggers and the cattle hands. The two exceptions to this masculinity are, potentially at least, the indigenous population and the colonist settlers, whose tribes and households are likely to have both men and women present, albeit not necessarily in equal numbers or positions of power and respectability.

While attention has been paid so far to the environmental and property dimensions of the sequential progression of land occupancy from, say, logging company to colonist to cattle ranch, people such as Lisansky (1979), Meertens (1993), Townsend (Townsend, 1993; Townsend and Wilson de Acosta, 1987) and Hecht (1985) have highlighted that it also has a gender dimension. Townsend and Wilson de Acosta (1987), for example, effectively identify four distinctive 'gender orders' corresponding to different stages in the colonisation process (see Connell (1987), Agg and Phillips (1998) and Phillips (forthcoming) for discussions on the notion of gender orders). Prior to colonisation one has low-density occupation by Indian groups, in which women play a major role in the economy of the tribe. Second, soon after the colonial contact, there is exploitation of the forest by transient groups in search of timber, skins or minerals. These groups are almost exclusively male and they act to displace or destroy the indigenous population either through violence or through disease (see discussion of ecological imperialism in Chapter 2). One has here a distinctly masculine gender order. The third phase sees colonisation by agricultural settler families, in which women play a significant economic and social role. Here one has a more balanced gender composition, although there are still likely to be more men than women, and some differentiation into distinct gender roles. The fourth phase is the penetration of capitalist cattle ranching, when family farms are destroyed. Men displaced from the farms may find employment on the ranches, but there are virtually no jobs available on these ranches for women. Women therefore tend to either become 'housewives', working in the home but with no income-generating employment, or else they resort to prostitution. The overall result is a highly masculinised gender order.

Overall it can be argued that deforestation is not exclusively driven by the capitalist world economy, but there are other, more localised contexts of social relations that need to be recognised. Many of these contexts are also capitalistic in nature, or at least help establish the 'conditions of existence' (Corbridge, 1986) for capitalist social relations, such as through forming private property rights that are a condition for a private market in land. There are, however, other social relations and social ordering processes at work, such as nationalism, militarism and masculinism, that are clearly of significance and require consideration alongside the social relations and orderings of capitalism and a capitalist world economy.

5.4.2 Environmental degradation in non-capitalist countries

Advocates of broadly Marxist political economy interpretations of environmental change as outlined in the section before last have often been criticised for ignoring the presence of severe environmental problems in seemingly 'non-capitalist' countries, such as existed prior to the late 1980s and early 1990s in the self-proclaimed 'socialist' or 'Communist' states such as the USSR and eastern bloc countries like East Germany, Poland, Czechoslovakia and Romania. The collapse of the 'Iron Curtain' associated with the unification of East and West Germany and the dissolution of the Union of Soviet Socialist Republics and the Warsaw Pact in the late 1980s and early 1990s has certainly brought some serious environmental problems to light (see Box 5.6), as indeed did the escape of radiation from the Chernobyl reactor in 1986. Much of the blame for these problems has been clearly laid at the door of the non-capitalist governments. Carter and Turnock, for example, argue that the presence of serious environmental problems in east-central Europe:

> can be laid at the feet of four decades of central planning by Communist parties, which failed to adhere to declared policies for sound environmental management; instead there was serious ecological damage of the sort previously attributed only to Western capitalist regimes (Carter and Turnock, 1993, p. 1).

Box 5.6 Examples of environmental degradation in eastern Europe

A series of studies of the environmental change in eastern Europe during and subsequent to Communist rule is provided in Carter and Turnock's (1993) book *Environmental problems in Eastern Europe*. Below are a selection of assessments from the book's various contributors relating to the extent of environmental degradation in the countries of eastern Europe:

> Leipzig was covered in 400,000 tons of SO_2 annually, life expectancy was six years less than the national average, and four-fifths of children aged below 7 years developed chronic bronchitis or heart problems. The dubious title of 'worst polluted city in Europe' was claimed by Bitterfeld, with its coal mines, thermal power station and chemical plant. The River Mulde directly received its untreated waste (eg

dioxin), while the nearby River Elbe had mercury levels 250 times the EC limits (F.W. Carter and D. Turnock, Chapter 1, p. 5)

With the advent of socialism (or what the Soviets called 'goulash Communism') Hungary's economy (largely dependent on agriculture) was forced to give more emphasis to heavy industry: steel mills, metal smelters, metallurgical plants, chemical, cement and mining. An ugly industrial crescent was created from Akja, Györ and Tatabánya in the west to Miskolc and Ozd in the east. Within this industrial corridor most of the country's heavy, polluting industries were concentrated ... virtually all industrial complexes were built without regard to pollution control measures or environmental repercussions (D. Hinrichsen and I. Láng, Chapter 5, p. 91)

Poland, along with Czechoslovakia and the former East Germany, is amongst the most polluted countries in Europe not only for air and water but also its soil and vegetation (F.W. Carter, Chapter 6, p. 107).

After a 16 per cent increase between 1937 and 1961 the forested area then decreased by a substantial 12 per cent from 10,300,000 ha in 1961 to 9,100,000 in 1976. The loss of forests was a primary factor in the spread of soil erosion which has now become critical as well (B. Jancar-Webster, Chapter 8, pp. 166-7).

Each year the chemical plant of Nováky released 6,000 tonnes of waste calcium carbide and generated 1.2 million tonnes of waste material ... about one hundred people in Nováky were found to be affected by diseases related to toxicity (D. Turnock, Chapter, 9, p. 193).

Caution, however, needs to be exercised in interpreting the unearthing of environmental problems in these formerly Communist countries. First, attention has often not been paid to examples where environmental considerations have been given high priority by the governments of these countries. O'Connor (1989), for example, highlights how massive investments and social and economic reorganisation were sometimes undertaken in the USSR for environmental reasons. For example, there were not only large state investments in pollution abatement technology in industries in areas such as Lake Baikal, but also decisions to completely remove industries from some areas, such as within the city limits of Moscow and in the area of Baikal. A second important consideration that needs to be borne in mind is that, as Carter and Turnock (1993) have noted, some of the environmental 'black spots' in these countries have a history that predates the establishment of Communist rule. The Silesia district of southern Poland, for example, which has been described as 'the most polluted part of Poland' (Carter, 1993, p. 126), was part of a Fascist-governed Germany until 1945. Third, as again Carter and Turnock demonstrate, when the 'serious' levels of pollution in eastern Europe are compared with those in western Europe the contrast may not be as great as expected, particularly if one disaggregates figures away from national boundaries. They comment, for example, that constructing zones of sulphur dioxide emission levels would place the 'black triangle' of southern Poland, south-east of the former East Germany and the southern part of the Czech Republic, alongside much of the rest of Germany, much of the Benelux countries and 'the whole of England apart from the West Country' (Carter and Turnock, 1993, p. 208). Finally, as O'Connor (1989) has argued, attention also needs to be paid to the structure of the economies of these countries and in particular to the predominance in them of industries that make highly extensive use of their environments, both in terms of providing natural resources and as a place to dispose of waste materials. As O'Connor (1989, p. 104) points out, many of these countries 'have "specialised" in polluting industries, eg paper and pulp, fossil fuel power production, oil refining, heavy chemical, gas processing, and other basic capital goods industries'. Many capitalist countries that today might be seen to specialise in less polluting 'post-industrial' activities had a period of earlier 'industrial' specialisation, which was shown in Chapter 3 to be highly polluting. It was also argued that 'post-industrial societies' are still highly dependent on industrial forms of production, albeit often occuring in places outside these societies. It may well be, therefore, that the environmental problems of the Communist regimes may be much more to do with relative economic specialisations than to do with anything peculiar to the presence of Communism.

Such arguments might justify suggestions that at least some of the attention paid to the environmental 'problems' of these 'post-Communist' societies is ideologically motivated: that is it is seeking to create an 'other' against which to posit the superiority of the West and its favoured forms of socio-political life such as democracy and capitalism (see Smith, G., 1993; Unwin, forthcoming). There are certainly grounds for suggesting that forms of socio-political life have been too starkly differentiated into the material and cultural spaces of East and West. This point has been made by ecocentrics like Jonathan Porritt who, as noted in Chapter 2, has suggested that communism and liberalism are just different forms of a broader 'super ideology' of industrialism. It is certainly evident that the majority of these Communist governments placed considerable stress on industrialisation and hence on the dominance and exploitation of nature, as indeed did the governments of the West and the Third World through much of the twentieth century.

It is also pertinent to note that there has been a long-running debate within Marxist political economy about how best to describe the political economies of the 'Communist' states. One area of debate has been over the political constitution of these societies and whether or not they fulfilled the criteria of communism and socialism as described by Marx and Engels in their *Manifesto of the Communist Party* (Marx and Engels, 1971). They argued that there were two stages in the emergence of Communist societies: the first they characterise as 'socialism', in which there is a strong state, and a second, 'advanced communism', in which the state has withered and died as people live by the principles of communism. In the former there was a centralised bureaucratic control of society and economic activity, in the latter power would be 'socialised', with people working co-operatively towards shared goals (see for example, O'Connor, 1989, pp. 96–7). This view that communism did not require a strong state was challenged by Lenin and more particularly Stalin who argued for the retention of a strong and centralised state within what was still to be characterised as a Communist society. As a result of such shifts distinctions have been drawn between 'communist' and 'socialist' or 'state-socialist' societies, with the latter being what was actually being established in the Soviet Union, eastern bloc and other countries such as China, Angola, Mozambique and Tanzania. These were 'the actually existing socialisms' (Altvater, 1993; Smith, G., 1993).

O'Connor (1989) has argued that the political structure of Communist countries, with a strong centralised and bureaucratic state and little socialisation of power, is an important constituent in the formation of their environmental problems. He argues, for instance, that although environmental policies were sometimes high on the agenda of the central state – in the USSR, for example, 'ecological science' was seen for over 20 years as being a central ingredient of government planning, which compares very favourably to the record of many western governments – its policies were often not implemented because 'managers, technicians, and workers' in localised organisations and enterprises often did not share this concern. There was in a sense an 'implementation gap' that stemmed quite directly from the structure of the state. Relatedly there was also, O'Connor (1989, p. 99) suggests, a problem in that there was often insufficient input of environmental concerns into the state bureaucracy from Soviet society because of 'an absence of freedom to organize and agitate independently around ... issues'. The state bureaucracy was often both slow and quite resistant to taking on board the knowledges and concerns of people who lived and worked in its own, highly formalised, channels of interaction.

The term political economy refers to how societies are organised not just politically, but also economically, and there has been a further area of debate around the economic character of what we might now characterise as 'state-socialist societies'. In particular, while Marx saw his communist and socialist societies as having 'modes of production' that were quite distinct from capitalism, a series of later writers began to highlight strong parallels between state-socialist and capitalist societies. In particular some theorisations developed to characterise capitalist societies – such as 'state capitalism' (Bukharin, 1972) or 'state monopoly capitalism' (Jessop, 1982) – may be seen to be applicable to state-socialist societies (see for example Peet, 1980). Wallerstein (1979) even goes so far as to consider 'post-socialist *states*' as being 'collective capitalist *firms*'.

There is also a need to recognise that these 'state-socialist' countries were themselves situated in the social context of the 'capitalist world economy'. Right from the onset of the Russian revolution this was seen as a site of considerable importance, with people like Trotsky arguing that it was not possible to establish communism in just one country but that there needed to be a sharing of resources between countries (see Short, 1982). This view was rejected by Stalin, and Trotsky was assassinated. However, time has arguably proved Trotsky right, in the sense that the demise of communist and/or state-socialist societies has widely been construed as evidence of economic failure in the face of competition from the 'capitalist West' (see for example, Fukuyama (1992) and Michalak and Gibb (1992), but compare with Smith (1994), G. Smith (1993) and Unwin (forthcoming)). According to Marxist theory, state-socialist societies should not have a perpetual growth dynamic built in, given that they are focused around fulfilling human need, not the establishment of profit through continued production. However, when surrounded by an economic system oriented to the creation of new wants in order to satisfy perpetually expanding production, it is arguably unsurprising that these consumer desires should become strong within, as well as without, state-socialist societies. To achieve a consumption level equivalent to that of the West – let alone achieve a goal of 'surpassing the West's ability to improve the material and social conditions of the producing classes' (O'Connor, 1989, p. 97) – within what were in many respects 'underdeveloped economies' (akin to many in the so-called Third World) was clearly a massive task, and one that it was openly acknowledged would necessitate making many sacrifices, including many environmental ones. The demands on nature were also heightened by a geopolitical contest between the USSR and USA that created, amongst other things, a massive armaments industry, which in turn produced its own particular growth logic (Altvater, 1993; Smith, 1994).

Overall one can suggest that while there were very specific demands placed on nature within 'state-socialist' societies many of these demands were either similar or indeed interconnected to those present within the contempory capitalist world economy. It is therefore important not to overemphasise the 'East'/'West' differences, although as Smith (1989) notes one should not erase them entirely. Once again one can suggest that environmental actions need to be situated in multiple social relational contexts, which range from the local, the national and the international to the global.

5.5 Capitalism: the cure to unsustainability?

In the previous section we outlined arguments that placed the blame for environmental degradation at the door of capitalism, or more precisely on the operation of capitalistic social relations. Although such arguments have a large number of broad adherents it is important to recognise that others have rejected these arguments, perhaps favouring the adoption of the more 'physical process' or 'individual mismanagement' focused interpretation discussed in Chapters 3 and 4; or else, as discussed in the preceding section, arguing for the centrality of social relationships other than capitalism, such as 'political ideology' or 'patriarchy'. There is also a third response, which is to suggest that capitalism has acted, or at least can act to protect and improve nature. It is on this argument that we will now focus attention, before considering whether some external, international or global force might be required to regulate people's use of natures.

5.5.1 Environmental agitation and business: from confrontation to symbiosis?

As discussed in this and several of the preceding chapters, modern production of agricultural and industrial products, and indeed services, can be seen to have considerable environmental impacts. As mentioned in Chapter 1, and as will be discussed in more detail in the following chapter, there have been through the course of the nineteenth and twentieth centuries increasingly visible and vocal expressions of environmental concern, many of them focused around the level of human impacts on natures and suggesting that they are reaching crisis proportions.

These views are often characterised as being ecocentric, and have been challenged both by people who adopt a technocentric viewpoint and those favouring a more social contextual perspective, such as was outlined in the previous section. Having said this, however, it is also clear that these views have had considerable impact. O'Riordan has amended his initial typology of environmental attitudes, claiming that today everyone is green, including the former bastions of cornucopianism, the business community (see O'Riordan, 1991; 1995e).

This is certainly the view of Elkington and Burke (1987), who argue that environmental pressure groups such as Greenpeace have forced capitalist businesses to take environmental issues seriously and to change the way they operate. They cite as an exemplar of this the oil and chemical industries, which in many people's mind are, for good reason (see Chapter 4), amongst the most notorious culprits of environmental destruction. Elkington and Burke (1987, p. 89) acknowledge this, noting that with the arrival of oil supertankers '"super spills" made the industry's problems headline news'. They argue that while these spills do have a significant local impact on the communities of people and organisms the oil companies have been amongst the pioneers of environmental science and management, being amongst the first organisations to start undertaking environment impact assessments (see Box 1.3). Likewise they claim that in the chemical industry:

> Because of the environmental pressure on hundreds of its products, from pesticides to aerosol propellants, the chemical industry has to carry out leading edge research in such abstruse fields as ecotoxiocology and stratospheric chemistry. Despite the contrary evidence of the Seveso and Bhopal disasters, the greening of the chemical industry has been proceeding at a spectacular rate (Elkington and Burke, 1987, p. 108).

More generally, Elkington and Burke suggest that capitalist industry has become increasingly 'green', that is increasingly concerned about its various impacts on people's environments and on the constitution of nature. Indeed Elkington and Burke claim that the extent of the 'greening' is becoming so great that it is reasonable to describe some capitalists as being 'green capitalists', that is capitalists who act to improve, rather than exploit, nature.

5.5.2 Are all capitalists bad? The case for 'green capitalism'

Elkington and Burke argue their case for the emergence of green capitalism by drawing upon examples where industry has designed new products that are less environmentally damaging than previous ones. Indeed they argue that across a whole range of products, 'from lighter cans to jet engines', industry is producing more environmentally sensitive products. They then cites examples (see Box 5.7), although more inclusive lists of green companies can be found in Elkington's later book *Green pages* (Elkington *et al.*, 1988) and in other similar 'green consumer guides', such as Button's (1990) *New green pages*. Furthermore, many companies are now actively advertising the beneficial relations between their products. As well as numerous advertisements extolling the fuel 'economy' of particular cars, one also finds the promotion of other environmental virtues, such as the use of unleaded petrol, catalytic converters and, in the case of *Audi* and *Volkswagen* cars, their 'recyclability' (for illustrations of environmental advertising by the car industry see Beaumont, 1993).

Box 5.7 Elkington and Burke's seven green capitalists

(1) *Castrol* because it began producing a biodegradable oil for use in outboard motors fitted to boats in lakes in Switzerland.
(2) *Du Pont* and other aerosol manufacturers because they conducted research on ozone depletion.
(3) *Philips* for developing 'long-life' light bulbs.
(4) Packaging companies for producing lightweight packaging that because it used less material also meant that less energy was used in transporting it.
(5) *Rolls-Royce* for developing lightweight and low-noise engines. The weight is important in terms of reducing the energy consumed in producing lift.
(6) Car manufacturers for making more fuel-efficient and less polluting engines.
(7) *ICI* for producing water-based, rather than solvent-based, metallic paints. Paint solvents are seen as an important 'greenhouse gas'.

Source: Elkington and Burke (1987).

Another industry in which there is a large, and perhaps surprising, emphasis on environmentalism is the petro-chemical industry. Shell has advertised that it is undertaking 'environmental research' and is, in the process, promoting local distinctiveness (in part because it is, one advert asks us to believe, not a large multinational corporation but simply 'a local company'). In another advert it is stated that Shell is not simply run by 'businessmen' but by people who are also 'environmentalists'. The company has also actively sponsored 'environmental campaigns', most notably its *Better Britain Campaign*. Such campaigns may go some way to explaining the results of a survey of the 'environmental reputation' of companies, reported on by Elkington *et al.* (1988, p. 18) which placed Shell at the top. This was, however, before the attempt by the company to sink the disused oil platform *Brent Spar* in the Atlantic Ocean, which as we will discuss in the next chapter, had major repercussions on the environmental reputation of the company.

Shell has not been alone in its use of environmental advertising and sponsorship. In Britain most of the major energy producers, for example, make use of environmental advertising, often incorporating both the rhetoric and people associated with environmentalist criticism of business (see Box 5.8). There are, however, serious reasons for questioning whether capitalist business has been greened. First, although people such as Elkington and Burke have documented particular instances of environmentally beneficial actions, it is important to note that these instances are relatively small given the total extent of capitalist business. As Beaumont (1993) has noted, empirical evidence suggests that many businesses have remained completely oblivious to calls to pay more attention to the environmental consequences of their actions. Furthermore, he notes that even businesses that are responding can do so in a variety of different ways, ranging from extremely localised actions through to complete redesign of products, production practices and indeed the arena of the firm's business. Given this there is a clear need to consider the extent to which the practices of companies engaging in green advertising actually correspond with their promoted 'greenness'. Third, some people have argued that the whole notion of businesses being green at all is nonsensical because they all rely on the exploitation of natural resources.

Box 5.8 British Gas and green advertising

In 1990 British Gas ran a series of 14 newspaper advertisements featuring, as they put it, a range of 'independent authorities on the environment'. These environmental 'experts' included John Elkington, who as author of *Green capitalists* might not have been a surprising choice, and Tim O'Riordan who, as discussed in Chapter 1, has linked business to a problematic technocentrism and therefore might be seen as a somewhat more surprising choice.

The environmental experts were, it was stated in the advertisement, allowed to expound their own viewpoints about particular environmental issues, although it was also stated that their views were 'not necessarily those of British Gas'. The advertisements also contained an 'explanation' of its format:

> For years we have taken the health of this planet for granted. Now it is under threat. The need for all of us to understand the issues and decide how best they should be tackled is vital ... By publishing the views of 14 independent authorities on the environment, we hope to stimulate debate. With the release of the White Paper this is particularly important. We want people like you to keep the ball rolling. To take an interest in conservation. It will be energy well spent. (Detail taken from advertisement.)

As Burgess (1990) outlines in the context of environmental advertising by the nuclear industry, advertisements can be interpreted in many different ways, sometimes even in ways that completely contradict the meanings the advertisers were seeking to promote. One therefore has to be careful about how one interprets the advertisements of British Gas. Here we would point to just two features. First, note the inclusion of a very brief reference to a 'White Paper', or prospective piece of legislation, which was proceeding through the British Parliament at the time of the advert. The nature of the White Paper, which was on the privatisation of the gas industry, was not mentioned, and the inference perhaps was that, whatever effects this legislation is going to bring about, it will not affect British Gas's concern for the environment. A second clear strand in the text is the use of a universalist construction of environment issues. Indeed the text makes use of the rhetorics of both the universal effect and universalistic cause. Hence it is the 'health of the planet', not the state of a specific area that is seen as being 'under threat'. Furthermore, because the text makes no attempt to specify where this threat may be seen to originate, the implication is perhaps that we are all to blame for problems in '*the* environment'. It is certainly implied that we all need to 'take an interest' and 'get the ball rolling'.

5.5.3 Green rhetoric – but what of the practice?

It is clear that many companies are promoting a green image. There are, however, a number of studies that have raised questions as to whether so-called 'green businesses' really live up to their promoted image. McCully (1991), for example, presents a damming assessment of the environmental record of two of the companies that figure on Elkington and Burke's list of 'green capitalist' companies outlined in Box 5.7. These two companies are Du Pont and ICI, and in clear contrast to Elkington and Burke's positive assessment of their environmental record McCully states:

> Over the next century and beyond, the ozone depleting chemicals produced by Du Pont and ICI will be responsible for death, disease and environmental damage on a massive scale. From their past record it appears unlikely that the executive of these companies are particularly concerned about this (McCully, 1991, p. 114).

The focus of McCully's claims about these companies is that both had been producing CFCs. As outlined in Chapters 1 and 3, these chemicals are now commonly recognised to have caused a depletion of atmospheric ozone, which in turn has been connected to skin cancer and other human ailments. McCully claims that both companies had known about the potential impacts of CFCs on the ozone layer since the 1970s, and suggests that both companies began research on alternative aerosol propellants at this time, albeit on a minor scale. McCully argues that, despite being aware of potential problems with CFCs, the firms continued to produce them on the grounds that replacements were deemed uneconomic. It was, he argues, only when political pressure was exerted on these companies in the mid-1980s that they began to seriously turn their attention towards developing alternatives. Indeed, as evidence of the role of political pressures, McCully states that Du Pont actually suspended its US-based research on alternative propellants when Ronald Reagan became president. Reagan was known to be against the introduction of legislative environmental control and hence companies like Du Pont felt that they had less cause to take any steps towards changing their production of CFCs. As a consequence this company labelled as being 'green capitalist' by Elkington and Burke had stopped a research programme on alternative propellants even though it accepted that its current products were depleting atmospheric ozone levels. It was only in 1986 that the company re-established its research programme, this resumption coinciding with government acceptance of ozone depletion over the Antarctic. As McCully comments (1991, p. 155): 'Six years of research into less or non-ozone depleting CFC substitutes was thus lost – years during which global production of the two most damaging CFCs increased from 737 to 825 million kilograms'.

Another critic of the greening of capitalism is Yearley (1991), who notes that there is some evidence to suggest that much of the green imagery was nothing more than a marketing ploy. He notes, for example, the emergence of so-called 'environmentally friendly', 'phosphate-free' washing-up liquids. The problem with these product promotions is that phosphates were not generally a constituent of washing-up liquids anyway, although they were used in some washing powders for clothes. As Yearley argues, there was 'no environmentally relevant difference' between the so-called 'green' or 'environmentally friendly' washing-up liquids and all the others. The claim to be phosphate free was truthful but 'the greenness was illusionary' (Yearley, 1991, p. 99).

Yearley also makes the more general claim that companies have responded to the rise of environmental concern and pressure merely by making 'the easiest and cheapest environmental improvements in their products', and only making these changes when they feel it is 'necessary to maintain their market share' (Yearley, 1991, p. 100). The greening of capitalism is therefore, at best, only partial, and in many instances may be more cosmetic than real. It is also something that is externally produced rather than reflecting a fundamental change in the philosophy of business owners and managers.

5.5.4 Social regulation as the foundation of green capitalism

This last point is an important one and very much in accord with Elkington and Burke's own analysis of green capitalism. As discussed earlier, Elkington and Burke argue that green capitalism emerged largely as a response to political pressures placed on companies as a result of the actions of environmental protest organisations such as Greenpeace. McCully also suggests that political pressures played the leading role in the greening of Du Pont, albeit seemingly in an overly late and limited manner. Other studies have, however, suggested that there are a variety of other mechanisms for greening businesses, including 'legislation' 'green consumerism' and 'environmental economics'.

The role of legislation in environmental management has already been extensively discussed in Chapter 4. It was highlighted how legislation can take 'hard' and 'soft' forms, and examples were given of both effective and ineffective legislation. A key condition in the legislative approach is the development of a political will to develop and implement laws and, as will be shown later in this chapter and also in Chapter 6, this is often hard to establish in the national, let alone international political arena.

Green consumerism refers to the pressure that consumers can bring to bear on companies, not so much through political pressure and fear of governmental regulation, but through the market pressure of their purchasing decisions. Mention has already been made of the production of 'environmentally friendly' or 'green' products such as phosphate-free washing up-liquids and CFC-free aerosols. These products emerged in the late 1980s when there was growing public concern over environmental problems such as water and air pollution but relatively little government response. As a result, environmental groups began to argue that individuals had to take action themselves, and that one of the most effective 'weapons' they had was their purchasing power. The broad argument was that if people began to demand environmental products then the producers would, by 'economic forces', be induced to produce such products.

While green consumerism focused on altering patterns of demand, environmental economics focuses on the cost structure of firms. As discussed in Chapter 1 in the context of cost–benefit analysis (see Box 1.3) and also in Chapter 4 in the context of 'environmental auditing' (see Table 4.5), it has been argued that environment degradation occurs because the elements of nature are used as a 'free commodity' and thereby people fail to recognise both the benefits and costs that are generated by their use of nature. Cost–benefit analysis and environmental audits are managerial attempts to increase environmental awareness by giving non-commodified resources a virtual monetary value. Environmental economics applies the same logic but seeks to give non-commodified nature a real monetary value: to create nature as a real commodity. As Eckersley puts it,

> [Environmental] economists have recommended that governments 'correct' market failure by imposing taxes or charges (or removing subsidies) on polluters and other environmental degraders to enable the 'environmental factors of production' to be 'internalised' by the market so that prices reflect the full costs of production (Eckersley, 1995b, p. 14).

The notion of 'debt-for-nature swaps' can be seen as illustrative of the arguments advanced for such a commodification of nature and some of the problems that may well be associated with them (see Box 5.9).

Box 5.9 Debt-for-nature swaps

A relatively new strategy to deal with environmental problems is 'debt-for-nature swaps', which involves linking the undertaking of particular courses of environmental action with the alleviation of debt. In the international arena this, at least potentially, takes the form of commerical lenders, the governments of developed countries and international development agencies reducing the debt burden of some underdeveloped countries in exchange for the enactment of measures that will conserve parts of nature that are seen to be 'globally significant', such as rain forests.

The idea of debt-for-nature swaps has emerged because of the high levels of debt in many countries that, as already discussed in the text, has meant that many struggle to pay even the interest of their loans, let alone repay the original loan. Some creditors faced with the risk of non-payment have preferred to sell their loans, at a substantial discount, to other agencies. Mahony (1992) calculates that some $20 billion of debt had been traded by 1989, and some of these discounted loans have been purchased by conservation organisations who have then cancelled the debt in return for some form of conservation. According to Klinger (1994), 22 debt-for-nature swaps have been agreed involving eight debtor nations and by 1991 approximately $85 million of debt had been reduced by such deals (Reilly, 1991). The Bolivian government, for example, agreed to safeguard and expand the 134,000 hectare Beni Biological Reserve in exchange for the cancellation of $650,000 in debt (Cartwright, 1989). Costa Rica swapped a $6.5 million debt to finance rain forest conservation projects, an exchange that reduced its total debt burden by some 5 per cent (Tokar, 1991; Klinger, 1994). Other countries that have participated in debt-for-nature swaps include the Dominican Republic, Ecuador, Madagascar, the Philippines, Poland and Zambia (Mahony, 1992).

While debt-for-nature swaps have been hailed as a way of harnessing the global economy to help address global environmental problems, they have also been heavily criticised. Debt-for-nature swaps raise issues of political control. Many of the governments of underdeveloped countries see these schemes as undermining their own sovereignty over their natural resources', in that they act as simply a mechanism by which control over land is sold to people and institutions of the developed world. Debt-for-nature swaps may hence be seen as a continuation of the processes of commodification and neo-imperialism (see Chapter 2). Significantly, many of the countries in which levels of environmental change are highest are also those that are most resistant to debt-for-nature swaps: two-thirds of all deforestation is said to be occurring in countries such as Brazil, Indonesia, Burma, Thailand and Colombia, which have not been involved in any swap deals (Klinger, 1994; Reilly 1991). Klinger (1994), on the other hand, argues that debt-for-nature swaps do not necessarily mean the loss of title interests over land, nor do they necessitate that the debtor has to abide by a foreign agenda. Swaps may hence be seen as allowing indebted countries to reallocate capital from debt servicing to conservation projects.

Critics such Rehmke (1991) and Tokar (1991) have, however, suggested that in practice these schemes may do little that acts to benefit the local economies of the recipient underdeveloped countries and, as argued earler in the text, there may be quite rational reasons for undertaking environmentally damaging actions. Debt-for-nature schemes may do little to address the underlying causes of environmental destruction such as poverty, landlessness and debt, and may indeed repoduce these problems. Mahony (1992), for example, suggests that the current structure of debt-for-nature swaps often does little to reduce debt problems, in that debtors are still expected to pay back similar amounts of money, even if the total amount of money owed is lower. In these circumstances, Mahony (1992) argues, the commercial lenders receive most of the benefits. There is also, as Afusa (1991) highlights, a question of internal imperialism, with resistance to debt-for-nature swaps being voiced by representatives of indigenous peoples who argue that their land is being used to pay off the debt of other, non-native, peoples (see also Tokar, 1991).

Another problem with debt-for-nature swaps is that they have largely been enacted by non-governmental organisations (NGOs) such as the *Nature Conservancy*, the *World Wide Fund for Nature* and *Conservation International*. Although these organisations have participated in $33 million worth of debt-for-nature swaps (Klinger, 1994), this constitutes a small percentage of the total debt problem: for example the total external debt of Nicaragua alone was $5929 million in 1996 (World Bank, 1999). Governments and international agencies appear unwilling to become involved in these schemes, which are currently undertaken only on commercial debts: some 60 per cent of international debt is owed to organisations or multilateral development banks that are currently prohibited from participating in debt-for-nature swaps.

All these mechanisms can be seen as attempts to regulate the structure of capitalist economic production, but without radically altering the nature of that economy. Critics of these approaches have tended to be of two types. First, attention has been drawn to the extent to which particular modes of regulation are effective and the extent to which capitalist businesses have sought to work around them. In the next section we will examine one particular aspect of this, namely the 'geography of greening'. In section 5.5.6 we will look at the second set of reactions to attempts to make capitalism green. For this second category of critics the whole notion of capitalism becoming green is a contradiction in terms.

5.5.5 The geography of greening

Although the notions of green capitalism have been the subject of some debate (eg see Johnston, 1989; Yearley, 1991; Eckersley, 1995a) there has been relatively little attention paid to its geography. It can be argued, however, that there is both quite a distinct empirical geography to 'green capitalism' and evidence to suggest that geography actually plays an important part in capitalist resistance to the mechanisms of greening identified above. We will begin with the observable geography of green capitalism.

In many western, so-called developed countries there is, as has been demonstrated here, some evidence to suggest a greening of business, at the very least in terms of the marketing of products and the promotion of companies, and also to some extent in the design of products and the execution of production processes. However, many of the companies that in the developed world might warrant some designation of 'greenness' are also undertaking actions in other parts of the world that could not really be described in such terms. Anderson (1992), for example, highlights that companies whose products and technologies are banned in the developed world because of their environmental impacts do not necessarily stop selling these products. Rather, in many cases, they are just shipped to a new market with less stringent environmental regulations. Anderson cites, as examples, how CFCs were being sold to industries in Indonesia, as also was the chemical DDT, which has been long banned in most western countries. Madeley (1986) and Rose (1986) have both highlighted how pesticides, including many that are banned from sale in Britain, are exported from Britain for sale in 'Third World' countries. John Murrel, writing in Irvine (1989a), suggests that Shell, which as noted earlier portrays itself in countries like Britain as being an environmentally responsible company, has been one of the companies involved in such a trade, exporting pesticides to Nigeria and Zambia that are banned in many western countries. It has also been the major petroleum exploiter in the Nigerian delta, and in the process a major contributor to what has been described by members of the European Parliament as 'an environmental nightmare' (cited in Watts, 1998). The character of this nightmare is clearly described by Watts:

> The heart of the ecological harm stems from oil spills – either from the pipelines which criss-cross Ogoniland (often passing directly through villages) or from blow-outs at the well-heads – and gas flaring. A staggering 76 per cent of natural gas in the oil producing areas is flared (compared to 0.6 per cent in the USA) ... Burning 24 hours per day at temperatures of 13–14,000 degrees Celsius, Nigerian natural gas produces 35 million tons of CO_2 and 12 million tons of methane, more than the rest of the world (and rendering Nigeria probably the biggest single cause of global warming). The oil spillage record is even worse ... there are roughly 300 spills per year in the delta and in the 1970s alone spillage was four times more than the much publicized Exxon Valdez spill in Alaska ... Between 1982 and 1992 Shell alone accounted for 1.6 million tonnes of spilled oil, 37 per cent of the company's spills world wide (Watts, 1998, pp. 257–8).

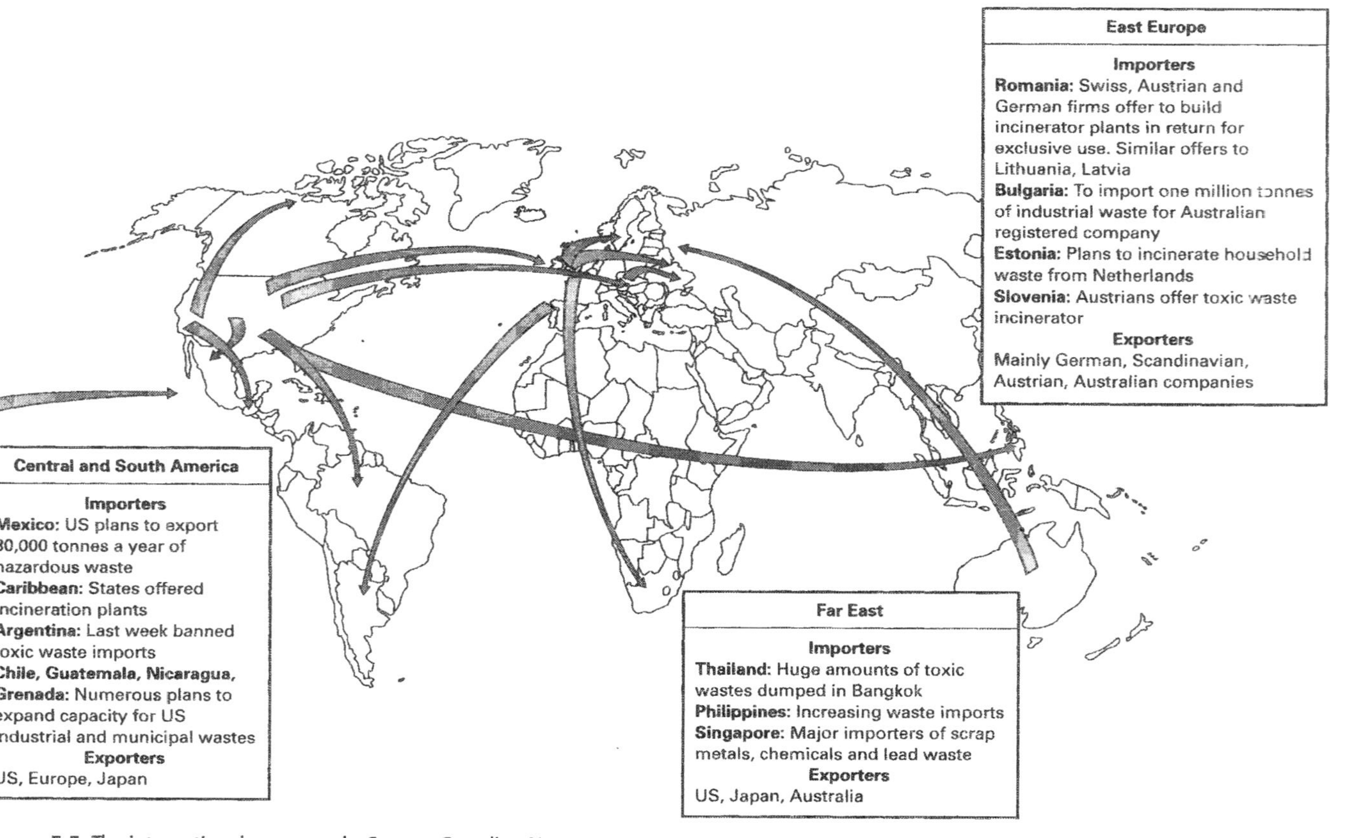

Figure 5.5 The international waste trade. Source: Guardian Newspaper.

With a record like this it is perhaps unsurprising that many geographers have come to question whether Shell deserves a 'green' epitaph and whether it should really be the sponsor of the Royal Geographical Society: perhaps a case of greening through geography as well as the more widespread practice of avoiding greening through geography.

It is not only 'environmentally unfriendly' products and technologies that are being exported from the 'green developed world', but also waste products. Vidal (1992), for example, has argued that a 'waste colonialism' has emerged whereby there is 'a massive export' of hazardous waste materials from 'the most 'environmentally aware' countries in the world. He adds:

> The US, Germany, Holland, Switzerland and the Nordic countries, who are introducing ever tighter environmental legislation on the back of consumer demands for clean industry, are now exporting millions of tonnes a year of hazardous waste to countries with weak laws or administrations that are unable to monitor or tell the difference between raw materials and toxic waste (Vidal, 1992, p. 29)

In the early 1990s this trade in waste and pollutants came to public attention when there were a number of highly publicised cases of ships being forbidden any port of call once their cargo was discovered, effectively marooning them at sea. The cases were, however, simply visible manifestations of a much broader trade (see Figure 5.5), which itself may be seen as a manifestation of attempts by firms to circumvent nationally based environmental regulations, often at the same time as seeking to present themselves as being green capitalists.

5.5.6 Can capitalism really be green?

'Green capitalism' and associated notions such as 'green economics' and 'green consumerism' can all be seen to be firmly embedded in the practices of capitalism, such as producing for exchange and for profit. These forms of 'greening' do not imply any radical change in how people perceive and act on what they take to be nature, nor does it imply any fundamental change in the organisation of capitalist production. As such it can be seen to exemplify what O'Riordan described as an 'accommodatory' perspective of contemporary society–nature relations: broadly technocentric but recognising that there may need to be some adjustments in society's exploitation of elements of nature. Not surprisingly for many environmentalists who may be seen to adopt more ecocentric viewpoints, minor adjustments in attitude and practice are insufficient.

One of the commonest lines of criticisms raised against green capitalism from adherents of more ecocentric viewpoints is that it perpetuates a level of exploitation of nature that is currently degrading nature and will, if continued though time or expanded through space, entail reaching natural limits. Irvine (1989b), for example, argues that attempts to reform industrial capitalism to achieve 'green capitalism' or 'sustainable development' are doomed to failure, essentially because this form of economy is geared towards a continued expansion in production and thereby a continued extraction of raw materials and energy out of the environment. Green consumerism, he argues, does not provide an ecological solution in that, as it preaches the virtue of consuming more environmentally friendly products, it is still preaching the virtue of consumption. As Irvine puts it:

> The Body Shop is certainly much more socially and environmentally enlightened than Boots the Chemist but they both want you to fill your bathroom with deodorants, perfumes and all the other paraphernalia of conspicuous consumption (Irvine, 1989b, p. 88).

The production of all goods involves the use of some resources and energy: so long as production expands, Irvine argues, so will the exploitation of the environment. The only way to prevent reaching a point of 'ecological disaster' is, Irvine suggests, to place a brake on production, although he recognised that this is very difficult to achieve given the present structure of the world economy.

Irvine focuses on the temporal dimension to capitalist exploitation of nature elements, but other ecocentrics have highlighted its current spatiality and the consequences if the practices of developed capitalist countries should spread, as is the hope of many developmentalists. People such as Daly (1973), Lappé and Collins (1980) and Hilyard (1993), for example, have all argued that the globalisation of a US-style economy is impossible given the demands it places on nature. As Hilyard notes:

> In the last 50 years, the US has single-handedly consumed more fossil fuels and minerals than the rest of humanity has consumed in all recorded history. The US beef industry alone consumes as much food as the population of India and China consumes (Hilyard, 1993, p. 30).

The only form of development that such ecocentrics would see as warranting the label 'green' is one in which there are limits placed on economic growth, particularly amongst the so-called 'developed countries'. The logic of this argument is clearly presented by Bahro (1982) who, as Redclift (1984) notes, basically argues that the solution to environmental degradation in both the developed 'North' and underdeveloped 'South' is for the 'North' to deindustrialise and for there to be some industrialisation of the 'South'.

The analysis of people such as Irvine and Bahro points out some of the limitations of the notion of green capitalism as advanced by the likes of Elkington and Burke. Whether their analysis leads them to propose a feasible way forward is, as will be discussed in Chapter 6, rather questionable. Furthermore their analysis does not specify when ecological limits will be met and, as discussed in Chapter 4, there are widely differing assessments of the extent of environmental degradation. However, one of the most problematic elements in such ecocentric analysis is that there is a tendency to omit any social explanation as to why there is this proclivity to growth and high levels of material exploitation. Often it appears just as a given of contemporary capitalist societies, or else it is seen as a consequence of a near-universal human tendency to consume as much as is feasible. In contrast, Marxist political economic analyses have focused considerable attention on the establishment of this tendency to growth and have argued that it is a logical outcome of the emergence of production oriented towards exchange and private capital accumulation. As Yearley (1991, p. 104) succinctly puts it, while expanding production may be 'attractive to the consumer, it is essential to the economic system'.

5.6 Global social regulation to protect global nature

In the preceding two sections we have highlighted some of the problems associated with notions that capitalism can be regulated to become green. For deep green and Marxist political economists capitalism may be inherently unregulatable in the sense of having all its negativities removed: to remove these would be to end capitalism. Even those advocating the regulation of capitalism via consumer pressure, market mechanisms or legislation have also foreseen problems in implementing regulation, not least in the ability of companies, particularly large multinational ones, to make use of differential geographies of regulation. One potential way of preventing this particular regulatory failure may be to move to a global system of regulation, and some of the attempts to do this have already been discussed in Chapter 4. A further rationale for a global scale of regulation may also be seen in the spatiality of environmental processes. As highlighted in Chapter 3, many environmental problems can have effects that are spatially quite extensive, and in the case of ozone depletion and the greenhouse effect might indeed be seen as being global in character. In this section we will examine attempts to forge some form of global environmental regulation, placing particular emphasis on the social contexts which these forms of regulation have emerged and whose interests they might serve. We will focus attention specifically on a series of conferences connected in some way with the United Nations that have examined the issues of environment and development and suggested a range of regulatory responses.

5.6.1 The emergence of environmental issues in the international political sphere

One of the principal institutions of international politics is the United Nations, and it is possible to chart a steady increase in its expressions of concern about environmental issues. One of the earliest of these was the United Nations' *Special Seminar on Development and the Environment* held in 1972. While this represented a clear recognition that environmental issues were worthy of some international discussion, the report of the seminar concluded that:

> the problems of environmental disruption are still a relatively small part of the development concern of the developing countries, and it may be premature for many of them to divert their administrative energies to the establishment of new institutions or machinery (United Nations, 1972, p. 27).

For the participants of this seminar environmental issues were low down their order of concern: they were much more concerned to maximise economic growth, as this was seen to be the measure of a country's development.

This should not, however, be taken to mean that environmental concerns were totally excluded from the international political arena at this time. In the same year, for instance, the United Nations convened another conference, officially entitled the *United Nations Conference on the Human Environment*, although more commonly referred to now as the *Stockholm Conference*. This conference was heavily influenced by the doom-laden predictions of neo-Malthusians such as Ehrlich and Hardin (see Chapter 3), and the focus was very much on issues of population growth and pollution (Hecht and Cockburn, 1992, p. 369). Although the conference was organised by the United Nations it was, as Hecht and Cockburn (1992, p. 369) point out, 'concerned primarily on First World concerns'. Indeed, McCormick (1992) argues that the conference stemmed largely from a growing concern within Europe about acid rain and the growing recognition that pollution was no respecter of national boundaries. The conference was initially viewed with considerable hostility by many governments of underdeveloped countries, who feared that environmental safeguards and regulations would act as a further brake on their economic development, largely by restricting their ability to export to the developed world. However, the conference was preceded by a series of preparatory meetings at which, according to the secretary-general of the conference, 'a clear recognition of the relationship between the issue of environment and development emerged' (Maurice Strong, quoted in McCormick, 1992, p. 93). This view is, however, generally seen as an overstatement, in that although the notions of development and environment were connected in the final conference it was in a series of distinct and rather contradictory ways. In many instances the 'developed worlds' and 'less-developed worlds' (to use the terminology of the time) were seen as quite disconnected from each other, each facing quite distinct and largely unrelated sets of environmental problems (see Box 5.10). There were also clear hints of environmental determinism in their accounts of developmental differences (compare with Figure 1.1). At other times the 'same boat' argument was very explicitly drawn upon, although the emphasis of change was often directed more at the less-developed rather than the developed countries (again see Box 5.10)

Box 5.10 Development and environmental degradation – the view from the Stockholm Conference

As part of the preparation for the *United Nations Conference on the Human Environment*, held in Stockholm in June 1972, the secretary-general of the conference, Maurice Strong, commissioned Barbara Ward and Réne Dubos to write, on the basis of consultations with 152 experts drawn from scientific, business, planning, economic and development organisations, a 'conceptual framework' for the conference delegates. This report was subsequently published as a book, *Only one earth: the care and maintenance of a small planet* (Ward and Dubos, 1972). This book outlines a very clear, albeit contestable, view of development and its association with environmental degradation:

> Taking a fairly arbitrary level of national income – say $400 per capita – as a rough dividing line, we can say that the countries below this level tend to share certain acute problems which directly affect their environmental outlook and which are not experienced on the same scale – or at all – in developed lands.
>
> Their populations are growing almost twice as fast as those of wealthier countries and twice as fast as did industrializing states in the nineteenth century.

> Their soils tend to be more fragile and their climates less reliable and moderate than in temperate zones, where the majority of high-income states are to be found. Moreover, much less is known in scientific terms about specific natural conditions and needs in developing lands.
>
> With few exceptions, urbanization has come ahead of industrialization .... [which] creates the risk of far greater social diseconomies than any known in the process of nineteenth-century industrialization ...
>
> All these pressures create environmental problems, both economic and social, which are sufficiently different to merit the most urgent consideration, apart from the issues which are the most urgent concern of high-income lands (Ward and Dubos, 1972, pp. 92–3).

The authors, however, also draw back from this argument to invoke the global consequence argument that we are all 'in the same boat', we all face a common, and singular, environmental crisis:

> However the distinction remains arbitrary. Rich and poor, developed and developing, industrialized and pre-technological – all are enmeshed in myriad webs of trade, communication and influence, all are struggling to adapt the technological order to truly human ends, all are involved in the welfare and survival of their fellow communities, all must inescapably share a single, vulnerable biosphere. Whatever their immediate differences, the environmental issue confronts them all with an ultimate challenge – the survival and good estate of their planetary home (Ward and Dubos, 1972, p. 93).

Furthermore, the book ends by suggesting a set of common solutions, or 'strategies for survival', which included: nations adopting 'a *collective* responsibility for discovering more – much more – about the natural system and how it is affected by man's activities and vice versa' (p. 290); the development of 'common policy and co-ordinated action' with respect to environmental problems (p. 295); and a recognition by 'all the inhabitants of Planet Earth' that they all 'belong to a single system, powered by a single energy, manifesting a fundamental unity under all its variation, depending for its survival on the balance and health of the total system' (p. 297).

However, while espousing common consequences and common solutions *Only one earth*, and the Stockholm Conference to which it was connected, tended to propose courses of action that impacted rather more on the underdeveloped than the developed world. In particular, although there was a call for some transfer of resources from rich to poor countries, there was no criticism directed at the level of economic activity and consumption of the developed world. Attention was restricted largely to measures deemed necessary to either raise the level of economic activity in the 'less developed' countries or to ameliorate its impacts in the 'more developed' ones. Overall, therefore, the Stockholm Conference can be seen to have adopted a broadly technocentric perspective on society–nature relations.

While the Stockholm Conference is often described now in quite negative terms (see Hecht and Cockburn, 1992; McCormick, 1992), a number of commentators have identified some positive features. One aspect of the conference worthy of note is how it came to involve not only governments and international agencies, but also formal and informal NGOs. The Stockholm Conference was in practice three conferences: the main conference of governmental representatives, a *Milföforum* for some 400 invited non-governmental environmental organisations, and a *Folket Forum* organised by the socio-environmental pressure group *Pow Wow*. There hence was some attempt to interconnect governments, formal pressure groups and a 'concerned public'.

The second feature that has been given prominence in many accounts of the Stockholm Conference was the *United Nations Environment Programme* (UNEP). UNEP was set up to help implement the ideas of the Stockholm Conference, in particular those contained within an action plan developed at the conference. The action plan had two basic principles: first that there was a need for increased knowledge of environmental trends and their human effects; and, second, to protect and improve the productivity of natural resources. UNEP very much adopted these technocratic principles, establishing *Earth Watch*: 'a UN sponsored network designed to research, monitor and evaluate environmental processes and provide early warning of environmental hazards and determine the status of selected natural resources' (McCormick, 1992, p. 107).

UNEP also sought to influence other UN agencies and make environmental considerations an element of all UN policy and management developments. Another element of UNEP was to bring countries together to exchange environmental information and develop environmental policies and regulations that were of mutual interest. A clear example of this was the *Regional Seas Programme*, in which ten action plans were drawn up by states adjoining common sea areas (see McCormick, 1992).

However, while UNEP is often seen as a useful outcome of the Stockholm Conference it has also itself been the subject of some criticism. McCormick (1992), for example, argues that it was beset by financial, organisational, locational and constitutional restrictions. UNEP was a small organisation within the UN in terms of budget and staff; it was seen at times to be unnecessarily bureaucratic and centralised; it was located in Nairobi, away from most of the rest of the UN and the important decision-making centre of New York; and it had no executive powers of its own, being able only to provide assistance to those who asked for environmental help rather than being able to initiate its own projects. More positive assessments of UNEP have argued, however, that despite its limitations it performed an important role as a catalyst, stimulating environmental awareness and debate within the institutions of international governance.

5.6.2 From environment and development to sustainable development: international discussions in the 1980s

Following on from UNEP there have been a whole series of United Nations-related reports and international conferences on environmental problems and development. One of the most widely heralded of the reports on environmental problems was the 1980 *World Conservation Strategy*, which was organised by the *International Union for the*

Conservation of Nature and Natural Resources (IUCN) with financial support from UNEP. The aim of the Strategy was to 'help advance the achievement of sustainable development through the conservation of living resources' (International Union for the Conservation of Nature and Natural Resources, 1980, p. iv). The focus was very much on living nature and how this might be rationally exploited. Conservation, for example, was defined as 'the management of human use of the biosphere so that it may yield the greatest sustainable benefit to the present generation while maintaining its potential to meet the needs and aspirations of future generations' (International Union for the Conservation of Nature and Natural Resources, 1980). The Strategy charted how living nature was impacted by cycles of development and underdevelopment.

In the same year as the World Conservation Strategy there was another major report on the issue of 'development', the so-called *Brandt Report* or, as it is more formally entitled, *North–South: a programme for survival* (Independent Commission on International Development Issues, 1980). Whereas the World Conservation Strategy started with examining environmental issues and ended up linking these to questions of development, the Brandt Report rather went in the other direction, starting with the development process and linking this to the environmental problems. It was argued, for instance, that basic development objectives were being undermined in many countries by 'irreversible destruction of the ecological systems of a number of poor countries' (Independent Commission on International Development Issues, 1980, p. 47). It was suggested in the 1980 report, and also in a follow-up report entitled *Common crisis* (Independent Commission on International Development Issues, 1983), that these environmental problems could be solved through greater incorporation of these poor countries into the world economy through expanding their trade.

There are, however, many people who would argue that incorporation into the world economy does not lead to an improvement in the environmental situation. There are at least two reasons for this. First, as discussed earlier, for many countries in the world incorporation into world trading systems does not lead to them increasing their economic prosperity, but rather sees their wealth being siphoned off into the already wealthy countries of the world. Second, unchecked economic growth can actively accelerate environmental degradation. As Redclift puts it:

> The pursuit of economic growth unchecked by environmental considerations, can accelerate, among other things, topsoil losses, the scarcity of fresh water, the deterioration of grassland and deforestation ... The irony is that these policies not only have negative environmental consequences, they also frequently fail to meet their economic objectives (Redclift, 1987, p. 59).

This idea that participation in the world economy leads to two distinct forms of environmental crisis – one related to poverty and one related to economic growth – was taken on board to some extent by an international commission and report, the so-called *Brundtland Report*, or more officially *Our common future* (World Commission on Environment and Development, 1987). The report, written by a commission chaired by the ex-Norwegian Prime Minister, Gro Harlem Brundtland, has become particularly well known through its definition of sustainable development as development that meets the needs of the present generation without compromising the ability of future generations to meet their own needs (see Chapter 4).

The Brundtland Report took up many of the themes of the Brandt Report, but like the earlier World Conservation Strategy focused particularly on the causes of environmental degradation. The report broadly concluded that environmental degradation was occurring for one of two reasons: either it was being created by the actions of the poor who were destroying their environments largely through their struggle simply to stay alive; or, in other areas, it was created through high consumption lifestyles and expanding production that largely ignored its environmental impacts. Hence, under this analysis, the world is facing two distinct types of environmental problem.

In a review of the Brundtland Report, Elkins (1992) argues that its analysis of the current environmental situation is essentially correct and marks a particular advance of the analysis of previous environmental/developmental reports. Elkins suggests (1992, p. 31), however, that the report contained a serious flaw: 'notwithstanding that its analysis reveals two very different causes of environmental destruction – over consumption by rich people and countries, poverty in poor countries – it recommends a single over-arching solution to the problem … further economic growth'.

The Brundtland Report hence follows the line of the earlier international reports, rather than perhaps the logic of its own analysis, in arguing for increased production and trade as a mechanism to environmental protection. As Finger (1993, p. 42) notes, the concept of sustainable development as established by the Brundtland Report effectively argues that 'ecological sustainability is good for economic development and economic development is good for ecological sustainabilty'. The Brundtland Report can be seen as providing what might be described, to use the vocabulary of Chapter 1, as an 'accommodatory' view of contemporary society–nature relations; that is it recognises that there might be some problems created by the exploitation of nature but sees these as being of manageable proportions. Overall it can be suggested that while international politics in the 1980s came to taken on board a concern for environmental as well as economic developmental issues, the resultant policy prescriptions were still very much technocentric in character.

5.6.3 The Earth Summit

The Earth Summit is the informal title given to the *United Nations Conference on Environment and Development* (or UNCED) that was held in Rio de Janeiro in June 1992. In many ways this conference can be seen as the culmination of the rise of environmental concern on to the international political agenda. The Earth Summit, for example, was described in the following terms by its organiser, Maurice Strong: 'no conference in history ever faced the need to take such an important range of decisions – decisions that will literally determine the fate of the earth' (quoted in Pearce, 1991a, p. 20).

The head of UNEP simply described the conference as 'our best, perhaps our last chance, to save the planet' (Mostafa Tobla, quoted in Pearce, 1991a, p. 20). However, while the Earth Summit was heralded with much high-sounding rhetoric, the reaction of many environmentalists and much of the media after it actually happened was essentially negative (see Box 5.11).

Box 5.11 A selection of assessments of the Earth Summit

We cannot permit the extreme in the environmental movement to shut down the United States. We cannot shut down the lives of many Americans by going to the extreme on the environment (George Bush, President of the United States of America, 1 June 1992).

We are not saving the earth: we are not even saving our own consciences (Carlo Ripa di Meana, European Commission, 1 June 1992).

This is the decisive first step (Fernando Collier, President of Brazil, 5 June 1992).

I came here with low expectations and all of them have been met (Jonathan Porritt, 13 June 1992).

They talk, they talk and they talk about poverty and then they don't do anything. I don't really see why we came here (Earnest Rukangira, Director, Rwanda National Environment Service, 15 June 1992).

We need a paradigm shift. I saw no sign of that happening in Rio. Of course we have to welcome any progress, but it has been microscopic (Jeremy Leggett, Scientific Director, Greenpeace, quoted in the *Independent*, 15 June 1992).

The hopes of Rio have run into the ground (André Simms, member of Christian Aid, 28 June 1997).

Sources: All quotes from the *Guardian* on the date shown unless otherwise stated.

Such divergent reactions to the Earth Summit raise a series of questions. The comments of Jonathan Porritt are, for example, suggestive that the summit may be interpreted as yet another instance of an implementation gap, although this time it is the politicians and governmental advisers at the summit who are to be blamed for failing to implement the rational course of action. Yet just as the actions of the peasant farmer need to be placed in their social contexts, so too do the actions of the agents of international politics. This point is well made by Hecht and Cockburn (1992) in a review of the *realpolitik* of the Earth Summit. They argue that the summit very much reflected the 'social and economic contours of the current environmental arena', and indeed encapsulated a clear 'social geometry' (Hecht and Cockburn, 1992, p. 367), or social geography.

The Earth Summit in international and national contexts

One of the most obvious reasons for the apparent failure of the Earth Summit is the existence of geographical differences in environmental concern, and the failure of the summit to overcome these. As Pearce argued in an article that appeared before the summit, there were clear 'North–South' divisions in the perception of environmental problems and solutions:

> The North – Europe, North America and Japan – is worried about global issues such as the greenhouse effect and the destruction of the planet's last rain forests. It says: we are all in the same boat and must pull together to protect our future prosperity.

> The South – most of Asia, Africa and South America – dismiss the 'same boat' argument. The North has the prosperity and has created the global crisis, it says. The North is responsible for most of the planet's pollution and consumes most of its natural resources. Moreover, the North's control of commerce is creating trade imbalances and poverty in the South that force poor nations to overuse their forests and soils, creating grave local environmental problems. The South says that if the North wants to help save the planet, then it must pay – by changing the rules of trade, providing more aid and allowing access to the latest technologies ... The governments of the North, and especially the US, will not admit that 'they are the problem' (Pearce, 1991, p. 29).

One of the hopes of those behind the Earth Summit, including Maurice Strong, was that political discussions could overcome these geographical differences. In practice, however, politics often meant that not even discussion, let alone agreement, could take place. President Bush of the United States of America, for example, persistently threatened not to attend the summit because of the inclusion of certain items on the agenda and, even when he did attend, he refused to sign the Biodiversity Convention and only signed the Climate Change Convention after specific levels and deadlines for reductions in CO_2 emissions had been removed (see Chapter 4). Many members of OPEC also opposed the Climate Change Convention, fearing it could diminish their oil revenues, while India and Malaysia led opposition to both a Forest Convention and a Forest Declaration on the grounds both that it interfered with their national sovereignty to exploit their resources as they wished, and because developed countries were seeking to use the forests of the underdeveloped world as a cheap 'carbon sink' for the disposal of CO_2 emissions rather than undertake a more costly programme of reducing the developed world's production of this and other greenhouse gases. As people such as Sachs (1993) have argued, many of the countries of the South effectively sought to use the summit to launch demands for more development assistance and sought to use environmental issues as 'a diplomatic weapon' to gain this assistance. The Earth Summit can, hence, be seen to have been caught up within existing geopolitical differences rather than to have broken them down.

The pre-existing geopolitical differences in which the Earth Summit was situated were themselves products of unequal power relations, and it should therefore be unsurprising that not only did the shape of the discussions often reflect these differences but that the outcome of these discussions tended to fall more into line with the view of the more dominant group, in this case the countries of the North. Hilyard (1993) and Shiva (1993), for example, both argue that the published outcome of the summit tended to be more in line with the interests of the North than the South. Hilyard, for instance, suggests that the underlying argument endorsed by the summit was that:

> environmental and social problems are primarily the result of insufficient capital (solution: increase Northern investment in the South); outdated technology (solution: open up the South to Northern technologies); a lack of expertise (solution: bring in Northern-educated managers and experts) and faltering economic growth (solution: push for an economic recovery in the North) (Hilyard, 1993 p. 31).

Differences in environmental attitudes do not, however, just relate to international geopolitical power relations. The attitudes of the British and US governments at the summit, for example, were subject to considerable criticism from within these two countries (see Plate 5.1), and equally stark 'intra-national' differences in attitudes could be found across the globe. For example, some of the governments of the South that were most vocal in calling on the countries of the North to take action on the environment actually had quite long records of both supporting projects that initiated considerable levels of environmental degradation within their own countries, and also of ignoring (and in such cases as India, Malaysia and Kenya actively suppressing) environmental movements within their own boundaries. Indeed one of the strongest elements of commonality within the summit was arguably that the governments of all the countries represented were hoping to ignore at least some of the demands of their own citizens. Hecht and Cockburn (1992) certainly suggest that the Earth Summit was an exercise in *realpolitik* dressed up in 'the popular language of environmentalism'. They note, by way of support for this claim, that although the G7 group of countries – that is the seven wealthiest countries in the world – were all very active at the Earth Summit, and invested considerable political energies in putting forward their environmental viewpoint, at a regular G7 meeting less than a month afterwards there was no mention of environment at all. In Hecht and Cockburn's (1992, p. 367) words, the G7 group had 'excised global ecological terms from their vocabulary'.

Plate 5.1 Intra-national differences over the Earth Summit: hanging of banner from Nelson's Column, London, by Greenpeace

This quick demise of environmental issues from the agenda of inter-state discussions so soon after the Earth Summit is clearly suggestive of an, at best, 'shallow' embrace of environmentalism. Indeed a number of commentators have argued that the governments at the Earth Summit were all essentially adopting a technocentric viewpoint. Sachs (1993, p. 11), for example, has argued that the summit was effectively 'a technocentric effort to keep development afloat against the drift of plunder and pollution', while Hecht and Cockburn argue that it was structured around just three views of environment and development:

(1) **Green consumerism**
A belief that 'perfecting markets (if the forests are correctly assessed in market terms they will not be destroyed) and deft marketing strategies (for the Body Shop, Ben and Jerry's Rainforest Crunch etc) create the conditions through which environmental results can most effectively be achieved' (Hecht and Cockburn, 1992, p. 374).

(2) **Scientific bureaucratic regulation**
A belief that policies based on extensive research and rational management techniques such as 'zoning' will provide the best solution.

(3) **International regulation**
A belief that international financial and regulatory bodies can steer the international economy away from disaster.

Although there are significant differences between the approaches, and these differences may go a long way to explain the acrimonious debate that went on at the summit, they can all be seen to be essentially technocentric. For Hecht and Cockburn all three strategies are inadequate to deal with current levels of environmental degradation. They argue, for instance, that the strategy of green consumerism fails to consider the 'historical and economic realities' of life in many countries; that the strategy of bureaucratic research lends itself to deferring uncomfortable decisions until it may well be too late; while the strategy of international regulation is idealistic in that:

> regulatory powers seem to be lacking at almost every level ... and are primarily tugged by international purse strings. Moreover, the long-standing resistance of the USA to the United Nations could compromise any enforcement powers. As the USA does not obey world court decisions that it finds distasteful, there is no a priori reason to believe that any kinds of enforcement or sanctions would have much effect, whether or not the USA is a signatory (Hecht and Cockburn, 1992, p. 375).

From Hecht and Cockburn's perspective the Earth Summit was not so much a failure in that it did not live up to its objectives, but rather was a failure in its choice of objectives in the first place. It is clear that inter-governmental discussions of environmental issues on the international political arena at the Earth Summit, as at the preceding international conferences, were associated with neither the enactment of ecocentric arguments nor the adoption of a technocentric notion of equitable sustainable development.

The social geography of the Earth Summit

So far we have talked about 'the' Earth Summit but, as in the case of the Stockholm Conference, it was in practice not one conference but three. There are indeed a series of parallels between the Earth Summit and the Stockholm Conference: both involved a series of preparatory meetings setting up an agreed agenda; both were chaired by Maurice Strong; both led to the publication of quite similar documents (see Table 5.5); and finally both conferences involved three distinct components, namely a formal governmental conference, a large arena for NGOs to present material and lobby governments, and a more informal, popular conference. The Stockholm Conference had the formal conference, the *Milföforum* and the *Folket Forum*. At the Earth Summit there was, first, the formal *Eco 92*, attended by representatives of national governments and international agencies such as the United Nations and the World Bank. Second, there were a series of meetings, lectures, conferences and press briefings organised by NGOs. These were organised pressure groups, ranging from large and identifiably 'environmental' organisations such as Friends of the Earth, Greenpeace and the World Wide Fund for Nature, through large and not clearly identifiably environmental organisations such as churches, trade unions and business corporations, and on into a range of smaller and diversified groups such as peasant organisations, women's co-operatives and youth groups. Finger (1993) argues that NGOs were involved in the summit both to establish a 'constituency' of supporters who would help establish public legitimacy for the conference and to help identify some NGOs or members of NGOs who might be incorporated into the administration of systems of environmental regulation. Hecht and Cockburn (1992) also point out that the presence of these organisations reveals much about the socio-political situation in many countries in the South, and in particular the rise of what they call the '*frazzled state*'. They use this term to suggest that the governments in many countries have virtually lost any control over their economies and have become reliant on external development organisations such as the World Bank. They argue that the people in these countries, with no effective government, have come to turn increasingly to NGOs for support. The participation of a large number of NGOs at the Earth Summit is hence seen by Hecht and Cockburn to be of great social significance.

Table 5.5 A comparison of the documents of the Stockholm Conference and the Earth Summit

Documents of the Stockholm Conference	Documents of the Earth Summit
Stockholm Declaration: an 'inspirational' text designed 'to put on record the essential arguments of human environmentalism and to act as a preface to the Principles'	*Earth Charter*: an 'inspirational expression of global environmental and development problems and the actions needed to solve them'
List of Principles: outlined 26 broad aims and goals	
Action Plan: '109 separate recommendations, ranging from the general to the specific'	*Agenda 21*: a 'blueprint of actions suggested as being necessary to ensure planetary survival in the 21st century'
	International Conventions: formal inter-governmental agreements on particular environmental problems (ie biodiversity and climate change treaties)

(Sources: descriptions of the Stockholm Conference documents are from McCormick, 1992; those of the Earth Summit documents from Pearce, 1992)

The third element of the Earth Summit was the so-called '*Earth Parliament*', which was 'largely composed of modest grass roots and indigenous people's organisations' (Hecht and Cockburn, 1992, p. 367). Again Hecht and Cockburn argue that this Parliament was of great significance, and they highlight two reasons for this. First, they suggest that the people in the Earth Parliament actually represented the people who 'bear the brunt of environmental degradation most directly' (Hecht and Cockburn, 1992, p. 367). If you like, these people acted as a sign that environmental degradation is not just an issue for a global 'us' to think about for future generations, but is already a critical issue for many people living, and dying, today. Second, Hecht and Cockburn suggest that this group, although perhaps the least powerful group at the Earth Summit, are arguably the most knowledgeable about resource use. The Earth Summit meetings at Rio de Janeiro reveal quite clearly the contradictions of the current socio-economic organisation of the exploitation of the environment, whereby many of those with the most knowledge of the environment are also the ones experiencing the most dire consequences of environmental mismanagement.

A final point we would make in the context of the Earth Summit is that, as Hecht and Cockburn note, these three elements of the Earth Summit had a very clear 'social geometry' or geography. The formal conference, Eco 92 – which on their reading was the most irrelevant in that it represented no change from the ill-thought-out and ill-fated conceptions of environment and development employed in the past – was located in RioCentro, a built-environment of opulence, luxury and wealth. It was also surrounded by, but separated off from the 'outside world' by a cordon of Brazilian and UN soldiers. On the furthest margins of the conference site was the Earth Parliament that, in Hecht and Cockburn's reading at least, involved the people who are most at the centre of contemporary environmental crises. In the middle, although still located some 40 kilometres from the RioCentro, were the NGOs, which can be seen as being in part connected to a failed past, but as yet without the power to create significantly different global futures. The Earth Summit, even in its physical layout, was largely an expression of current economic, social and political geographies of the environment rather than somewhere in which new configurations of society and nature relations were being created. Overall, as Shiva has put it, the Earth Summit was largely a re-creation of a very particular political-economic space:

> the global as it emerged in discussions and debates around the UN Conference on Environment and Development ... was not about universal human interests or about a planetary consciousness. The life of all people, including the poor of the Third World, or the life of the planet, are not at the centre of concern ... The 'global' in the dominant discourse is the political space in which a particular dominant local seeks global commons, and frees itself of local, national and international constraints. The global does not represent the universal human interest, it represents a particular local and parochial interest which has been globalised through the scope of its reach (Shiva, 1993, pp. 149–50).

The particular global that Shiva argues predominated at the Earth Summit was effectively that of the world capitalist system and contemporary successful agents within this, such as multinational corporations and the governments of the so-called developed world.

5.7 Conclusion

In this chapter we have explored the role that social contexts play in the establishment of and persistent failure to ameliorate actions that have seemingly serious environmentally damaging consequences. It has been argued that, contrary to some interpretations of the presence and continuation of environmental problems, it is not reasonable to place the blame on particular individuals nor see them as an inherent feature of human nature: a consequence of all human activity. Rather, it is argued, attention needs to be centred on elucidating the social contexts in which people act and, in many instances, on the world of capitalistic social relations, which arguably have inherent tendencies that encourage people to undertake environmentally damaging actions. This argument has been advanced by looking at both the creation of, and the failure to prevent, actions that are widely seen to have environmentally damaging consequences, such as deforestation, famine and industrial pollution. The need to recognise the global reach of capitalistic social relations is highlighted, but it is also argued that attention needs to be paid to differential positioning in these relations and the role of local agencies such as culture, political institutions and the environments of nature.

The chapter has also addressed attempts to regulate the global economy and its uses of nature. The notion of green capitalism and green consumerism was discussed and a series of problems associated both with their implementation and overall scope were outlined. While green capitalism and green consumerism may be more environmentally sustainable than non-green versions serious questions remain in the longer term, given that they retain the logic of expanding environmental exploitation. Furthermore, important contemporary issues relating to social justice are also apparent.

The final section of the chapter looked at a range of international conferences and initiatives relating to the regulation of the capitalist world economy. It was argued that these conferences and initiatives, although raising the issue of the use and degradation of natures, have been limited in their impact, in part through their largely technocentric conceptions of society–nature relations and also very clearly by geopolitical differences in social interest, which themselves relate in large part to the contemporary structuring of the capitalist world economy.

Chapter 6

Alternative Relations between Society and Nature

6.1 Introduction: alternatives to technocentrism

This chapter draws on some of the ideas presented in Chapter 1 to explore in more detail the arguments for non-exploitative relationships between society and nature. Through the course of the preceding five chapters a series of problems surrounding exploitative relationships with nature have been outlined, and it has been highlighted how environmental exploitation creates problems that are extremely difficult to manage and/or appear unlikely to be addressed given current modes of social organisation. Such problems may suggest a need to adopt a radical new attitude to the environment and to the organisation of society, and indeed, as noted in Chapter 1, this has been the argument of people who might be said to adopt an ecocentric perspective on society–nature relations. It was, however, also highlighted in that chapter that such calls have been rejected by some who do not accept the validity of the arguments put forward by ecocentrics. In this chapter we want to return to these issues and consider arguments about new societal–environmental relationships. In particular the chapter will seek to outline something of the extensive history of calls for the establishment of non-exploitative relationships with nature, the forms of argument made and the social context in which they have been placed.

In Chapter 1, it was suggested that social attitudes towards nature and the environment can usefully be considered using the ecocentric/technocentric distinction proposed by O'Riordan (1976). It was argued that technocentrism refers to ideas that encourage and attempt to justify an 'exploitative' relationship to those things that are taken to be 'nature', while ecocentrism seeks to enact and legitimate a non- or less-exploitative relationship to 'nature'. Technocentrics tend to see society–nature relations in terms of human control and use, while ecocentrics see nature both as a source of material and cultural/emotional/spiritual value and as something that has material limits and is under threat from human activity.

As was also outlined in Chapter 1, technocentrism has been seen to be a feature of both contemporary and historical lines of Occidental thought, including Judaeo Christianity, classical and modern science, liberalism and communism. Furthermore, as has been demonstrated in Chapter 5, it can be seen as very influential in contemporary global politics. It is, however, important to recognise that ecocentric perspectives have also had quite an extensive history and geography of expression. Pointers to this have, indeed, been variously made through the preceding chapters. For example, reference was made in Chapter 1 both to medieval animalistic thought and to suggestions that 'environmental stewardship' was an important strand of Christian thought. It should also be noted that Pepper (1984) has discussed how the idea of 'natural magic' and its notions of people as a microcosm of a universal cosmological nature were important within the scientific revolution that, as discussed in Chapter 1, has also been suggested as an important moment in the rise of technocentrism. Furthermore, in contrast to the hierarchy of societies and nature that, as discussed in Chapter 2, characterised much of European thought during European imperialism, much aboriginal thought viewed people as part of nature (see Pawson, 1992; Phillips, forthcoming-b), although this is not to say that aboriginal people have not impacted nature adversely.

In the present chapter attention will first focus on Occidental thought, which has sought to challenge technocentric values and that may therefore be seen to promote a broadly ecocentric perspective on society and nature. Although such thoughts do, as noted above and as more clearly spelt out by Pepper (1984), have 'roots' that extend well back in time, our attention will focus on environmental attitudes in Britain and North America since the eighteenth century and the emergence of what is commonly termed the 'Romantic movement'. The spatial and temporal starting point is largely pragmatic: the Romantic movement has been the starting point or early focus for several other studies on which we will be drawing (eg Lowe and Goyder, 1983; Evans, 1992; MacNaghten and Urry, 1998) and provides a manageable period of time to cover within the confines of this book. It has, however, also been seen by people such as Short (1991) as marking a key point of transition in ideas about nature. Short argued, for instance, that between the seventeenth and nineteenth centuries there was a shift from a 'classical view' of landscapes, which valued those that bore the clear imprint of human activity, towards a 'romantic view' in which 'wilderness' areas, seemingly unaffected by human activity, were seen to have particular appeal. We feel that Short rather overdraws the speed and direction of change – Thomas (1984) suggests a similar shift but presents it as being more gradual and complex – but it is certainly the case that between the seventeenth and nineteenth centuries there were important changes in how some people in Britain and North America viewed landscapes and nature. It is with these changes that we will begin, before proceeding to examine subsequent challenges to technocentric attitudes towards nature and the environment in Britain and North America and then, towards the end of this chapter, in other areas of the world.

6.2 Ecocentric environmentalisms in Britain and North America

6.2.1 Romanticism and the rejection of social exploitation of nature

Romanticism is commonly used to refer to a cluster of ideas adopted by many artists, musicians, poets and writers in Europe and North America from the late eighteenth century. Romanticism is particularly associated in Britain with people such as Edmund Burke, Thomas Carlyle, Coleridge, Shelley and, perhaps most famous of all, William Wordsworth. In the USA key figures of Romanticism were Alexander Wilson, Ralph Waldo Emerson and Henry David Thoreau.

Eighteenth- and nineteenth-century Romanticism can be see to have at least four key features. First, as evidenced by the list of names associated with it Romanticism was an 'artistic movement'. It was artistic not only in the sense of being propounded by artists, poets and writers, but also in the sense that it legitimated artistic endeavour, that is it seemed to provide a justification for the activities of painting and writing. It did this by placing higher value on creative, thoughtful, aesthetic activities than it did on manual, mechanistic labour.

A second feature of Romanticism worthy of some comment is its individualism. As Pepper (1984) argues, Romantics placed great stress on the freedom of the individual. Creativity, seen as central to the Romantic way of life, was seen to be encouraged by, indeed generally required, freedom for the individual to do as they pleased. Creativity was seen as being a moment unique to the individual, not least because artistic creativity required the expression of personal feelings and emotions. This emphasis on individual creativity is in sharp contrast to both the increasing routinisation of labour within the factory and within mechanised agriculture (see Chapter 2) and also notions of science that, following the ideas of such philsophers as August Comte and Francis Bacon, was seen to involve the application of a standard set of procedures – the so-called scientific method – and steady and gradual accumulation of knowledge on the foundations of knowledge already established by others.

A third feature of Romanticism was that it was profoundly 'anti-materialistic'. Romanticism rejected the material measures of progress, again favoured alike by economic commentators and philosophers of science such as Bacon. As noted in Chapter 1, Bacon argued that the aim of science was to 'command nature' for human action and welfare (see Pepper, 1984, p. 55). By contrast, consider the following extract from one of William Wordsworth's poems, *The Tables Turned:*

> One impulse from a vernal wood
> May teach you more of man,
> Of moral evil and of good,
> Than all the sages can.
> Sweet is the lore which Nature brings;
> Our meddling intellect
> Misshapes the beauteous forms of things –
> We murder to dissect (Wordsworth, 1798).

Nature in this extract is not, as it was for classical scientists like Francis Bacon, an object of minor significance. Instead, it is a source of more knowledge, and more wisdom, than any person.

The extract also illustrates a fourth feature of Romanticism, namely the way it often claimed that the desire to exploit or dominate nature was not desirable. Instead, Wordsworth and other Romantics constructed the desire to tame nature as a meddlesome habit that misshaped the natural, the inherent, form of nature.

The idea of nature as an inherent state was crucial to Romanticism, as also was the idea of nature as external to society. Romantics, such as Wordsworth, saw nature as being areas outside the ravages of modern human activity. In particular they came to value wild and seemingly barren areas in Britain such as the Lake District, the Highlands of Scotland, Snowdonia and, slightly later, Dartmoor and

Exmoor. The appreciation of such areas, which today may seem such a 'natural' viewpoint, was in practice a dramatic change from the way such landscapes were looked at in the sixteenth and seventeenth centuries. The Lake District, for instance, had been variously described in the seventeenth century as 'desert and barren', with 'very terrible' or 'hideous' mountains and 'dreadful fells, hideous wastes, horrid waterfalls, terrible rocks and ghastly precipices' (Thomas, 1984, p. 258; see also Urry, 1995a). In the sixteenth and seventeenth centuries the preferred landscapes were not those that appeared unaffected by human activity, but rather those where could be seen the mark of human activity, of human control. This desire was most clearly seen in the creation of formal gardens that, as one contemporary writer puts it, were created to: 'deviate from it [nature] as much as possible. Our trees rise in cones, globes and pyramids. We see the marks of scissors upon every plant and bush' (Joseph Addison, 1712, quoted in Passmore, 1974, p. 36). Other instances of this desire for evidence of tamed nature included the preference for regular, cultivated, landscapes over uncultivated, seemingly wilderness, areas.

The ideas of the Romantic poets, writers and painters found a wider audience, particularly among the upper middle class. Together they started to engage in a new form of activity: the touristic search for the Romantic experience, which was discussed in Chapter 2 in relation to the 'post-industrial economy'. Green (1990, p. 6) argues that nature became increasingly seen as related to 'leisure and pleasure', that is as a resource for 'tourism, spectacular entertainment, visual refreshment'. MacNaghten and Urry (1998, p. 115) concur, arguing that in the nineteenth century there was 'a general growth in viewing nature as spectacle', an argument that is also made by Daniels (1993) who argues that 'wilderness', or untrammelled nature, became in effect a commodity produced for a 'spectacular' consumer culture: that is a culture that valued visual brilliance and design. As Daniels (1993, pp. 47–8) notes, wilderness areas such as Snowdonia and the Lake District: 'hitherto avoided by polite society, were packaged as scenic attractions with guidebooks, marked paths and viewpoints. A range of optical gadgets was on sale for the tourists to enhance their views'.

Many of the guide books to such areas clearly expounded on the virtues of a romantic wilderness experience. Here, for instance, is a passage from William Adam's guide book to the Peak District, *Gem of the Peak,* where he describes Dovedale:

> There is an indescribable and overpowering majesty in nature, especially in mountain scenery, that is difficult to account for; it seems not so much to arise from a minute examination of the parts of which it is composed, as from the combined effect of the whole as it is rapidly traversed by the eye, until the mind is completely filled with its vast dimensions, and inspired with a deep sense of its own insignificance and nothingness, as compared with the monuments of creative wisdom and omnipotent power; and that power reigning in supreme though silent majesty around us. We experienced the full effect of this as we made our way from the smiling fields and busy haunts of man to plunge into and examine the deep recesses of Dovedale, which still retains all its ancient simplicity and beauty, uninjured by the art of man, and its magic charm still unbroken by the intrusion of his dwellings – it would seem, therefore, that its solitude is that which, combined with its romantic scenery, speaks so impressively to the heart, and which has elicited the admiration of all the lovers of nature (Adam, 1851, p. 211).

Similar developments can be observed within North America where, for example, an Act of Congress in 1864 ceded the Yosemite Valley and the Mariposa Grove of giant sequoias to the state of California for the purpose of establishing them as public parks, and in 1872 the world's first designated 'national park' was established at Yellowstone. As MacNaghten and Urry (1998, p. 183) note, such designations were very much bound up with a growth in symbolic representation of these areas as various 'preachers, painters, photographers and writers' began to 'describe and lyricise Yosemite and especially its stupendous sequoias'. Daniels (1993) and Short (1991) similarly record how in the second half of the nineteenth century there emerged a growing sensibility over 'wild nature' that was fuelled by, for example, the paintings of George Catlin, Albert Bierdstadt and Thomas Moran; the photographs of W.H. Jackson, Charles Leader Weed, Carlton E. Watkins and, somewhat later, of Ansel Adams; and the writings of George Perkins Marsh, Ralph Waldo Emerson, Henry David Thoreau and John Muir. As discussed in Chapter 2, the portrayal of areas as 'awesome nature' were then often accompanied by a growth in Romantic tourism.

While the landscapes of areas such as the Lake and Peak Districts in Britain and Niagara and Yosemite in the USA came to be increasingly portrayed in the nineteenth century as places of majestic and sublime nature, as MacNaghten and Urry (1998, p.115) argue there were a series of landscapes that were seen to hold particular visual approach or act as 'nature as spectacle': 'in England, the awesome sublime landscapes of the Lakes; the sleek, well-rounded beautiful landscapes of the Downs; and the picturesque, irregular and quaint southern villages full of thatched roofs'.

Howkins (1986) has similarly noted, for example, how from the 1880s through into the 1920s and 1930s there was an increase in representations of what he terms 'south country', that is 'a unified landscape type' composed of 'rolling and dotted with woodlands', has hills that are 'smooth and bare, but never craggy', and '[a]bove all ... is a cultivated and it is post-enclosure countryside'. Cosgrove *et al.* (1996, p. 536) describe this landscape as one of 'English domesticity, agrarian productivity and secure, conservative, stable social order', or what Wright (1985) described as 'deep England'. Howkins adds that while this landscape may have had its referent, 'roughly speaking' in the country south of the Thames and Severn and east of Exmoor, the referent rarely matched the symbol exactly, but rather the image came to act as a 'set of yardsticks of 'rurality' by which the observer judged landscapes'. Some landscapes, such as the Cotswolds, seemed to be positioned at the height of the yardstick and in the heart of the 'south country'. Other areas outside the geographical contingent 'south country', such as Shropshire, were adjudged to match up to the definition of preferred rurality and hence were effectively incorporated into the 'south country'. Other areas seemingly within the boundaries of the 'south country', such as East Anglia, were seen to be wanting, lacking such necessary features as a rolling topography or a village architecture of half-timbered or mellow sandstone buildings. The landscapes of these areas were often described in disparaging or lukewarm terms, or else in ways that draw on differing values from those associated with the Romanticisms of the 'south country'.

The numbers of people travelling to spectacularised nature and other countrysides raised, in the minds of many, the possibility that their Romantic experience would be undermined by the opening up of such areas to an increasing number of people. Concern began to be raised by leading Romantics such as William Wordsworth that the arrival of 'the masses' in areas such as the Lake District would ruin the 'natural character' of the area. The problem of 'the masses' for people such as Wordsworth was not simply one of numbers, but also their social character. The masses were, so Wordsworth argued, simply incapable of fully appreciating a romantic experience:

> A vivid perception of romantic scenery is neither inherent in mankind nor a necessary consequence of even a comprehensive education ... Rocks and mountains, torrents and wild spread waters ... cannot in their finer relations to the human mind, be comprehended without opportunities of culture in some degree habitual ... Persons in that condition, when upon holiday, or on a Sunday after having attended divine worship (would be better off making) little excursions with their wives and children among the neighbouring fields within reach of their own dwellings (William Wordsworth, quoted in Glyptis, 1991, p. 27).

Wordsworth also talked of the Lake District becoming 'a sort of national property, in which every man has a right and interest who has an eye to perceive and a heart to enjoy' (Wordsworth, 1952, p. 127). While people such as McCormick (1992) have interpreted such comments in terms of a discourse of establishing parks for the nation – a democratising of access to the countryside and nature – these comments also sustain more exclusionary interpretations. In other words, if you do not have the necessary perceptual eye and emotional heart, then by implication you have no rights or interests with regard to the Lake District.

Wordsworth was not alone in his scorn of the masses. Here is one nineteenth-century description of Ambleside on Lake Windermere:

> A great steam monster ploughs up our lake and disgorges multitudes upon the pier; the excursion trains bring thousands of curious, vulgar people ... the donkeys in our streets increase and multiply a hundred fold ... our hills are darkened by swarms of tourists; our lawns are picnicked upon twenty at a time (James Payne, quoted in Glyptis, 1991, p. 27).

Such fear of the masses and the destruction of valued countryside led to the establishment of a number of 'preservationist' groups. In 1883, for example, the *Lake District Defence Society* was formed to oppose the extension of the railway into Borrowdale. In this it was successful, and there emerged in the late nineteenth and early twentieth centuries a series of other organisations concerned with the preservation of a range of natures, the general characteristics of which are summed up by the later part of the *Selbourne Society's* title: for the 'protection of birds, plants and pleasant places' (see Table 6.1). The Lake District Defence Society itself formed one of the nuclei for the establishment of one of the most significant of conservation organisations in Britain, the *National Trust,* or as it is more correctly known, *The National Trust for Places of Historic Interest and Natural Beauty.*

Table 6.1 Dates of formation of British nature conservation organisations

Organisation	Year founded
Society for the Protection of Animals	1824
The Commons, Open Spaces and Footpaths Preservation Society	1865
East Riding Association for the Protection of Seabirds	1867
Association for the Protection of British Birds	1870
Thirlmere Defence Association	1878
The Lake District Defence Society	1883
The Selbourne Society for the Protection of Birds, Plants and Pleasant Places	1885
The Royal Society for the Protection of Birds	1889
The National Trust for Places of Historical Interest and Natural Beauty	1895
The Coal Smoke Abatement Society	1899
Society for the Preservation of the Wild Fauna of the Empire	1903
The British Vegetation Committee	1904
The British Association for Shooting and Conservation	1908
The Society for the Promotion of Nature Reserves	1912
The British Ecological Society	1913
The British Correlating Committee	1924
The Council for the Preservation of Rural England	1926
The Council for the Preservation of Rural Wales	1928
National Trust for Scotland	1931
The Standing Parks Committee	1936
Civic Trust	1957
The Council for Nature	1957
British Trust for Nature Conservation	1965
Conservation Society	1966

Today the National Trust is very much a preservationist and protectionist organisation, seeking to purchase land or buildings seen to be under threat of new and (in the view of the Trust) unwanted development. This objective of purchasing land or buildings to ensure a particular use has led to the National Trust becoming the largest landowner in Britain, owning more than 1 per cent of the country's land surface and over 200 historic buildings (Lowe and Goyder, 1983, p. 139). However, at its inception, the National Trust had a rather anti-landownership influence, in that another of its points of origin was *The Commons Preservation Society* (compare Blunden and Curry, 1990; Lowe and Goyder, 1983). This was essentially a middle-class reaction to the enclosure of land by landowners. It was feared by many in the middle class that dispossessed labourers would form a rabble that might overthrow the existing structure of society. Poverty was seen to breed social unrest and the landowners, by taking over formerly common land, were seen to be creating further poverty. Poverty was seen not solely in economic or monetary terms but also in a lack of access to 'open space'. The openness of the countryside was seen to be an important influence on people's health, and the poor health of the working class in towns attributed to a lack of access to the invigorating powers of open space, arguably to the exclusion of any recognition of the influence of poverty on diet and living conditions. Open space was seen as a basic amenity to which all people should have access, not least because participation in open-air recreation was seen by many members of the middle class to have a positive effect on people's morals and behaviour: positive that is in the sense that they were seen to accord more with the moral behaviour of the middle class. Hence, as Short (1991, p.77) comments, the aim of organisations such as the Commons Preservation Society was: 'to bring the town dweller more frequently into contact with the beauty of nature, to help forward the ideal of the simpler life, plain living and high thinking'.

The middle-class concern for recreational provision for the poor stood in some tension with the Romantic concern for the conservation of wilderness. The tensions between these two concerns can be seen to have exerted a continuing pressure upon the countryside right through to the present day. For example, many of the organisations founded by people who adopted either a Romantic or a public amenity viewpoint on the countryside now play major roles in the current organisation of the countryside. Second, the desire to use the countryside as either a place for Romantic wilderness experiences or as a site for public amenity also led to one of the most significant ideas affecting the countryside today, namely the idea of 'National parks'.

Another important protectionist organisation in England is the *Council for the Protection of Rural England* (CPRE), or as it was originally termed, the *Council for the Preservation of Rural England.* This organisation was founded in the 1920s, a period that saw a resurgence in the number of environmental groups being formed and that has been called by Lowe and Goyder (1983) the 'second wave of environmentalism'.

6.2.2 The second wave: conservation and amenity in places of nature

According to Lowe and Goyder (1983) it is possible to discern four waves of environmental concern within a period running from the late nineteenth century through to the 1970s and 1980s (see MacNaghten and Urry, 1998, for a classification that is both slightly different and that also runs through into the 1990s). Lowe and Goyder's first wave ran from the late nineteenth and into the early twentieth century and involved the formation of essentially upper-middle-class preservation organisations. Lowe and Goyder date the 'second wave' to the inter-war years, and suggest that the organisations formed in this period were constituted rather more broadly in society. Many of the organisations formed in this period did not subscribe to the elitist implications of the Romantic movement, and indeed the inter-war years saw the growth of many 'amenity groups', that is organisations oriented much more towards expanding access to the countryside than protecting a Romantic experience of the countryside. The *British Field Sports Society* was founded in 1930 to promote the capture and/or killing of various fauna in nature, the *British Canoe Union* in 1936 to encourage recreational use of water, while the *Youth Hostels Association,* which was founded in 1930, had over 83,000 members by 1939 (Harrison, 1991). There was also considerable growth in amenity organisations formed before the First World War: the *Cyclists' Touring Club* founded in 1878 had 36,000 members by 1939; while the *Federation of Rambling Clubs,* which was formed in 1905 and was to become the *Ramblers Association* in 1935, had 40,000 members by 1931. Furthermore, even more traditional 'protectionist' bodies, such as the CPRE, tended to favour new arguments and develop new strategies for the countryside. In particular, the CPRE strongly favoured a 'planned' system of land use: some areas particularly 'the countryside', should be protected from new developments; other areas of the country, such as industrial and urban areas, should see the encouragement of new development. The CPRE can hence be seen as an illustration of what Porritt and Winner (1988) see as the characteristic environmental movement of the second age of environmentalism: technocratic movements oriented towards the manipulation of nature in order to maximise potential social benefits.

Technocratic environmentalism can be seen to have a number of important features. First, it very much promoted its own values as being somehow universal. The CPRE, for instance, has frequently promoted the preservation of the countryside as being of general benefit, although as Matless (1990a; 1990b) has recorded the precise grounds of argument were both complex and changeable. In the 1920s and 1930s, for example, considerable emphasis was placed on educational and social benefits of the countryside with Cyril Joad, a leading member of both the CPRE and the Ramblers Association, for instance, arguing that 'familiarity with nature in walking and riding, in swimming and climbing, is an essential contribution' in the development of personality and that '[c]onfined to the towns we need the country to enable us to develop to the full individualities that ours might be' (Joad, 1938, p. 65). He further suggested that:

> country sights and sounds are the best cures for the neurosis of the mind, as fresh air and exercise are the saving antidotes against ailments of the body, that freedom, physical movement, and the stimulating self-help of open-air life are the best aids to companionships, so that a man will know his friend better after a week in the country than after a year in the town (Joad, 1938, p. 68).

Nature in the form of the sights, sounds and fresh air of the countryside were seen as being of benefit to people's physical body, the mind and spirituality (Matless, 1995). This view of nature can be seen to produce 'othering' (see Box 5.3), both by establishing rather contestable images of difference – why should living in the countryside make one more spiritual than living in the town? – but also privileging some people over others. In particular many of the commentaries of preservationists of the 1920s and 1930s, and indeed since (see Chapter 2), show clear echoes of the elitist concerns of nineteenth-century Romantics such as Wordsworth. The urban 'masses', for instance, were again singled out for repeated criticism. Joad, for example, complained about:

> the hordes of hikers cackling insanely in the woods, or singing raucous songs as they walk arm in arm at midnight down the quiet village street. There are people, wherever there is water, upon sea shores or upon river banks, lying in every attitude of undressed and inelegant squalor, grilling themselves, for all the world as if they were steaks, in the sun. There are tents in meadows and girls in pyjamas dancing beside them to the strains of the gramophone, while stinkingly disorderly dumps of tins, bags and cartons bear witness to the tide of invasion for weeks after it has ebbed; there are fat girls in shorts, youths in gaudy ties and plus fours, and a roadhouse round every corner and a cafe on top of every hill for their accommodation (Joad, 1938, p. 73).

The urban in-migrants were seen to lack the 'cultural competence' (Cloke *et al.*, 1998) to live in the countryside, and hence had to be educated to act in the correct manner, that is, effectively, broadly in line with the Romantic or countryside aesthetic (see Chapter 2). As Harrison (1991) has demonstrated, although this view has been widely promoted by many of the people and agencies who own and regulate access to the countryside, it is not a view shared by everybody (see also Phillips, 1998a, 1998b; Urry, 1995b). Hence, while the second wave of environmentalism may have seen a widening of the social base away from a very small artistic group of the upper middle class, it certainly did not represent the creation of a mass environmental movement comprising membership spread across all social groups.

A second feature of 'second wave' environmentalism is that it 'localised' nature. The countryside was seen to be a particularly important place of nature: people were encouraged to experience nature by going to the countryside, and environmental problems were seen largely in terms of urban growth and pollution that might lead to loss of nature in the countryside. Although this line of argument stood in some contrast to governmental plans and policies, which saw 'environmental issues' and changes in rural space as being of marginal significance in societal development, and instead focused primarily on 'economic issues' such as employment, wealth and balance of payments and on the development of 'urban spaces', preservationist thought still posited quite clear separations of nature and society. MacNaghten and Urry (1998), and more particularly Matless (1990a; 1990b), have argued that the rural preservation movement was not so much anti-modernisation as concerned to regulate the influence of modernism: to order modernisation and to present clear boundaries between this and the pre-modern countryside and the spaces of nature; to create, as Matless (1994) puts it, 'reserves of nature'. These reserves took a large variety of forms including the 'reserves of wilderness' such as the national parks in the USA and, to a lesser extent, in Britain; 'landscape reserves', such as the protected landscapes of British national parks and *Areas of Outstanding Natural Beauty* (AONBs); and the specifically designated *'nature reserves'* that, as Livingstone (1995) and Adams (1997) note, effectively drew highly restricted definitions of what constituted nature, excluding from its gaze both what was happening in the spaces of industrialisation and urbanisation, and also in the landscapes of modernising agriculture.

6.2.3 The third wave: conservation and technocratic environmentalism

A third 'wave of environmentalism', beginning in the late 1950s and continuing through the 1960s, has been identified by Porritt and Winner (1988). In Britain this period saw the establishment of conservation organisations such as the *Civic Trust,* the *Council for Nature,* the *Noise Abatement Society* and the *British Trust for Conservation Volunteers.* With the exception of the last named, all these organisations were effectively watch-dog institutions: they were concerned with the establishment and/or supervision of some form of environmental regulation – be these relating to historic buildings (as in the case of the Civic Trust), the protection of wildlife and nature (in the case of the Council for Nature) or noise pollution by the Noise Abatement Society. Porritt and Winner (1988, p. 20) have called this phase of environmentalism, the 'age of institution building and environmental regulation' and suggest that it was characterised by 'scientifically respectable, politically articulate and litigious pressure groups'. One of the clearest examples of the use of scientific practice and political rhetoric was Rachel Carson's book *Silent spring* (1962), the publication of which McCormick (1992, p. 55) argues has been 'the single event most frequently credited as signifying the environmental revolution', or the emergence of what has often been termed 'the new environmentalism'. As MacNaghten and Urry remark, the book details:

> with relentless detail the effects of pesticides (most notably organochloride insecticides such as DDT, aldrin, isodrin, entrin and dieldrin) on the wildlife of the countryside, and their potentially catastrophic long-term effects on life support systems as toxic poisons worked their way up the food chain (MacNaghten and Urry, 1998, p. 45).

By the early 1970s many of the issues identified by the pressure groups and environmental activists such as Rachel Carson had come to be recognised by national governments, and some of the regulatory functions formerly pursued by these groups became part of governmental activities. In 1970, for example, the *Department of the Environment* (DOE) and *Environmental Protection Agency* (EPA) were established in Britain and the USA respectively.

As well as having a heightened focus on environmental regulation and monitoring, the third wave of environmentalism differed from the second in that it tended to see nature in a rather less localised sense. Indeed, MacNaghten and Urry (1998, p. 49) have argued that a series of major pollution incidents – such as the spillage of oil from the Torrey Canyon off the coast of the United Kingdom and the *Union Oil Company* platform off the coast of Santa Barbara, California; mercury poisoning at Minamata Bay in Japan; and evidence of widespread chemical and oil pollution in the Great Lakes of the USA and Canada – began to generate 'an awakening sense of ... environmental bads ... [moving] ... across national borders and potentially invading everyone's body'. They also add, however, that although within countries like the United Kingdom there was something of a widening conception of what might be seen as nature and the environment, 'concerns over global and systemic environmental issues did not find a vehicle for public expression, since both environmental groups and government agencies still conceived of nature within the older vocabularies of rural preservation and amenity' (MacNaghten and Urry, 1998, p. 50). In other words, a cultural localisation of nature and environment was still very strong. There was also a strong 'compartmentalisation' of nature in that, as MacNaghten and Urry (1998, p. 34) comment, concern over nature was conceived of in terms of 'distinct and largely unrelated sets of issues', such as amenity and access to the countryside, the control of pollution, the preservation of beautiful landscapes and resources of commercial value or scientific interest. There was no widespread acceptance of the notion of 'an environmental crisis', or of global threats to the form of nature.

6.2.4 The fourth wave: deep green environmentalism

In the late 1960s and 1970s there was, however, the establishment of a range of new environmental organisations within Europe, North America and beyond. McCormick (1992, p. 101) states that by 1982 there were some 2230 environmental non-governmental organisations in the so-called 'less developed countries' and 13,000 in the 'more developed countries'. Many of these organisations were similar to those of the 'third wave', in that they were 'technocratic' groups concerned to monitor and supervise use of environmental resources. Robinson (1992), for example, suggests that while in 1970 there were only 15 countries with environmental management institutes, by 1980 this had grown to 115. The majority of the environmental groups formed in the 1970s and 1980s were, on the other hand, more squarely 'pressure groups', oriented towards undertaking campaigns to influence public opinion and politics. These later organisations included Friends of the Earth, Greenpeace and a range of 'green' political parties.

Friends of the Earth was initially established in San Francisco in 1969, with further branches being established in London and Paris in 1970. Today there are 'Friends of the Earth' in most western European countries, in Australia and New Zealand, in the USA and Canada, and in an array of other countries including Argentina, Bangladesh, Ghana, Hong Kong, Japan, Malaysia, Papua New Guinea and South Africa. Greenpeace also originated in North America, being initially set up as the *Don't make a Wave Committee* in Vancouver in 1969. In 1971 the name had changed to Greenpeace. Shortly afterwards offices were set up in Britain, France and The Netherlands, and by 1983 Greenpeace had offices 17 countries, a global membership of 1.2 million and there was a *Greenpeace International,* located in Lewes, Sussex, Britain.

The 1970s and 1980s also saw the growth of 'green' political parties. The first to use the epithet 'green' was the *New Zealand Green Party,* established in 1972. In 1979 the most renowned of the green parties, *die Grünen* (or 'the Greens') was created, and by 1983 had gained its first seat in the Bundestag, the national parliament of the then German Federal Republic. By January 1987 die Grünen had 8.5 per cent of the national vote and 42 seats in the Bundestag (see Sallnow and Arlett, 1989). In Britain there was a dramatic rise in the vote for the 'Green Party' in the 1989 elections for the European Parliament, when it received 15 per cent of the votes cast. This compared to a mere 0.6 per cent cast for the Green Party in the 1987 British General Election. McCormick (1992, p. viii) records that by 1990: 'nearly 20 countries had functioning green parties, ten had returned members to their national assembles, and 29 green members sat in the European Parliament'.

In the 1980s not only was there a rise in votes for 'green' political parties, but there was also some indication that politicians from other parties were starting to take the concerns of environmentalists seriously, at least in their rhetoric. One of the starkest illustrations of this was the then Conservative British Prime Minister, Margaret Thatcher. In her first eight years in office she neither made a speech on environmental matters nor visited an environmental project (Bristow, 1993). Many public pronouncements by government ministers and closely associated organisations were quite antagonistic to the concerns of fourth wave environmentalists. The *Centre for Policy Studies*, which acted as a 'think tank' for the Conservative government, argued, for example, that conservation and preservation movements constituted the 'real green' movement:

> To be British and to be green is far more accurately reflected in our national passions for fishing, gardening and country walking, than in any radical, ecological fervour. We are indeed a green nation, but not in the way the Left would like. We care about the details of our immediate surroundings, street corners and hedgerows, our parks and fens, our hillsides (Centre for Policy Studies, quoted in Porritt and Winner (1988), p. 82).

However in 1988 Margaret Thatcher pronounced in a speech to the *Royal Society* that protecting the 'balance of nature' was one of the 'greatest challenges' of the late twentieth century and that urgent action was required to deal with such 'global' environmental problems as ozone depletion, global warming and acid rain (quoted in McCormick, 1992, p. 200).

In the 1980s there was also evidence that in the general populace of many western countries there was a growing recognition of environmental issues and a growing adoption of an environmentalist identity. Dunlap and Scarce (1991), for example, record that public opinion polls in the USA from the early 1970s through to 1990 record generally rising levels of awareness and concern about environmental issues such as acid rain, waste disposal and the greenhouse effect, and while often it was not a principal object of concern 'growing majorities see environmental problems as serious, worsening, and increasingly threatening to human well-being' (Dunlap and Scarce, 1991, p. 657). Kessell (1985) conducted questionnaire surveys in the USA, England and the former Federal German Republic and asked, for example, whether people thought that pollution was 'rising to dangerous levels'. The proportion of people who did was apparently 79, 70 and 91 per cent respectively. Surveys have also asked people whether they engage in particular 'green activities', such as using 'ozone friendly' aerosols, buying products made from recycled material, and lowering their energy usage, and also whether they belong to environmental organisations or participate in environmental campaigns. The results of many of these surveys suggested that people through the 1980s were joining environmental organisations and self-consciously adopting green behaviour in increasing numbers (see MacNaghten and Urry, 1998).

According to Evans (1992) these environmental organisations of the 1970s and 1980s represent a marked break with the more conservation-oriented bodies that had preceded them. At least seven features stand out.

First, the organisations tended to adopt markedly more holistic views of nature and environment than had the organisations of the preceding wave. Elements of the environment were seen as interconnected to, and interdependent upon, other elements. As Buttel (1993, p. 21) says, environmentalism since the late 1960s has often come to be 'packaged' within 'a comprehensive ... framework'. Buttel (1993, p.21) adds that many of these frameworks tend to: 'accede to the well institutionalised instrumental–scientific rationality of the modern world ... by transforming scientific concepts into value or ethical claims ... which are in turn legitimated by the authority of science'. In other words notions of science and scientific evidence are used to construct a notion of nature that has very particular, but often easily overlooked, social and political implications. A very clear instance of such framework is Gaianism (see Box 6.1).

Box 6.1 Lovelock and the science of Gaia

Although the notion of Gaia and the work of James Lovelock is used by people such as O'Riordan (1989) to encapsulate the total character of 'deep green' environmentalism, the term was first used by Lovelock in a very scientific context, namely as a term to summarise an hypothesis about how life came to be established on earth. The starting point of this hypothesis was Lovelock's observation that the elements of the earth the atmosphere, the biosphere and the hydrosphere are interconnected and have remained at a steady state of equilibrium that sustains life, even though the output from the sun has been steadily increasing over time and the atmosphere is composed of an unstable mixture of reactive gases. Lovelock suggests that this life-supporting equilibrium is maintained by 'the process of life', whereby living organisms effectively 'regulate' their environments so that life continues. Life, so Lovelock claimed, did not just depend on the earth, 'but also adapts the earth to make it and keep it a home' (Lovelock, 1988, p. 66). Once life had become established on earth there emerged, Lovelock suggested, 'a self-organising and self-regulating system' and it was this he termed *'Gaia'* (Lovelock, 1979).

Simplifying Lovelock's arguments greatly, one can suggest that his Gaia hypothesis has three distinct elements. First, it is a 'holistic' theory: it seeks to look at the overall pattern of things rather than break things down into small elements. Gaia is a theory of the earth in all its constituent parts: geochemical and biotic elements. In this it differs, so Lovelock claims, from traditional sciences. Specifically, Lovelock suggest that to understand Gaia it is necessary to break down the distinction between 'earth' and 'life' sciences: between for example, geography/geology and biology. Lovelock argues, for instance, that:

> geologists have tried to persuade us that the earth is just a ball of rock, moistened by the oceans; that nothing but a tenuous film excludes the hard vacuum of space; and that life is merely ... a quiet passenger that happens to have hitched a ride on this rock ball in its journey through time and space. Biologists have been no better. They have asserted that living organisms are so adaptable that they have been fit for any material changes that have occurred during the earth's history (Lovelock, 1988, pp. 11–12).

Lovelock coins the term *'geophysiology'* to describe a science that integrates the 'earth' and 'life' sciences.

To counterbalance the division of earth and life processes and to develop 'geophysiology', Lovelock attempts to formulate a new theory of evolution, one that builds on a 'Darwinian' notion of the 'survival of the fittest' with a geological concern for environmental change. Traditional 'evolutionary' theory, such as that of Darwin, presupposes, so Lovelock claims, that evolution of species is 'independent' of the evolution of their environment (Lovelock, 1988, p. xvi). Lovelock claims, however, that biological life is 'tightly coupled' with, and fully integrated into the biochemical environment, and that species evolve in connection with changes in the environment. Lovelock attempts to develop this argument by providing a new 'evolutionary history' of the earth: one that rests equally upon changes in the geochemical environment as it does on changes in biological species (see Table B6.1).

Table B6.1 Lovelock's evolutionary history of Gaia

Period	Time span (eons)	Geochemical features	Biological features
Pre-Gaia	Pre 4.58	Atmosphere dominated by methane – oxygen only a trace gas	No life
Archean	4.58–2.5	Atmosphere dominated by methane – oxygen only a trace gas	'Cyanobacteria' – photosynthesisers
Proterozoic	2.5–0.57	Oxygen became dominant atmospheric gas	'Eukaryotes' emerged – organisms composed of a collection of cells that consumed organic matter and oxygen to live rather than photosynthesisers
Phanerozoic	0.57–now	Further rise in atmospheric oxygen	Plants and animals (and eventually *Homo sapiens*) emerge

Note: 1 eon = 100 million years

(Source: based on Lovelock, 1988)

In this new evolutionary history Lovelock distinguishes four distinct phases: first there is the period 'pre-life', pre Gaia; second there is the *Archean;* third the *Proterozoic*; and fourth the *Phanerozoic*. As Table B6.1 illustrates, each of these phases, bar the first, is characterised by distinct geochemical and biological features.

A third element of scientific Gaia, and one that has caught the attention of the non-scientists, is Lovelock's characterisation of the process by which life and environment interact. Lovelock argues that the emergence of biotic life so transforms geochemical processes that it becomes meaningless to talk about them separately from life processes: the environment, the earth, is so caught up in the processes of constituting life that one should see it as being 'alive'. The 'earth' can be seen as being as central to animal life as the heart or lungs of an animal: it is as much a part of an animal as the heart or lungs.

This is arguably the most distinct element of Lovelock's Gaia. From it he draws a number of further contentious arguments, including:

(1) 'Life is a planetary phenomenon. On this scale it is near immortal.'
(2) 'There can be no partial occupation of a planet by living organisms ... The presence of sufficient living organisms on a planet is needed for the regulation of the environment. Where there is incomplete occupation,

the ineluctable [inescapable] forces of physical and chemical evolution would make it uninhabitable'.

(3) 'Increased diversity among the species leads to better regulation' (Lovelock, 1988, p. 63).

These three claims have all been taken up within non-scientific debates, particularly by those wishing to provide a justification for deep ecological values. Advocates of a non-scientific interpretation of Gaia include Porritt and Winner, and Goldsmith, all leading activists in the Green Party and writers in the *Ecologist* magazine. All see Lovelock's Gaia as a crucial element of a new philosophy for the current age of environmental crisis.

Lovelock's Gaia hypothesis and related arguments have been the subject of considerable criticism, originating both within and beyond 'scientific' discourse. Restricting ourselves to the broadly scientific debate, at least three lines of criticism can be discerned. First, objections have been raised against the originality of Lovelock's theory, and Lovelock himself has been forced to admit that many of his ideas have had their forerunners, not least in the writings of the famous geologist James Hutton, who talked about the idea of *'planetary physiology'* in much the same manner as Lovelock talks about 'geophysiology' (Hutton, 1788).

A second line of criticism is the way Lovelock links findings from very disparate sources. For some people this is the power of Lovelock's work: it breaks down, or appears to break down, rigid disciplinary boundaries. For others the borrowing of terms is largely a rhetorical device: it does not demonstrate the interdependence that Lovelock would like. Lovelock, for instance, attempts to define the term 'life' by borrowing concepts from mathematics, physics and molecular biology. The result is a definition of life as 'a self-organising system characterised by an actively sustained low entropy' (Lovelock, 1988, p. 27). For many such a definition is meaningless. Other ambiguous concepts include Lovelock's suggestion that Gaia acts 'unconsciously' to regulate the earth. Applying the term 'unconscious' to a biogeochemical process seems close to 'anthropomorphism': the transference of concepts used to explain human experience on to non-human processes and objects. Lovelock (1988, p. 212), for example, suggests that: 'Gaia ... is stern and tough, always keeping the world warm and comfortable for those who obey the rules but ruthless in her destruction of those who transgress. Her unconscious goal is a planet for life'. There is here also a very clear gendering of Gaia.

A third line of criticism that has been directed at the Gaia hypothesis, and indeed the 'deep green' movement as a whole, is the degree to which one can really have a 'non-anthropocentric' philosophy. I have already quoted Lovelock's argument about the Gaianists being concerned with 'the health of the planet rather than that of some individual species', such as the human race. Many people have found this argument both distasteful and illogical. Dobson (1990, pp. 64–6) has argued that a 'weak form of anthropocentrism', one that is concerned about people but that does not see the environment merely as an object to be used, is 'a necessary feature of the human condition'.

As Tim O'Riordan has pointed out, 'Man's conscious actions are anthropocentric by definition. Whether he [*sic*] seeks to establish a system of biotic rights or transform a forest into a residential suburb, the act is conceived by man in the context of his social and political culture' ... 'It is this factor that links even the search for intrinsic [environmental] value with anthropocentrism. The search is a *human search,* and although it may be successful in displacing the human being from the centre stage in terms of value, one will always find a human being at the centre of the enterprise, asking the questions ... any human undertaking will be (weakly) anthropocentric, including the Green movement itself' (Dobson, 1990, pp. 65–6).

As well as frequently utilising holistic science frameworks, some fourth wave environmentalists often constructed holistic images of the environmentalism itself. For example Porritt and Winner, in their provocatively titled book *The coming of the Greens,* have likened the growth of environmentalism to the formation of a river delta:

> The way the British Green Movement has been growing is a bit like the slow but steady geological process by which new land takes shape at the mouth of a giant river such as the Mississippi. In the upper reaches of the river, the weight and power of rushing water pounding tears away the river bed and surrounding topsoil, which is carried downstream as sediment. Likewise, in society, ecological anxieties and an increasingly critical view of the quality of modern life are persuading people to question the values and assumptions of industrialism. As the river slows and deposits the sediment when it approaches the sea, so the doubts and fears are leading to new attitudes and ideas. Over time, small islands begin to appear in the delta. From above the water, they appear to be separate, but they are in fact interconnected parts of the emerging land mass. The disparate 'islands' of the Green movement – environmentalism, holistic health, peace campaigning, green spirituality, organic farming, alternative economics, green consumerism, animal rights activists and all the rest – may not look like they have much in common, but potentially they are all manifestations of fundamental green principles (Porritt and Winner, 1988, p. 16).

Under this interpretation environmental concern is a reflection of the destruction of the environment caused by industrialisation that emerges once the turbulent phase of initial industrialisation has been navigated and that grows gradually, and seemingly inevitably, into a single all-encompassing 'green movement'. Quite similar analyses, albeit adopting more 'detached' sociological perspectives, can be found in the work of people such as Oloffson (1988).

A second feature of the fourth wave of environmentalism was that human activity was very much included within the scope of its concern. As McCormick (1992, p. 48) argues, while for earlier 'protectionists' the key issue of concern was 'wildlife and habitat' largely within the margins of modern society, for the 'New Environmentalists, human survival itself was at stake'. Environmental problems such as global warming, the destruction of the ozone layer or the disposal of waste were seen as human creations and, in accordance with the holism of the new environmentalism, as having reinforcing impacts that created a global environmental crisis that threatens the existence of every part of nature, including people. Rather than argue for the protection of localised environments – for the creation of nature or wildlife reserves – or for the careful monitoring and regulation of human activity, these fourth wave environmentalists tended to see the solution to these problems as requiring major transformations in the way people feel, think about, use and manage 'the environment' and in the way societies are organised. As discussed in Chapter 1, the term 'deep green' was often used to refer to these calls for radical change, spanning people's beliefs, emotional and experiential connection with nature.

Third, and arguably in some contradiction to the above point about the centrality given to human activity, the 'deep green' fourth wave tended to espouse a quite anti-anthropocentric perspective. This is quite clearly expressed in Lovelock's Gaia theory (see Box 6.1), but can also be found in, for example, animal liberationism, spiritual ecology and a range of philosophical and scientific 'deep ecologies', such as Varela *et al.'s* (1974) *'autopoiesis'* thesis, which claims all entities that are capable of 'self-reproduction' should be considered to have moral value and therefore rights, and Fox's (1990) *'transpersonal ecology'*, which argues for a psychological identification with as many entities as possible (for more details of these ecocentric philosophies see Eckersley, 1992).

A fourth feature of the fourth wave of environmentalism was a shift in the way the environmental groups worked. As mentioned earlier, the focus for many groups in the third wave was a technocratic one, concerned with establishing and enforcing particular environmental standards. In part this focus was continued into the fourth wave: Rose (1993), for example, has suggested that the environmental movements of the 1970s and 1980s were essentially engaged in a 'struggle for proof', that is they were continually seeking to find scientific evidence relating to the state of particular environmental issues. However, a key difference was that now the focus was rather less on seeking to establish new governmental regulations and rather more on changing public attitudes towards these environmental issues. In particular, and in line with this wave's more holistic perspective, there was a concern to move people's awareness from what we have described as the localised and compartmentalised environmentalism of the third wave on to more globalised and holistic concerns. There was also a movement towards 'direct action' as a means to change public opinion. As McKay (1996, p. 130) notes, the term 'direct action' encompasses a large range of activities that have, as a common denominator, an abandonment of formal political lobbying.

Two other important organisational changes in the fourth wave were the growing 'globalisation' and 'mediaisation' of environmentalism. The notion of globalisation of environmentalism can be seen to refer not only to the growing number of countries containing environmental organisations but also to the spatial extent and impact of these organisations. Environmental organisations such as Friends of the Earth, Greenpeace and the World Wide Fund for Nature have become 'trans-national', with international offices, and frequently organising campaigns spanning several nations, such as in the case of Greenpeace's *'Brent Spar* protest' that, although oriented at changing the actions of Shell UK and the United Kingdom government (see Box 6.2), critically involved actions in European countries such as Germany, where a boycott of petrol stations was organised.

Greenpeace's Brent Spar campaign is also a very clear example of the 'mediaisation of environmentalism'. The centrepiece of the campaign was an occupation of the Brent Spar oil platform executed by protesters climbing up the platform from

Box 6.2 Greenpeace, Shell and the Brent Spar

The Brent Spar was an oil storage and tanker loading platform that from 1976 had been in the Brent oil field in the North Sea, which lies to the north-east of Shetland. In 1991 the operators of the field, Shell and Esso, decided to take it out of operation. The platform, which was some 141 metres in height and weighed some 14,500 tonnes, had six large oil storage tanks that together displaced some 66,500 tonnes of water. After operations ceased these six tanks were flushed out with sea water, although some oil was acknowledged to have remained in the tanks together with a bituminous sludge at their base. A survey commissioned by Shell as part of proposals to dispose of the platform also found evidence of some 19 ml of toxic polychlorine biphenyls (PCBs) that had been in two electricity transformers on the platform. Shell had to decide what to do with the rig, and it was seen that there were three possible courses of action. First, the platform could have been left unused in its original position. However this option would have involved the company in leaving a sizeable navigational hazard and potential source of pollutants in the sea. A second option was to tow the platform close to shore and dismantle it there, but it was suggested that this might impose stresses on the platform that might damage its structural integrity and lead to the release of its contents in more ecological sensitive coastal environments. The third option, which Shell decided was the 'best practicable environmental option' (or BPEO), was to tow the rig to a deeper part of the North Sea and sink it. It was argued that sinking it in the North Feni Ridge area of the North Atlantic, which was some 2000 metres deep, would avoid serious pollution because the pollutants would be contained close to the sea-bed by water density and pressure. The company sought, and received, permission from the United Kingdom government to undertake this course of action.

Shell had already arranged to tow the platform to its proposed site for sinking when Greenpeace decided to launch its 'direct action protest' in an operation it termed 'Bee's Knees'. A ship, the Embla, was hired for some £20,000, equipped with provisions for a six-month operation, and set sail for the Brent Spar platform. On 30 April 1995 Greenpeace protesters climbed aboard the oil platform and started occupying it. As MacNaghten and Urry (1998, p. 69) remark, once it had occupied the platform Greenpeace employed a double strategy 'of producing scientific evidence in favour of disposing the oil platform on shore and in mobilising public opinion'.

Members of Greenpeace took samples of the contents of the platform and announced that it held 5000 tonnes of hydrocarbons, as against Shell's claim that it contained some 50 tonnes. It was subsequently calculated that the platform held some 75–100 tonnes, and Greenpeace wrote to Shell in September 1995 apologising for its miscalculations. However, by this time Greenpeace had effectively beaten Shell via its media campaign.

Once Greenpeace had occupied the platform media reporters were invited to come aboard, and then given access to the communication facilities set up there. On 12 May Shell gained a court order requiring the protesters to leave the platform, and on 19 May it gained another authorising their eviction. At just after 6.00 am on 23 May Shell employees, police and sheriff officials boarded the plat-

Plate B6.2 Water cannon is sprayed round Brent Spar to stop Greenpeace from boarding the North Sea oil platform

form, and by shortly after 8.00 pm they had removed just over 20 protesters from the platform. However, even after their removal, Greenpeace was able to produce further publicity for its campaign through a series of 'spectacular' attempts to return to the platform, which were all recorded by a watching media given facilities on board the Embla.

With this publicity there developed a substantial boycott of petrol stations across Europe, particularly in Germany. It is suggested that at its height this boycott was costing Shell some £5 million a day. Furthermore, many European politicians announced public support for Greenpeace's campaign. On 20 June Shell suddenly announced that a decision had been made to abandon plans to sink the platform. This change of mind provoked considerable reaction, not least from the then British Prime Minister, John Major, who was reported to have described the company as behaving like 'wimps', and to have stressed that the government might not give permission for the platform to be disposed onshore.

Following the abandonment of the plan for deep sea sinking the Brent Spar was towed to Erfjord in Norway while a new decision about how to dispose of it was reached. To help it make this decision, in October 1995 Shell invited firms to bid for designing and executing a new BPEO. In August 1996 it was announced that 19 contractors had altogether submitted 30 different proposals, and after three dialogue seminars held between November 1996 and May 1997 in London, Copenhagen and Rotterdam some six contractors were asked to submit formal bids. These were received by Shell in early June 1997, and on 29 January 1998 the company announced that the base of the Brent Spar would be used to construct a quay extension at Mekharvick, near Stavanger, Norway and its superstructure would be dismantled and recycled on land.

inflatable boats. Such tactics of occupation had been used many times before by Greenpeace, notably in campaigns against whaling (see Pearce, 1991c). However, as Pilkington *et al.* (1995, p. 5) note, while the action embodied a traditional format of protest – 'simple, direct and with maximum publicity potential' – it was also an operation of 'hi-tec and skilful PR':

> The protesters had satellite telephones and a Mac computer that downloaded photographs and video footage to a media base in Frankfurt. Greenpeace employed its own photographer and cameraman to capture the images that ensured the story was splashed across the world (Pilkington *et al.*, 1995, p. 5).

As Anderson (1997) records clearly, a major feature of Greenpeace is its use of the global media:

> Greenpeace Communications, a division of Greenpeace International ... has a full in-house film, video and photographic capability incorporating a small television studio, three editing suites, a digital sound studio and a commercial film and television archive. These facilities also include compressed digital satellite encoders and decoders and three-dimensional computer graphics. The Greenpeace press desk operates on a 24-hour basis (Anderson, 1997, p. 85).

Greenpeace, through its staging and packaging of protest actions, can be seen to be engaging in what Truett-Anderson (1990) describes as 'theatrical politics': that is they are trying to change public opinion, and thereby politics and policy making, through the staging of 'action-events' that provide a 'spectacle' that is both attractive to the media and sticks in people's minds. They may be seen to have been successful in at least one case, given that one of the leading texts of 'green capitalism' argues 'If any one image sums up the collision between market values and radical environmentalism it is a newswire picture of a Greenpeace inflatable slicing towards an "enemy ship"' (Elkington *et al.*, 1988, p. 12). It is certainly the case that *Greenpeace's* inflatables frequently appear on the pages of many newspapers (see Plate 6.1).

Plate 6.1a Greenpeace protest at HMS Vanguard Trident submarine missile test off Cape Canaveral, USA

Plate 6.1b Greenpeace action against USS Eisenhower, Palma, Spain

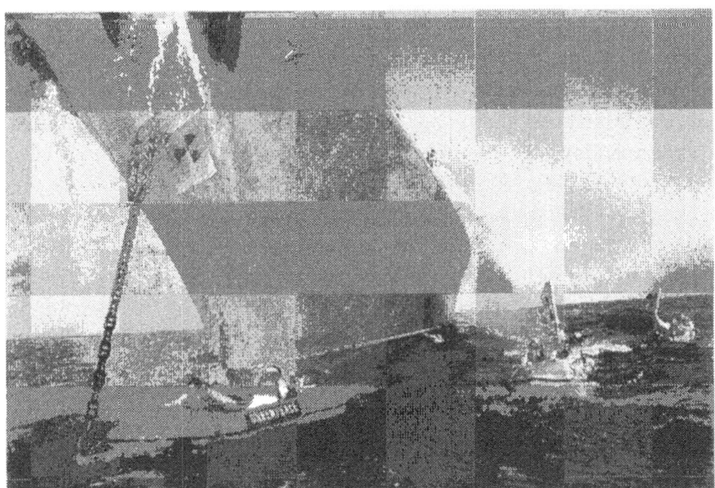

Plate 6.1c Greenpeace inflatable darting in front of Japanese whaler, Northern Ross Sea

A final, and rather contested, feature of the fourth wave of environmentalism relates to its social origins and make-up. People such Porritt and Winner (1988) who adopt an evolutionary and holistic perspective on environmentalism tend to construct fourth wave environmentalism very much as a logical outgrowth from earlier forms of environmentalism, related both to a deepening environmental crisis and to continuing failure to persuade people to change attitudes and actions with regard to nature. There is certainly some evidence to support an evolutionary perspective on environmentalism, in that many of the fourth wave activists had previously been members of third wave institutions. In 1969 David Brower was forced to resign as executive director of the American conservation organisation the *Sierra Club*, a position he had held since 1952, on the grounds that he had neglected his organisation duties in favour of political campaigns and had effectively brought the stated apolitical status of the group into doubt. Brower's response was to form a more explicitly politically active organisation, namely *Friends of the Earth,* the first branch of which he set up in San Francisco in 1969. Furthermore, as Pearce (1991c, p. 48) notes, a couple of years later several other 'disaffected members' of the Sierra Club were involved in the establishment of the Greenpeace organisation in Vancouver.

Not only was there continuity in terms of some key personnel between fourth wave environmentalism and organisations set up in earlier periods of environmentalism, there was also continuity in terms of a continuing use of scientific argument or discourse, a point that is again well illustrated by the Greenpeace's 'Brent Spar campaign' (see Box 6.2). Environmental groups such as Friends of the Earth and Greenpeace very much sought to engage in the 'politics of science and reason' – that is they tried to persuade people to take certain actions through the presentation of scientific arguments and evidence – as well as engage in theatrical politics. They differed from the earlier generation of environmental groups by giving greater recognition to the value of other forms of persuasion, and, as argued previously, by tending to adopt more holistic views of science. Hence, as Sandbach (1980) has commented, many of the new environmentalists of the 1970s did not question the status of scientific knowledge but saw a 'desperate need' for a new, 'less reductionist' form of science. Sandbach notes how this frequently translated into the espousal of some form of ecological science.

A number of authors, including Sandbach, have argued that sciences such as ecology were not, however, the only significant influence on the environmentalism of the 1970s and 1980s. MacNaghten and Urry (1998, p.47) have argued, for instance that: 'to assume that the energy and dynamism of contemporary environmentalism originated in "science" is to overlook the wider cultural critique of which New Environmentalism was merely one part'. They go on to argue that many of the environmentalists of the 1970s had been active in student protest movements in the 1960s, and that from the 1970s environmentalism became a major element of a *'counter-culture'* that identified a series of ills with modern, organised capitalist, societies and its dominant cultural values:

> its embedded values of materialism, individual achievement and technological progress; its emphasis on fixed roles, borders and boundaries, especially of sex and gender; its valuation of dull and predictable sources of pleasure mainly housed within the family; and its prioritisation of the interests and concerns of the generation that had 'won the war' and which was determined to exploit the earth's resources to enjoy the victory (MacNaghten and Urry, 1998, p. 48).

A 'green lifestyle' became seen as 'a way of resisting an alienating and destructive' mainstream society (MacNaghten and Urry, 1998, p. 47), and environmentalism became linked to a burgeoning peace and anti-nuclear movement. Indeed, Greenpeace was originally established, under the name of the *Don't make a Wave Committee,* to protest against nuclear explosion testing, while the German 'green party', die Grünen, was established in the 1980s by a number of anti-nuclear power protesters (see Sallnow and Arlett, 1989).

6.2.5 The fifth wave

The 1980s appeared, at least to people such as Porritt and Winner, as the point where environmental issues were at last coming to the fore in public awareness and where environmental organisations were coming together to form a mass social movement. However, in the 1990s a series of further changes have occurred that cast some doubt on this interpretation.

First, there is some evidence to suggest that environmental issues have declined in importance for many people. The electoral success of green political parties, for example, has proved quite short lived. In 1990, for example, the vote for die Grünen fell to 3.9 per cent of the total vote, and they thereby failed to achieve the 5 per cent of the vote required under the system of proportional representation to get a member in the Bundestag. In Britain the vote of the *Green Party* plummeted after the 1989 European election; likewise in France, where green and ecological parties that had 'once been trumpeted as the great new force in ... politics' (Gumbell, 1993, p. 8) failed to gain any parliamentary seats in the 1992 national elections. In the early 1990s a series of divisions were reported to exist within many of the European green parties, and also within environmental organisations such as Greenpeace and Friends of the Earth, and there have been suggestions of declining public support for these organisations. MacNaghten and Urry, for example, argue that 'much of the public has become disappointed with and sceptical about' environmental groups such as Greenpeace and Friends of the Earth. Opinion polls spanning many counties have further suggested a decline in public concern about environmental issues (Anderson, 1997), although membership levels of environmental groups is generally much greater in the 1990s than it was in the 1980s and earlier (see Table 6.2).

Criticisms have, however, been levelled at the use of opinion polls and green voting behaviour as measures of environmental concern (for a clear example see MacNaghten and Urry, 1998). It has also been suggested that green voting behaviour is not a good indicator of environmental concern, in large part because many of the issues and policies espoused by green parties have become co-opted by other parties. MacNaghten and Urry (1998, pp. 60–2) argue that, since the late 1980s, environmental groups have struggled to find a clear role and identity, largely because there has been 'normalising [of] the environment', that is notions of 'the environment' and 'environmental problems' have become widely accepted across a whole range of discourses, including those of national and international governance (see the discussion in Chapter 5). Environmental pressure groups, such as Friends of the Earth and Greenpeace, have been invited to develop policy, not merely campaign about issues.

The overall picture relating to levels of environmental concern and their influence on the formation of government policies is quite difficult to discern. In many ways the 1990s can be seen as either a continuation of the fourth wave or its slight dissipation. However, what is more clear, and can be seen as a basis of designating the 1990s as a new wave of environmentalism, is the emergence of yet another set of environmental organisations, although the term 'organisation' arguably is too strong to accurately reflect the nature of this new wave. Terms such as 'cells' (Vidal, 1993a), 'networks' (Anderson, 1997) or 'tribes' (Maffesoli, 1996) are perhaps more appropriate. As Anderson remarks:

Table 6.2 Membership of environmental organisations in Britain

Organisation	Membership figures ('000)			
	Mid-1960s	Mid-1970s	Mid-1980s	Mid-1990s
National Trust	165	549	1,460	2,400
Royal Society for the Protection of Birds	31	204	340	925
Royal Society for Nature Conservation	35	109	180	260
Council for the Protection of Rural England	15	28	40	45
Greenpeace	–	–	100	400
Friends of the Earth	–	5	60	250

(Sources: Sandbach, 1980; Porritt and Winner, 1988; Evans, 1992; McCormick, 1992; Council for the Protection of Rural England and the Countryside Commission, 1995; MacNaghten and Urry, 1998)

> Unlike organizations such as Greenpeace and Friends of the Earth, these grassroots movements do not have a fixed fee-paying membership. Distinct leaders are absent; they come and go. There is no hierarchical structure ... These networks ... encompass a varied mixture of individuals united only at times seemingly by a deeply held political cynicism .... [and a tendency] to emphasise individual subjectivity and expressionism rather than collective responsibility (Anderson, 1997, pp. 87-88).

Maffesoli (1996) has argued that the late twentieth-century 'post-modern societies' is a 'time of the tribes', in which people live in 'elective socialities' or 'neo-tribes' and join up with others to pursue a particular activity or interest. This participation may be only fleeting, and people may well participate in a whole series of activities, many of which may at times appear somewhat contradictory. It is now argued that many people do not adopt stable and rationally consistent 'world views', but rather have multiple and fragmented social identities and activities (see Pile and Thrift, 1995; Urry, 1995c; Phillips, 1998a). These identities and activities may often form in resistance to established social relations, practices, institutions and meanings, and according to Routledge (1997) involve the formation of 'autonomous' or 'liminal spaces', that is spaces where established social conventions and rules are not seen to apply.

A number of studies have argued that such 'spaces' are important in the new environmentalism of the 1990s. McKay (1996), for example, argues that from the 1960s onwards one can see the emergence of a series of activities and spaces that seek effectively to negate the rules and resources of predominant forms of economic power. Illustrations of liminal spaces of significance to environmentalism include the emergence of 'free festivals' and 'fairs' in the 1970s, the convoys of the 'New Age travellers' of the late 1970s and 1980s, the creation of communities such as *Tippee Valley* in Dyfed, Wales and the *Tinkers Bubble* in Somerset, and the politically charged communities of the anti-road protest movement (see Box 6.3).

Box 6.3 1990s road protest in the United Kingdom

According to Vidal (1993d), July 1993 saw one of Britain's largest 'mass trespasses', when some 500 people broke down six-feet high chain-link fences and walked on to a 400-feet wide and 100-feet deep cutting that was being constructed to contain an extension to the M3 motorway. This mass trespass was one of the last acts taken at Twyford Down in protest at the building of the motorway. Although this protest failed, in the sense that the M3 now does run through the cutting, it was a protest that has been seen by many as marking the start of a new phase of environmental protest in Britain (eg see MacNaghten and Urry, 1998; McKay, 1996), what we have termed the 'fifth wave'. We will here briefly outline the nature of the protest at Twyford Down, and also at three other major sites of anti-road protest – Pollok Estate near Glasgow, Wanstead and Leytonstone in London and Newbury in Berkshire – to give you some illustrations of this type of environmentalism.

Twyford Down and the M3 extension

The proposal to extend the M3 motorway across Twyford Down was originally made in the 1970s, but had been objected to by local residents and national environmental groups. The area had two scheduled ancient monuments, two areas that had been designated by the government's own *Nature Conservancy Council* as *Sites of Special Scientific Interest* (or SSSIs) because they contained several rare and protected species, and it was part of a designated *Area of Outstanding Natural Beauty.* The area was also the site of a seventeenth-century burial ground. Calls to conserve the Downs and not build the road had led first to a public inquiry in 1985 and, in 1991, to the then European Commissioner on the Environment, Carlo Ripa di Mena, ordering a halt to the development because an environmental impact assessment had not been carried out as required by European law (see Box 1.3 on environmental impact assessments).

The British government argued that it had carried out assessments that were equivalent to environmental impact assessments and after some political disputation this view prevailed. In 1992 the Department of Transport determined that construction should proceed and the majority of the local and national pressure groups stopped their protests and were resigned to witness the destruction of this part of nature. However, in March of that year two of the protesters decided to try to protect the Downs 'just by living on it' (Vidal, 1994, p. 4). They were soon joined by some 40 or so other people, most of whom were apparently not members of any established environmental groups (Vidal, 1994). This group came to refer to themselves as *Indigenous Englanders* and as the *Donga Tribe,* the later being a Matabele name that had been given in the nineteenth century to a series of ancient trackways that criss-crossed the Down. Many of the protesters claimed allegiance to *Earth First!,* an environmental movement or network that had first emerged in Britain in 1991, drawing its name and inspiration from a group formed in the USA in 1980 that adopted a range of direct action tactics. At ▶

Twyford protesters began to adopt non violent direct action, such as putting themselves in the way of the bulldozers and other earth-moving equipment being used to construct the road cutting. They were joined by a number of local residents, many of whom held positions of responsibility in the local community (see Davies, 1992). On 9 December, however, the road construction company hired private security guards to remove the protesters. The result was a violent confrontation between the protesters on the one hand and the security guards and the police on the other, which lasted three days and become widely known as the *Battle of Twyford Down.*

The response of the Department of Transport was to hire further security guards (the total cost of policing Twyford was an estimated £1.7 million (MacNaghten and Urry, 1998, p. 65)), to hire private detectives to trace and photograph protesters (at an estimated cost of £250,000 (Vidal, 1994)) and to try, unsuccessfully, to sue 76 protesters for £1.2 million to cover the costs of the delay in road construction. The protests did not diminish; instead more protesters came to visit the site, even after the cutting for the motorway had been built. Hence in July 1993 there was the mass trespass of the construction site.

Pollok Estate and the M77 extension

In 1965 a proposal was made to extend the M77 to link it to the A736 in order to reduce traffic congestion on the A77 Glasgow to Ayr road (see Figure B6.3a). The proposed new road ran through part of the Pollok Estate, which was an area of farmland and woods given to the 'people of Glasgow' as an area of open space. Objections were raised that the construction of the road would be environmentally, economically, socially and politically damaging (see Routledge, 1997). However in 1992 it was decided that construction of the road would commence. In April 1994 the *Stop the Ayr Road Route (STARR) Alliance* was formed, with members being drawn from local residents and environmental groups, including the recently formed *Glasgow Earth First!* A variety of forms of protest were engaged in, including the long-established tactics such as lobbying of local politicians, staging of mass demonstrations, flyposting and graffiti; more direct forms of action such as the disabling and the 'locking on' of personnel on to earth-moving equipment to prevent their use; and symbolic actions such as establishing the *Pollok Free State* and *Carhenge.*

Pollok Free State was an encampment in woods on the Pollok Estate. In part this can be seen as a standard process of protest through occupation. It also provided a site from which other forms of protest could be planned and initiated. However, as Routledge (1997) details, it also acted as an 'autonomous', 'liminal' space, where established social rules were seen not to apply. The autonomy and liminality of the community was actively stated on 20 August 1994 when protesters declared that the area was to be an 'independent free state' and starting issuing 'Pollok Free State' passports. The community remained until late 1996, by which time most of the rest of the area had been cleared for the new road.

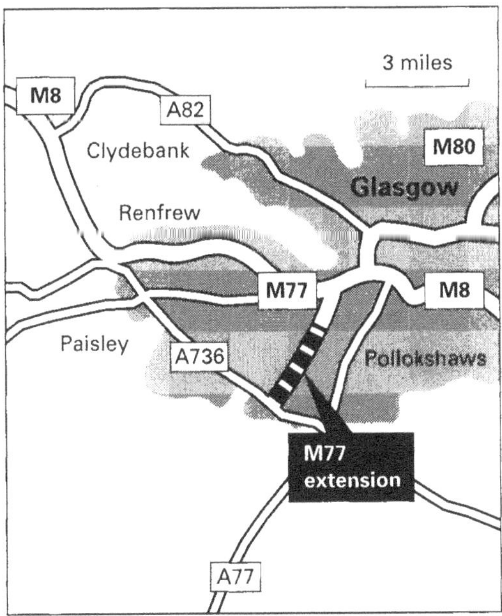

Figure B6.3a M77 link road

Carhenge was built in early 1995 as part of a series of symbolic protests. It consisted of nine cars buried in the ground in a pattern similar to the stones at Stonehenge. As Routledge (1997, p. 369) argues, Carhenge is a clear instance of theatrical politics, acting to parody society's valuation of certain creations over others and in particular the obsession with the car.

Despite these novel forms of protest the M77 Link, like the extension to the M3 at Twyford, was built, as protesters were gradually evicted or voluntarily left the site of the road.

Wanstead, Leytonstone and the M11 link road

For much of the same period of time as the M77 protest in Glasgow there was an environmental protest in London. This protest shared many common characteristics with the M77 protest. First, it was a protest about a similar sort of road development: it was suggested that a link road was needed, this time between the M11 and Hackney and central London (see Figure B6.3b on page 349). It was argued that this would reduce the travel time between Hackney and the M11 by some seven minutes and would also ease traffic circulation in the general area.

Once again there were local protests, related to the destruction of both 'nature' in the form of 'green spaces' such as areas of parkland and the more human environments of local housing. Once again these complaints were ignored and constructors moved in to start clearing the route in September 1993. This led to the start of direct actions by the *No M11 Link Campaign,* initially in the form of a squat in a house in

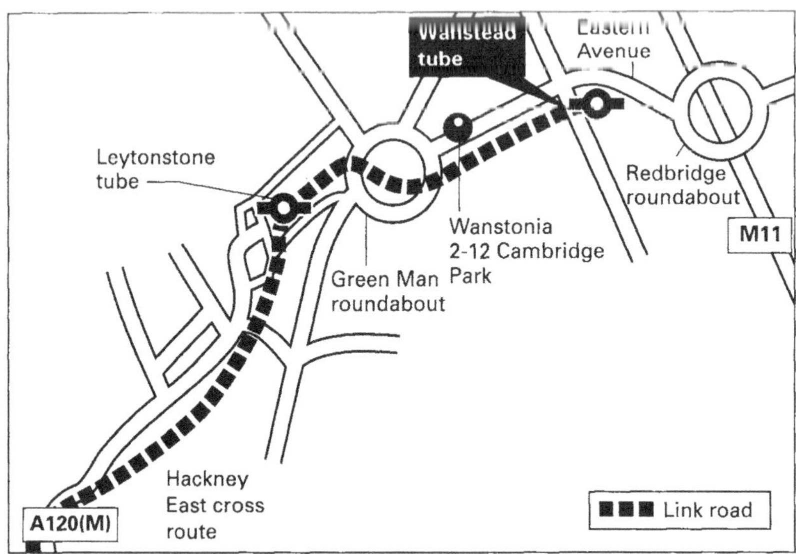

Figure B6.3b M11 link road

line for demolition and attempts to disrupt the felling of trees. A 250-year-old sweet chestnut tree in George Green, Wanstead became a particular focus of attention, perhaps because it could be seen to symbolise the series of concerns that underlay the protest: 'residents' health, their past, nature more generally, even simply an effective local voice' (McKay, 1996, p. 149). A tree-house was built in the tree and a successful legal campaign launched to get this tree-house designated as a legal dwelling (Vidal, 1993b), which then required the contractors to obtain a court order to remove the protesters, or the 'residents' as they were now legally defined. However in December 1993 they were removed by bailiffs and police with the aid of a high-lift hydraulic platform or 'cherry-picker'.

The removal of these protesters from the chestnut tree did not, however, signal the end of the protest. Rather the focus of protest moved to a line of Edwardian houses in Wanstead that were scheduled for demolition, some of which were still occupied by long-term residents. These houses became the site of the *Independent State of Wanstonia,* a territorial designation that was declared in January 1994, some eight months before the declaration of the Pollok Free State. Passports and a flag were also produced here.

Occupation of the houses lasted only until 16 February 1994 (as opposed to Pollok, which remained occupied for over two years) when the protesters were evicted. Protest once more moved on, however, with a second 'independent state' being declared, this time entitled *Leytonstonia.* This became the site for a series of protest actions, including the construction of *'kerbhenge'* and *'brickhenge',* which were 'Stonehenge'-like creations built out of kerbstones and bricks respectively. Eviction from Leytonstonia also followed, this time in November 1994, when some 200 bailiffs, 700 police and 400 security guards removed protesters, many of whom had buried their arms in tarmac or concrete or climbed up a greased scaffolding tower (McKay, 1996). It has been estimated that the cost of dealing with the first year of the M11 protest was some £16 million (MacNaghten and Urry, 1998, p. 66).

The Newbury by-pass

In July 1995, as one of his last acts as Transport Secretary before becoming Chair of the Conservative Party, Brain Mawhinney approved a £70 million by-pass scheme for Newbury. The proposed scheme was to ease traffic in the town of Newbury by building a by-pass to the west of the town. An earlier proposal to build a by-pass on the eastern side had been rejected, largely on the grounds of financial cost. However the newly approved route had major environmental costs, as it ran across three SSSIs, an historic battlefield and a series of other archaeological sites (see Figure B6.3c) . Nature conservationists also became concerned about loss of habitat, particularly for a rare species of snail, the Desmoulin Whorl Snail.

A series of protests were made about the decision to accept this proposal, but in November 1995 the government announced that it wanted the construction of the road to begin 'as soon as possible in 1996' (John West quoted in Harper, 1995). Rather ironically, given that over 360 acres of land were set to be consumed by the new road (Bellos, 1996b), the contractors had to move fast because a European Union Directive sought to conserve nature by banning the felling of trees during the breeding season for birds, between mid-March and August.

In August 1994 a lone protester had set up a tree-house in Snelsmore Woods at the northern end of the proposed by-pass route. A year later some six camps had been set up; by the end of 1995 this had grown to nine and from January 1996, following the government's announcement of its intention to carry on with the road scheme, the number of camps 'mushroomed' to 29 (Vidal, 1996). Another 'free independent state' was declared, this one called *The Isle of Kennet Free Independent State.* Many of these camps were composed of tree-houses, linked together by overhead walkways. Protesters undertook a long series of direct actions such as 'tree hugging' (see Plate B6.3b), 'locking-on' to equipment, trees and concrete blocks, and the blocking of access roads through standing at the top of 'tripods'. In February 1996 there was what has been described as Britain's largest ever anti-road protest (Bellos, 1996a), when an estimated 5000 people marched along the proposed route of the by-pass (note that this is a tenfold increase on the estimated number involved in the Twyford Down trespass). In March there were a series of (at times violent) confrontations in the trees between protesters and climbers hired to evict them. It is estimated that between January and April 1996 some 900 people were arrested in what came to be known as the *Third Battle of Newbury* (the first two battles were held in the English Civil War of the seventeenth century). Construction of the road then began and the road was opened, without ceremony, in the autumn of 1998.

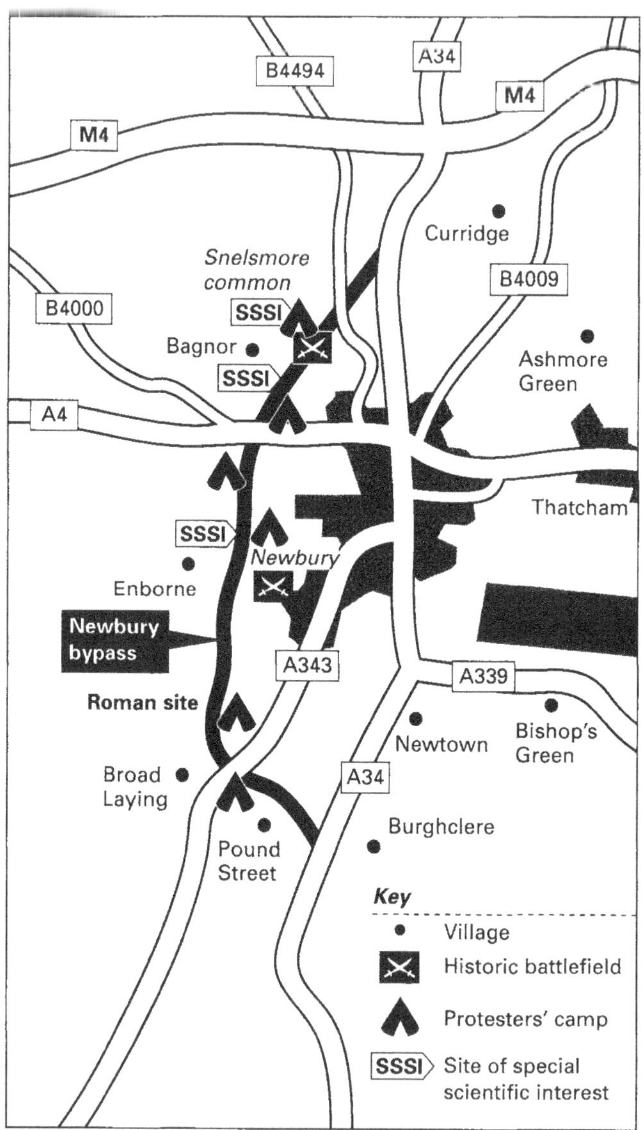

Figure B6.3c Newbury by-pass

Plate B6.3 Tree hugger, Newbury by-pass protest

The formation of such networks and spaces has been the source of some debate. Some people, such as Habermas (1987), Harvey (1989) and Buttel (1993), have argued that they constitute 'neo-conservative' social movements in that they act to sustain rather than transform prevailing social orders. Routledge (1997), for example, documents how at least some of the conventions of prevailing social orders, such as the privileging of male voices and a gendered division of labour, may be actively reproduced within some liminal spaces. More generally, such spaces may simply provide a place where relatively small populations can temporarily escape from the predominant structurings of social life. As such they may pose a very limited challenge to the social order.

Furthermore, as Buttel (1993) notes, many of these protest groups, although at times 'constituted as a critique of industrial capitalism' may also be highly tolerant of it, for example by accepting the presence of social inequalities as part of an 'expressive individualism' or, as Eder (1993) calls it, an 'individualistic ethos of personal identity', in which social inequalities are seen to stem from people having inherent differences: frequently expressed in terms of everybody is 'naturally' different. This ethos contrasts with others, such as the view that people are basically the same or that they come to establish differences because of the way they act (see Eder, 1993; Phillips, forthcoming-a). This can lead to a socially conservative 'politics of identity' in which currently recognised differences are seen to be due respect whatever their consequences; it can also, however, lead to a socially liberating 'politics of identity', whereby being seen as different is not seen as a reason for condemnation, control and reform.

The environmentalism of the 1990s is a complex and quite hybrid entity. Much of it may well be quite accurately described as neo-conservative or individualistic, giving primacy to personal experiences over social concern. On the other hand much of it is quite clearly political, with many environmental groups, for example, adopting explicitly radical epithets. The *Sea Shepherd Conservation Society,* for example, was founded by Paul Watson, who had been turned down by Greenpeace for being 'too radical' (Pearce, 1991c, p. 28). Sea Shepherd undertook 'direct action' to conserve nature, principally in terms of endangered marine fauna such as whales. Their actions have often been focused around the same issues as Greenpeace. However, while Greenpeace has a policy of refraining from actions that endanger human life or property, Sea Shepherd's activities have included the ramming of whaling ships and, allegedly, the sinking of ships using mines (see Pearce, 1991c, p. 31). In Britain a plethora of environmental protest organisations has emerged in the late 1980s and 1990s, including the *Animal Liberation Front, Earth First!,* the *Earth Liberation Front,* the *Hunt Saboteurs Association, Surfers against Sewage, Small World, The Land is Ours,* the *'Dongas', 'Flowerpots', 'Rainbow'* and *'Space Goats'* tribes, as well as a range of local road protest groups. Vidal (1993c) and Chaudhary (1994) estimated that there were between 220 and 250 groups in the United Kingdom protesting against road developments associated with a £23 billion road-building programme announced by the Conservative government in 1993.

Many of these groups espousing direct action are extremely critical of earlier generation environmental organisations. As Grant (1995, p. 21) remarks, organisations like Greenpeace and Friends of the Earth are 'perceived as sell-outs, mired in the bureaucratic process, in the pockets of the Government after the Earth Summit'. As mentioned earlier, both these organisations have been acting as policy advisers, and while this may be welcomed by many people in these organisations as a means of 'having an influence', it may well have the negative consequences observed by Grant. MacNaghten and Urry (1998), for example, have observed that many people are quite concerned about the state of 'the environment, particularly the local one in which they live, but that they are extremely fatalistic as to whether they personally, or indeed anybody else, can do anything about these problems'. MacNaghten and Urry (1998, p. 244) claim that people seemed to have faith neither in any form of environmental collective action, that is 'participation in political parties, consumer or environment groups, or trade unions', nor in political representation through voting. In such a situation of 'desocialisation' and 'depoliticisation' it is not surprising that direct actions by committed individuals have come to generate both widespread attention and public support.

Actions taken have been directed at a large range of environmental issues, including the disposal of nuclear and non-nuclear waste, the destruction of 'natural' landscapes and ecosystems as a result of open cast mining or transport developments such as new road and airport developments, and the hunting and live export of animals or their use in laboratory experiments. Vidal and Bellos (1996) have suggested that in 1996 there were over 500 instances of 'direct action' (see Figure 6.1). Particularly significant in terms of the public attention given to them have been protests related to new road developments (see Box 6.3), but there have also been more general anti-road protests organised, for example, by groups such as *Reclaim the streets,* which has organised street parties with up to 8000 blocking a section of motorway in London (Vidal and Bellos, 1996).

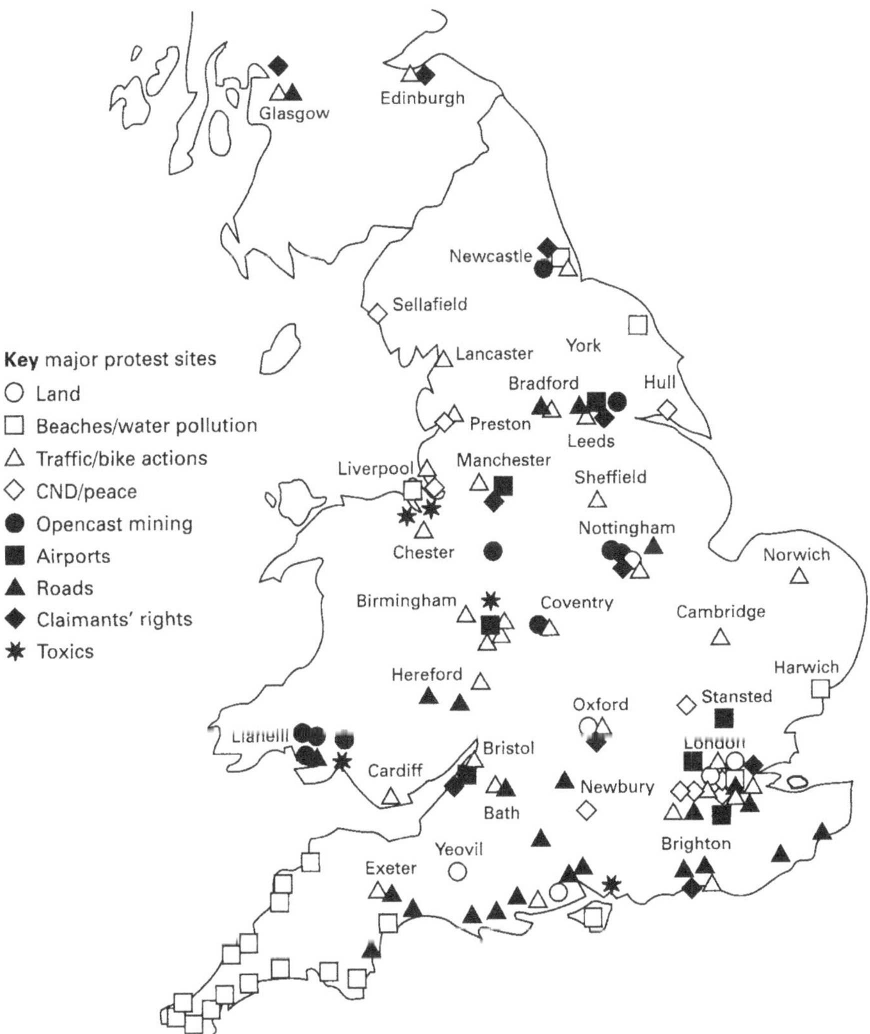

Figure 6.1 Environmental protests in 1990s' Britain

Many of these 'direct actions', like those of earlier international environmental organisations such as Greenpeace, are undertaken as spectacles that have an appeal to the media. While for Greenpeace the spectacle often involved small inflatables in the world's seas, new images have come to signify the latest group of environmentalism. Hence one has images of the 'tree hugger', first associated with the Chipko movement in India (see Box 6.4) but now very much a globalised image (cf. Bandyopadhyay, 1992; Guha, 1989; MacNaghten and Urry, 1998); the 'tunneller'; the encampment of 'benders'; and 'the youth, the dreadlocks, the crusty dresses and the vegetarian boots' (Grant, 1995; quoted in Routledge, 1997, p. 370). As Routledge (1997) outlines, these images are a complex, negotiated and conflictual creation: in part being elements of a 'theatrical politics' originated by the protesters – people consciously look and dress in particular ways and live in a different manner in order to unsettle people living in the mainstream – and in part being a negative imagery constructed by an unsympathetic media that see these environmental protestors as variously comic, disreputable and dangerous. With reference to the former, Routledge makes use of Melucci's (1989) suggestion that there are three forms of theatrical politics: prophecy, reversal and representation. In the first, environmentalists act to promote an alternative framework of meaning to those which are socially dominant or 'hegemonic': it is hence frequently suggested that these protesters adopt an 'alternative culture'. Second, it is suggested that many of the practices of these groups seek to reverse the values of the hegemonic culture, in order to demonstrate contradictions and arbitrariness. Third, Melucci argues that protest groups can construct representations that, although appealing to mainstream media, also contain 'counter-hegemonic messages', that is they seek to promote different values and ways of acting.

Box 6.4 The Chipko tree huggers of India

The Chipko movement emerged in the Garhwal district of the Himalayas (see Figure B6.4) during the 1970s, when the State Forest Department in India took control of all forested and open land owned by the national government and local states and was charged with expanding timber production. As part of its strategy to do this, the State Forest Department started planting quick-growing chir pines and imposed new access and felling restrictions on forested areas, allowing local populations only to cut trees for household consumption, at the same time as auctioning off rights for commercial felling to wealthy merchants and large-scale timber companies. This caused considerable resentment and economic hardship for local people, particularly those working in local timber-based artisanal industries. In the early 1970s there was also major flooding in the area, which some people argued stemmed from increased surface water runoff related to the clearance of large areas of land by the new large commercial foresting concerns and to the planting of the chir pine, which created forests that were, unlike the pervious banj or Himalayan oak forest, virtually devoid of water retentive undergrowth.

In 1971 in the village of Gosephwar demonstrations began demanding the restoration of communal rights to the forests. A series of other protests followed,

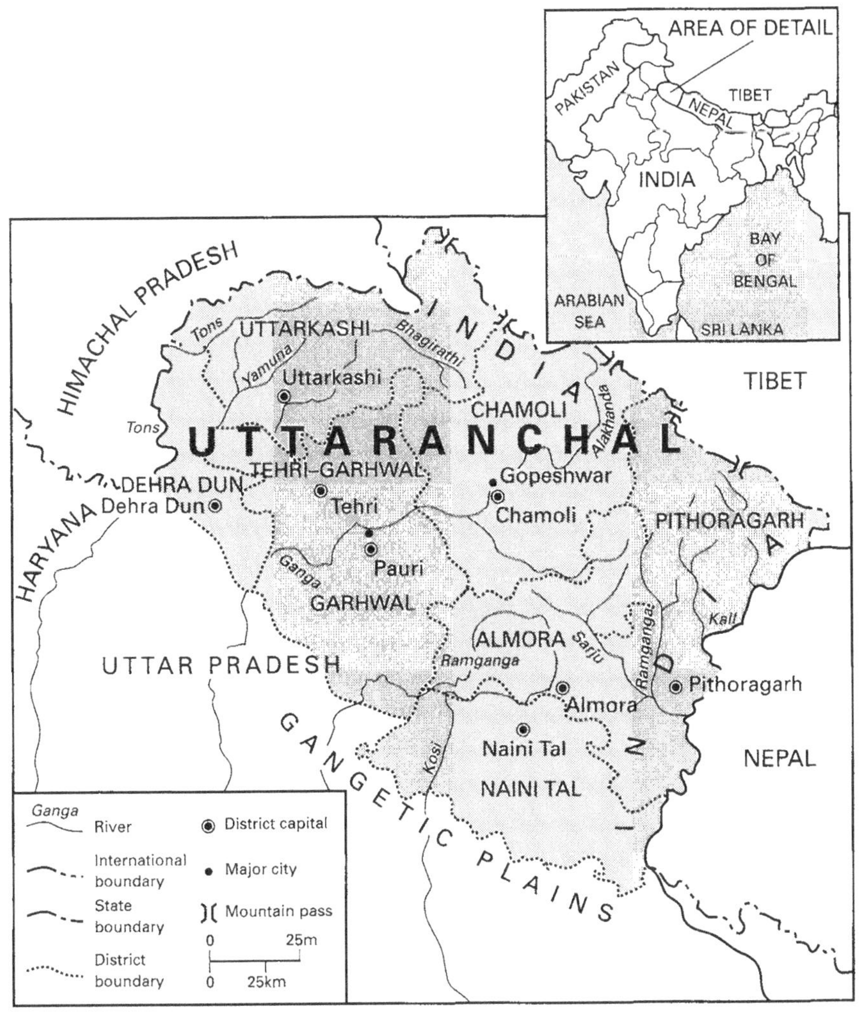

Figure B6.4 Indian Himalayan region (Source: Peet and Watts 1996)

particularly in the felling seasons from 1973 through to 1974. In 1973 one particularly significant protest centred around the granting of a timber concession in the Alaknada river to a tennis-racket manufacturer from Allahabad, a city some 800 kilometres away. This concession was granted in preference to an application from a local co-operative venture that had sought to fell trees to make agricultural implements. In protest, villagers sought to prevent the felling of trees by loggers hired by the tennis-racket manufacturer by literally hugging the trees. The term *'chipko'* means to stick or adhere.

The protest at Alaknada river is frequently seen as marking the birth of the Chipko movement, although the act of tree hugging has a long history in the

In 1731 Amrita Devi and her three daughters, who were all members of the Bishnois sect of tree worshippers, were hacked to death when they hugged trees being felled to build a local maharajan palace. For Amrita Devi, her three daughters and 355 people who were also killed trying to stop felling, trees had a spiritual importance. By contrast, for many of the 1970s' Chipkos the motive was more pragmatic: they were trying to stop large commercial felling by organisations located outside the region so that local artisanal producers could continue to fell trees. As one of the leaders of the Chipko movement, Prasad Bhatt, put it, 'Saving our trees is only the first step in the Chipko movement. Saving ourselves is the real goal' (quoted in Pearce, 1991c, p. 268). However, others in the 1970s' movement did take a rather less instrumental view, most notably Sunderlal Bahuguna who, although he had once run a craft co-operative using timber products, came to favour a complete ban on tree felling.

The presence of two divergent rationales for the Chipko has become a matter of some debate. Commentators such as Guha (1989), Pearce (1991c), Bandyopadhyay (1992) and Rangan (1996) have, for example, focused on documenting shifts in emphasis in the Chipko movement and tensions and conflicts between various representatives therein. Shiva (Shiva, 1987a and b; 1989; Shiva and Bandyopadhyay, 1986) has taken a slightly different tack, developing an 'ecofeminist' interpretation of the Chipko, arguing that not only were the majority of tree huggers women, but also tensions within the movement can be seen to revolve around a masculinist concern to tame and control the forests and an ecofeminist concern to live within nature. Shiva argues, for example, that women such as Mirabeh and Saralabehn, who had both been associates of Mahatma Gandhi in the 1930s and 1940s, were prominent in developing the ecocentric strand of the Chipko movement and effectively act to educate Bahugana into his later views. Bandyopadhyay (1992, pp. 267-9), who had earlier written with Shiva (Shiva and Bandyopadhyay, 1986), claims, however, that her analysis is a 'decadent and outdated western model of gender conflict' and that the Chipko movement is 'a Gandhian movement characterized by unique gender collaboration'.

6.3 Modern environmentalism: holistic social movement or class ideologies?

The previous section has outlined a range of different forms of environmentalism and when they were of particular significance within Britain. It has also been noted on various occasions that for people such as Porritt and Winner the various forms of environmentalism have some overall coherence, in that they can be seen to care about nature and reject an instrumental, technocentric, attitude towards it. For Porritt and Winner the various 'waves' of environmentalism represent new stages of growth of an environmental movement that is growing, delta like, within the modern world.

There are, however, rather different interpretations of modern environmentalism that highlight the differences, rather than the commonalities, between various environmentalists. These interpretations tend to suggest that modern environmentalism is quite fragmented, not only in terms of utilising different forms of action, but also being concerned with different natures and having distinct social origins. In this section we will examine this argument in more depth and see whether it may help explain the wave-like character of environmental concern.

6.3.1 Political differences in environmental concern

It is frequently suggested by advocates of environmentalism that a concern for the environment lies outside and generally above social conflict and traditional political distinctions. The German green party, die Grünen, for example, uses the slogan, 'not left, not right but on', while the British *Ecology Party,* the forerunner of the British Green Party, frequently portrayed itself as being outside traditional party political divisions (see Figure 6.2). Furthermore, Cotgrove (1982), in an examination of the political affiliations of members of environmentalist groups, found that environmentalists frequently identify themselves as being in the political 'centre' or with having 'no party political position'.

Figure 6.2 The Green Party's position on conventional party politics (The Green Party)

Pepper (1984; 1993) has, however, argued strongly that this notion of being beyond existing political struggles is wrong and that environmental concerns are not 'above' or unrelated to traditional political concerns, but rather can be said to both run alongside and even stem from traditional socio-political interests.

In using the term 'running alongside traditional political concerns' Pepper is suggesting that most environmentalist positions can be connected to existing political concerns: that is, environmental concern can be 'bolted on' to existing political concerns and leave these political perspectives largely unchanged. Pepper suggests, for example, that many of the arguments of 'deep ecologists' can be bolted on to right-wing political arguments, to produce what he terms 'ecofascism', while other arguments of the deep ecologists can be 'bolted on' to traditional left-wing politics, to produce what Pepper terms 'ecosocialism'. Cotgrove (1982, p. 91) has similarly argued that environmentalists share many of the same values of supporters of the 'traditional left' which: 'like the environmentalists, see the need for public interest to override market mechanisms, emphasise participation against authority, and collective provision of welfare and the re-assertion of community'. Cotgrove (1982, p. 91), however, also suggests that there were crucial differences between environmentalists and the 'traditional left', particularly in relation to their views on economic growth: 'the traditional left at least shares with the new environmentalists a commitment to non-economic values. But it differs in supporting economic growth which environmentalists actively oppose'.

Clearly there may well be issues and concerns that are specific to environmentalism. However, as both Cotgrove and Pepper argue, many of their arguments are common to other political positions.

Pepper, and also Cotgrove, further suggest that environmental concern, or at least some elements of it, can be connected to existing political positions. They also suggest that environmental concern is rooted in existing socio-political interests and struggles. Put at its most basic, Cotgrove and Pepper suggest that environmental concern is an expression of specific class interests, in particular it is said to be associated with middle-class groups.

6.3.2 Class and stratifications in environmental concern

There is some clear evidence to support such a claim. Lowe and Goyder (1983), for example, have shown that in Britain in the late 1970s members of environmental groups were drawn predominantly from middle-class households (see Table 6.3). They also found that expressions of concern about environmental problems such as pollution tended to be higher amongst members of the middle class than among those in working-class households (see Figure 6.3). Similar results have been found in the more recent study of Pattie *et al.* (1991) who found, for instance, that salaried middle-class groups were rather more concerned about the impact of house building in the countryside and we are more willing to pay for pollution control than were more working or *petit bourgeois* class groups (see Figures 6.4 and 6.5). Clearly this willingness to pay may reflect the differential incomes of these groups as well as, or perhaps more than, differences in degree of environmental concern. Indeed Lowe and Goyder note that if one asks whether people are concerned about the environment one finds a high level of concern across all classes (see Figure 6.6), while Pattie *et al.* found that the working class were amongst the most concerned about nuclear power (see Figure 6.7).

Table 6.3 The social composition of environmental groups in Britain in the 1970s and early 1980s

Occupation of 'head of household'	Social structure of environmental groups (%)			
	Bedfordshire & Huntingdon Naturalist Trust [1]	National Trust [2]	Royal Society for Protection of Birds [3]	Social structure of England and Wales (%)
Managerial and professional	78	72	25	14
Technical and clerical	9	24	41	22
Skilled manual	11	3	20	31
Unskilled manual	2	1	14	33

Notes: Survey data on which the table is based were published at the following dates: [1]1980; [2]1973; [3]1979

(Source: after Lowe and Goyder, 1983)

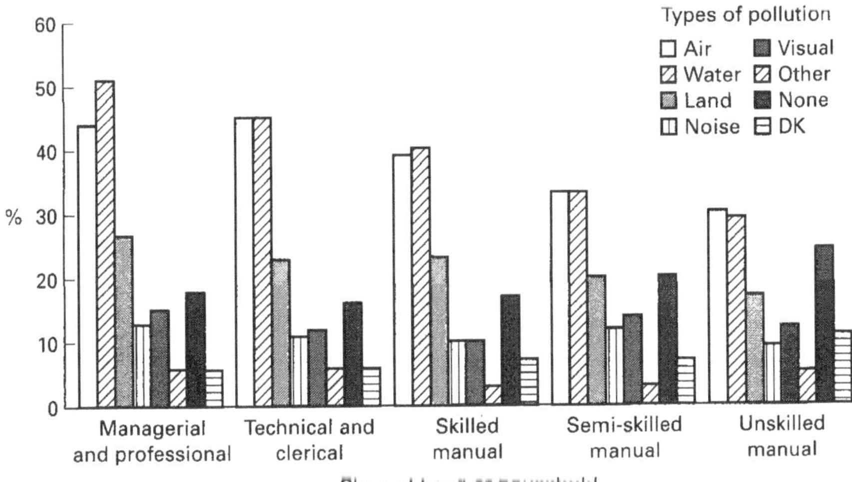

Figure 6.3 Concern about pollution, 1976 (source: after Lowe and Goyder, 1983)

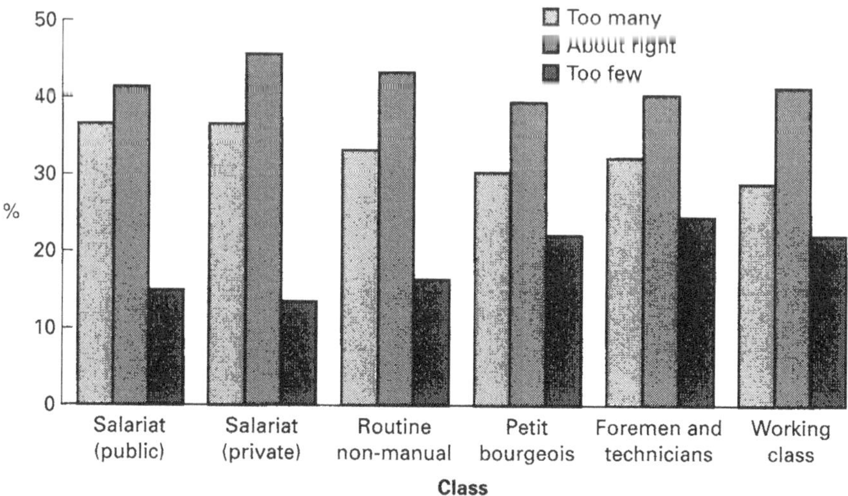

Figure 6.4 Attitudes in the late 1980s towards house building in the British countryside (source: Pattie *et al.*, 1991)

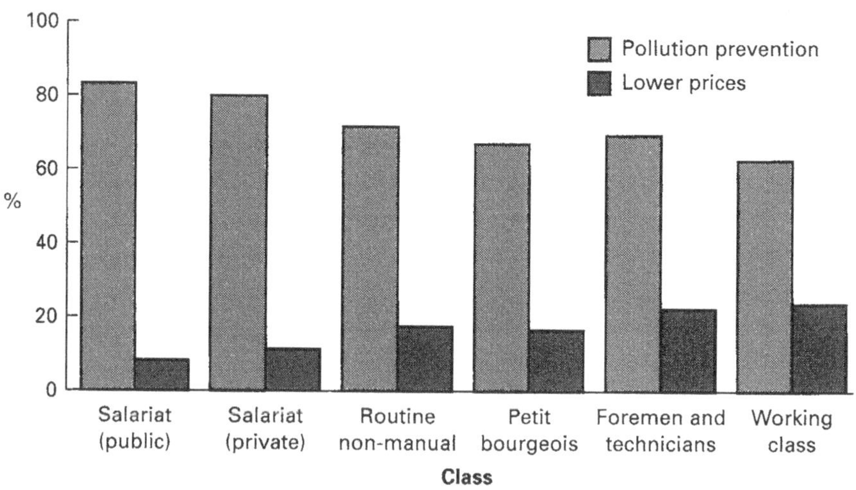

Figure 6.5 Preference for pollution prevention versus lower prices (source: Pattie *et al.*, 1991)

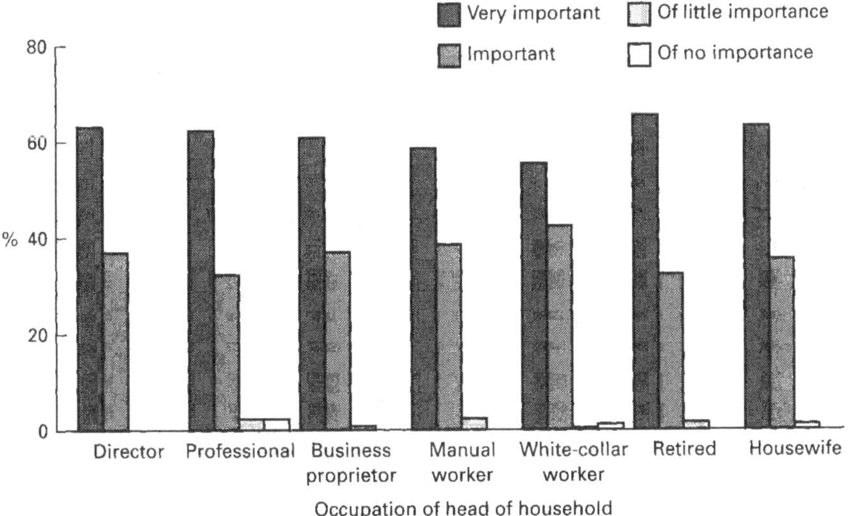

Figure 6.6 Concern about pollution, 1979 (source: after Lowe and Goyder, 1983)

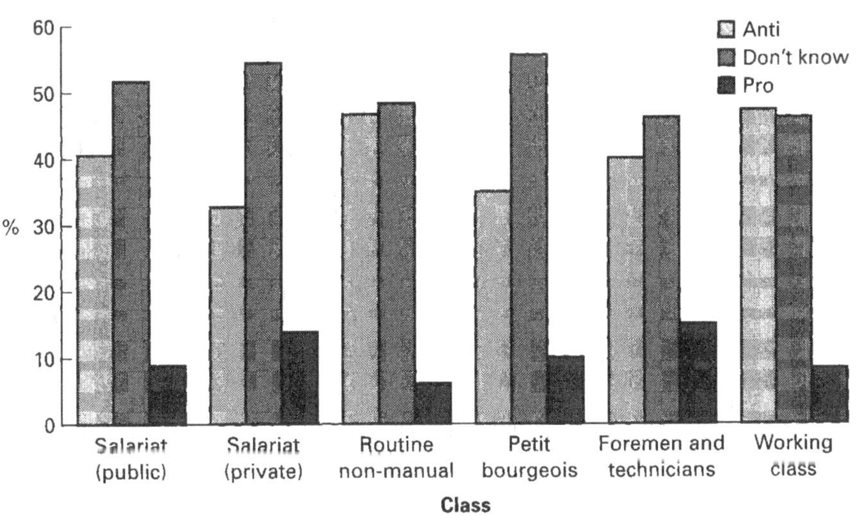

Figure 6.7 Attitudes towards nuclear power in late 1980s' Britain (source: Pattie *et al.*, 1991)

The graph of attitudes towards nuclear power is also interesting, in that Pattie *et al.* distinguish between a public sector middle class, or the public salariat, and a private sector middle class, or the private salariat. They note that the public sector salariat is much more anti-nuclear power than is the private sector salariat, a finding that might suggest that a simple middle/working class distinction cannot account for all the variations in environmental attitudes. This point is made most clearly by Cotgrove:

> To say that environmentalists are middle class is true, but misleading. Industrialists too are well educated and in the upper income bracket. By any definition they too are middle class. Yet they are opposed to the environmentalist movement (Cotgrove, 1982, p. 19).

Since the early 1970s there has been a steady debate about the relationship between class and environmentalism (see for example Inglehart, 1971a; Buttel and Flinn, 1974; Cotgrove and Duff, 1980; Cotgrove, 1982; Morrison and Dunlap, 1986; Eckersley, 1989; Beck, 1992; Eder, 1993). At least three positions can usefully be distinguished. First, there are some people who argue that expressions of environmental concern stem from the self-interest of the middle class, or parts of the middle class. A second viewpoint is that environmental concern stems from the differential experiences particular classes have of nature. A third argument that has been advanced is that contemporary environmental concern is essentially unconnected with class but rather stems from broad economic and cultural shifts within contemporary capitalist societies. Some of the arguments of each of these perspectives will now be outlined.

Within these debates about class and environmentalism there have been several attempts to develop the point made by Cotgrove above, and by Cotgrove and Duff (1980), which suggests that 'environmentalists' appear to be drawn predominantly from a specific element, or 'fraction' of the middle class. In particular, they argued that people's views about environmental issues are linked to their occupational position with regard to the private market economy and the public, non-market or state sector, and they suggest that environmental concern is a specific attribute of public sector, middle-class professional and semi-professional groups.

Similar arguments have been made by a variety of other authors. Eckersley for example, has suggested that environmentalism may be associated with what, following people like Gouldner (1979), she terms the 'new class' of a 'professional intelligentsia' (Eckersley, 1989, p. 208). Eder (1993) links environmentalism with a 'new *petit bourgeoisie*', while as discussed in Chapter 2, Thrift (1989), Harrison (1991) and Urry (1990) have talked in terms of the 'service class'. Despite the variety of titles and subtle differences in argument, these authors can be said to be referring to a broadly similar set of occupations, focused particularly around professional occupations.

A key issue is why is there such an association. Two distinct, albeit interrelated lines of interpretation focus on class 'interests' and class 'experience'. The former suggests that the 'new middle class' is involved in environmentalism for self–centred reasons. Two clear examples of the class-interest argument can be found in Pepper (1993):

> the ruling and affluent classes expend great time and energy on the protection of humanised animals rather than the welfare of brutalised children in home or factory workers or adult workers reduced to the level of animals. Their concern for animals is a displacement for their class position (Parsons, 1990, quoted in Pepper, 1993, pp. 150–1).

> greens, as a new social movement, can be seen as a largely third generation displaced working-class group struggling for status and recognition (Scott, 1990, pp. 145–7). From relatively privileged and educated, but not over wealthy, backgrounds, they lack political power ... real power still lies with industrialists or trade unions, and is based on a stable, technocentric set of values, around which capital and labour form a consensus ... Excluded middle class, professional groups therefore mobilise ... struggling for the political power to match its social position (Pepper, 1993, p. 151).

The arguments of Parsons are problematic in that they adopt a very general view of class and ignore the findings of people such as Cotgrove and Duff (1980) that many environmentalists are drawn from social welfare professions and that their concern for nature very much overlaps with a concern for society. However, Cotgrove and Duff can be seen to be in broad agreement with the line of argument advanced in the second extract from Pepper, in that they argue that environmentalists are drawn from a specific fraction of the middle class:

> whose interests and values diverge markedly from other groups in industrial society ... environmentalism is an expression of the interests of those whose class position locates them at the periphery of the institutions and processes of industrial capitalist societies. Hence their concern to win greater participation and influence (Cotgrove and Duff, 1980, pp. 340–1).

In other words, environmentalism is seen to be attractive to members of a 'new middle class' because it enables them to gain access to social and political power.

Class interest-based interpretations have been criticised as being overly 'structuralist' and ignoring the values, knowledgeability and agency of people (see Chapter 1 on structuralist approaches and their critics). To say that people do something because it is in their class interest implies that they know what these interests are, a claim that is often hard to substantiate (see Lockwood, 1989). People such as Buttel and Flinn (1974) and Cotgrove (Cotgrove, 1982; Cotgrove and Duff, 1980; 1981) have, however, sought to develop more 'experience close' interpretation of environmentalism and class interests. Buttel and Flinn (1974, p. 58), for example, suggest that a concern with environmental issues might be connected to the degree to which people experience 'exploitative or non-exploitative relations with natural resources'. People such as farmers or semi-skilled or unskilled work, who are working to 'exploit' natural resources are found to express little support for environmentalism, while those in professional and technical work, 'more or less insulated from dominant economic interests', and thereby the capitalist exploitation of nature, were 'more likely to support environmental reform' (Buttel and Flinn, 1974).

While Buttel and Flinn tie environment concern very directly to a position within, and thereby experience of, society–nature relations, Cotgrove (1982) very much saw environmentalism as being connected to more general social experiences and relations. In particular he argued that the new class espoused environmentalism as a result of the political 'alienation' this class was experiencing in modern societies: this class was seen to be 'at the periphery of the institutions of industrial capitalist societies' and thereby excluded from 'the processes of decision-making' (Cotgrove, 1982, p. 95). However, as Eckersley (1989) has argued, this alienation thesis can be seen to overemphasise the marginality of the 'new middle class':

> While most of is members do not work in the core institutions of capitalism, their occupations can hardly be described as peripheral ... many members of the new class are often closely connected with such institutions and participate 'to an above average extent in the cognitive culture of society (ie in the knowledge and information available in society)'. Likewise .... [they] are relatively experienced in, and often frustrated with, conventional practices and their limitations (Eckersley, 1989, p. 213).

Eckersley argues that the new class is not so much 'alienated' from modern society as engaged in a 'moral and intellectual disagreement' (p. 213) with the dominant, 'technocentric' value systems and forms of organisation. She argues that there is a need for a more 'comprehensive' interpretation of the new class and its advocacy of environmentalism, an interpretation that addresses both the structure of class and also gives some agency to the values and opinions expressed by environmentalists.

A concern with giving agency to people's values and opinions can, indeed, be seen to constitute a third focus of discussion about class and environmentalism. As stressed in Chapter 1, and in various places through this chapter, environmentalists often place great stress on changing people's attitudes and perceptions of nature. As also highlighted here and in Chapter 3, and even more particularly in Chapter 5, this view can be seen as being 'idealistic' and paying insufficient attention to the social context and consequences of action. It may, in line with our discussion of the relationship of environmentalism with more socially focused ideologies, be criticised for overemphasising the distinctiveness of environmental ideas. This point has been picked up in several studies of class and environmentalism that have suggested that environmentalism can be seen as a feature of a broader cultural shift in ideas. A key exponent of this idea is Inglehart (1971a; 1971b; 1981), who suggested that environmentalism is just one strand of what he terms variously 'post-bourgeois' or 'postmaterial' values, which he suggests have become adopted by particular sections of the middle classes in western countries as a result of growing affluence and security since the Second World War. Basically Inglehart argued that as people become more affluent they move from a concern with securing 'material needs' – such as gaining access to food, water and shelter – towards other, less material, 'wants', including a concern with their environments and with nature.

Inglehart's ideas have been widely utilised in interpretations of environmentalism: indeed the studies of Cotgrove and Cotgrove and Duff all couch their discussions of environmentalism within Inglehart's broader notion of postmaterial ideas. Studies by people such as Eckersley (1989) have, however, identified a series of problems with Inglehart's arguments, ranging from its very narrow focus on the impact of affluence on values, its temporisation of environmentalism and, relatedly, a neglect of the specificity and agency of environmental ideas of an 'ecocentric perspective'.

Eckersley makes the first point very clearly when she states:

> Inglehart has said much about the unparalleled economic affluence of the post-1945 years. He has said very little, however, about the significance of new technologies (Nuclear, genetic, Chemical), the international communications revolution (computers and television), the considerable expansion in higher education, and the growth in our biological and ecological knowledge and understanding (Eckersley, 1989, p. 221).

Eckersley goes on to argue that there is a need for a much wider analysis of the influences that such changes might have on people's values, including the espousal of environmental attitudes. Dickens (1996) has argued that contemporary societies are extremely complex, and that this very complexity is a key influence on the rise of environmentalism, which he suggests is 'about alienation ... the failure of people to understand the stratified world' (p. 9).

With reference to the issue of temporality, Inglehart very much utilises a linear history of environmentalism. In particular he argues that a period of prosperity in the

West that lasted from 1945 until the 1970s produced a steady growth in adherents to postmaterialism. He argued that people have a tendency to retain their value systems throughout their adult life, and that the switch to postmaterialism hence occurred when people experienced affluence through their childhood. So long as people were experiencing such a childhood, so the number of people adopting these values would grow, even if they were to experience periods of economic hardship through their later life. Eckersley (1989), however, identifies this as a highly problematic claim, and even goes so far as to suggest that a rather greater influence on the rise of environmentalism is higher education. Certainly a series of studies have highlighted that members of many environmental groups have high levels of education (e.g. Buttel and Flinn, 1974). However, it is also important to note that there have been important changes in the form of environmentalism and that expressions of support for these various forms have fluctuated over time. Indeed, people such as Lowe and Goyder (1983) have suggested that there is very much a cyclical, rather than linear, character to the history of environmentalism. Support for this argument can be found in Sandbach's (1980) 'content analysis' of the coverage of environmental issues in major US newspapers and magazines such as the *New York Times,* the *Wall Street Journal, Time* and *Newsweek* in the 1960s and early 1970s, in the growth and then levelling of membership of environmental organisations in the USA in the 1960s and 1970s (see Anderson, 1997), as well as the decline in the late 1980s and 1990s of the 'green vote' and public opinion poll measures of 'environmental concern' discussed earlier in this chapter.

Lowe and Goyder suggest that two factors underlie the wave-like pattern of the development of the environmental movement. The first is one internal to the movement and is related to its organisational structure. They suggest that environmental pressure groups have effectively a fixed lifespan: 'a particular configuration of groups prevails until a new generation perceives new environmental problems, or wishes to create its own institutions to express its separate social identity and style of participation' (Lowe and Goyder, 1983, p. 24). In other words, Lowe and Goyder see the separate environmental 'waves' as representing the influx of a new generation of people into the environmental movement rather than a spread on environmental concern into more and more elements of society.

The second factor Lowe and Goyder identify as a cause for the cyclical development of the environmental movement is the cyclical nature of capitalist world economic development. As they put it:

> It is perhaps no coincidence that each of the periods of sudden growth of the new environmental groups in the 1890s, the late 1920s, the late 1950s and the early 1970s occurred at similar phases in the world business cycle – towards the ends of periods of sustained economic expansion. This helps to explain why, despite different levels of economic activity, the advanced capitalist countries have simultaneously experienced heightened environmental concern. We would suggest that environmental groups arose at these times as more and more people turned to count the mounting material cost of unbridled economic growth and sought to reassert non-material values (Lowe and Goyder, 1983, p. 25).

Lowe and Goyder amend Inglehart's analysis in two significant ways. First, although they see environmentalism very much as a 'post' or 'non-material' value, and see levels of economic prosperity as being a key motivator in its adoption, they recognise that modern capitalist economies are highly cyclical, moving from growth to depression and back again. As a result they suggest there are fluctuations in levels of environmental concern. A second important feature is that they also recognise that there may be good reasons for people being concerned about the environment, and that the condition of the environment may itself be a significant contributor to levels of environmental concern. As Lowe and Rudig (1984, p. 219) comment, Inglehart looked on 'the environment' as simply 'one among many "postmaterialist issues" which arose to prominence unrelated to any change in the environment'. By contrast Lowe and Goyder, and also Eckersley (1989) and Anderson (1997) argue for a need to consider the 'reality', or rather better, the 'agency' of nature (see Dickens, 1996 and Fitzsimmons, 1989); an agency that is not beyond society given that many environmental problems relate to economic production and consumption.

A third criticism that can be levelled at the work of Inglehart is that he ignores the agency of ideas. Eckersley (1989, p. 233), for example, argues that environmentalists 'genuinely want a better world', a point also made by Cotgrove and Duff (1981, p. 99), who argue that environmentalism is about the values that people, or at least some of them, 'want to see made real in some future'. Cotgrove and Duff (1981, p. 99) add: 'Thus it is not present need which generates values but their utopia: not their conditions of existence *per se,* but what they would like to see as their conditions of existence'. More recently, Anderson (1997) has argued that:

> while social class positioning undoubtedly plays a part in collective action, it would be rather simplistic to suggest that a determinate relationship exists. Additionally we need to look at the role of values, identity and knowledge in understanding the growth of environmentalism (Anderson, 1997, p. 96).

6.3.3 Environmentalism: an ideology of the advanced capitalism?

In this chapter we have so far been largely focusing our attention on environmentalism within the so-called developed world, and in particular on Britain. However, as noted in Chapter 5 and mentioned several times in the course of this chapter, environmental concern is now globalised, not least in the sense of being found across the globe. Furthermore, many western environmentalists have come to look to environmental protests in other parts of the world for ideas and inspiration. The practice of 'tree hugging', for example, has been adopted from the Chipko movement and as Guha (1989, p. 178) has commented, the Chipko movement has become 'one of the most celebrated environmental movements in the world'.

The presence of environmental movements in the 'underdeveloped world' poses some problems for those, such as Inglehart, who suggest that environmental concern appears as a consequence of material well-being. Dunlap *et al.* have explicitly criticised such arguments, claiming on the basis of a survey of environmental concern conducted in 24 countries that environmental concern cannot simply be viewed as 'a luxury affordable only by those who have enough economic security to pursue quality-of-life goals' (Dunlap *et al.,* 1993, p. 81). On the other hand, people such as Routledge (1995) have argued that there are important differences between expressions of environmental concern in 'developed' and 'underdeveloped' countries, differences that very much reflect the levels of material affluence:

> In the North, new social movements have often concentrated upon 'quality of life issues', whereas in the South, movements have often focused upon access to economic resources. An example is the issues faced by the ecological movements. In the North, the ecology movement has taken much of the industrial economy, working to preserve nature as an item of 'consumption', as a haven from the world of work. In the South, however, those affected by environmental degradation – poor and landless peasants, women and tribals – are involved in struggles for economic and cultural survival rather than quality of life (Routledge, 1995, p. 273).

As highlighted in Chapter 5, it is all too easy for the views and interests of the so-called 'developed' countries to displace other opinions and concerns, and too often discussions of environmentalism and development have sought to understand situations in other areas of the world through concepts that in many senses have applicability only to the 'developed world'. Hence, rather too much of the discussion of environmentalism in the non-western world draws a sharp contrast between the conservation of nature and issues of economic development. While in the developed countries a critical issue for many people may be the conservation of wilderness areas and other environments that are in one way or another distinctively different from mainstream modern life, in many areas of the world the crucial environment is the economic one.

6.4 Conclusion

This chapter has sought to detail, principally though examples drawn from Britain and North America, the presence in varying forms and to varying extents of claims for ecocentric environmentalism. Ecocentric environmentalism is understood to involve calls for the establishment of some form of non-exploitative relationships between society and nature. Attention is paid to the emergence of the Romantic movement in Britain and North America and its valuation and 'spectacularisation' of 'wild nature', or landscapes where seemingly society has had little or no impact. As noted in Chapter 1, such constructions of nature are highly problematic, and through the course of the twentieth century four other 'waves' of environmentalism are identified. In each nature is represented rather differently, and social activities oriented towards this nature have varied. In the 1930s, for instance, a 'technocratic environmentalism' emerged that saw nature in terms of a space that could be clearly differentiated from the modern spaces of industry and urbanity and that should, at the same time, be made accessible to people for recreation, providing that its distinctiveness could be retained. In the 1960s this form of environmentalism was joined by another view of nature, which saw it in a more ecological and less localised sense, although it was still highly compartmentalised, with concern being expressed about the state of quite a range of seemingly distinct natures, such as natural landscapes, natural resources and the flora and fauna of nature. By the 1980s this comparmentalisation had broken down for many people, and there was a concern for 'the environment' or the 'future of nature'. For many commentators this is seen to represent the emergence of a full environmental consciousness, although as mentioned in Chapter 5 such holism may occlude significant differences in both perceptions of and interests in nature. In the final, current phase of environmentalism, some of these issues have been recognised and there has been a turn to, once again, a much more localised vision of nature. Such a vision, however, is not without its problems, in that it may fail to address both global geochemical environmental changes and the presence of localised social inequalities.

Chapter 7

Conclusion

7.1 So where have we got to?

At the start of this book questions were raised about 'What is nature?', 'What is society?' and 'How are they and how should they be interrelated?' Five other questions were also highlighted:

(1) How has the exploitative relationship between society and nature changed over time?
(2) Is the exploitative relationship facing problems of sustainability in the contemporary world?
(3) Is it possible to manage this relationship to overcome or alleviate the problems it is creating?
(4) Why is this relationship established and who benefits from it?
(5) Is it possible to have alternative relationships between society and nature?

In this final chapter we want to return to these questions and give our own views as to how they might be answered. These are simply our views, and you may well have different views that contradict and contest ours. If this is the case, good! In writing this book we wanted to encourage people to reflect on what they take to be nature, how they come to see things as being of nature, what the consequences of so viewing it might be, and how do you, and should you, interact with parts of nature and with parts of society. If you feel the need to take issue with what we have said then we may have gone some way to delivering what we intended.

As we have progressed through writing this book we have come to realise how varied and complex are the potential answers to the seemingly simple questions we set ourselves. This makes writing this conclusion quite daunting, but we feel it is useful, certainly to us and hopefully to you, to consider whether we can draw any general points from our text. We will therefore run briefly through the general issues of society and nature we discussed in Chapter 1, and then look at each of the five questions we set ourselves to answer.

7.2 What is nature?

In Chapter 1 it was argued that the term nature should not be looked on as a singular, static, discrete entity but should be seen as something that is defined, enacted and acted upon in a myriad of different ways. In a sense we have adopted what in

Chapter 1 was described as a 'social constructionist' perspective, although we would add not a 'strict social constructionist' perspective, in that like Castree and Braun (Castree, 1995; Castree and Braun, 1998), and also like many ecocentrics, we do feel that there are parts of our world that are not currently socially recognised, nor amenable to social organisation or control, but that do have important effects and that might reasonably be described as constituting a natural agency. However, we would emphasise that these agents have multiple and changeable characteristics or 'natures': hence we have talked of the 'agencies of natures' rather than 'the agency of nature'.

In the first chapter the concept of society was also discussed. It was noted that society is often seen either in terms of an aggregation of individuals, analysable solely in terms of patterns of individual beliefs and actions, or as some supra- individual entity, setting conditions of how people act. We would resist both these constructions in favour of a more interrelational, perhaps even dialectical, view of the constitution of social life. In this book we have therefore used the terms society and the social to variously refer to social groups or aggregates, social practices, social representations and social relationships. We would also note the need for spatial and historical specificity in social groups, practices and relations: hence we talk of societies rather than society. We would however also suggest that there is value in recognising common elements in and between these social forms.

The overall aim of this book, established in Chapter 1, is therefore to consider the extent to which the elements of the social – social groups, social practices, social representations and social relations – interrelate to each other and with agencies of nature in an exploitative manner or in some other way. In Chapter 1 we highlighted two ways in which we are using the term exploitation: first to mean use and transformation, and second, to imply injustice. We highlighted how exploitative and non–exploitative relations can be seen to underlie O'Riordan's classification of society-nature relations, with his notion of 'technocentrism' being effectively an attempt to justify an exploitative social relation with agencies of nature, while his 'ecocentricism' involves attempts to explain the potentialities of other relationships with such agencies. In the rest of the book we then seek to both illustrate the significance of these differences and to understand both their formation and consequences.

7.3 How has the exploitative relationship between society and nature changed through time?

In the second chapter we focused attention on the history of exploitative relations between societies and natures. It can be argued that people must 'exploit' parts of nature in order to, for example, meet their basic needs such as accessing food, water and heat sources. By the very act of breathing people are both using nature – taking in oxygen – and transforming it – breathing out carbon dioxide. In a sense, therefore, people will always have an 'exploitative' relationship with agencies of nature. However, it has frequently been suggested that there has been a gradual increase in the level of human exploitation of nature over time.

In Chapter 2 we explored four episodes in the history/prehistory of society–nature relations – the emergence of agriculture and industrialisation, imperialism and post-industrialism. We argued that while it is possible to see more and more agencies of nature being socially exploited it is important to, first, recognise that this expansion of exploitation is not the only social relation in operation, in that there are both social relations of exploitation between social agencies (that is between people through their actions, products, representations) and also there are what might be seen as non-exploitative relations between social agencies and nature (and between social agencies). Non-exploitative here means 'non-instrumental', or not interacting with the intention of gaining something, and 'just', in the sense of not gaining something, either intentionally or non-intentionally, at the expense of others. Many ecocentrics have sought to highlight how people interact emotionally, experientially, aesthetically and cognitively with what is seen as nature, and this is sometimes seen as something quite modern, emerging in the Romantic movement as a sort of perverse reaction to heightened exploitation of natures that characterised industrialisation. We would, however, suggest that such non-exploitative relations have a much longer history, extending back into prehistory.

A second cautionary remark we make about constructing a history of growing social exploitation of nature is that this history should not be seen as a clear linear one. At various points, for example, we highlighted how mineral extraction did not simply emerge in the industrial revolutions of the eighteenth and nineteenth century, but had been around for much longer. Similarly we emphasised that the energy sources of the 'organic economy' – plants, animals and humans – did not cease to be of significance with the emergence of industrialisation, nor indeed did the power sources of this 'mineral-based economy' cease to be of significance to post-industrial societies. We would also add that increasing exploitation does not necessarily equate, as technocentric interpretations would imply, greater control over and freedom from the agencies of nature. As Chapter 3 highlighted, there are important questions to be raised as to whether contemporary forms of exploitation can be sustained into the future.

7.4 Is the exploitation of natures sustainable in the contemporary world?

Chapter 3 focused on the notion that societies are heading towards environmental crises. It was argued that two schools of thought have dominated this debate. First, there are the so-called 'neo Malthusians', who firmly believe that humanity is heading for an environmental catastrophe because people are over exploiting the earth's resources (which are finite) and causing widespread environmental damage. The 'cornucopians', however, believe the polar opposite. Chapter 3, therefore, considered whether societies were running out of resources, cannot feed themselves or provide sufficient energy to meet their future demands.

With regard to the question 'is it *possible* for a growing human population to continue to feed itself?' the answer, at a global level, has to be yes. There is the land and the technology to grow sufficient food and there are also the technology and managerial methods to tackle problems in the agricultural sector such as soil erosion, acidification, and pollution from fertilisers, pesticides and insecticides. However, if you ask the question 'will the human population be fed adequately', then the answer is probably no. The so-called developed countries will continue to produce enough food to feed their own people and will probably continue to produce surplus supplies. The situation for some underdeveloped countries is much less positive. There are clearly local problems relating to food production and distribution and there are other worrying trends, especially with renewable resources. Soil erosion, deteriorating water quality and the possibility of regional water shortages need to be addressed in many areas, and the depletion of fish stocks shows how there can be a collapse of seemingly renewable resources. Efforts to substitute this resource with aquacultural methods at present can only produce a minor proportion of ocean fish catches. Despite these concerns, it is too early to decide whether Brown's (1997; 1998) assertion that there is currently a global food crisis is correct. Trends like global grain production have levelled out over the past few years and it is difficult to comment on whether these represent short-term fluctuations or long-term trends. However, they clearly need to be monitored.

With regard to energy, the answer to the question 'is it *possible* to produce sufficient energy for an ever-increasing human population?' is, again, at a global level yes. In the short term there are sufficient supplies of oil, gas and coal, and there is also clearly the potential to exploit other sources of these conventional fuels. Renewable energy technology is also rapidly becoming more efficient and cost effective. Nuclear power can also meet any shortfalls in energy from fossil fuels and/or renewables, although the future expansion of nuclear power is limited by environmental factors rather than its inability to generate vast amounts of energy. To avoid a radical transformation in the type of fuels used to generate energy and alleviate energy demand, societies could make considerable savings by introducing energy efficiency and conservation measures. With regard to non-fuel minerals it is also evident that these resources are, at a global level, not scarce.

The third major theme of Chapter 3 was to review the extent of environmental degradation of local, regional and global scales. It is evident that human action is responsible for creating a series of problems such as pollution, loss of biodiversity and changes to the atmosphere. Thus, so far, there has been an environmental cost from human exploitation. However, while the evidence for human actions contributing to environmental problems such as acid rain and ozone depletion is compelling, some research has begun to question whether the extent of human influence is as great as previously thought, and has argued that natural agencies may be equally or more significant. The result of some human action may also prove beneficial. For example some of the predicted changes to terrestrial ecosystems as a result of global warming may be advantageous to some societies, with increasing temperatures and atmospheric carbon dioxide levels leading to increase plant growth and crop yields.

Overall there is no convincing evidence to suggest that at a global level society is close to reaching its limits to growth. However, this does not mean people can afford to be complacent about the environment and its resources, especially soil and water. Surely the most sensible option is to try and alleviate pressure on the natural environment and its resources when it is possible to do so. People have developed the technology and the managerial methods to exploit natures more efficiently and sustainably.

7.5 Is it possible to manage this relationship to overcome or alleviate the problems it is creating?

It is possible that society can alleviate many of the contemporary environmental problems outlined in Chapter 3. Chapter 4 demonstrated that people have produced a range of methods to manage agencies of nature. It is currently possible, for example, to prevent the release of acid rain pollutants by the use of filters, clean coal technology, low sulphur coal, renewable energy or nuclear power, energy efficiency measures or liming acidified lakes. There is evidence that measures to combat acid rain are having some success (Montzka *et al.*, 1996). Similar mitigation options are available to lower greenhouse gas emissions, lower rates of soil erosion and reduce other forms of pollution. The success of the protocols phasing out CFCs and other ozone-destroying chemicals shows that environmental problems can be tackled at a global level.

On the basis of the evidence presented in Chapters 3 and 4 contemporary societies are reasonably good at reacting to environmental problems but poor at being proactive, of taking action to prevent a problem from occurring in the first place. People are seemingly often quite prepared to release chemicals into the physical environment without considering the environmental consequences of that action, even though scientists and environmental managers have developed strategies such as environmental impact assessments and risk assessment to forecast and/or predict environmental impacts. Furthermore, Chapter 4 was littered with examples where agreements, regulations, laws, taxes, fines and bans have failed to stop environmentally damaging behaviour.

Such inactivity in implementing environmental policy and management strategies – which we describe as 'the implementation gap' in Chapter 5 – clearly introduces serious doubt as to whether the possibilities of dealing with environmental degradation will be fully realised. Key questions are therefore 'why is this?' and 'can anything be done to change it?'. As we mentioned in Chapter 5 there are two common responses to perceived implementation gaps: first, some people argue that implementation failures stem from the design of policies, while others argue that the problems stem less from the design of the policies and more from how these policies are implemented.

In Chapter 4 we highlighted examples of both sets of problems. So far, for example, policies dealing with environmental problems have begun to move away from a simple technology-based approach to a more holistic approach. There are two key elements to this holism. First, nature is seen as a set of interacting elements, often operating across a variety of scales. Second social, economic, political and natural factors often operate in combination (Oppenheimer, 1995). For many the recognition of these two sets of factors has led to improvement in policy. On the other hand holistic policy development may have a tendency to slide into 'scientism', in that it may involve the creation of a series of expert knowledges that are not available to everybody. This knowledge can have powerful effects when integrated into policy making, but as we argued in Chapter 4 it may contain a high degree of uncertainty and contradiction. So, for example, scientists do not fully understand all the effects of oil spills or other pollutants.

In Chapter 4 we also highlighted how environmental policies may flounder because of conflicts of interest. Many politicians in underdeveloped countries are reluctant to adopt environmental regulations because they are worried about losing international trade and the perceived economic benefits. So in India, for example, the government has allowed multinational corporations to draw up agreements that absolve them from any liability in the event of an industrial accident, such as the release of poisonous gases as occurred at Bhopal. Unless environmental protection forms part of international trade agreements like the General Agreement on Tariffs and Trade (GATT), it seems inevitable that poorer countries will capitulate to the demands and environmentally damaging actions of multinational corporations and foreign investors. International and global organisations such as the European Union, World Bank and United Nations may be seen to be instrumental in ensuring co-operation between countries to mitigate against environmental exploitation. However, as was shown in Chapter 5, these policies have often been highly technocentric in focus and structured to benefit the interests of powerful governments and institutions at the expense of weaker ones.

7.6 Why do societies exploit nature unsustainably and who benefits?

In Chapter 5 we argued that both the design and implementation of environmental policy making, and responses to it, need to be understood in social context. In particular we suggested that capitalistic social relations may have inherent tendencies that encourage people to over-exploit the agencies of nature. We illustrated this point by looking at the reasons for the creation of such environmental problems as deforestation, famine and industrial pollution, and suggested that much of their origin and persistence stems from the predominance of capitalistic social relations.

This account clearly raises a further question, namely why capitalist relations should be so exploitative of nature. A range of reasons for this can be advanced. First it can be argued that capitalist social relations condense and augment the transformative capacity people have on agencies of nature. This point is made by Marxists such as N. Smith (1984, pp. 35–60), who argues that capitalism should be seen as a specific form of production and that to understand society–nature relations one needs to appreciate both the characteristics of 'production in general' and the 'specific characteristics of capitalist production'. More specifically, Smith argues that Marx sees production as fundamentally involving the transformation of natures:

> Labour is, first of all, a process between man [*sic*] and nature, a process by which man, through his own actions, mediates, regulates and controls the metabolism between himself and nature. He confronts the materials of nature as a force of nature. He sets in motion the natural forces which belong to his own body, his arms, legs, head and hands, in order to appropriate nature's production in a form adapted to his own needs. Through this movement he acts upon external nature and changes it, and in this way he simultaneously changes his own nature (Marx, 1976, p. 177).

In other words, and omitting the gendered language of Marx, people adopt an instrumental relationship with things beyond them and in doing so they simultaneously appropriate elements of that world – gaining energy, motion, foodstuffs, clothing and so on – change that world, and impact their own selves. Three forms of nature are hence created through the labour process: that which is appropriated as being of value, that which is created as a material consequence of human labour, and nature of people themselves.

These three aspects of produced nature are all still seen to be of significance within capitalist societies, but they take on particular forms. As we have argued in Chapter 2, the modern societies have seen an expansion in the agencies of nature that are socially appropriated: societies no longer just appropriate the air people breathe or the plants or animals they eat, but now their appropriation extends from minerals far underground through to the chemicals high up in the stratosphere and into the imagery of nature as well. Furthermore the scale of appropriation has grown immensely too: much of nature is consumed at places far from where it was first appropriated. This expansion in the appropriation of nature is not, as noted in Chapter 5, coincidental to the emergence of capitalism but can be seen as a concomitant consequence of it: capitalism is production for accumulation, and if accumulation is to continue then production has to continue. As we noted in Chapter 5 under capitalism things, including such basic requirements as foodstuffs, are produced for exchange, not for need; and even if needs are satisfied production needs to continue if capitalist social relations or capital accumulation are to continue. This means that there is both a tremendous draw on appropriating nature as a productive resource and also a tendency for innovation in the uses to which the agencies of nature are put. This latter tendency can in part be seen as the consequence of the success of capitalist-oriented production in appropriating nature and fulfilling human 'needs'. In other words, once societies are able to appropriate the agencies of nature to produce food to satisfy people's hunger, they may then wish to use the agencies of nature to produce other things as well although, as was shown in Chapter 5, given the unequal distribution of social power, agencies of nature are often appropriated for new uses at the expense of the production of foodstuffs. Indeed innovations in the use of nature may more generally be seen to relate to the significance of competition with capitalism: greater levels of accumulation may well be had in finding new uses for nature rather than trying to compete with a large number of other producers involved in long-established uses of nature.

As we have already argued in a previous section, along with the growing material appropriation of nature has often come increased material transformation. The material transformations may act as a constraint on the future of capitalist production, although as we have argued in the previous section we would suggest that globally capitalism is far from reaching any 'natural limits'. We would also, however, concur with Castree and Braun's (1998, p. 9) assessment that: 'The competitive and accumulative imperatives of capitalism bring all manner of natural environments and concrete labour processes ... together in an abstract framework of market exchange which, literally, produces nature(s) anew'. In other words, natures are socially constructed by labour in general and by capitalism in particular, arguably to the extent that there is no earthy nature untouched by this (see Smith, N., 1984; 1998).

According to Neil Smith (1984) the transformation of human nature under capitalism can be seen to centre on the forms of 'alienation' it creates. In particular Smith highlights how growing divisions of labour associated with the expansion of capitalist social relations leads to changing relationships with nature in the practices of labour; people no longer utilise a range of skills to appropriate nature and create a finished product, but rather perform only a small part of the labour put into manufacturing a product, which is finished and sold off without any involvement of those who made it. These are clearly important changes, but there is also clearly a danger of over-generalisation. Labour processes come in a myriad of forms and have not necessarily moved from undifferentiation to specialisation, as Smith's arguments might be seen to suggest. Furthermore, as Smith (p. 53) himself notes, the rising differentiation of labour can have seemingly opposing consequences, leading for example both to the 'homogenisation' and a 'differentiation' of cultures, while there is also further cultural differentiation through other lines of social difference such as gender, sexuality, race and locality. It is therefore very difficult to trace clearly the impact that the exploitation of external natures may have on human natures, although in the course of this book we have highlighted some points of connection. In Chapter 1, for example, we highlighted the arguments of ecofeminists who have suggested that gender differences may stem from, and act to create, different ways of relating with nature. So, for example, it has been argued that women are 'more natural' and 'closer to nature' due to experiences of nurturing children, animals and plants, and because of this experience they adopt a more intimate and caring, ecocentric, attitude towards nature than men. In Chapter 2 we also highlighted how tourists might relate differently to nature: for some tourism is about getting in touch with nature – experiencing it physically – while others may be said to travel around in an 'environmental bubble' that isolates them from everything but carefully staged spectacles of nature, and yet others immerse themselves in symbolic images of nature. In Chapter 2, we also highlighted how European imperialism had severe impacts on the biological nature of plants, animals and people, leading in many cases to the destruction of whole communities, of both human and non-human kinds.

In Chapters 2 and 5 we also highlighted how the exploitation of people is often conducted through notions of nature. Examples mentioned included the exploitation of people labelled as 'savage' and thereby seen as behaving like animals or being less than human (see Chapter 2). Sometimes these people were exploited directly, as in the case of the slave trade; in other instances they were exploited indirectly, in that they were removed from areas of nature that were then appropriated by others. The removal could occur physically or through economic mechanisms such as taxation or the land market (see Chapter 5), or indeed through the agency of disease (see Chapters 2 and 5). Another illustration of social exploitation through the valuing of nature discussed in Chapter 5 was the way that the rhetoric and practices of contemporary 'global environmentalism' involves a prioritisation of particular constructions and uses of nature at the expense, both symbolically and practically, of others. There are, as Chapter 6 highlighted, also more localised empowerments and dis-empowerments of constructions of nature with there being, in Britain and North America at least, five discernible 'waves' of environmentalism, each of which values and enacts particular natures. Hence, there was a 'Romantic' concern with places of wild nature; a 'technocratic' concern with creating reserves immune from modernisation and yet accessible for people living in and seeking temporary refuge from modernity; 'ecological' constructions of natures that sought to create spaces for the reproduction of a range of valued natures such as natural landscapes, natural resources and the flora and fauna of nature; a concern for 'globalised' nature that argued for a unified nature under threat of world-wide change; and 'localised' visions of nature under threat from mainstream and increasingly globalised society. These constructions of nature have emerged in contestation both with a technocentric/capitalist concern to exploit natures simply for privatised accumulation, and also, to some extent, with each other. As particular constructions of nature have become more widely recognised and acted upon, so other constructions of nature have declined in their effects. At present there is some plurality of ways in which natures are being defined and enacted, although the technocentric/capitalist focus on using nature as a productive force often prevails.

7.7 Is it possible to have alternative relationships between societies and natures?

The dominance of capitalist technocentrism in constructions of nature raises the issue of whether or not it is possible to have alternative relations between people and natures. We have in part already answered this question, in that we have just highlighted the presence of a range of ways of considering and enacting nature. Even in countries that are quite explicitly capitalist in character there are many people who have long argued against simply seeing nature as a resource for private accumulation. In Chapter 1 we criticised people who saw nature, and indeed societies, as having a singular dimension, and we would restate that argument now. Indeed, while we have previously broadly endorsed a Marxist political economy interpretation of society–nature relations, we would also concur with those such as Eckersley (1992) who have criticised Marxism for having an overly reductionist view of society–nature relations.

As noted above, Marx sees labour as the centre of society–nature relations: it is labour that allows both the use and exploitation of nature, it is labour that produces environmental transformations and it is labour that creates human consciousness. As Eckersley comments, Marx views humans as being effectively *homo faber* (working man). She also notes, however, that this concept has been criticised by both later neo-Marxist writers and also by ecocentrics. Hence, for example, critical theorists such as Horkheimer, Adorno, Marcuse and Habermas argued that human life was, or at least should be, about more than performing labour. Habermas (1979), for example, argues for a reconstruction of Marxist theory away from an exclusive emphasis on labour towards a recognition of the communicative aspects of social life. He argues, contra to Marx, that the performance of labour does not mark out humans from animals but rather it is the performance of communicative action through language. Hence, for Habermas, the basis of human life is both labour, or instrumental action, and language, or communicative action. He goes on to argue in a series of other writings (eg Habermas, 1984, 1987) that in modern, capitalist society, instrumental activity is exercising too much influence and is 'colonising' areas of social life at the expense of expressive and communicative actions. As a consequence people are becoming treated in a manner analogous to the way agencies of nature are treated in technocentrism: as a legitimate object of exploitation. One might hence say that there is a 'naturalisation' of social exploitation.

Eckersley, however, argues that critical theorists such as Habermas do not go far enough in their critique of the productivist focus of Marx. Eckersley (1992, p. 110) argues, for instance, that while Habermas is concerned to challenge the naturalisation of social exploitation he is happy to endorse the social exploitation of 'non-human nature'. She continues by arguing that there is a 'need to revise and extend Habermas's communicative ethics to a full-blown ecocentric ethics' (Eckersley, 1992, p. 112) which involves:

> the recognition of the *relative* autonomy and unique mode of being of the myriad life-forms that make up the non-human world ... To ecocentric theorists, freedom or self-determination is recognised as a legitimate entitlement of *both* human and non-human life forms... an ecocentric political theory is 'emancipation writ large' – the maximisation of the freedom of *all* entities to unfold or develop in their own ways (Eckersley, 1992, p. 91).

Like Pepper (1993) we have some doubts as to whether Eckersley's 'full-blown ecocentric ethic' is realisable or desirable. It seems to us, for example, to beg such questions as do 'we', where the 'we' does include humanity and parts of non-human nature such as plants and animals, really want entities such as viruses that cause disease to have a maximal ability to unfold in their own ways? And, relatedly, would the unfolding of one agency of nature not have detrimental impacts on the emancipation of others? On the other hand, we would also argue that *purely* instrumental relations towards the agencies of nature are neither necessary, beneficial nor fully enacted. Rather, drawing on Eckersley (1992, p. 114), we would suggest there is room for, and many benefits to be gained from, a range of agencies of nature, if societies managed to infuse and temper their instrumental relations with natures with 'normative considerations concerning human well-being and respect for other life forms'.

References

Abramovitz, J.N. (1997) Valuing nature's services. In L.R. Brown *et al. State of the world, 1997* (London: Earthscan), 95–114.

Abrams, E.M., Freter, A.-C., Rue, D.J. and Wingard, J.D. (1996) The role of deforestation in the collapse of the Late Classic Maya State. In L.E. Sponsel, T.N. Headland and R.C. Bailey (eds) *Tropical deforestation* (New York: Columbia University Press), 55–75.

Adam, W. (1851) *Gem of the Peak* (Derby: J. & C. Mozley, 5th edn).

Adams, W. (1990) *Green development: environment and sustainability in the third world* (London: Routledge).

Adams, W. (1997) Rationalization and conservation: ecology and management of nature in the United Kingdom. *Transactions, Institute of British Geographers* 22(3), 277–91.

Addiscott, T.M., Whitmore, A.P. and Powlson, D.S. (1991) *Farming, fertilisers and the nitrate problem* (Wallingford: CAB International).

Adelman, M.A. (1995) Trends in the price and supply of oil. In J.L. Simon (ed.) *The state of humanity* (Oxford: Blackwell), 287–93.

Afusa, S.A. (1991) Indigenous people can save rainforests. In M. Polesetsky (ed.) *Global resources: opposing viewpoints* (San Diego: Greenhaven Press), 156–61.

Agg, J. and Phillips, M. (1998) Neglected gender dimensions of rural social restructuring. In M. Boyle and K. Halfacree (eds) *Migration into rural areas: theories and issues* (London: Wiley), 252–79.

Agnew, C. and Anderson, E. (1992) *Water resources in the arid realm* (London: Routledge).

Aks, S.E., Erickson, T.B., Branches, F.J.P. and Hryhorczuk, D.O. (1995) Blood-mercury concentrations and renal biomarkers in Amazonian villagers. *Ambio* 24(2), 103–5.

Albury, D. and Schwartz, J. (1982) *Partial progress: the politics of science and of technology* (London: Pluto Press).

Allen, E. (1992) Jungle geopolitics. *Geographical Magazine* April, 28–32.

Allott, T.E.H., Harriman, R. and Battarbee, R.W. (1992) Reversibility of lake acidification at the Round Loch of Glenhead, Galloway, Scotland. *Environmental Pollution* 77, 219–25.

Altvater, E. (1993) *The future of the market: an essay on the regulation of money and nature after the collapse of actually existing socialism* (London: Verso).

Alvares, C. and Billorey, R. (1987) Damning the Narmada: the politics behind dam construction. *The Ecologist* 17(2), 62–73.

American Petroleum Institute (1994) Petroleum companies work to reduce pollution. In C.P. Cozic (ed.) *Pollution* (San Diego: Greenhaven Press), 104–12.

Ames, B.N. (1995) Pesticides, cancer and misconceptions. In J.L. Simon (ed.) *The state of humanity* (Oxford: Blackwell), 588–94.

Ammerman, A.J. and Cavalli-Sforza, L.L. (1979) The wave of advance model for the spread of agriculture in Europe. In C. Renfrew and K.L. Cooke (eds) *Transformations: mathematical approaches to culture change*, (London: Academic Press), 275–94.

Ammerman, A.J. and Cavalli-Sforza, L.L. (1984) *The Neolithic transition and the genetics of population in Europe* (Princeton: Princeton University Press).

Anderson, A. (1997) *Media, culture and the environment* (London: UCL Press).

Anderson, C.K. (1995) Interim spent fuel management: 1995 update. *Nuclear Engineering International* 40(488), 34.

Anderson, I. (1992) Dangerous technology dumped on Third World. *New Scientist*, 7 **March**, 9.

Anderson, J.G., Toohey, D.W. and Brune, W.H. (1991) Free radicals within the Antarctic vortex: the role of CFCs in Antarctic ozone loss. *Science* 251, 39–46.

Anderson, R.N. (1998) Oil production in the 21st century. *Scientific American* 278(3), 68–73.

Anderson, T.L. (1995) Water, water everywhere but not a drop to sell. In J.L. Simon (ed.) *The state of humanity* (Oxford: Blackwell), 425–33.

Andres, R.J., Marland, T., Boden, T. and Bischoff, S. (1994) Carbon dioxide emissions from fossil fuel consumption and cement manufacture 1751 to 1991 and an estimate for their isotopic composition and latitudinal distribution. In T.M.L.Wigley and D. Schimel (eds) *The carbon cycle* (Cambridge: Cambridge University Press).

Anisimov, O.A. and Nelson, F.E. (1996) Permafrost distribution in the northern hemisphere under scenarios of climate change. *Global and Planetary Change* 14, 59–72.

Anisimov, O.A., Shiklomanov, N.I. and Nelson, F.E. (1996) Global warming and active-layer thickness: results from transient general circulation models. *Global and Planetary Change* 15, 61–77.

Anon. (1992) The best global warming treaty yet. *Nature* 357, 97–8.

Anon. (1995a) Making nuclear power usable again. *Nature* 375, 91–2.

Anon. (1995b) Berlin and global warming policy. *Nature* 374, 199–200.

Anon. (1995c) The great Berlin greenhouse compromise. *Nature* 374, 749–50.

Anon. (1995d) Disposing of nuclear waste. *Scientific American* 273(3), 143.

Anon. (1997a) Frankly, my dear, I don't want a dam. *Scientific American* 277, 11.

Anon. (1997b) Japan set to miss target to cut carbon emissions. *Nature* 387, 447.

Anon. (1997c) Seizing global warming as an opportunity. *Nature* 387, 637.

Antarctic Treaty (1959) *Text of the Antarctic Treaty*.

Antarctic Treaty (1991a) *Convention on the Regulation of Antarctic Mineral Activities*.

Antarctic Treaty (1991b) *Protocol on Environmental Protection*.

Antarctic Treaty Consultative Committee (1991) *Report of the XVIth Antarctic Consultative Committee Meeting*.

ApSimon, H.M. and Warren, R.F. (1996) Transboundary air pollution in Europe. *Energy Policy* 24(7), 631–40.

Archer, J. (1994) Policies to reduce nitrogen losses from water from agriculture in the United Kingdom. *Marine Pollution Bulletin* 29, 444–9.

Ardö, J., Lambert, N., Henzlik, V. and Rock, B.N. (1997) Satellite-based estimation of coniferous forest cover changes: Krušné Hory, Czech Republic 1972–1989. *Ambio* 26(3), 158–66.

Arieh, E. and Merzer, A.M. (1974) Fluctuations in oil-flow before and after earthquakes. *Nature* 247, 534–5.

Arnold, D. (1996) *The problem of nature: environment, culture and the European expansion* (Oxford: Blackwell).

Arnold, J. (1987) Deforestation. In D. McLaren and B. Skinner (eds) *Resources and world development* (Chichester: Wiley), 711–25.

Arntzen, J. (1995) Economic instruments for sustainable resource management: the case of Botswana's water resources. *Ambio* 24(6), 335–42.

Asian Development Bank (1982) *Nepal agricultural sector* (Manila: Asian Development Bank).

Avery, D. (1995) The world's rising food productivity. In J.L. Simon (ed.) *The state of humanity* (Oxford: Blackwell), 378–91.

Aweto, A.O. and Adejumbobi, D.O. (1991) Impact of grazing on soil in the southern Guinea savannah zone of Nigeria. *The Environmentalist* 11, 27–32.

Bache, B.W. (1985) Soil acidification and aluminium mobility. *Soil Use and Management* 1(1), 10–14.

Bagarinao, T. (1998) Nature parks, museums, gardens and zoos for biodiversity conservation and environmental education. *Ambio* 27(3), 230–7.

Bahro, R. (1982) *Socialism and survival* (London: Heretic Books).

Bai, H. and Wei, J.-H. (1996) The CO_2 mitigation options for the electric sector. *Energy Policy* 24(3), 221–8.

Bailey, R. (1994) Scientists exaggerate the extent of ozone depletion. In C.P. Cozic (ed.) *Pollution* (San Diego: Greenhaven Press), 50–5.

Baker, C. (1991) Tidal power. *Energy Policy* 19(8), 792–7.

Baker, C.V. (1998) Forest resource scarcity and social conflict in Indonesia. *Environment* 40(4), 5–9 and 28–37.

Baker, R. (1987) Linking and sinking: economic externalities and the persistence of destitution and famine in Africa. In Glantz, M. (ed.) *Drought and hunger in Africa* (Cambridge: Cambridge University Press), 149–68.

Balling, Jr., R.C. (1994) Interpreting the global temperature record. *Economic Affairs* 14(3), 18–21.

Bandyopadhyay, J. (1992) From environmental conflicts to sustainable mountain transformation: ecological action in the Garhwal Himalaya. In D. Ghai and J. Vivian (eds) *Grassroots environmental action: people's participation in sustainable development* (London: Routledge), 259–78.

Barbier, F. (1989) *Economics, natural resources scarcity and development* (London: Earthscan).

Barker, G.W. (1985) *Prehistoric farming in Europe* (Cambridge: Cambridge University Press).

Barker, T. (1995) Taxing pollution instead of employment: greenhouse gas abatement through fiscal policy in the UK. *Energy and Environment* 6(1), 1–29.

Barker, T. (1997) Taxing pollution instead of jobs. In T. O'Riordan (ed.) *Ecotaxation* (London: Earthscan), 163–200.

Barnes, N. (1996) Conflicts over biodiversity. In P. Sloep and A. Blowers (eds) *Environmental policy in an international context: conflicts* (London: Arnold), 217–41.

Barnett, H.J., van Muiswinkel, G.M., Shechter, M. and Myers, J.G. (1984) Global trends in non-fuel minerals. In J.L. Simon and H. Kahn (eds) *The resourceful earth* (Oxford: Basil Blackwell), 316–38.

Barratt-Brown, M. (1974) *The economics of imperialism* (London: Penguin).

Barrett, F. (1989) On the Algarve's road to ruin. *The Independent*, 22 July, 45.

Barrow, C.J. (1995) *Developing the environment: problems and management* (Harlow: Longman Scientific and Technical).

Bartlett, K.B. and Harriss, R.C. (1993) Review and assessment of methane emissions. *Chemosphere* 26, 261–320.

Bateman, I. (1993) Valuation of the environment, methods and techniques: the contingent valuation method. In R.K. Turner (ed.) *Sustainable environmental economics and management: principles and practice* (London: Belhaven), 192–267.

Bateman, I. (1995) Environmental and economic appraisal. In T. O'Riordan (ed.) *Environmental Science for environmental management* (Harlow: Longman), 45–65.

Battarbee, R.W. (1989) Geographical research on acid rain. I. The acidification of Scottish lochs. *The Geographical Journal* 155, 353–60.

Battarbee, R.W., Flower, R.J., Stevenson, A.C., Jones, V.J., Harriman, R. and Appleby, P.G. (1988) Diatom and chemical evidence for reversibility of acidification of Scottish lochs. *Nature* 332, 530–2.

Battarbee, R.W., Allott, T.E.H., Juggins, S., Kreiser, A.M., Curtis, C. and Harriman, R. (1996) Critical loads of acidity to surface waters: an empirical diatom-based palaeolimnological model. *Ambio* 25(5), 366–9.

Baumol, W.J. and Oates, W.E. (1995) Long-term trends in environmental quality. In J.L. Simon (ed.) *The state of humanity* (Oxford: Blackwell), 444–60.

Bawa, K.S. and Dayanandan, S. (1997) Socioeconomic factors and tropical deforestation. *Nature* 386, 562–3.

Bayliss-Smith, T. and Wanmali, S. (1984) (eds) *Understanding green revolutions: agrarian change and development planning in South Asia* (Cambridge: Cambridge University Press).

Bean, M.J. (1994) Legislative and public agency initiatives in ecosystem and biodiversity conservation. In Ke Chung Kim and R.D. Weaver (eds) *Biodiversity and landscapes: a paradox of humanity* (Cambridge: Cambridge University Press), 381–9.

Beardsley, T. (1995) Rio redux. *Scientific American* 272(6), 19.

Beaumont, J. (1993) The greening of the car industry. *Environment and Planning A* 25, 909–22.

Beck, P. (1990) Antarctica enters the 1990s: an overview. *Applied Geography* 10(4), 247–63.

Beck, U. (1992) *Risk society: towards a new modernity* (London: Sage).

Beck, U. (1995) *Ecological politics in an age of risk* (Cambridge: Polity Press).

Beck, U. (1996) When experiments go wrong. *The Independent*, 26 March.

Bedoya, E. and Klein, L. (1996) Forty years of political ecology in the Peruvian upper forest: the case of upper Huallaga. In L.E. Sponsel, T.N. Headland and R.C. Bailey (eds) *Tropical deforestation: the human dimension* (New York: Columbia University Press), 163–86.

Beerling, D.J. and Woodward, F.I. (1996) In situ gas exchange responses of boreal vegetation to elevated CO_2 and temperature: first season results. *Global Ecology and Biogeography Letters* 5, 117–27.

Behre, K.-E. (1981) The interpretation of anthropogenic indicators in pollen diagrams. *Pollen et Spores* 23, 225–45.

Behre, K.-E. (1988) The role of man in European vegetation history. In B. Huntley and T. Webb III (eds) *Vegetation history* (Kluwer Academic Publishers), 633–72.

Beisner, C. and Simon, J.L. (1995) Editor's appendix. In J.L. Simon (ed.) *The state of humanity* (Oxford: Blackwell), 469–75.

Bekki, S., Tourmi, R. and Pyle, J.A. (1993) Role of sulphur photochemistry in tropical ozone changes after the eruption of Mount Pinatubo. *Nature* 362, 331–3.

Beland, P. (1996) The Beluga whales of the St Lawrence river. *Scientific American* 274(5), 58–65.

Bell, D. (1973) *The coming of the post-industrial society* (London: Heinemann).

Bell, M. (1994) *Childerley: nature and morality in a country village* (Chicago: Chicago University Press).

Bell, M. and Boardman, J. (eds) (1992) *Past and present soil erosion* (Oxford: Oxbow Books).

Bell, M. and Walker, M.J.C. (1992) *Late quaternary environmental change* (Harlow: Longman).

Bell, P.R.F. and Elmetri, I. (1995) Ecological indicators of large-scale eutrophication in the Great Barrier Reef lagoon. *Ambio* 24(4), 208–15.

Bellos, A. (1996a) By-pass march claims protest record. *The Guardian*, 2 February.

Bellos, A. (1996b) By-pass protest stops digger in new victory. *The Guardian*, 11 January.

Bender, W.H. (1997) How much food will we need in the 21st century? *Environment* 39(2), 7–11 and 27–8.

Bennetts, M. (1995) Modelling climate change, 1860–2050. *The Globe* 24, 2–4.

Benton, L.M. (1996) The greening of free trade? The debate about the North American Free Trade Agreement (NAFTA) and the environment. *Environment and Planning A* 28, 2155–77.

Berglund, B.E. (1985) Early agriculture in Scandinavia: research problems related to pollen-analytical studies. *Norwegian Archaeological Review* 18, 77–105.

Berglund, B.E. (1986) The cultural landscape in a long-term perspective. Methods and theories behind research on land use and landscape dynamics. *Striae* 24, 79–87.

Bergmann, K.H., Hecht, A.D. and Schneider, S.H. (1981) Climate models. *Physics Today* October, 44–51.

Bernstein, H., Hewitt, T. and Thomas, A. (1992a) Capitalism and the expansion of Europe. In T. Allen and A. Thomas (eds) *Poverty and development in the 1990s* (Oxford: Clarendon Press), 168–84.

Bernstein, H., Johnson, H. and Thomas, A. (1992b) Labour regimes and social change under colonialism. In T. Allen and A. Thomas (eds) *Poverty and development in the 1990s* (Oxford: Clarendon Press), 185–203.

Best, J. (1989) Afterword: extending the constructionist perspective – a conclusion and an introduction. In J. Best (ed.) *Images of issues: typifying contemporary problems* (New York: Aldine de Gruyter).

Betsill, M.M., Glantz, M.H. and Crandall, K. (1997) Preparing for El Niño. What role for forecasts? *Environment* 39(10), 6–13 and 26–9.

Bevington, R. and Rosenfeld, A.H. (1990) Energy for buildings and homes. *Scientific American* 263(3), 77–86.

Bhagwati, J. (1993) The case for free trade. *Scientific American* 269(5), 17–23.

Binns, T. (1990) Is desertification a myth? *Geography* 75, 106–13.

Bird, C. (1991) Medicines from the rainforest. *New Scientist* 131, No 1782, 34–9.

Birks, H.J.B. (1998) Long term ecological change in the British Isles. In B.M. Usher and D.B.A. Thompson (eds) *Ecological change in the uplands* (Oxford: Blackwell), 37–56.

Birks, H.J.B. (1989) Holocene isochrone maps and patterns of tree spreading in the British Isles. *Journal of Biogeography* 16, 503–40.

Birks, H.J.B. and Birks, H.H. (1980) *Quaternary palaeoecology* (London: Arnold).

Bishop, K.H. and Hultberg, H. (1995) Reversing acidification in a forested ecosystem: the Gärdsjon covered catchment. *Ambio* 24(2), 98–102.

Blaikie, P. (1985) *The political economy of soil erosion in developing countries* (Harlow: Longman).

Blaikie, P. and Brookfield, H. (1987a) Colonialism, development and degradation. In P. Blaikie and H. Brookfield (eds) *Land degradation and society* (London: Methuen), 100–21.

Blaikie, P. and Brookfield, H. (1987b) (eds) *Land degradation and society* (London: Methuen).

Blaikie, P. and Brookfield, H. (1987c) Management, enterprise and politics in the development of the tropical rainforests. In P. Blaikie and H. Brookfield (eds) *Land degradation and society* (London: Methuen), 157–76.

Blaustein, A.R. and Wake, D.B. (1995) The puzzle of declining amphibian populations. *Scientific American* 272(4), 56–61.

Bleken, M.A. and Bakken, L.R. (1997) The nitrogen cost of food production: Norwegian society. *Ambio* 26(3), 134–42.

Blowers, A. (1996) Transboundary transfers of hazardous and radioactive wastes. In P.B. Sloep and A. Blowers (eds) *Environmental policy in an international context: environmental problems as conflicts of interest 2* (London: Arnold), 151–86.

Blowers, A. and Glasbergen, P. (1996) The search for sustainable development. In P. Glasbergen and A. Blowers (eds) *Environmental policy in an international context: perspectives* (London: Arnold), 31–58.

Blowers, A. and Lowry, D. (1987) Out of sight: out of mind: the politics of nuclear waste in the UK. In A. Blowers and D. Pepper (eds) *Nuclear power in crisis* (London: Croom Helm), 129–63.

Blunden, J. and Curry, N. (1990) *A people's charter? Forty years of the National Parks and Access to the Countryside Act, 1949* (London: HMSO).

BMA (British Medical Association) (1991) *Hazardous wastes and human health* (London: BMA).

Boardman, J. (1991) Land use, rainfall and erosion risk on the South Downs. *Soil Use and Management* 7, 34–8.

Bock, Y., Agnew, D.C., Fang, P., Genrich, J.F., Hager, B.H., Herring, T.A., Hudnut, K.W., King, R.W., Larsen, S., Minster, J.B., Stark, K., Wdowinski, S. and Wyatt, F.K. (1993) Detection of crustal deformation from the Landers earthquake sequence using continuous geodetic measurements. *Nature* 361, 337–40.

Boehmer-Christiansen, S. (1986) An end to radioactive waste disposal at sea? *Marine Policy* 10(2), 119–31.

Boerner, C. and Chilton, K. (1994) False economy: the folly of demand-side recycling. *Environment* 36(1), 6–15 and 32–3.

Bongaarts, J. (1996) Population pressure and the food supply system in the Developing World. *Population and Development Review* 22(3), 483–503.

Bookchin, M. (1979) Ecology and revolutionary thought. *Antipode* 10/11, 3/1, 21–32.

Boorstin, D. (1964) *The image: a guide to pseudo events in America* (New York: Harper).

Boraiko, A. (1986) Earthquake in Mexico. *National Geographic* 165, 655–75.

Boserup, E. (1965/1993) *The conditions of agricultural growth: the economics of agrarian change under population pressure* (Chicago: Aldine).

Boserup, E. (1981) *Population and technological change: a study of long-term trends* (Chicago: Chicago University Press).

Boyd, R., Niskavaara, H., Kontas, E., Chekushin, V., Pavlov, V., Often, M. and Reimann, C. (1997) Anthropogenic noble-metal enrichment of topsoil in the Monchegorsk Area, Kola Peninsula, north-west Russia. *Journal of Geochemical Exploration* 58, 283–90.

Boyle, S. (1989) More work for less energy. *New Scientist* No 1676, 37–40.

Bradley, P.N. and Carter, S.E. (1989) Food production and distribution – and hunger. In R.J. Johnston and P.J. Taylor (eds) *A world in crisis* (Oxford: Blackwell), 101–25.

Brady, N.C. and Weil, R.R. (1999) *The nature and properties of soils* (London: Macmillan, 12th edn).

Bramwell, A. (1989) *Ecology in the twentieth century: a history* (London: Yale University Press).

Brasseur, G. (1992) Volcanic aerosols implicated. *Nature* 359, 275–6.

Brazaitis, P., Watanabe, M.E. and Amato, G. (1998) The Caiman trade. *Scientific American* 278(3), 52–8.

Brenner, R. (1977) The origins of capitalist development: a critique of neo-Smithian Marxism. *New Left Review* 104, 25–92.

Brewer, A. (1980) *Marxist theories of imperialism: a critical survey* (London: Routledge and Kegan Paul).

Bridges, E.M. (1991) Dealing with contaminated soils. *Soil Use and Management* 7(3), 151–8.

Bridges, E.M. and Morgan, H. (1990) Dereliction and pollution. In Griffiths, R.A. (ed.) *The city of Swansea: challenges and change* (Stroud: Allan Sutton Publishing), 270–90.

Briguglio, L., Butler, R., Harrison, D. and Filho, W.L. (eds) (1996) *Sustainable tourism in islands and small states: case studies* (London: Pinter).

Bristow, T. (1993) Environmental awareness and societal change in the UK. In S. Harper (ed.) *The greening of rural policy: international perspectives* (London: Belhaven Press), 42–63.

Broecker, W.S. (1992) *The glacial world according to Wally (draft), III. Records* (Broecker: Columbia University, Palisades).

Bromley, R. and Humphrys, G. (eds) (1979) *Dealing with dereliction: the redevelopment of the Lower Swansea Valley* (Swansea: University College of Swansea).

Brookfield, H. (1975) *Interdependent development* (London: Methuen).

Brookfield, H. and Padoch, C. (1994) Appreciating agrodiversity. A look at the dynamism and diversity of indigenous farming practices. *Environment* 36(5), 6–11 and 37–44.

Brookfield, H., Lian, F.J., Kwai-Sim, L. and Potter, L. (1990) Borneo and the Malay Peninsula. In B.L. Turner II, W.C. Clark, R.W. Kates, J.F. Richards, J.T. Matthews and

W.B.Meyer (eds) *The earth as transformed by human action* (Cambridge: Cambridge University Press), 496–512.

Brooks, D. (1994) US factories in Mexico cause toxic pollution. In C.P. Cozic (ed.) *Pollution* (San Diego: Greenhaven Press), 79–84.

Brown, A.G. (1997) *Alluvial geoarchaeology*. (Cambridge: Cambridge University Press).

Brown, A.G. and Barber, K.E. (1985) Late Holocene palaeoecology and sedimentary history of a small lowland catchment in central England. *Quaternary Research* 24, 87–102.

Brown, E. (1996) Deconstructing development: alternative perspectives on the history of an idea. *Journal of Historical Geography* 22(3), 333–9.

Brown, Jr., G.E. (1987) US nuclear waste policy: flawed but feasible. *Environment* 29(8), 6–7 and 25.

Brown, H.S., Kasperson, R.E. and Raymond, S. (1990) Trace pollutants. In B.L.Turner II, W.C. Clark, R.W. Kates, J.F. Richards, J.T. Matthews and W.B.Meyer (eds) *The earth as transformed by human action* (Cambridge: Cambridge University Press), 437–54.

Brown, J.L. (1987) Hunger in the US. *Scientific American* 256(2), 21–6.

Brown, L. and Wolf, E. (1984) *Soil erosion: quiet crisis in the world economy. Worldwatch paper 60* (Washington DC: Worldwatch Institute).

Brown, L.R. (1991) Global resource scarcity is a serious problem. In M. Polesetsky (ed.) *Global resources: opposing viewpoints* (San Diego: Greenhaven Press), 17–23.

Brown, L.R. (1994) Facing food insecurity. In L.R. Brown *et al. State of the world, 1994* (London: Earthscan), 177–97.

Brown, L.R. (1997) Facing the prospect of food scarcity. In L.R. Brown *et al. State of the world, 1997* (London: Earthscan), 23–41.

Brown, L.R. (1998) The future of growth. In L.R. Brown *et al. State of the world, 1998* (London: Earthscan), 3–20.

Brune, W. (1996) There's safety in numbers. *Nature* 379, 486–7.

Bryson, R.A. (1973) Drought in the Sahel: who or what is to blame? *The Ecologist* 3, 366–71.

Buckland, P., Dugmore, A. and Sadler, J. (1991) Faunal change or taphonomic problem? A comparison of modern and fossil insect faunas from south-east Iceland. In J.K. Maizels and C.J. Caseldine (eds) *Environmental change in Iceland: past and present* (Dordrecht: Kluwer), 127–46.

Buckley, R. and Araujo, G. (1997) Green advertising by tourism operators on Australia's Gold Coast. *Ambio* 26(3), 190.

Budiansky, S. (1996) The environmental crisis is exaggerated. In A. Sadler (ed.) *The environment: opposing viewpoints* (San Diego: Greenhaven Press), 25–31.

Buffin, D. (1992) Calls to phase out methyl bromide: a major ozone depleter. *Pesticide News* 18, 5–11.

Bukharin, N. (1972) *Imperialism and the world economy* (London: Merlin).

Bunting, M.J. (1994) Vegetation history of Orkney, Scotland. Pollen records from small basins in west Mainland. *New Phytologist* 128, 771–92.

Bunyard, P. (1987) The significance of the Amazon Basin for global climatic equilibrium. *The Ecologist* 17(4/5), 139–41.

Bunyard, P. and Morgan-Grenville, F. (eds) (1987) *The green alternative* (London: Methuen).

Burgess, J. (1990) The production and reception of environmental meanings in the mass media: a research agenda for the 1990s. *Transactions, Institute of British Geographers*, New series 15(2), 139–61.

Burt, T. and Haycock, N.E. (1991) Farming and the nitrate problem. *Geography* 76(1), 60–3.

Bush, M.B. (1988) Early Mesolithic disturbance: a force in the landscape. *Journal of Archaeological Science* 15, 453–62.

Bush, M.B., Piperno, D.R. and Colinvaux, P.A. (1989) A 6000-year-old history of Amazonian maize cultivation. *Nature* 340, 303–5.

Butler, J.H., Elkins, J.W., Hall, B.D., Cummings, S.O. and Montzka, S.A. (1992) A decrease in the growth rates of atmospheric halon concentrations. *Nature* 359, 403–5.

Butler, J.N., Burnett-Herkes, J., Barnes, J.A. and Ward, J. (1993) The Bermuda fisheries: a tradegy of the commons averted. *Environment* 35(1), 7–15 and 25–33.

Buttel, F. (1993) Environmentalization and greening: origins, processes and implications. In S. Harper (ed.) *The greening of rural policy: international perspectives* (London: Belhaven Press), 12–26.

Buttel, F. and Flinn, W. (1974) The structure and support of the environmental movement, 1968–1970. *Rural Sociology* 39, 56–69.

Buttel, F. and Taylor, P. (1994) Environmental sociology and global environmental change: a critical assessment. In M. Redclift and T. Benton (eds) *Social theory and the global environment* (London: Routledge), 228–55.

Button, J. (1990) *New green pages: a directory of natural products, services, resources and ideas* (London: Optima, 2nd edn.).

Byrd, B.F. (1994) From early humans to farmers and herders – recent progression key transitions in south-west Asia. *Journal of Archaeological Research* 2, 221–53.

Caincross, F. (1994) Chemical companies are reducing pollution. In C.P. Cozic (ed.) *Pollution* (San Diego: Greenhaven Press), 98–103.

Caldwell, M.M., Teramura, A.H., Tevini, M., Bornmann, J.F., Björn, L.O. and Kulandaivelu, G. (1995) Effects of increased solar ultraviolet radiation on terrestrial plants. *Ambio* 24(3), 166–73.

Campbell, C.J. and Laherrere, J.H. (1998) The end of cheap oil. *Scientific American* 278(3), 60–5.

Campbell, K., McWhir, J., Ritchie, W. and Wilmut, I. (1996) Sheep cloned by nuclear transfer from a cultured cell line. *Nature* 380, 7 March, 64–6.

Caputo, R. (1993) Tradegy stalks the horn of Africa. *National Geographic* 184(2), 88–122.

Carless, J. (1994) Recycling should be a high priority. In C.P. Cozic (ed.) *Pollution* (San Diego: Greenhaven Press), 172–8.

Carls, E.G., Fenn, D.B. and Chaffey, S.A. (1995) Soil contamination by oil and gas drilling and production operations in Padre Island National Seashore, Texas, USA. *Journal of Environmental Management* 45, 273–86.

Carson, R. (1962) *Silent spring* (Boston: Houghton Mifflin).

Carter, F.W. (1993) Poland. In F.W. Carter and D. Turnock (eds) *Environmental problems in Eastern Europe* (London: Routledge).

Carter, F.W. and Turnock, D. (1993) *Environmental problems in Eastern Europe* (London: Routledge).

Cartwright, J. (1989) Conserving nature, decreasing debt. *Third World Quarterly* 11(2), 114–27.

Carvallo, M. (1994) Antarctic tourism must be managed, not eliminated. *Forum for Applied Research and Public Policy* 9, 76–9.

Caseldine, C.J. and Hatton, J. (1993) The development of high moorland on Dartmoor: fire and the influence of Mesolithic activity on vegetation change. In F.M. Chambers (ed.) *Climate change and human impact on the landscape* (London: Chapman & Hall), 119–31.

Castree, N. (1995) The nature of produced nature. *Antipode* 27, 12–48.

Castree, N. and Braun, B. (1998) The construction of nature and the nature of construction. In B. Braun, and N. Castree, (eds) *Remaking Reality* (London: Routledge), 3–42.

Cater, E. (1995) Environmental contradictions in sustainable tourism. *The Geographical Journal* 161(1), 21–28.

Cathcart, B. (1995) Where fishes swim, men will fight. *Independent on Sunday*, 19 March, 24.

Catt, J. (1985) Natural soil acidity. *Soil Use and Management* 1(1), 8–10.

Chadwick, D. (1991) Elephants – out of time, out of space. *National Geographic* 179(5), 2–49.

Chadwick, D.H. (1995) The Endangered Species Act. *National Geographic* 187(3), 2–41.

Chambers, F.M. (1993) Late Quaternary climatic change and human impact: commentary and conclusions. In F.M. Chambers (ed.) *Climate change and human impact on the landscape* (London: Chapman & Hall), 247–59.

Chambers, F.M. (1995) *Peatlands and palaeoclimates: putting global warming in perspective.* Inaugural lecture, Cheltenham and Gloucester College of Higher Education.

Chambers, F.M. (1998) Global warming: new perspectives from palaeoecology and solar science. *Geography* 83(3), 266–77.

Champion, A. (1989) *Counterurbanization: the changing pace and nature of population deconcentration* (London: Edward Arnold).

Champion, A. (1992) Urban and regional demographic trends in the developed world. *Urban Studies* 29, 461–82.

Charles, D. (1993) In search of a better burn. *New Scientist,* 23 January, 28–31.

Charters, W.W.S. (1991) Solar energy: current status and future prospects. *Energy Policy* 19(8), 738–41.

Chaudhary, V. (1994) Anti-road lobby stirs nationwide. *The Guardian*, 14 June.

Cherfas, J. (1986) What price whales? *New Scientist,* 5 June, 36–40.

Claussen, E. (1994) EPA programs reduce air pollution. In C.P. Cozic (ed.) *Pollution* (San Diego: Greenhaven Press), 127–32.

Clifford, F. (1996) Curtailing environmental regulations may prove harmful. In A. Sadler (ed.) *The environment: opposing viewpoints* (San Diego: Greenhaven Press), 193–200.

Cloke, P. (1992) 'The countryside': development, conservation and an increasingly marketable commodity. In P. Cloke (ed.) *Policy and change in Thatcher's Britain* (Oxford: Pergamon).

Cloke, P. (1993) The countryside as commodity: new rural spaces for leisure. In S. Glyptis (ed.) *Leisure and the environment* (London: Belhaven), 53–67.

Cloke, P. and Little, J. (1990) *The rural state? Limits to planning in rural society* (Oxford: Oxford University Press).

Cloke, P. and Park, C. (1985) *Rural resource management* (London: Croom Helm).

Cloke, P. and Phillips, M. (forthcoming) *The myth and persistence of rural culture* (London: Edward Arnold).

Cloke, P. and Thrift, N. (1987) Intra-class conflict in rural areas. *Journal of Rural Studies* 3, 321–33.

Cloke, P., Goodwin, M., Milbourne, P. and Thomas, C. (1998) Inside looking out: outside looking in. Different experiences of cultural competence in rural lifestyles. In M. Boyle and K. Halfacree (eds) *Migration to rural areas* (Chichester: John Wiley & Sons), 134–50.

Cloke, P., Phillips, M. and Thrift, N. (forthcoming) *Moving to rural idylls* (London: Paul Chapman).

Clunies Ross, T. (1993) Taxing nitrogen fertilizers. *The Ecologist* 23(1), 13–16.

Cobb, Jr., C.E. (1987) The Great Lakes. Troubled waters. *National Geographic* 172(1), 2–31.

Cobb Jr., C.E. (1996) Eritrea wins the peace. *National Geographic* 189(6), 82–105.

Coghlan, A. (1991) Fresh water from the sea. *New Scientist* No 1784, 37–40.

Cohen, B. (1984) Statement of dissent. In J.L. Simon and H. Kahn (eds) *The resourceful earth* (Oxford: Basil Blackwell), 566.

Cohen, B.L. (1995a) The costs of nuclear power. In J.L. Simon (ed.) *The state of humanity* (Oxford: Blackwell), 294–302.

Cohen, B.L. (1995b) The hazards of nuclear power. In J.L. Simon (ed.) *The state of humanity* (Oxford: Blackwell), 576–87.

Cohen, P. (1996) Brazil acts on the incredible shrinking rainforest. *New Scientist* 149, No 2041, 4.

Coles, R.W. and Taylor, J. (1993) Wind power and planning. *Land Use Policy* 10(3), 205–26.

Collins, L. (1995) Eco-labelling. *Geography Review* 8(3), 24–27.

Collis, J. (1984) *The European Iron Age* (London: Batsford).

Conacher, A.J. (1990) Salt of the earth. *Environment* 32(6), 5–9 and 40–2.

Connell, R. (1987) *Gender and power: society, the person and sexual politics* (Cambridge: Cambridge University Press).

Constanza, R. and Cornwell, L. (1992) The 4-P approach to dealing with scientific uncertainty. *Environment* 34, 12–17 and 40–2.

Cook, R.M., Sinclair, A. and Stefansson, G. (1997) Potential collapse of North Sea cod stocks. *Nature* 385, 521–2.

Cook-McGuail, J. (1978) Effects of hikers and horses on mountain trails. *Journal of Environmental Management* 6, 209–12.

Cooper, W.J. and Holloway, L.A. (1996) Recipes for cleaner air. *Nature* 384, 313–14.

Cope, C. (1991) Gazelle hunting strategies in the southern Levant. In O. Bar-Yosef and F.R.Valla (eds) *The Natufian culture in the Levant* (Michigan: Ann Arbor), 341–58.

Corbridge, S. (1986) *Capitalist world development: a critique of radical development geography* (London: Macmillan).

Corcoran, E. (1991) Cleaning up coal. *Scientific American* 264(5), 70–80.

Corry, S. (1993) The rainforest harvest: who reaps the benefit? *The Ecologist* 23(4), 148–53.

Cosgrove, D. (1994) Contested global visions: one-world, whole-earth, and the Apollo space photographs. *Annals, Association of American Geographers* 84(2), 270–94.

Cosgrove, D., Roscoe, B. and Rycroft, S. (1996) Landscape and identity at Ladybower Reservoir and Rutland Water. *Transactions, Institute of British Geographers* 23(3), 534–51.

Cotgrove, S. (1982) *Catastrophe or cornucopia* (Chichester: John Wiley & Sons).

Cotgrove, S. and Duff, A. (1980) Environmentalism, middle class radicalism and politics. *Sociological Review* 28, 333–51.

Cotgrove, S. and Duff, A. (1981) Environmental values and social change. *British Journal of Sociology* 32(1), 92–110.

Coull, J. (1984) Canada's Atlantic fisheries: new opportunities and problems. *Geography* 69, 353–6.

Cozic, C.P. (ed.) (1994) *Pollution* (San Diego: Greenhaven Press).

Craddock, P.T. (1995) *Early metal mining and production* (Edinburgh: Edinburgh University Press).

Crang, M. (1998) *Cultural geography* (London: Routledge).

Croll, B.T. (1991) Pesticides in surface and groundwaters. *Journal of the Institution of Water and Environmental Management* 5, 389–95.

Croll, B.T. and Hayes, C.R. (1988) Nitrate and water supplies in the United Kingdom. *Environmental Pollution* 50, 163–87.

Crosby, A. W. (1986) *Ecological imperialism: the biological expansion of Europe, 900–1900* (Cambridge: Cambridge University Press).

Cross, M. (1993) A very dirty business. *New Scientist*, 23 January, 28–31.

Crosson, P. (1997) Will erosion threaten agricultural productivity? *Environment* 39(8), 5–9 and 29–31.

Crow, B. (1992) Understanding famine and hunger. In T. Allen and A. Thomas (eds) *Poverty and development in the 1990s* (Oxford: Oxford University Press), 15–33.

Crush, J. (ed.) (1995) *Power and development* (London: Routledge).

Cutter, S. (1993) *Living with risk* (London: Edward Arnold).

D'Elia, C.F. (1987) Nutrient enrichment of Chesapeake Bay. *Environment* 29(2), 6–11 and 30–3.

Dabas, M. and Bhatia, S. (1996) Carbon sequestration through afforestation: role of tropical industrial plantations. *Ambio* 25(5), 327–30.

Daly, H. (1973) *Towards a steady state economy* (New York: W.H. Freeman).

Daly, H. (1993) The perils of free trade. *Scientific American* 269(5), 24–9.

Daniels, S. (1993) *Fields of vision: landscape imagery and national identity in England and the United States* (Cambridge: Polity Press).

Danielsen, O. (1994) Large-scale wind power in Denmark. *Energy Policy* 12(1), 60–2.

Darby, H.C. (1976) The age of the improver: 1600–1800. In H.C. Darby (ed.) *A new historic geography of England after 1600* (Cambridge: Cambridge University Press), 1–88.

Davidson, K. (1994) Predicting earthquakes. Can it be done? *Earth*, May, 56–63.

Davies, D.M. and Williams, P.J. (1995) The effect of nitrification inhibitor Dicyandiamide on nitrate leaching and ammonia volatilization: a UK nitrate sensitive areas perspective. *Journal of Environmental Management* 45, 263–72.

Davies, N. (1992) A road too far. *The Guardian*, Weekend Section, 18–19 July, 14–15.

Davis, G. (1990) Energy for Planet Earth. *Scientific American* 263(3), 20–7.

Davis, O.K. and Turner, R.M. (1986) Palynological evidence for the historic expansion of juniper and desert shrubs in Arizona. *Review of Palaeobotany and Palynology* 49, 177–93.

Davison, P. (1995) Hell's highway set to rip through Amazon jungle. *Independent on Sunday*, 9 April, 18.

Dawkins, R. (1976) *The selfish gene* (Oxford: Oxford University Press).

Dayan, T. (1994) Early domesticated dogs of the Near East. *Journal of Archaeological Science* 21, 633–40.

Dayan, T. and Simberloff, D. (1995) Natufian gazelles: proto-domestication reconsidered. *Journal of Archaeological Science* 22, 671–5.

De Alessi, M. (1996) Tender loving hunters. *New Scientist*, 22 June, 47.

De Laet, S.J. (1994) Europe during the Neolithic. In S.J. De Laet, A.H. Dani, J.L. Lorenzo and R.B. Nunoo (eds) *History of humanity, volume 1: Prehistory and the beginnings of civilisation* (London: Routledge), 490–500.

Degens, E.T., Kempe, S. and Richey, J.E. (1991) *Biogeochemistry of major world rivers* (Chichester: John Wiley).

Delcourt, H.R. (1987) The impact of prehistoric agriculture and land occupation on natural vegetation. *Trends in Ecology and Evolution* 2, 39–44.

Delcourt, H.R. and Delcourt, P.A. (1991) *Quaternary Ecology: a palaeocological perspective.* (London: Chapman & Hall).

Demeritt, D. (1998) Science, social constructivism and nature. In B. Braun and N. Castree, (eds) *Remaking Reality* (London: Routledge), 173–93.

Dennell, R. (1983) *European economic prehistory* (London: Academic Press).

Devarakonda, M.S. and Hickox, J.A. (1996) Radioactive wastes. *Water Environment Research* 68(4), 608–16.

Devolpi, A. (1995) Fast finish to Plutonium Peril. *Bulletin of Atmospheric Sciences* 51(5), 20.

Dickens, P. (1996) *Reconstructing nature: alienation, emancipation and the division of labour* (London: Routledge).

Dickenson, R.E. (1986) The climate system and modelling of future climate. In B. Bolin, B.R. Doos, J. Jäger and R.A. Warrick, (eds) *The greenhouse effect, climate change and ecosystems* (Chichester: John Wiley & Sons), 207–70.

Dickson, M.H. and Fanelli, M. (1994) Small geothermal resources: a review. *Energy Sources* 16, 349–76.

Dietz, F.J., Van der Straaten, J. and Van der Velde, M. (1991) The European market and the environment: the case of the emission of NO_x by motorcars. *Review of Political Economy* 3(1).

Dillon, P.J., Reid, R.A. and Girard, R. (1986) Changes in the chemistry of lakes near Sudbury, Ontario, following reductions in SO_2 emissions. *Water, Air and Soil Pollution* 31, 59–65.

Dillon, P.J., Reid, R.A. and Grobois, E. (1987) The rate of acidification of aquatic ecosystems in Ontario, Canada. *Nature* 329, 45–8.

DiPippo, R. (1991) Geothermal energy: electricity generation and environmental impact. *Energy Policy* 19(8), 798–807.

Dixon, J.A. and Fallon, L.A. (1989) The concept of sustainability: origins, extensions, and usefulness for policy. *Society and Natural Resources Journal* 2, 73–84.

Dixon, R.K., Sathaye, J.A., Meyers, S.P., Masera, O.R., Makarov, A.A., Toure, S., Makundi, W. and Wiel, S. (1996) Greenhouse gas mitigation strategies: preliminary results from the US Country Studies Program. *Ambio* 25(1), 26–32.

Dlugokencky, E.J., Masarie, K.A., Lang, P.M. and Tans, P.P. (1998) Continuing decline in the growth rate of the atmospheric methane burden. *Nature* 393, 447–50.

Dobson, A. (1990) *Green political thought: an introduction* (London: HarperCollins).

Dobson, A. (1996) Environment sustainabilities. *Environmental Politics* 5(3), 401–28.

Dodds, K. (1996) To photograph the Antarctic: British polar exploration and the Falkland Islands and Dependencies Aerial Survey Expedition (FIDASE). *Ecumene* 3(1), 63–89.

Dodgshon, R.A. (1987) *The European past: social evolution and spatial order* (London: Macmillan).

Donkin, R.A. (1976) Changes in the early middle ages. In H.C. Darby (ed.) *A new historical geography of England before 1600* (Cambridge: Cambridge University Press), 75–135.

Doughty, R. (1981) Environmental theology: trends and prospects in Christian thought. *Progress in Human Geography* 5(2), 234–48.

Dovers, S.R. and Handmer, J.W. (1995) Ignorance, the precautionary principle, and sustainability. *Ambio* 24(2), 92–7.

Downing, T.E., Hecht, S.B., Pearson, H.A. and Garcia-Downing, C. (eds) (1992) *Development or destruction: the conversion of tropical forest to pasture in Latin America* (Boulder, Colorado: Westview).

Doyle, R. (1996a) The changing quality of life. *Scientific American* 275(1), 18.

Doyle, R. (1996b) Soil erosion of cropland in the US, 1982 to 1992. *Scientific American* 275(4), 23.

Doyle, R. (1996c) Global forest cover. *Scientific American* 275(5), 22.

Doyle, R. (1997a) Access to safe drinking water. *Scientific American* 277(5), 19.

Doyle, R. (1997b) Plants at risk in the US. *Scientific American* 277(1), 14.

Doyle, R. (1997c) Threatened mammals. *Scientific American* 276(1), 25.

Drake, F. (1995) Stratospheric ozone depletion – an overview of the scientific debate. *Progress in Physical Geography* 19, 1–17.

Dreze, J. and Sen, A. (1993) *The political economy of hunger. Volume 2: Famine prevention* (Oxford: Clarendon Press).

Driver, F. (1997) Bodies in space: Foucault's account of disciplinary power. In T. Barnes and D. Gregory (eds) *Reading human geography* (London: Arnold), 279–89.

Dudley, N. (1986) Acid rain and pollution control policy in the UK. *The Ecologist* 16(1), 18–23.

Duffus, D.A. and Dearden, P. (1990) Non-consumptive wildlife-orientated research: a conceptual framework. *Biological Conservation* 53(3), 213–31.

Dumayne, L. (1993) Invader or native? – vegetation clearance in northern Britain during Romano-British time. *Vegetation History and Archaeobotany* 2, 29–36.

Dumayne, L. and Barber, K.E. (1994) The impact of the Romans on the environment of northern England: pollen data from three sites close to Hadrian's Wall. *The Holocene* 4(2), 165–73.

Dumayne-Peaty, L. (1998) Human impact on the environment during the Iron Age and Romano-British times: palynological evidence from three sites near the Antonine Wall, Great Britain. *Journal of Archaeological Science* 25, 203–14.

Dunlap, R. and Scarce, R. (1991) The polls-poll trends: environmental problems and protection. *Public Opinion Quarterly* 55, 651–72.

Dunlap, R., Gallup, G. and Gallup, A. (1993) The health of the planet: a global concern. *Environment* 35, 7–15 and 33–9.

Dusseault, M.B. (1997) Flawed reasoning about oil and gas. *Nature* 386, 12.

Eckersley, R. (1989) Green politics and the new class: selfishness or virtue. *Political Studies* XXXVII, 205–23.

Eckersley, R. (1992) *Environmentalism and political theory: toward an ecocentric approach* (London: UCL Press).

Eckersley, R. (ed.) (1995a) *Markets, the state and the environment* (Basingstoke: Macmillan).

Eckersley, R. (1995b) Markets, the state and the environment: an overview. In R. Eckersley (ed.) *Markets, the state and the environment* (Basingstoke: Macmillan), 7–45.

Eckholm, E. (1976) *Losing ground: environmental stress and world food prospcets* (New York: Norton).

Eden, S. (1998) Environmental issues: knowledge, uncertainty and the environment. *Progress in Human Geography* 22(3), 425–32.

Eder, J.F. (1996) After deforestation: migrant lowland farmers in the Philippines Uplands. In L.E. Sponsel, T.N. Headland and R.C. Bailey (eds) *Tropical deforestation: the human dimension* (New York: Columbia University Press), 253–71.

Eder, K. (1993) *The new politics of class* (London: Sage).

Eder, K. (1996) *The social construction of nature* (London: Sage).

Edge, G. and Tovey, K. (1995) Energy: hard choices ahead. In T. O'Riordan (ed.) *Environmental science for environmental management* (Harlow: Longman), 317–34.

Edwards, K.J. (1988) The hunter-gatherer/agricultural transition and the pollen record in the British Isles. In H.H. Birks, H.J.B. Birks, P.E. Kaland and D. Moe (eds) *The cultural landscape: past, present and future* (Cambridge: Cambridge University Press), 255–66.

Edwards, K.J. (1989) The cereal pollen record and early agriculture. In A.D. Milles, D. Williams and N. Gardner (eds) *The beginnings of agriculture* (Oxford: British Archaeological Reports, International Series S496), 113–35.

Edwards, K.J. (1990) Fire and the Scottish Mesolithic: evidence from microscopic charcoal. In P.M. Vermeersch and P. Van Peer (eds) *Contributions to the Mesolithic in Europe*. Leuven: Leuven University Press, 71–9.

Edwards, K.J. (1993) Models of mid-Holocene forest farming for north-west Europe. In F.M. Chambers (ed.) *Climate change and human impact on the landscape* (London: Chapman & Hall), 133–45.

Edwards, K.J. and Hirons, K.R. (1984) Cereal pollen grains in pre-elm decline deposits: implications for the earliest agriculture in Britain and Ireland. *Journal of Archaeological Science* 11, 71–80.

Edwards, K.J. and MacDonald, G.M. (1991) Holocene palynology: II Human influence and vegetation change. *Progress in Physical Geography* 15(4), 364–91.

Edwards, K.J. and Whittington, G. (1997) Vegetation change. In K.J. Edwards and I.B.M. Ralston (eds) *Scotland: environment and archaeology 8000BC–AD1000* (Chichester: John Wiley & Sons), 63–82.

Edwards, R. (1997) Nuclear firms want special treatment. *New Scientist* No 2086, 7.

Eger, H., Fleischhauer, E., Hebel, A. and Sombroek, W.G. (1996) Taking action for sustainable land-use: results from 9th ISCO conference in Bonn, Germany. *Ambio* 25(8), 480–3.

Ehrlich, P.R. (1968) *The population bomb* (New York: Ballantine).

Ehrlich, P.R. (1982) Human carrying capacity, extinctions and nature reserves. *BioScience* 32, 331–3.

Ehrlich, P.R. and Ehrlich, A.H. (1990) *The population explosion* (London: Arrow).

Eisenbud, M. (1990) The ionizing radiations. In B.L. Turner II, W.C. Clark, R.W. Kates, J.F. Richards, J.T. Matthews and W.B.Meyer (eds), *The earth as transformed by human action* (Cambridge: Cambridge University Press), 455–66.

El Asswad, R.M. (1995) Agricultural prospects and water resources in Libya. *Ambio* 24(6), 324–7.

Elinder, C.-G. and Järup, L. (1996) Cadmium exposure and health risks: recent findings. *Ambio* 25(5), 370–3.

Elkington, J. and Burke, T. (1987) *The green capitalists* (London: Gollancz).

Elkington, J., Burke, T. and Hailes, J. (1988) *Green pages: the business of saving the world* (London: Routledge).

Elkins, P. (1992) *A new world order: grassroots movements for global change* (London: Routledge).

Ellis, D. (1989) *Environments at risk* (London: Springer-Verlag).

Ellis, S. and Mellor, A. (1995) *Soils and environment* (London: Routledge).

Ellis, W.S. (1987) Africa's stricken Sahel. *National Geographic* 172(2), 140–79.

Elmsley, J. (ed.) (1996) *The global warming debate* (London: European Science and Environment Forum).

Elsworth, S. (1984) *Acid rain* (London: Pluto Press).

Emel, J. and Peet, R. (1989) Resource management and natural hazards. In R. Peet and N. Thrift (eds) *New models in geography, volume 1: the political economy perspective* (London: Unwin Hyman), 49–76.

Emerson, J.W. (1971) Channelization: a case study. *Science* 173, 325–6.

Environmental Protection Agency (1994) The EPA responds effectively to pollution emergencies. In C.P. Cozic (ed.) *Pollution* (San Diego: Greenhaven Press), 121–6.

Escobar, A. (1996) Constructing nature: elements for a poststructural political ecology. In R. Peet and M. Watts (eds) *Liberation ecologies: environment development, social movements* (London: Routledge), 46–68.

Evans, D. (1992) *A history of nature conservation in Britain* (London: Routledge).

Evans, J. (1972) *Land snails and archaeology* (London: Seminar Press).

Evans, L.T. (1993) *Crop evolution, adaptation and yield* (Cambridge: Cambridge University Press).

Evans, R. (1990) Soil erosion: its impact on the English and Welsh landscape since woodland clearance. In J. Boardman, I.D.L. Foster and J.A. Dearing (eds) *Soil erosion on agricultural land* (Chichester: John Wiley & Sons), 231–54.

Fairlie, S. (ed.) Overfishing: its causes and consequences. *The Ecologist* 25(2/3), 42–125.

Fajer, E.D. and Bazzaz, F.A. (1992) Is carbon dioxide a 'good' greenhouse gas? *Global Environmental Change* 2(4), 301–10.

Falnes, J. and Lovseth, J. (1991) Ocean wave energy. *Energy Policy* 19, 768–75.

Fan, S.-M. and Jacob, D.J. (1992) Surface ozone depletion in Arctic spring sustained by bromine reactions on aerosols. *Nature* 359, 522–4.

Farman, J.C., Gardiner, B.G. and Shanklin, J.D. (1985) Large seasonal losses of total ozone in Antarctica reveal seasonal ClO_x/NO_x interaction. *Nature* 315, 207–10.

Farmer, B. (ed.) (1977) *Green revolution? Technology and change in rice-growing areas of Tamil Nadu and Sri Lanka* (Macmillan: London)

Farrer, S. (1998) Death Threat to 7000 tree species. *The Sunday Times* 10th May, p.8.

Fearnside, P. (1987) Deforestation and international development projects in Brazilian Amazonia. *Conservation Biology* 1(3), 214–20.

Fearnside, P. (1989) A prescription for slowing down deforestation in Amazonia. *Environment* 31(4), 17–20 and 39–40.

Feifer, M. (1985) *Going places* (London: Macmillan).

Fiedel, S. (1987) *Prehistory of the Americas* (Cambridge: Cambridge University Press).

Filho, S.R. and Maddock, J.E.L. (1997) Mercury pollution in two gold mining areas of the Brazilian Amazonia. *Journal of Geochemical Exploration* 58, 231–40.

Findlay, A. and Findlay, A. (1985) *Population and development in the Third World* (London: Methuen).

Finger, M. (1993) Politics of the UNCED process. In W. Sachs (ed.) *Global ecology: a new arena of political conflict* (London: Zed Books), 36–48.

Fischhoff, B. (1991) Science of politics in the midst of environmental disaster: report from Poland. *Environment* 33(2), 12–17 and 37–8.

Fischmann, J. (1992) Falling into the gap: a new theory. *Discover* October, 56–63.

Fisher, L.J. (1994) The EPA helps prevent toxic pollution. In C.P. Cozic (ed.) *Pollution* (San Diego: Greenhaven Press), 33–137.

Fisher, M. (1994) Conservation through commercialisation. *Economic Affairs* 14(3), 15–17.

Fitzsimmons, M. (1989) The matter of nature. *Antipode* 21, 106–20.

Flavin, C. and Durning, A.B. (1991) Sweeping energy policies are needed to conserve global resources. In M. Polesetsky (ed.) *Global resources: opposing viewpoints* (San Diego: Greenhaven Press), 247–53.

Fleming, A. (1988) *The Dartmoor Reaves: investigating prehistoric land divisions* (London: Batsford).

Flower, R.J. and Battarbee, R.W. (1983) Diatom evidence for recent acidification of two Scottish lochs. *Nature* 305, 130–3.

Fogel, R.W. (1995) The contribution of improved nutrition to the decline in mortality rates in Europe and North America. In J.L. Simon (ed.) *The state of humanity* (Oxford: Blackwell), 61–71.

Foresta, R.A. (1992) Amazonia and the politics of geopolitics. *The Geographical Review* 82, 128–42.

Forestier, K. (1989) The degreening of China. *New Scientist* 1 July, 52.

Forman, D., Al-Dabbagh, S. and Doll, R. (1985a) Nitrates, nitrites and gastric cancer in Great Britain. *Nature* 313, 620–5.

Forman, D., Al-Dabbagh, S. and Doll, R. (1985b) Nitrates and gastric cancer risks. *Nature* 317, 675–6.

Foster, I.D.L., Harrison, S.H. and Clark, D. (1997) Soil erosion in the West Midlands: an act of God or agricultural mismanagement? *Geography* 82(3), 231–9.

Foster, I.D.L., Mighall, T.M., Wotton, C., Owens, P. and Walling, D. (in press) Evidence of medieval soil erosion in the South Hams region of Devon, *The Holocene*.

Foucault, M. (1977) *Discipline and punish: the birth of the prison* (London: Allen Lane).

Fouda, S.A. (1998) Liquid fuels from natural gas. *Scientific American* 278(3), 60–5.

Fowler, D., Cape, J.N. and Leith, I.D. (1985) Acid inputs from the atmosphere in the United Kingdom. *Soil Use and Management* 1(1), 3–5.

Fox, W. (1990) *Toward a transpersonal ecology: developing foundations for environmentalism* (Boston: Shambhala).

French, C.A.I. (1988) Aspects of buried prehistoric soils in the lower Welland Valley and fen margin north of Peterborough, Cambridgeshire. In W. Groenman-van Waateringe and M. Robinson (eds) *Man-made soils* (Oxford: British Archaeological Reports International Series 410), 115–28.

French, H.F. (1994) Making environmental treaties work. *Scientific American* 271(6), 62–5.

Friis-Christensen, E. and Lassen, K. (1991) Length of the solar cycle: an indicator of solar activity closely associated with climate. *Science* 254, 698–700.

Fröbel, F., Heinrichs, J. and Kreye, O. (1980) *The new international division of labour* (Cambridge: Cambridge University Press)

Fukuyama, F. (1992) *The end of history and the last man* (London: Penguin).

Fulkerson, W., Judkins, R.R. and Sanghvi, M.K. (1990) Energy from fossil fuels. *Scientific American* 263(3), 83–9.

Furth, H. (1995) Fusion. *Scientific American* 273(3), 140–2.

Futang, W. and Zong-Ci, Z. (1995) Impact of climate change on natural vegetation in China and its implication for agriculture. *Journal of Biogeography* 22, 657–64.

Gäbler, H.-E. (1997) Mobility of heavy metals as a function of pH of samples from an overbank sediment profile contaminated by mining activities. *Journal of Geochemical Exploration* 58, 157–72.

Garcia, S., Gulland, J.A. and Miles, E. (1984) The new law of the sea, and the access to surplus fish resources. *Marine Policy* 10(3), 192–200.

Geller, H. and McGaraghan, S. (1998) Successful government–industry partnership: the US Department of Energy's role in advancing energy-efficient technologies. *Energy Policy* 26(2), 167–77.

Geller, R.J. (1991) Shake–up for earthquake prediction. *Nature*, 352, 275–6.

George, R.L. (1998) Mining for oil. *Scientific American* 278(3), 66–7.

George, S. (1988) *A fate worse than debt* (London: Penguin).

Gerrard, S. (1995) Environmental risk assessment. In T. O'Riordan (ed.) *Environmental science for environmental management* (Harlow: Longman), 296–316.

Giddens, A. (1990) *The consequences of modernity* (Cambridge: Polity Press).

Giddens, A. (1991) *Modernity and self-identity* (Cambridge: Polity Press).

Giddens, A. (1997) *Sociology* (Cambridge: Polity Press, 3rd edn).

Girling, M. (1988) The bark beetle *Scolytus scolytus* (Fabricius) and the possible role of elm disease in the early Neolithic. In M. Jones (ed.) *Archaeology and the flora of the British Isles*. Oxford University Committee for Archaeology Monograph, 14, 34–8.

Glacken, C. (1967) *Traces on the Rhodian shore* (Berkeley: University of California Press).

Gleick, P.H. (1994) Water, war and peace in the Middle East. *Environment* 36(3), 5–15 and 35–42.

Glyptis, S. (1991) *Countryside recreation* (Harlow: Longman).

Godlewska, A. (1995) Map, text and image. The mentality of enlightened conquerors: a new look at the Description de l'Egypte. *Transactions, Institute of British Geographers* 20(1), 5–28.

Goeller, H.E. (1995) Trends in nonrenewable resources. In J.L. Simon (ed.) *The state of humanity* (Oxford: Blackwell), 313–22.

Goldsmith, E., Allen, R., Allaby, M., Davoll, J. and Lawrence, S. (1972) *Blueprint for survival* (Harmondsworth: Penguin).

Goodall, B. (1995) Environmental auditing: a tool for assessing the environmental performance of tourism firms. *The Geographical Journal* 161(1), 29–37.

Goodland, R. and Irwin, H. (1975) *Amazon jungle: green hell to red desert* (Amsterdam: Elsevier).

Goodwin, H. (1960) Prehistoric wooden trackways of the Somerset Levels: their construction, age and relation to climate change. *Proceedings of the Prehistoric Society* 26, 1–36.

Goodwin, H. (1996) In pursuit of ecotourism. *Biodiversity and Conservation* 5, 277–91.

Göransson, H. (1983) När börjar neolitikum? *Popular Arkeologi* 1, 4–7.

Göransson, H. (1986) Man and forests of nemoral broad-leaved trees during the Stone Age. *Striae* 24, 143–52.

Gouldner, A. (1979) *The future of intellectuals and the rise of the new class* (New York: Continuum).

Grant, L. (1995) Just say no. *The Guardian*, Weekend Section, 3 June, 13–22.

Gray, Jr., C.L. and Alson, J.A. (1989) The case for methanol. *Scientific American* 261(5), 86–92.

Green, N. (1990) *The spectacle of nature* (Manchester: Manchester University Press).

Greenhalgh, G. (1995a) Mental health consequences of Chernobyl. *Nuclear News* 38, 50.

Greenhalgh, G. (1995b) Chernobyl health effects: radiation or stress? *Nuclear Engineering International* 40(497), 14.

Greer, J. (1993) The price of gold: environmental costs of the new gold rush. *The Ecologist* 23(3), 91–6.

Gregory, D. (1990) 'A new and differing face in many places': three geographies of industrialisation. In R.A. Dodgshon and R.A. Butlin (eds) *An historical geography of England and Wales* (London: Academic Press, 2nd edn), 351–400.

Gregory, K., Webster, C. and Durk, S. (1996) Estimates of damage to forests in Europe due to emissions of acidifying pollutants. *Energy Policy* 24(7), 655–64.

Griffin, K. (1969) *Underdevelopment in Spanish America: an interpretation* (London: Allen & Unwin).

Griffin, K. (1979) *The political economy of agrarian change* (London: Macmillan).

Groenman-van Waateringe, W. (1983) The early agricultural utilization of the Irish landscape: the last word on the elm decline. In T. Reeves-Smyth and F. Hammond (eds) *Landscape archaeology in Ireland* (Oxford: British Archaeological Reports, British Series 116), 217–232.

Groenman-van Waateringe, W. and Robinson, M. (eds) (1988) *Man-made soils* (Oxford: British Archaeological Reports, International Series 410).

Grubb, M.J. (1988) The potential for wind energy in Britain. *Energy Policy* 16(6), 594–607.

Guha, R. (1989) *The unquiet woods: ecological change and peasant resistance in the Himalaya* (New Dehli: Oxford University Press).

Gujja, B. and Finger-Stich, A. (1996) What price prawn? Shrimp aquaculture's impact in Asia. *Environment* 38(7), 12–15 and 33–9.

Gulland, J. (1998) The end of whaling? *New Scientist*, 29 October, 42–7.

Gumbell, A. (1993) Ecology party switches leader and takes on a shade of pink. *The Guardian*, 15 November, 8.

Gunnarsson, J., Broman, D., Jonsson, P., Olsson, M. and Rosenberg, R. (1995) Interactions between eutrophication and contaminants: towards a new research concept for the European aquatic environment. *Ambio* 24(6), 383–5.

Gwyer, D. (1991) Fishing for trouble. *Geographical Magazine*, July, 20–4.

Haas, P. (1991) Policy responses to stratospheric ozone depletion. *Global Environmental Change* 1(3), 224–34.

Habermas, J. (1975) *Legitimation crisis* (London: Heinemann).

Habermas, J. (1978) *Knowledge and human interests* (London: Heinemann).

Habermas, J. (1979) *Communication and the evolution of society* (London: Heinemann).

Habermas, J. (1984) *The theory of communicative action, volume 1: reason and the rationalization of society* (London: Heinemann).

Habermas, J. (1987) *The theory of communicative action, volume 2: the critique of functionalist reason* (Cambridge: Polity Press).

Hacon, S., Rochedo, E.R.R., Campos, R.R.R. and Lacerda, L.D. (1997) Mercury exposure through fish consumption in the urban area of Alta Floresta in the Amazon basin. *Journal of Geochemical Exploration* 58, 157–72.

Häder, D.-P., Worrest, R.C., Kumar, H.D. and Smith, R.C. (1995) Effects of increased solar ultraviolet radiation on aquatic ecosystems. *Ambio* 24 (3), 174–80.

Hadfield, P. (1997) Raining acid on Asia. *New Scientist* 153, No 2069, 16–17.

Haines, M.R. (1995) Disease through the ages. In J.L. Simon (ed.) *The state of humanity* (Oxford: Blackwell), 51–60.

Halfacree, K. (1994) The importance of 'the rural' in the constitution of counterurbanization: evidence from England in the 1980s. *Sociologica Ruralis* XXXIV (2–3), 164–89.

Halfacree, K. (1996) Out of place in the country: travellers and the rural idyll. *Antipode* 28(1), 42–72.

Halfacree, K. (1997) Contrasting roles for the post-productivist countryside: a postmodern perspective on counterurbanisation. In P. Cloke and J. Little (eds) *Contested countryide: otherness, marginalisation and rurality* (London: Routledge), 109–22.

Hall, D.O. and House, J.I. (1995) Biomass energy in western Europe to 2050. *Land Use Policy* 12(1), 37–48.

Hall, D.O., Rosillo-Calle, F., Williams, R.H. and Woods, J. (1993) Biomass for energy: supply prospects. In T.B. Johansson, H. Kelly, A.K.N. Reddy and R.H. Williams, *Renewable energy* (Washington DC: Island Press).

Halliday, F. (1977) Migration and the labour force in the oil producing states of the Middle East. *Development and Change* 8, 263–92.

Hansen, J., Fung, I., Lacis, A., Rind, D., Lebedeff, S., Ruedy, R., Russell, G. and Stone, P. (1988) Global climate changes as forecast by Goddard Institute for Space Studies Three Dimensional Model. *Journal of Geophysical Research* 93, 9341–64.

Haraway, D. (1990) A manifesto for cyborgs: science, technology, and socialist feminism in the 1980s. In L. Nicholson (ed.) *Feminism and postmodernism* (London: Routledge).

Hardin, G. (1968) The tragedy of the commons. *Science* 162, 1234–48.

Hardy, T. (1978) *Tess of the D'Urbervilles* (Harmondsworth: Penguin).

Harlan, J.R. (1994) Plant domestication: an overview. In S.J. De Laet, A.H. Dani, J.L. Lorenzo and R.B. Nunoo (eds) *History of humanity, volume 1: prehistory and the beginnings of civilisation* (London: Routledge), 377–88.

Harper, K. (1995) Newbury by-pass approval angers anti-road lobby. *The Guardian*, 23 November, 5.

Harris, D. (1986) The mackerel massacre. In E. Goldsmith and N. Hildyard N. (eds) (1986) *Green Britain or industrial wasteland?* (London: Polity Press), 108–19.

Harris, D.R. (1996) *The origins and spread of agriculture and pastoralism in Eurasia* (London: UCL Press).

Harris, J. and Harris, B. (1989) Agrarian transformation in the Third World. In D. Gregory and R. Welford (eds) *Horizons in human geography* (London: Macmillan), 258–78.

Harrison, C. (1991) *Countryside recreation in a changing society* (London: TMS Partnership).

Harrison, P. (1990) Too much life on earth? *New Scientist* 1717, 28–9.

Harvey, D. (1982) *The limits to capital* (Oxford: Blackwell).

Harvey, D. (1989) *The condition of postmodernity* (Oxford: Blackwell).

Hasselmann, K. (1997) Are we seeing global warming? *Science* 276, 914–15.

Hauhs, M. and Wright, R.F. (1988) *Acid deposition: reversibility of soil and water acidification – a review* (Brussels: Commission of the European Communities).

Havens, K.E., Aumen, N.G., James, R.T. and Smith, V.H. (1996) Rapid ecological changes in a large subtropical lake undergoing cultural eutrophication. *Ambio* 25(3), 150–5.

Healy, P.F., Lambert, J.D.H., Arnason, J.T. and Hebda, R.J. (1983) Caracol, Belize: evidence of Ancient Maya agricultural terraces. *Journal of Field Archaeology* 10, 397–410.

Hecht, S. (1985) The Latin American livestock sector and its potential impact on women. In J. Monson and M. Kalb (eds) *Women as food producers in developing countries* (Los Angeles: University of California Press).

Hecht, S. and Cockburn, A. (1990) *The fate of the forest: developers, destroyers and defenders of the Amazon* (Harmondsworth: Penguin).

Hecht, S. and Cockburn, A. (1992) Realpolitik, reality and rhetoric in Rio. *Environment and Planning D: Society and Space* 10(4), 367–76.

Hedin, L.O. and Likens, G.E. (1996) Atmospheric dust and acid rain. *Scientific American* 275, 56–60.

Hellden, U. (1986) Desertification monitoring: remotely sensed data for drought impact studies in the Sudan. *European Space Agency SP–248*, Noordwijk, 417–28.

Hennepin, L. (1698) *A new discovery of a vast country in America* (London: British Library).

Henriksen, A., Lien, L,. Traen, T.S., Sevaldrud, I.S. and Brakke, D.F. (1988) Lake acidification in Norway – present and predicted chemical status. *Ambio* 17(4), 259–66.

Herrington, P. (1997) Pricing water properly. In T. O'Riordan (ed.) (1997) *Ecotaxation* (London: Earthscan), 263–86.

Hildyard, N. (1989) Adios Amazonia? *The Ecologist* 19(2), 53–62.

Hildyard, N. (1993) Foxes in charge of the chickens. In W. Sachs (ed.) *Global ecology: a new arena of political conflict* (London: Zed Books), 22–35.

Hill, K. (1995) The decline in child mortality. In J.L. Simon (ed.) *The state of humanity* (Oxford: Blackwell), 37–50.

Hillman, G.C. and Davies, M.S. (1990) Measured domestication rates in wild wheats and barley under primitive cultivation, and their archaeological significance. *Journal of World Prehistory* 4, 157–222.

Hinrichsen, D. (1997) Feeding a future world. *People and the Planet* 7(1), 6–9.

Hirons, K.R. and Edwards, K.J. (1990) Pollen and related studies at Kinloch, Isle of Rhum, Scotland, with particular reference to possible early human impacts on vegetation. *New Phytologist* 116, 715–27.

Hirst, E. (1991) Boosting US energy efficiency through federal action. *Environment* 33(2), 7–11 and 32–6.

HMSO (1986) *Radioactive waste. First report of the House of Commons Environment Committee Session 1985–6* (London: HMSO).

Hoagland, W. (1995) Solar energy. *Scientific American* 273(3), 136–9.

Hobson, J. (1902) *Imperialism – a study* (London: George Allen and Unwin).

Hodder, I. (1990) *The domestication of Europe* (Oxford: Blackwell).

Hodell, D.A., Curtis, J.H. and Brenner, M. (1995) Possible role of climate change in the collapse of Classic Maya civilisation. *Nature* 375, 391–4.

Hodgson, B. (1990) Alaska's big spill – can the wilderness heal? *National Geographic* 177(1), 5–43.

Hodgson, G. and Dixon, J.A. (1989) Logging versus fisheries in the Philippines. *The Ecologist* 19(4), 139–43.

Hofmann, D.J., Deshler, T., Aimedieu, P., Matthews, W.A., Johnston, P.V., Kondo, Y., Sheldon, W.R., Byrne, G.J. and Benbrook, J.R. (1989) Stratospheric clouds and ozone depletion in the Arctic during January 1989. *Nature* 340, 117–21.

Hofmann, D.J., Oltmans, S.J., Harris, J.M., Soloman, S., Deshler, T. and Johnson, S.J. (1992) Observation and possible causes of new ozone depletion in Antarctica in 1991. *Nature* 359, 283–7.

Hohenemser, C. (1996) Chernobyl, 10 years later. *Environment* 38(3), 3–5.

Holdgate, M.W. (1979) *Perspectives of environmental pollution* (Cambridge: Cambridge University Press).

Hollister, C. and Nadis, S. (1998) Burial of radioactive waste under the seabed. *Scientific American* 278(1), 40–5.

Holloway, M. (1991) Soiled shores. *Scientific American* 265(4), 80–94.

Holloway, M. (1996) Sounding out science. *Scientific American* 275(4), 82–8.

Holmes, I. (1997) Self-help in the jungle. *Independent on Sunday*, Review Section, 5 January, 54–5.

Holt, S. (1985) Whale mining, whale saving. *Marine Policy* 7(3), 192–213.

Hong, S., Candelone, J.-P., Patterson, C.C. and Boutron, C.F. (1994) Greenland ice evidence of hemispheric lead pollution two millennia ago by Greek and Roman civilisations. *Science* 265, 1841–3.

Hong S.M., Candelone, J.-P., Patterson, C.C. and Boutron, C.F. (1996) History of ancient copper smelting pollution during Roman and Medieval times recorded in Greenland ice. *Science* 272, 246–9.

Hornocker, M. (1997) Siberian tigers. *National Geographic* 191(2), 100–9.

Hornung, M. (1985) Acidification of soils by trees and forests. *Soil Use and Management* 1(1), 24–8.

Horton, T. (1993) Chesapeake Bay – hanging in the balance. *National Geographic* 183(6), 2–35.

Houghton, J.A. (1994) Emissions of carbon from land use change. In T.M.L. Wigley and D. Schimel (eds) *The carbon cycle* (Cambridge: Cambridge University Press).

Houghton, J.T., Meira Filho, L.G., Bruce, J., Hoesung Lee, Callander, B.A., Haites, E., Harris, N. and Maskell, K. (eds) (1995) *Climate change (1994) Radiative forcing of climate and an evaluation of the IPCC IS92 emission scenarios* (Cambridge: Cambridge University Press).

Houghton, J.T., Meiro Filho, L.G., Callender, B.A., Harris, N., Kattenburg, A. and Maskell, K. (eds) (1996) *Climate change (1995) The science of climate change* (Cambridge: Cambridge University Press).

Houghton, R.A. (1996) Converting terrestrial ecosystems from sources to sinks of carbon. *Ambio* 25(4), 267–72.

Howell, D.G., Bird, K.J. and Gautier, D.L. (1993) Oil: are we running out? *Earth* 2(2), 26–33.

Howkins, A. (1986) The discovery of rural England. In R. Colls and P. Dodd (eds) *Englishness: politics and culture, 1880–1920* (London: Croom Helm), 62–88.

Hoyt, E. (1993) Saving whales by watching them. *New Scientist,* 8 May, 45–6.

Hubbard, K.G. and Flores-Mendoza, F.J. (1995) Relating United States crop land use to natural resources and climate change. *Journal of Climate Change* 8, 329–35.

Huggett, R.J. (1998) *Fundamentals of biogeography* (London: Routledge).

Hughes, L., Cawsey, E.M. and Westoby, M. (1995) Climatic range sizes of Eucalyptus species in relation to future climate change. *Global Ecology and Biogeography Letters* 5, 23–9.

Hulme, M. (1989) Is environmental degradation causing drought in the Sahel? An assessment from recent empirical research. *Geography* 74(1), 38–46.

Hulme, M. (1997) Global warming. *Progress in Physical Geography* 21(3), 446–53.

Hulme, M. and Kelly, M. (1993) Exploring the links between desertification and climate change. *Environment* 35(6), 5–11 and 39–45.

Humborg, C., Ittekkot, V., Cociasu, A. and Bodungen, B.V. (1997) Effect of Danube River dam on Black Sea biogeochemistry and ecosystem structure. *Nature* 386, 385–8.

Huntley, B. (1993) Rapid early-Holocene migration and high abundance of hazel (*Corylus avellana* L.): alternative hypotheses. In F.M. Chambers (ed.) *Climate change and human impact on the landscape* (London: Chapman & Hall), 205–16.

Hutchings, G.J., Heneghan, C.S., Hudson, I.D. and Taylor, S.H. (1996) Uranium-oxide-based catalysts for the destruction of volatile chloro-organic compounds. *Nature* 384, 341–3.

Hutton, J. (1788) Theory of the Earth; or an investigation of the laws observable in the composition, dissolution, and restoration of land upon the globe. *Transactions, Royal Society of Edinburgh* 1, 209–304

IAEA (1992) *Radioactive waste management* (Vienna: IAEA).

IBRD (1998) *World Bank Atlas* (Washington: IBRD).

Idso, S.B. and Kimball, B.A. (1993) Tree growth in carbon dioxide enriched air and its implications for global carbon cycling and maximum levels of atmospheric CO_2. *Global Biogeochemical Cycles* 7, 537–55.

Igarashi, G., Saeki, S., Takahata, N., Sumikawa, K., Tasaka, S., Sasaki, Y., Takahashi, M. and Sano, Y. (1995) Groundwater radon anomaly before the Kobe earthquake in Japan. *Science* 269, 60–1.

Independent Commission on International Development Issues (1980) *North South: a programme for survival* (London: Pan Books).

Independent Commission on International Development Issues (1983) *Common crisis. North South: co-operation for world recovery* (London: Pan Books).

Inglehart, R. (1971a) The silent revolution in Europe: intergenerational change in post-industrial societies. *American Political Science Review* 65, 991–1017.

Inglehart, R. (1971b) *The silent revolution: changing values and political styles among western publics* (Princeton, New Jersey: Princeton University Press).

Inglehart, R. (1981) Post-materialism in an age of environment of insecurity. *American Political Science Review* 75, 880–900.

Ingram, R.G., Wang, J., Lin, C., Legendre, L. and Fortier, L. (1996) Impact of fresh water on a sub-Arctic coastal ecosystem with seasonal sea ice (south-eastern Hudson Bay,

Canada) I: Interannual variability and predicted global warming on river plumes dynamics and sea ice. *Journal of Marine Systems* 7, 221–31.

Innes, J.B. and Simmons, I.G. (1988) Disturbance and diversity: floristic changes associated with pre-elm decline woodland recession in north-east Yorkshire. In M. Jones (ed.) *Archaeology and the flora of the British Isles*. Oxford University committee for Archaeology Monograph number 14, 7–20.

International Labour Office (1997) *Yearbook of labour statistics* (Geneva: International Labour Office).

International Union for the Conservation of Nature and Natural Resources (1980) *World Conservation Strategy* (Gland, Switzerland: IUCN, UNEP and WWF).

Irvine, S. (1989) Changing ourselves or changing society? The limits of green consumerism. *The Ecologist* 19(1), 33–6.

Irvine, S. and Ponton, A. (1988) *A green manifesto: policies for a green future* (London: MacDonald Optima).

IUCN (1996) *IUCN red list of threatened animals* (Switerland: Gland).

Iversen, J. (1941) Landnam i Denmarks stenalder (Land occupation in Denmark's Stone Age). *Danmarks Geologiske Undersøgelse II*, 66, 1–68.

Iversen, J. (1973) The development of Denmark's nature since the last glacial. *Danmarks Geologiske Undersøgelse Series V*, 7–C, 1–26.

Jacobs, M. (1991) *The green economy* (London: Pluto Press).

Jardine, K. (1994) Finger on the carbon pulse: climate change and the boreal forests. *The Ecologist* 24(6), 220–4.

Jaworowski, Z. (1996) Damage assessment. *Scientific American* 275(4), 6–8.

Jeffreys, K. (1994) Free market environmentalism. Can it save the planet? *Economic Affairs* 14(3), 6–9.

Jeffreys, A. (1998) Ecotourism in north-west Ecuador. *Geography Review* 11(3), 26–9.

Jennings, S. and Polunin, N.V.C. (1996) Impacts of fishing on tropical reef systems. *Ambio* 25(1), 44–9.

Jensen, K.W. and Snekvik, E. (1972) Low pH levels wipe out salmon and trout populations in southernmost Norway. *Ambio* 1, 223–5.

Jessop, B. (1982) *The capitalist state* (Oxford: Blackwell).

Joad, C. (1938) The people's claims. In C. Williams-Ellis (ed.) *Britain and the beast* (London: J.M. Dent & Sons), 64–85.

Johansson, T.B.J., Kelly, H., Reddy, A.K.N. and Williams, R.H. (1993) (eds) *Renewable energy: sources for fuels and electricity* (Washington DC: Island Press).

Johnston, R.J. (1989) *Environmental problems: nature, economy and state* (London: Belhaven).

Jolly, A. (1988) Madagascar's lemurs – on the edge of survival. *National Geographic* 174(2), 132–61.

Jones, A. (1987) Green tourism. *Tourism Management,* December, 354–6.

Jones, A.E. and Shanklin, J.D. (1995) Continued decline of total ozone over Halley, Antarctica, since 1985. *Nature* 376, 409–11.

Jones, P.D. and Wigley, T.M.L. (1990) Global warming trends. *Scientific American* 263, 84–91.

Jones, P.D., Raper, S.C.B. and Wigley, T.M.L. (1986a) Southern hemispheric surface air temperature variations: 1851–1984. *Journal of Climatology and Applied Meteorology* 25, 213–30.

Jones, P.D., Raper, S.C.B., Bradley, R.S., Diaz, H.F., Kelly, P.M. and Wigley, T.M.L. (1986b) Northern hemispheric surface air temperature variations: 1851–1984. *Journal of Climatology and Applied Meteorology* 25, 161–79.

Jones, P.D., Wigley, T.M.L. and Wright, P.B. (1986c) Global temperature variations: between 1851 and 1984. *Nature* 322, 430–4.

Jones, D.K.C. (1991) Environmental hazards. In R. Bennett and R. Estall (eds) *Global change and challenge: geography for the 1990s* (London: Routledge), 27–56.

Juniper, C. (1994) Efficient recycling programs can reduce waste. In C.P. Cozic (ed.) *Pollution* (San Diego: Greenhaven Press), 184–7.

Kaland, P.E. (1986) The origin and management of Norwegian coastal heaths as reflected by pollen analysis. In K.-E. Behre (ed.) *Anthropogenic indicators in pollen diagrams* (Rotterdam: A.A. Balkema), 19–36.

Kammen, D.M. (1995) Cookstoves for the developing world. *Scientific American* 273, 64–7.

Karp, B. (1996) Sustainable development is impracticable. In A.E. Sadler (ed.) *The environment: opposing viewpoints* (San Diego: Greenhaven Press), 163–70.

Karr, J. (1994) Landscapes and management for ecological integrity. In Ke Chung Kim and R.D. Weaver (eds) *Biodiversity and landscapes: a paradox of humanity* (Cambridge: Cambridge University Press), 229–51.

Kasahara, K. (1981) *Earthquake mechanics* (Cambridge: Cambridge University Press).

Kattenburg, A., Giorgi, F., Grassl, H., Meehl, G.A., Mitchell, J.F.B., Stouffer, R.J., Tokioka, T., Weaver, A.J. and Wigley, T.M.L. (1996) Climate models – projections of future climate. In J.T. Houghton, L.G. Meiro Filho, B.A. Callender, N. Harris, A. Kattenburg and K. Maskell (eds) *Climate change (1995) The science of climate change* (Cambridge: Cambridge University Press).

Keeling, C.D., Whorf, T.P., Wahlen, M. and van der Pilcht, J. (1995) Interannual extremes in the rate of rise of atmospheric carbon dioxide since 1980. *Nature* 375, 666–70.

Keller, E.A. (1975) Channelization: a search for a better way. *Geology* 3, 246–8.

Keller, E.A. (1976) Channelization: environmental, geomorphic, and engineering aspects. In D.R. Coates (ed.) *Geomorphology and engineering* (Stroudsburg, Pennsylvania: Dowden, Hutchinson & Ross), 115–40.

Kelly, M. (1998) Into the greenhouse century. *People and the Planet* 7(1), 29.

Kelly, S. (1997) Wind power. *Geography Review* 10(4), 22–5.

Kemp, D.D. (1994) *Global environmental issues: a climatological approach* (2nd edn) (London: Routledge).

Kemp, T. (1967) *Theories of imperialism* (London: Dobson).

Kennedy, I.R. (1992) *Acid soil and acid rain* (Chichester: John Wiley & Sons).

Kerr, R.A. (1988) Evidence of Arctic ozone destruction. *Science* 240, 1144–5.

Kerr, R.A. (1996a) A new way to ask the experts: rating radioactive waste risks. *Science* 274, 913–4.

Kerr, R.A. (1996b) Ozone-destroying chlorine tops out. *Science* 271, 32.

Kerr, R.A. (1998) A hint of unrest at Yucca Mountain. *Science* 279, 2040–1.

Kessell, H. (1985) Changes in environmental awareness: a comparative study of the FRG, England and the USA. *Land Use Policy* 2, 103–13.

Keyfitz, N. (1989) The growing human population. *Scientific American* 261(3), 71–8.

Khalil, M.A.K. and Rasmussen, R.A. (1985) Causes of increasing atmospheric methane: depletion of hydroxyl radicals and the rise of emissions. *Atmospheric Environment* 19(3), 397–407.

King, C.Y. (1988) Earthquake prediction techniques. *Encyclopedia of Earth System Science* 2, 111–8.

King, R. (1973) Geographical perspectives on the green revolution. *Tijdschrift voor Economische en Sociale Geografie* 64, 237–44.

Kirby, S. (1993) The thick black line. *New Scientist*, 30 January, 24–5.

Kleemola, J., Pehu, E., Peltonen-Sainio, P. and Karvonen, T. (1995) Modelling the impact of climate change on growth of spring barley in Finland. *Journal of Biogeography* 22, 581–90.

Kleiner, K. (1995) Billion-dollar drugs are disappearing in the forest. *New Scientist* No 1985, 5.

Klinger, J. (1994) Debt-for-nature swaps and the limits to international cooperation on behalf of the environment. *Environmental Politics* 3(2), 229–46.

Kliskey, A.D. (1994) A comparative analysis of approaches to wilderness perception mapping. *Journal of Environmental Management* 41, 199–236.

Knight, D,. Mitchell, B. and Wall, G. (1997) Bali: sustainable development, tourism and coastal management. *Ambio* 26(2), 90–6.

Knopman, D.S. and Smith, R.A. (1993) 20 years of the clean water act. *Environment* 35(1), 17–20 and 34–41.

Knox, P. and Agnew, J. (1989) *The geography of the world economy* (London: Edward Arnold).

Kohl, L. (1989) Heavy hands on the land. *National Geographic* 174, 633–52.

Kropotkin, P. (1899) *Fields, factories and workshops tomorrow* (London: Freedom Press).

Kubiakmartens, L. (1996) Evidence for possible use of plant foods in Palaeolithic and Mesolithic diet from the site of Calowanie in the central part of the Polish Plain. *Vegetation History and Archaeobotany* 5, 33–8.

Kulp, J.L. (1995) Acid rain. In J.L. Simon (ed.) *The state of humanity* (Oxford: Blackwell), 523–35.

Kurien, J. (1993) Ruining the commons: coastal overfishing and fishworkers' actions in south India. *The Ecologist* 23(1), 5–11.

Kwai-Cheong, C. (1995) The Three Gorges Project of China: resettlement prospects and problems. *Ambio* 24(2), 98–102.

Lacerda, L.D. and Marins, R.V. (1997) Anthropogenic mercury emissions to the atmosphere in Brazil: the impact of gold mining. *Journal of Geochemical Exploration* 58, 223–9.

Lance, A., Baugh, I. and Love, J. (1989) Continued footpath widening in the Cairngorm Mountains, Scotland. *Biological Conservation* 49, 201–14.

Langedal, M. (1997a) Dispersion of tailings in the Knabeåna–Kvina drainage basin, Norway 1. Evaluation of overbank sediments as sampling medium for geochemical mapping. *Journal of Geochemical Exploration* 58, 157–72.

Langedal, M. (1997b) Dispersion of tailings in the Knabeåna–Kvina drainage basin, Norway 2. Mobility of Cu and Mo in tailings-derived fluvial sediments. *Journal of Geochemical Exploration* 58, 173–84.

Lappé, F. and Collins, J. (1980) *Food first: a new action plan to break the famine trap* (London: Abacus).

Larrain, J. (1989) *Theories of development: capitalism, colonialism and development* (Cambridge: Polity).

Lash, S. and Urry, J. (1987) *The end of organised capitalism* (Cambridge: Polity).

Lashof, D.A. and Ahuja, D.R. (1990) Relative contributions of greenhouse gas emissions to global warming. *Nature* 344, 529–31.

Lean, G. (1995a) The food runs out. *Independent on Sunday*, 12 November, 17.

Lean, G. (1995b) Assault on green laws endangers Newt Brigade. *Independent on Sunday*, 23 April, 15.

Lean, G. (1996) And still the children go hungry. *Independent on Sunday*, 10 November, 12–13.

Lee, J.A. and Tallis, J.H. (1973) Regional and historical aspects of lead pollution in Britain. *Nature* 245, 216–18.

Lee, R. (1986) Post-industrial society. In R.J. Johnson, D. Gregory and D. Smith (eds) *The dictionary of human geography* (Oxford: Blackwell, 2nd edn), 362.

Lee, R. (1994) Development. In R.J. Johnson, D. Gregory and D. Smith (eds) *The dictionary of human geography* (Oxford: Blackwell 3rd edn), 128–30.

Leffell, D.J. and Brash, D.E. (1996) Sunlight and skin cancer. *Scientific American* 275(1), 38–43.

Lele, S. (1991) Sustainable development: a critical review. *World Development* 19(6), 607–21.

Lenihan, J.M. and Neilson, R.P. (1995) Canadian vegetation sensitivity to projected climate change at three organisation levels. *Climate Change* 30, 27–56.

Lenin, V. (1915) *Imperialism, the highest form of capitalism* (Moscow: Foreign Language Publishing House).

Lewis, D. (1991) The rape of the rainforest. *The Guardian* Environment Section, 1 November, 33.

Lewis, S. (1991) Selective logging can save rainforests. In M. Polesetsky (ed.) *Global resources: opposing viewpoints* (San Diego: Greenhaven Press), 169–75.

Ley, D. (1980) Liberal ideology and the post-industrial city. *Annals, Association of American Geographers* 70, 238–58.

Ley, D. (1987) Styles of the times: liberal and neo-conservative landscapes in inner Vancouver, 1968–1986. *Journal of Historical Geography* 31(1), 40–56.

Leyden, B., Brenner, M. and Dahlin, B.H. (1998) Cultural and climatic history of Cobà, a lowland Maya city in Quintana Roo, Mexico. *Quaternary Research* 49, 111–22.

Liansky, J. (1979) Women in the Brazilian frontier. *Latinamericanist* 15(1).

Likens, G.E., Driscoll, C.T. and Buso, D.C. (1996) Long-term effects of acid rain: response and recovery of a forest ecosystem. *Science* 272, 244–5.

Likhtarev, I.A., Sobolev, B.G., Kairo, I.A., Tronko, N.D., Bogdanova, T.I., Olenic, V.A., Epshtein, E.V. and Beral, V. (1995) Thyroid cancer in the Ukraine. *Nature* 375, 365.

Line, L. (1993) Silence of the songbirds. *National Geographic* 183(6), 68–91.

Lipton, M. (1977) *Why poor people stay poor: a study of urban bias in world development* (London: Temple Smith).

Lister, A. and Bahn, P. (1994) *Mammoths* (New York: Macmillan).

Livingstone, D. (1992) *The geographical tradition* (Oxford: Blackwell).

Livingstone, D. (1995) The polity of nature: representation, virtue, strategy. *Ecumene* 23, 353–77.

Lloyd Parry, R. (1997) A wet, warm, unhappy Christmas. *Independent on Sunday*, 7 December, 14.

Lockeretz, W., Shearer, G. and Kohl, D.H. (1981) Organic farming in the corn belt. *Science* 211, 540–7.

Lockwood, D. (1989) The weakest link in the chain? Some comments on the Marxist theory of action. In D. Rose (ed.) *Social stratification* (London: Hutchinson).

Lockwood, J.G. (1984) The southern oscillation and El Niño. *Progress in Physical Geography* 8, 102–10.

Løkke, H., Bak, J., Falkengren-Grerup, U., Finlay, R.D., Ilvisniemi, H., Holn Nygaard, P. and Starr, M. (1996) Critical loads of acidic deposition for forest soils: is the current approach adequate? *Ambio* 25(8), 510–16.

Lomnitz, C. (1995) *Fundamentals of earthquake prediction* (Chichester: John Wiley & Sons).

Longstreth, J.D., de Gruijl, F.R., Kripke, M.L., Takizawa, Y. and van de Leun, J.C. (1995) Effects of increased solar ultraviolet radiation on human health. *Ambio* 24(3), 153–65.

Loske, R. (1991) Ecological taxes, energy policy and greenhouse gas reductions: a German perspective. *The Ecologist* 21(4), 173–6.

Lottermoser, B.G. (1995) Noble metals in municipal sewage sludges of south-eastern Australia. *Ambio* 24(6), 354–7.

Lovejoy, A.O. (1935) Some meanings of 'Nature'. In A.O. Lovejoy and G. Boas (eds) *Primitivism and related ideas in antiquity* (Baltimore: Johns Hopkins University Press).

Lovelock, J. (1979) *Gaia* (Oxford: Oxford University Press).

Lovelock, J. (1988) *The ages of Gaia: a biography of our living planet* (Oxford: Oxford University Press).

Lowe, J.J. and Walker, M.J.C. (1997) *Reconstructing past environments* (Harlow: Longman).

Lowe, P. and Goyder, J. (1983) *Environmental groups in politics* (London: Allen & Unwin).

Lowe, P. and Rudig, W. (1984) Political ecology and the social sciences – the state of the art. *British Journal of Political Science* 16, 513–50.

Lund, M.W. (1990) Energy galore. *Geographical Magazine* September, 40–4.

Luo, Y., TeBeest, D.O., Teng, P.S. and Fabellar, N.G. (1995) Simulation studies on risk analysis of rice leaf blast epidemics associated with global climate change in several Asian countries. *Journal of Biogeography* 22, 673–8.

Luoma, J.R. (1988) The human cost of acid rain. *Audubon* 90(4), 16–27.

Lutz, W., Sanderson, W. and Scherbov, S. (1977) Doubling of world population is unlikely. *Nature* 387, 803–5.

Luxemburg, R. (1951) *The accumulation of capital* (London: Routledge and Kegan Paul).

Luxemburg, R. and Bukharin, N. (1972) *Imperialism and the accumulation of capital* (London: Allen Lane).

Lynch, A. (1981) *Man and the environment in south-west Ireland* (Oxford: BAR British Series 85).

Mabogunje, A.L. (1980) *The development process: a spatial perspective* (London: Hutchinson).

Mabogunje, A.L. (1995) The environmental challenges in sub-Saharan Africa. *Environment* 37(4), 4–9 and 31–5.

MacCannell, D. (1989) *The tourist: a new theory of the leisure class* (New York: Schoken Books).

MacDonald, I.R. (1998) Natural oil spills. *Scientific American* 279(5), 30–5.

MacIlwain, C. (1997) Clinton sets a careful course on climate. *Nature* 387, 640–1.

Mackay, R.M. and Probert, S.D. (1996) Iceland's energy and environmental strategy. *Applied Energy* 53, 245–81.

MacKenzie, D. (1996) Captain Birdseye plans award for friendly fisheries. *New Scientist* 149 No 2019, 11.

MacKenzie, D. and Joyce, C. (1990) Antarctica: a tale of two treaties. *New Scientist*, 17 November, 16.

Macklin, M. (1992) Metal pollution of soils and sediments: a geographical perspective. In M.D. Newson (ed.) *Managing the human impact on the natural environment: patterns and processes* (London: Belhaven Press), 172–95.

Macmillan, B. (1991) Famine: The unnatural disaster. *Geography Review*, September, 18–23.

MacNaghten, P. and Urry, J. (1998) *Contested natures* (London: Sage).

Madeley, J. (1986) Britain and the Third World. In E. Goldsmith and N. Hildyard (eds) *Green Britain or industrial wasteland?* (Cambridge: Polity), 322–7.

Maffesoli, M. (1996) *The time of the tribes* (London: Sage).

Mahat, T., Griffin, D. and Shepherd, K. (1986a) Human impact on some forests of the Middle Hills of Nepal, 1. Forestry in the context of the traditional resources of the state. *Mountain Research and Development* 6, 223–32.

Mahat, T., Griffin, D. and Shepherd, K. (1986b) Human impact on some forests of the Middle Hills of Nepal, 2. Some major human impacts before 1950 on the forests of Sindhu Palchok and Kabhre Palanchock. *Mountain Research and Development* 6, 325–34.

Mahat, T., Griffin, D. and Shepherd, K. (1987a) Human impact on some forests of the Middle Hills of Nepal, 3. Forests in the subsistence economy of Sindhu Palchok and Kabhre Palanchok. *Mountain Research and Development* 7, 53–70.

Mahat, T., Griffin, D. and Shepherd, K. (1987b) Human impact on some forests of the Middle Hills of Nepal, 4. A detailed survey of south-east Sindhu Palchok and north-east Kabhre Palanchok. *Mountain Research and Development* 7, 111–34.

Mahony, R. (1992) Debt-for-nature swaps: who really benefits? *The Ecologist* 22(3), 97–103.

Malle, K.-G. (1996) Cleaning up the River Rhine. *Scientific American* 274(1), 54–9.

Malmer, N. (1976) Acid precipitation: chemical changes in the soil. *Ambio* 5, 231–4.

Malthus, T. (1798) *An essay on the principle of population as it affects the future improvement of society with remarks on the speculations of Mr Godwin, M. Condorcet, and other writers* (London: J. Johnson).

Malthus, T. (1800) The present high price of provision. In *The pamphlets of Thomas Malthus* (1961 reprinted) (New York: Kelly).

Mann, C. and Plummer, M. (1995) Is endangered species act in danger? *Science* 267, 1256–8.

Manney, G.L., Froidevaux, L., Waters, J.W., Zurek, R.W., Read, W.G., Elson, L.S., Kumer, J.B., Mergenthaler, J.L., Roche, A.E., O'Neill, A., Harwood, R.S., Mackenzie, I. and Swinbank, R. (1994) Chemical depletion of ozone in the Arctic lower stratosphere during winter 1992–93. *Nature* 370, 429–34.

Mannion, A.M. (1991a) Aquatic ecosystem development in Scotland: a review based on evidence from diatom assemblages. *Scottish Geographic Magazine* 103, 13–20

Mannion, A.M. (1991b) *Global environmental change* (Harlow: Longman).

Mannion, A.M. (1995) *Agriculture and environmental change: temporal and spatial dimensions* (Chichester: John Wiley & Sons).

Mannion, A.M. (1997) *Global environmental change* (Harlow: Longman, 2nd edn).

Mannion, A.M. (1999) Domestication and the origins of agriculture: an appraisal. *Progress in Physical Geography*, 23(1), 37–56.

Manshard, W. (1985) The world's food and energy requirements – a global conflict for resources. *Natural Resources and Development* 21, 120–7.

Martin, M.N., Coughtrey, P.J. and Ward, P. (1979) Historical aspects of heavy metal pollution in the Gordano Valley. *Proceedings of the Bristol Naturalists Society* 37, 91–7.

Martin, P.H. and Lefebvre, M.G. (1995) Malaria and climate: sensitivity of Malaria potential transmission to climate. *Ambio* 24(4), 200–7.

Marx, K. (1976) *Capital: a critique of political economy* (Harmondsworth: Penguin)

Marx, K. and Engels, F. (1971) *Manifesto of the Communist Party* (Moscow: Progress Publishers).

Masood, E. (1997a) Britain seeks leadership role with ambious greenhouse-gas targets. *Nature* 387, 640.

Masood, E. (1997b) US seeks greenhouse gas cuts from the Third World. *Nature* 386, 103.

Masood, E. (1997c) Europe seeks to head off oil-exporters' veto on climate treaty. *Nature* 387, 537.

Masood, E. (1997d) Error re-opens scientific whaling debate. *Nature* 374, 587.

Massonnet, D., Rossi, M., Carmona, C., Adragna, F., Peltzer, G., Feige, K. and Rabaute, T. (1993) The displacement field of the Landers earthquake mapped by radar interferometry. *Nature* 364, 138–42.

Matless, D. (1990a) Ages of English design: preservation, modernism and the tales of their history. *Journal of Design History* 3(4), 203–12.

Matless, D. (1990b) Definitions of England 1928–1989. *Built Environment* 16, 179–91.

Matless, D. (1994) Moral geography in Broadland. *Ecumene* 1(2), 127–55.

Matless, D. (1995) The art of right living: landscape and citizenship, 1918–39. In S. Pile and N. Thrift (eds) *Mapping the subject: geographies of cultural transformation* (London: Routledge).

Maugh, T.H. II (1984) Acid rain's effects on people assessed. *Science* 226, 1408–10.

McAndrews, J.H. (1988) Human disturbance of North American forests and grassland: the fossil pollen record. In B. Huntley and T. Webb III (eds) *Vegetation history* (Dordrecht: Kluwer Academic Publishers), 673–97.

McAndrews, J.H. and Boyko-Diakonow, M. (1989) Pollen analysis of varved sediment at Crawford Lake, Ontario: evidence of Indian and European farming. In R.J. Fulton (ed.) *Geology of Canada and Greenland*. Ottawa: Geological Survey of Canada, 528–30.

McCormick, J. (1992) *The global environmental movement* (Chichester: Wiley, 2nd edn).

McCormick, J. (1998) Acid pollution: the international community's continuing struggle. *Environment* 40(3), 17–20 and 41–5.

McCorriston, J. and Hole, F.A. (1991) The ecology of seasonal stress and the origins of agriculture in the Near East. *American Anthropologist* 93, 46–9.

McCracken, D.V., Scott-Smith, M., Grove, J.H., Mackown, C.T. and Blevins, R.L. (1994) Nitrate leaching as influenced by cover cropping and nitrogen source. *Soil Science Society America Journal* 58, 1476–83.

McCully, P. (1991) A message to the executive and shareholders of D.I. Dupont du Nemours and Co. and Imperial Chemical Industries Ltd. *The Ecologist* 21(3), 114–16.
McCully, P. (1996) *Silenced rivers: the ecology and politics of large dams* (London: Zed Books).
McEvoy, A.F. (1987) Towards an interactive theory of nature and culture: ecology, production and cognition in the Californian fishing industry. *Environmental Review* 11(4), 289–305.
McGowan, F. (1991) Controlling the greenhouse effect: the role of renewables. *Energy Policy* 19(2), 110–18.
McGregor, D.F.M. and Barker, D. (1991) Land degradation and hillside farming in the Fall River basin, Jamaica. *Applied Geography* 11, 143–56.
McKay, G. (1996) *Senseless acts of beauty* (London: Verso).
McLaughlin, B. (1986) Rural policy in the 1980s: the revival of the rural idyll. *Journal of Rural Studies* 2(2), 81–90.
Meadows, D.H., Meadows, D.L., Randers, J. and Behrens, W.W. III (1972) *The limits to growth* (New York: Universe Books).
Meadows, D.H., Meadows, D.L. and Randers, J. (1992) *Beyond the limits: global collapse or sustainable future?* (London: Earthscan).
Meertens, D. (1993) Housewifisation and colonisation in the Colombian rainforest. In J. Momsen and V. Kinnaird (eds) *Women's roles in colonisation: a Colombian case study* (London: Routledge), 256–69.
Melamed, R., Villas Boas, R.C., Goncalves, G.O. and Paiva, E.C. (1997) Mechanisms of physico-chemical interaction of mercury with river sediments from a gold region in Brazil: relative mobility of mercury species. *Journal of Geochemical Exploration* 58, 119–24.
Mellaart, J. (1967) *Catal Huyuk: a Neolithic town in Anatolia* (London: Thames and Hudson).
Mellor, M. (1997) *Feminism and ecology* (Cambridge: Polity).
Mellucci, A. (1989) *Nomads of the present* (London: Radius).
Mendels, F. (1972) Protoindustrialization: the first phase of industrialization. *Journal of Economic History* 32, 241–61.
Merryfield, D.L. and Moore, P.D. (1974) Prehistoric human activity and blanket peat initiation on Exmoor. *Nature* 250, 439–41.
Mestel, R. (1995) Tension mounts as the big one stalks LA. *New Scientist*, 145, 16.
Metcalfe, J.D. and Arnold, G.P. (1997) Tracking fish with electronic tags. *Nature* 387, 665–6.
Michalak, W. and Gibb, R. (1992) Political geography and Eastern Europe. *Area* 24(4), 341–9.
Michaels, P.J. (1995) The greenhouse effect and global change: review and reappraisal. In J.L. Simon (ed.) *The state of humanity* (Oxford: Blackwell), 544–64.
Michaels, P.J., Singer, F.S. and Knappenberger, P.C. (1994) Analyzing ultraviolet-B radiation: is there a trend? *Science* 264, 1341–2.
Mighall, T.M. (in press) Geochemical monitoring of heavy metal pollution and prehistoric mining: evidence from Copa Hill, Cwmystwyth, mid-Wales and Mount Gabriel, Co. Cork, south-west Ireland. In P.T. Craddock and J. Lang (eds) *Aspects of early mining and extractive metallurgy* (London: British Museum).
Miller, G.T. (1990) *Living in the environment* (Belmont, California: Wadsworth Publishing Company).
Miller, J.R. (1997) The role of fluvial geomorphic processes in the dispersal of heavy metals from mine sites. *Journal of Geochemical Exploration* 58, 101–18.
Miller, K. (1998) Weather. *Life* August, 38–52.
Miller, R.V. (1998) Bacterial gene swapping in nature. *Scientific American* 278(1), 46–51.
Millington, A.C., Mutiso, S.K., Kirkby, S.J. and O'Keefe, P. (1989) African soil erosion – nature undone and the limitations of technology. *Land Degradation and Rehabilitation* 1, 279–90.
Milne, R. and Brown, W. (1991) Electricity giant axes studies on acid rain. *New Scientist*, 13 July, 15.

Mishan, E. (1969) *The costs of economic growth* (Harmondsworth: Penguin).

Mitchell, B. (1989) *Geography and resource analysis* (Harlow: Longman).

Mitchell, B. and King, P. (1984) Resource conflict, policy change and practice in Canadian fisheries management. *Geoforum* 15(3), 419–32.

Mitchell, J.F. and Johns, T.C. (1997) On the modification of global warming by sulphate aerosols. *Journal of Climate* 10, 245–67.

Mitchell, J.G. (1996) Our polluted runoff. *National Geographic* 189(2), 106–25.

Mitchell, R.B. (1995) Compliance with international treaties: lessons from international oil pollution. *Environment* 37(4), 10–15 and 36–41.

Molebatsi, C.O. (1996) Towards a sustainable city: Gaborone, Botswana. *Ambio* 25(2), 126–33.

Molina, M.J. and Rowland, F.S. (1974) Stratospheric sink for chlorofluoromethanes: chlorine atom-catalysed destruction of ozone. *Nature* 249, 810–12.

Molleson, T. (1994) The eloquent bones of Abu Hureyra. *Scientific American* 271, 60–5.

Monk, M. (1993) People and environment: in search of the farmers. In E. Shee Twohig and M. Ronayne (eds) *Past perceptions: the prehistoric archaeology of south-west Ireland* (Cork: Cork University Press), 35–52.

Montzka, S.A., Butler, J.H., Myers, R.C., Thompson, T.M., Swanson, T.H., Clarke, A.D., Lock, L.T. and Elkins, J.W. (1996) Decline in the tropospheric abundance of halogen from halocarbons: implications for stratospheric ozone depletion. *Science* 271, 1318–22.

Moore, A.T.M. and Hillman, G. (1992) The Pleistocene to Holocene transition and human economy in south-west Asia: the impact of the Younger Dryas. *American Antiquity* 57, 482–94.

Moore, P.D. (1975) Origin of blanket mires. *Nature* 256, 267–69.

Moore, P.D. (1987) Man and mire, a long and wet relationship. *Transactions of the Botanical Society of Edinburgh* 45, 77–95.

Moore, P.D. (1988) The development of moorlands and upland mires. In M. Jones (ed.) *Archaeology and the flora of the British Isles*, Oxford University Committee for Archaeology Monograph number 14, 116–22.

Moore, P.D. (1993) The origin of blanket mire, revisited. In F.M. Chambers (ed.) *Climate change and human impact on the landscape* (London: Chapman & Hall), 217–24.

Moran, E.F. (1996) Deforestation in the Brazilian Amazon. In L.E. Sponsel, T.N. Headland and R.C. Bailey (eds) *Tropical deforestation: the human dimension* (New York: Columbia University Press), 149–64.

Morehouse, W. (1994) Unfinished business: Bhopal ten years after. *The Ecologist* 24(5), 164–68.

Morgan, R.P.C. (1995) *Soil erosion and conservation* (Harlow: Longman).

Morris, E.M. and Thomas, A.G. (1986) Transient acid surges in an upland stream. *Water, Air and Soil Pollution* 34, 429–38.

Morrison, D. and Dunlap, R. (1986) Environmentalism and elitism: a conceptual and empiririical analysis. *Environmental Management* 10, 581–9.

Motluk, A. (1995) Paralysis stalls progress on whaling. *New Scientist* 146, No 1980, 4.

Motluk, A. (1996) Blood on the water. *New Scientist*, 22 June, 12–13.

Mukerjee, M. (1995a) Persistently toxic. *Scientific American* 272(6), 9–10.

Mukerjee, M. (1995b) Toxins abounding. *Scientific American* 273(1), 15–16.

Muller, F. (1996) Mitigating climate change: the case for energy taxes. *Environment* 38(2), 13–20 and 36–42.

Munksgaard, J. and Larsen, A. (1998) Socio-economic assessment of wind power – lessons from Denmark. *Energy Policy* 26(2), 85–93.

Murdoch, J. and Marsden, T. (1991) *Reconstituting the rural in an urban region: new villages for old? ESRC Countryside Change Project, working paper* (University of Newcastle).

Murdoch, J. and Marsden, T. (1994) *Reconstituting rurality: class, community and power in the development process* (London: UCL Press).

Murphy, P.E. (1985) *Tourism: a community approach* (London: Routledge).

Murphy, R. (1994) *Rationality and nature: a sociological enquiry into a changing relationship* (Oxford: Westview Press).

Myers, J.G., Moore, S. and Simon, J.L. (1995) Trends in the availability of non-fuel minerals. In J.L. Simon (ed.) *The state of humanity* (Oxford: Blackwell), 303–12.

Myers, N. (1979) *The sinking ark: a new look at the problem of disappearing species* (Oxford: Pergamon).

Myers, N. (1986) Environmental repercussion of deforestation in the Himalayas. *Journal of World Forest Resource Management* 2, 63–72.

Myers, N. (1989) The future of forests. In L. Friday and R. Laskey (eds) *The Fragile Environment. The Darwin Lectures* (Cambridge: Cambridge University Press).

Myers, N. (1992) *The primary source: tropical forests and our future* (New York: W.W. Norton).

Myers, N. (1993a) Biodiversity and the precautionary principle. *Ambio* 22(2/3), 74–9.

Myers, N. (1993b) Tropical rainforests: the main deforestation fronts. *Environmental Conservation* 20, 9–16.

Myers, R.A., Barrolman, N., Hoenig, J.M. and Qu, Z. (1996) The collapse of cod in eastern Canada: the evidence from tagging data. *Journal of Marine Science* 53, 629–40.

Naess, A. (1973) The shallow and the deep, long-range ecology movements. *Inquiry* 16, 95–100.

Naess, A. (1989) *Ecology, community and lifestyle* (Cambridge: Cambridge University Press).

Nash, J. and Ehrenfeld, J. (1996) Code green: business adopts voluntary environmental standards. *Environment* 38(1), 16–20 and 36–45.

Nef, J. (1964) *The conquest of the material world* (Chicago: University of Chicago Press).

Nesmith, C. and Radcliffe, S. (1997) Remapping Mother Nature: a geographical perspective on environmental feminism. *Environment and Planning D: Society and Space* 11, 79–94.

New Forests Project (1991) Changing farmers lifestyles can save rainforests. In M. Polesetsky (ed.) *Global resources: opposing viewpoints* (San Diego: Greenhaven Press), 143–9.

Newson, L. (1996) The Latin American colonial experience. In D. Preston (ed.) *Latin American development: geographical perspectives* (Harlow: Longman), 11–40.

Nicholson, S.E. (1978) Climatic variations in the Sahel and other African regions during the past five centuries. *Journal of Arid Environments* 1, 3–24.

Nicholson, S.E. (1989) Long term changes in African rainfall. *Weather* 44, 47–56.

Nisbett, E.G. (1989) Some northern sources of atmospheric methane: production history and future implications. *Canadian Journal of Science* 26, 1603–11.

Nriagu, J.O. (1983) *Lead and lead poisoning in antiquity* (Chichester: John Wiley & Sons).

Nriagu, J.O. (1996) A history of global metal pollution. *Science* 272, 223–4.

Nukl, P. (1996) Pesticide levels in vegetables and fruit cause alarm. *Sunday Times*, 3 November, 11.

O'Brien, W.F. (1994) *Mount Gabriel: Bronze age mining in Ireland* (Galway: Galway University Press).

O'Connor, J. (1989) Political economy of ecology in socialism and capitalism. *Capitalism, Nature, Socialism* 3, 93–107.

Odell, P. (1963) *An economic geography of oil* (London: G Bell & Sons).

Odum, E. (1975) *Ecology* (London: Holt, Rinehart & Winston).

OECD (1988) *Decision of the Council on Transfrontier Movements of Hazardous Wastes*, 685th Session, C(88) (Final) (Paris: OECD).

Oechel, W.C., Hastings, S.J., Vourlitis, G., Jenkins, M., Riechers, G. and Grulke, N. (1993) Recent change of Arctic tundra ecosystems from a net carbon dioxide sink to a source. *Nature* 361, 520–3.

Oechel, W.C., Cowles, S., Grulke, N., Hastings, S.J., Lawrence, B., Prudholme, T.,

Reiechers, G., Strain, B., Tissue, D. and Vourlitis, G. (1994) Transient nature of CO_2 fertilization in Arctic tundra. *Nature* 371, 500–3.

Oehme, M., Schlabach, M. and Boyd, I. (1995) Polychlorinated dibenzo-p-dioxins, dibenzofurans and coplanar biphenyls in Antarctic fur seal blubber. *Ambio* 24(1), 41–6.

Office for National Statistics (1997) *Social trends*, 27 – 1997 (London: HMSO).

Ohta, S., Uchijima, Z. and Oshima, Y. (1995) Effect of 2 × CO_2 climatic warming on water temperature and agricultural potential in China. *Journal of Biogeography* 22, 649–55.

Oldeman, R., Hakkeling, R. and Sombroeck, W. (1992) *World map of the status of human-induced soil degradation: an explanatory note* (Wageningen, The Netherlands: International Soil Information and Reference Center, and Nairobi, Kenya: United Nations Environment Programme, 1990).

Oldenburg, K.U. and Hirschhorn, J.S. (1987) Waste reduction: a new strategy to avoid pollution. *Environment* 29(2), 16–20 and 39–45.

Oliver, D., Elliott, D. and Reddish, A. (1991) Sustainable energy futures. In J. Blunden and A. Reddish (eds) *Energy, resources and environment* (London: Hodder & Stoughton), 180–232.

Oloffson, G. (1988) After the working-class movement? An essay on what's 'new'and what's 'social' in new social movements. *Acta Sociologica* 31, 15–34.

Olson, G.W. (1981) Archaeology: lessons on future soil use. *Journal of Soil and Water Conservation* 36, 261–4.

Olsson, A. and Bergmann, Å. (1995) A new persistent contaminant detected in Baltic wildlife: bis(4-chlorophenyl) sulfone. *Ambio* 24(2), 119–23.

Olsson, M., Karlsson, B. and Ahnland, E. (1992) Seals and seal protection: a presentation of a Swedish research project. *Ambio* 8, 494–6.

Oppenheimer, M. (1995) Context, connection and opportunity in environmental problem solving. *Environment* 37(5), 10–15 and 34–8.

O'Riordan, T. (1976) *Environmentalism* (London: Pion).

O'Riordan, T. (1988) The politics of sustainability. In R.K. Turner (ed.) *Sustainable environmental management* (London: Belhaven Press), 29–50.

O'Riordan, T. (1989) The challenge for environmentalism. In R. Peet and N. Thrift (eds) *New models in geography, volume 1: the political economy perspective* (London: Unwin Hyman), 77–104.

O'Riordan, T. (1991) The new environmentalism and sustainable development. *The Science of the Total Environment* 108, 5–15.

O'Riordan, T. (ed.) (1995a) *Environmental science for environmental management* (Harlow: Longman).

O'Riordan, T. (1995b) Frameworks for choice: core beliefs and the environment. *Environment* 37(8), 4–9 and 25–9.

O'Riordan, T. (1995c) Environmental science on the move. In T. O'Riordan (ed.) *Environmental science for environmental management* (Harlow: Longman), 1–11.

O'Riordan, T. (1995d) Managing the global commons. In T. O'Riordan (ed.) *Environmental science for environmental management* (Harlow: Longman), 347–60.

O'Riordan, T. (1995e) Section A. In T. O'Riordan (ed.) *Environmental science for environmental management* (Harlow: Longman).

O'Riordan, T. (ed.) (1997a) *Ecotaxation* (London: Earthscan).

O'Riordan, T. (1997b) Ecotaxation and the sustainability transition. In T. O'Riordan (ed.) *Ecotaxation* (London: Earthscan), 7–20.

Orme, T.W. (1994) The EPA wastes taxpayers' money. In C.P. Cozic (ed.) *Pollution* (San Diego: Greenhaven Press), 156–61.

Oskarsson, A., Nordberg, G., Block, M., Rasmussen, F., Petterson, R., Skerfving, S., Vahter, M., Wicklund Glynn, A., Öborn, I., Heikensten, M.-L. and Thuvander, A. (1996) Adverse health effects due to soil and water acidification: a Swedish research program. *Ambio* 25(8), 527–31.

Ott, W.R. and Roberts, J.W. (1998) Everyday exposure to toxic pollutants. *Scientific American* 278(2), 72–7.

Owens, M. and Cornwell, J.C. (1995) Sedimentary evidence for decreased heavy-metal inputs to the Chesapeake Bay. *Ambio* 24(1), 24–7.

Pacyna, J.M. (1986) Emission factors of atmospheric elements. In J.O. Nriagu and C.I. Davidson (eds) *Toxic metals in the atmosphere*. Advances in Environmental Science and Technology, volume 17 (Chichester: John Wiley & Sons), 2–32.

Pain, S. (1993) Valdez spill wasn't so bad, claims Exxon. *New Scientist* No 1858, 8 May, 4.

Pan, J.H. and Hodge, I. (1994) Land use permits as an alternative to fertiliser and leaching taxes for the control of nitrate pollution. *Journal of Agricultural Economics* 45, 102–12.

Pan, Y., McGuire, A.D., Kicklighter, D.W. and Melillo, J.M. (1996) The importance of climate and soils for estimates of primary production – a sensivity analysis with the terrestrial ecosystem model. *Global Change Biology* 2, 5–23.

Parenti, M. (1996) Against Empire. http://www.vida.com/parenti/ imperialism101.html.

Parfit, M. (1993) Water, the power, promise and turmoil of North America's fresh water. *National Geographic* special edition, 184(5), 1–119.

Parfit, M. (1995) Diminishing returns. *National Geographic* 188(5), 2–37.

Parson, E.A. and Greene, O. (1995) The complex chemistry of the international ozone agreements. *Environment* 37(2), 15–20 and 35–42.

Parsons, J. (1990) Too many people: too much taboo. *New Ground Spring*, 10.

Passmore, J. (1974) *Man's responsibility for nature: ecological problems and Western traditions* (London: Duckworth).

Pasztor, J. (1991) What role can nuclear power play in mitigating global warming? *Energy Policy* 19(2), 98–109.

Pattie, C., Russell, A. and Johnston, R. (1991) Going green in Britain? Votes for the Green Party and attitudes to green issues in the late 1980s. *Journal of Rural Studies* 7(3), 285–99.

Pawson, E. (1990) British expansion overseas c. 1730–1914. In R.A. Dodgshon and RA. Butlin (eds) *An historical geography of England and Wales* (London: Academic Press, 2nd edn), 521–44.

Pawson, E. (1992) Two New Zealands: Maori and European. In K. Anderson and F. Gale (eds) *Inventing places: studies in cultural geography* (Harlow: Longman), 15–36.

Pearce, D., Markandya, A. and Barbier, E. (1989) *Blueprint for a green economy* (London: Earthscan).

Pearce, F. (1991a) North–South rift bars path to Summit. *New Scientist*, 23 November, 20–1.

Pearce, F. (1991b) Building a disaster: the monumental folly of India's Tehri Dam. *The Ecologist* 21(3), 123–8.

Pearce, F. (1991c) *Green warrior: the people and politics behind the environmental revolution* (London: Bodley Head Press).

Pearce, F. (1993) What turns an oil spill into a disaster? *New Scientist*, 30 January, 11–13.

Pearce, F. (1995) Acid fall-out hits Europe's sensitive spots. *New Scientist* No 1985, 6.

Pearce, F. (1996) After the Falkland's bonzana. *New Scientist* 149 No 2017, 32–5.

Pearce, F. (1997a) Aircraft wreak havoc on ozone layer. *New Scientist* 153 No 2069, 18.

Pearce, F. (1997b) Environment body goes to pieces. *New Scientist* 153 No 2069, 11.

Peet, R. (1980) On geographic inequality under socialism. *Annals, Association of American Geographers* 70, 280–6.

Peglar, S. (1993) The mid-Holocene *Ulmus* decline at Diss Mere, Norfolk, UK: a year by year pollen stratigraphy from annual laminations. *The Holocene* 3, 1–13.

Peglar, S. and Birks, H.J.B. (1993) The mid-Holocene *Ulmus* fall at Diss Mere, south-east England – disease or human impact? *Vegetation History and Archaeobotany* 2, 61–8.

Pepper, D. (1984) *The roots of modern environmentalism* (London: Routledge).

Pepper, D. (1993) *Eco-socialism: from deep ecology to social justice* (London: Routledge).

Pepper, D. (1996) *Modern environmentalism* (London: Routledge).

Perera, J. (1993) Why Russia still wants nuclear power. *New Scientist*, 8 May, 29–30.

Perera, J. (1995) Studying the long-term effects. *Nuclear Engineering International* 40(496), 38.

Perry, A.H. (1985) The nuclear winter controversy. *Progress in Physical Geography* 9, 76–81.

Pestana, M.H.D., Formoso, M.L.L. and Teixeria, E.C. (1997) Heavy metals in stream sediments from copper and gold mining areas in southern Brazil. *Journal of Geochemical Exploration* 58, 133–44.

Peters, C.M. (1991) Harvesting rainforests. Native plants can stop deforestation. In M. Polesetsky (ed.) *Global resources: opposing viewpoints* (San Diego: Greenhaven Press), 162–8.

Peters, C.M., Gentry, A.H. and Mendelsohn, R.O. (1989) Valuation of the Amazonian rainforest. *Nature* 339, 655–6.

Peters, R.L. (1994) Conserving biological diversity in the face of climate change. In Ke Chung Kim and R.D. Weaver (ed.) *Biodiversity and landscapes: a paradox of humanity* (Cambridge: Cambridge University Press), 105–32.

Phillips, M. (1998a) The restructuring of social imaginations in rural geography. *Journal of Rural Geography* 18(2), 121–53.

Phillips, M. (1998b) Rural change: social perspectives. In B. Ilbery (ed.) *The geography of social change* (Harlow: Longman), 31–54.

Phillips, M. (forthcoming) Gender relations and identities in the colonisation of rural Middle England. In P. Boyle and K. Halfacree (eds) *Gender and migration in Britain* (London: Routledge).

Phillips, M. (forthcoming-a) Class, collective action and the countryside: a perspective on class formation in late twentieth century Britain. In J. Hooge and G.Van Gyes (eds) *Trade unions and late modernity: class action and the differentiated workforce* (HIVA, Leuven, Belgium).

Phillips, M. (forthcoming-b) Ecological colonialism: imperialism through nature. In H. Matthews and M. Phillips (eds) *Contested worlds: an introduction to human geography* (London: Routledge).

Phillips, M. (forthcoming-c) Philosophical arguments in human geography. In H. Matthews and M. Phillips (eds) *Contested worlds: an introduction to human geography* (London: Routledge).

Phillips, M., Fish, R. and Agg, J. (forthcoming) Putting together ruralities: towards a symbolic analysis of rurality in the British mass media. *Journal of Rural Studies*.

Pickering, K.T. and Owen, L.A. (1994) *An introduction to global environmental issues* (London: Routledge).

Pile, S. and Thrift, N. (1995) *Mapping the subject: geographies of cultural transformation* (London: Routledge).

Pilkington, W., Clouston, E. and Traynor, I. (1995) Battle of giants, large and small, fought at sea with hi-tech and skilful PR. *The Guardian*, 22 June, 4–5.

Pimental, D., Herdendorf, M., Eisenfeld, G., Carroquino, M., Carson, C., McDado, J., Chung, Y., Cannon, W., Roberts, J., Bluman, L. and Gregg, J. (1994) Achieving a secure energy future: environmental and economic issues. *Ecological Economics* 9(3), 201–19.

Pimental, D., Harvey, C., Resosudarmo, P., Sinclair, K., Kurz, D., McNair, M., Crist, S., Shpritz, L., Fitton, L., Saffouri, R. and Blair, R. (1995) Environmental and economic costs of soil erosion and conservation benefits. *Science* 267, 1117–23.

Plochl, M. and Cramer, G. (1995) Coupling global models of vegetation structure and ecosystem processes: an example from Arctic and boreal ecosystems. *Tellus* 47b, 240–50.

Plucknett, D.L. and Winkelmann, D.L. (1995) Technology for sustainable agriculture. *Scientific American* 273(3), 148–52.

Plumwood, V. (1988) Women, humanity and nature. *Radical Philosophy* 48, 16–24.

Plumwood, V. and Routley, R. (1982) World rainforest destruction – the social factors. *The Ecologist* 12, 4–22.

Polesetsky, M. (ed.) (1991) *Global resources: opposing viewpoints* (San Diego: Greenhaven Press).

Porritt, J. (1984) Industrialism in all its glory. In J. Porritt (ed.) *Seeing green: the politics of ecology explained* (Oxford: Basil Blackwell), 43–53.

Porritt, J. (1986) Beyond environmentalism. In E. Goldsmith and N. Hildyard (eds) *Green Britain or industrial wasteland?* (Cambridge: Polity).

Porritt, J. and Winner, D. (1988) *The coming of the greens* (London: Fontana).

Postel, S. (1994) Carrying capacity: earth's bottom line. In L.R. Brown and others, *State of the world, 1994* (London: Earthscan), 3–21.

Poteat, E. (1995) To minimise radwaste at McGuire, imagination went a long way, *Nuclear Engineering International* 40 (492), 47.

Poten, C.J. (1991) America's illegal wildlife trade. *National Geographic* 180(3), 106–32.

Potter, D. (1995) Environmental problems in their political context. In P. Glasbergen and A. Blowers (eds) *Environmental policy in an international context: perspectives* (London: Edward Arnold), 85–110.

Powell, J. and Craighill, A. (1997) The UK landfill tax. In T. O'Riordan (ed.) (1997) *Ecotaxation* (London: Earthscan), 304–20.

Powlson, D.S., Goulding, K.W.T. and Christian, D.G. (1994) *Nitrate leaching from arable land*. Agriculture and Food Research Council.

Prather, M., Derwent, R., Ehhalt, D., Fraser, P., Sanhueza, E. and Zhou, X. (1995) Other trace gases and atmospheric chemistry. In J.T. Houghton, L.G. Meira Filho, J. Bruce, Lee Hoesung, B.A. Callander, E. Haites, N. Harris and K. Maskell (eds) *Climate change (1994) Radiative forcing of climate and an evaluation of the IPCC IS92 emission scenarios* (Cambridge: Cambridge University Press), 77–126.

Pretty, J. (1995) *Regenerating agriculture: policies and practice for sustainability and self-reliance.* (London: Earthscan).

Pretty, J.N., Thompson, J. and Klara, J.K. (1995) Agricultural regeneration in Kenya: the catchment approach to soil and water conservation. *Ambio* 24(1), 7–15.

Price, T.D. (1987) The Mesolithic of western Europe. *Journal of World Prehistory* 1, 225–305.

Prigogine, I. and Stangers, I. (1995) *Order out of chaos: man's new dialogue with nature* (London: Flamingo).

Pringle, J. (1996) Strong quake kills 240 in Chinese town. *The Times*, 5 February.

Profitt, M.H., Margitan, J.J., Kelly, K.K., Loewenstein, M., Podolske, J.R. and Chan, K.R. (1990) Ozone loss in the Arctic polar vortex inferred from high altitude aircraft measurements. *Nature* 347, 31–6.

Prosser, K. (1994) The EPA fails to assist states' environmental efforts. In C.P. Cozic (ed.) *Pollution* (San Diego: Greenhaven Press), 149–55.

Prosterman, R.L., Hanstad, T. and Ping, L. (1996) Can China feed itself? *Scientific American* 275(5), 70–7.

Pushnik, J.C., Demaree, R.S., Houpis, J.L.J., Flory, W.B., Bauer, S.M. and Anderson, P.D. (1995) The effect of elevated carbon dioxide on a Sierra-Nevadan dominant species: *Pinus ponderosa*. *Journal of Biogeography* 22, 249–54.

Pye-Smith, C. (1997) Mimicking nature to grow more. *People and the Planet* 7(1), 12–14.

Rackham, O. (1986) *A history of the countryside* (London: Dent).

Ramanaiah, Y.V. and Reddy, N.B.K. (1983) Carrying capacity of land in Andhra Pradesh. *The Indian Geographical Journal* 58, 107–18.

Rangan, H. (1996) From Chipko to Uttaranchal: development, environment and social protest in the Garhwal Himalayas, India. In R. Peet and M. Watts (eds) *Liberation ecologies: environment development, social movements* (London: Routledge), 205–26.

Ray, G.C., Hayden, B.P., Bulger, jun., A.J. and McCormick-Ray, M.G. (1991) Effects of global warming on the biodiversity of coastal marine zones. In R.L. Peters and T.E. Lovejoy (eds) *Global warming and biological diversity* (New Haven: Yale University Press), 91–104.

Rayner, C., Roberts, S., Willett, P. and Willett, B. (1998) *Philip's geographical digest* (Oxford: Heinemann Educational).

Redclift, M. (1984) *Development and the environmental crisis: red or green alternatives* (London: Methuen).

Redclift, M. (1987) *Sustainable development: exploring the contradictions* (London: Methuen).

Redclift, M. (1992) The meaning of sustainable development. *Geoforum* 23(3), 395–403.

Reddy, A.N.K. and Goldemberg, J. (1990) Energy for the developing world. *Scientific American* 263(3), 62–73.

Rees, J. (1990) *Natural resources: allocation, economics and policy* (London: Routledge, 2nd edn).

Rees, J. (1997) Towards implementation realities. In T. O'Riordan (ed.) *Ecotaxation* (London: Earthscan), 287–303.

Rehfuess, K.E. (1985) On the causes of decline of Norway spruce (*Picea abies* Karst.) in central Europe. *Soil Use and Management* 1(1), 30–1.

Rehmke, G.F. (1991) Eliminating government support for deforestation can save rain forests. In M. Polesetsky (ed.) *Global resources: opposing viewpoints* (San Diego: Greenhaven Press), 150–5.

Reichhardt, T. (1995) Protecting wildlife becomes endangered act. *Nature* 374, 9.

Reid, J.W. and Bowles, I.A. (1997) Reducing the impacts of roads on tropical forests. *Environment* 39(8), 11–13 and 32–5.

Reid, T.R. (1998) Feeding the planet. *National Geographic* 194(4), 56–75.

Reid, W.V. and Goldemberg, J. (1998) Developing countries are combating climate change. *Energy Policy* 26(3), 233–7.

Reij, C., Scoones, I. and Toulmin, C. (1996) *Sustaining the soil: soil and water conservation in Africa* (London: Earthscan).

Reilly, W. (1994) The EPA is effective at reducing hazardous waste. In C.P. Cozic (ed.) *Pollution* (San Diego: Greenhaven Press), 115–20.

Reilly, W.K. (1991) Debt-for-nature swaps promote conservation. In M. Polesetsky (ed.) *Global resources: opposing viewpoints* (San Diego: Greenhaven Press), 262–6.

Reisenweber, R.L. (1995) Making environmental standards more reasonable. *Environment* 37(2), 15 and 32.

Renner, M. (1987) A critical review of tropical forests: a call for action. *The Ecologist* 17(4/5), 150.

Repetto, R. (1987) Population, resources, environment: an uncertain future. *Population Bulletin* 42(2), 2–43.

Repetto, R. (1992) Accounting for environmental assets. *Scientific American* June, 64–70.

Reus, J.O., Cosby, B.J. and Wright, R.F. (1987) Chemical processes governing soil and water acidification. *Nature* 329, 27–32.

Rhoades, R.E. (1991) The world's food supply at risk. *National Geographic* 179(4), 74–105.

Rice, J.O. (1996) Efforts to protect forests endanger loggers livelihood. In A. Sadler (ed.) *The environment: opposing viewpoints* (San Diego: Greenhaven Press), 124–8.

Rice, R.E., Gullison, R.E. and Reid, J.W. (1997) Can sustainable management save tropical rainforests? *Scientific American* 276(4), 34–9.

Ridley, M. (1993) Cleaning up with cheap technology. *New Scientist* No 1857, 26–7.

Ritter, D. (1995) Challenging current environmental standards. *Environment* 37(2), 11–12.

Roberts, N. (1989) *The Holocene* (Oxford: Blackwell).

Robinson, D. and Rasmussen, P. (1989) Botanical investigations at the Neolithic lake village at Weier, north-east Switzerland: leaf hay and cereals as animal fodder. In A. Milles, D. Williams and N. Gardner (eds) *The beginnings of agriculture* (Oxford: British Archaeological Reports, International Series S496), 149–63.

Robinson, M. (1992) *The greening of British party politics* (Manchester: Manchester University Press).

Robinson, M.H. (1985) Alternatives to destruction: investigations into the use of tropical rainforest resources with comments on repairing the effects of destruction. *The Environmental Professional* 7, 232–9.

Rodhe, H. and Svensson, B. (1995) Impact on the greenhouse effect of peat mining and combustion. *Ambio* 24(4), 221–5.

Rodrigues Filho, S. and Maddock, J.E.L. (1997) Mercury pollution in two gold mining areas of the Brazilian Amazon. *Journal of Geochemical Exploration* 58, 231–40.

Rogers, P. (1993) The value of co-operation in resolving international river basin disputes. *Natural Resources Journal* 17(2), 117–32.

Ronald, P.C. (1997) Making rice disease-resistant. *Scientific American* 277(5), 68–73.

Roodman, D.M. (1997) Reforming subsidies. In L.R. Brown *et al. State of the world, 1997* (London: Earthscan), 132–150.

Roosevelt, A.C. (1984) Population, health and the evolution of subsistence: conclusions from the conference. In M.N. Cohen, and G.A. Armelagos (eds) *Paleopathology and the origins of agriculture* (Orlando: Academic Press), 559–83.

Rosa, L.P. and Schaeffer, R. (1995) Global warming potentials: the case of emissions from dams. *Energy Policy* 23(2), 149–58.

Rose, C. (1986) Pesticide exports: Britain's record. In E. Goldsmith and N. Hildyard (eds) *Green Britain or industrial wasteland?* (Cambridge: Polity Press), 328–31.

Rose, C. (1993) Beyond the struggle for proof: factors changing the environmental movement. *Environmental Values* 2, 285–98.

Rosegrant, M.W. and Livernash, R. (1996) Growing more food, doing less damage. *Environment* 38(7), 6–11 and 28–31.

Rosenberg, N.J. (1982) The increasing CO_2 concentration in the atmosphere and its implication on agricultural productivity: II. Effects through human-induced climate change. *Climatic Change* 4, 239–54.

Rosenzweig, C. and Parry, M.L. (1994) Potential impact of climate change on world food supply. *Nature* 367, 133–38.

Ross, D. (1990) On the crest of a wave. *New Scientist* No 1717, 50–2.

Rothbard, D.M. and Rucker, C.J. (1994) Urban recycling has failed. In C.P. Cozic (ed.) *Pollution* (San Diego: Greenhaven Press), 202–7.

Rothenberg, D. (1989) Introduction: ecophilosophy – from intuition to system. In A. Naess, *Ecology, community and lifestyle* (Cambridge: Cambridge University Press).

Routledge, P. (1995) Resisting and reshaping the modern. In R. Johnston, P. Taylor and M. Watts (eds) *Geographies of global change* (Oxford: Blackwell), 263–79.

Routledge, P. (1997) The imagineering of resistance: Pollok Free State and the practice of postmodern politics. *Transactions, Institute of British Geographers* 22(3), 359–76.

Rowell, D.L. and Wild, A. (1985) Causes of soil acidification: a summary. *Soil Use and Management* 1(1), 32–3.

Rowell, G. (1991) Falcon rescue. *National Geographic* 179(4), 106–15.

Roy, S.K. (1987) The Bodhghat project and the World Bank. *The Ecologist* 17(2), 73–4.

Royle, S.A. and Phillips, D.R. (1997) China's population policy and its consequences. *Geography Review* 11(1), 2–7.

Rue, D.J. (1987) Early agriculture and early postclassic Maya occupation in western Honduras. *Nature* 326, 285–6.

Rusk, R. (ed.) (1939) *The letters of Ralph Waldo Emerson, volume 6* (New York: Columbia University Press).

Russell, D. (1996) Vacuuming the seas. *The Environmental Magazine* 7(4), 28–35.

Russell, J.M., Luo, M., Cicerone, R.J. and Deaver, L.E. (1996) Satellite confirmation of the dominance of chlorofluorocarbons in the global stratospheric chlorine budget. *Nature* 379, 526–9.

Russell, M. (1995) Environmental policy's great dilemma. *Environment* 37(2), 13–14.

Ryan, C. (1991) *Recreational tourism: a social science perspective* (London: Routledge).

Sabloff, J.A. (1987) New perspectives on the history of ancient Maya civilisation. *Nature* 326, 242–3.

Sabloff, J.A. (1995) Drought and decline. *Nature* 375, 357.

Sachs, W. (1993) Global ecology and the shadow of development. In W. Sachs (ed.) *Global ecology: a new arena of political conflict* (London: Zed Books), 3–21.

Sadler, A. (ed.) (1996) *The environment: opposing viewpoints* (San Diego: Greenhaven Press).

Safina, C. (1995) The world's imperiled fish. *Scientific American* 273(5), 30–7.

Safina, C. (1998) The world's imperiled fish. *Scientific American* 9(3), 58–63.

Sage, C. (1996) Population, poverty and land in the south. In P. Sloep and A. Blowers (eds) *Environmental policy in an international context: conflicts* (London: Edward Arnold), 97–125.

Said, E. (1992) *Culture and imperialism* (London: Chatto & Windus).

Sallnow, J. and Arlett, S. (1989) Green today, gone tomorrow? *Geographical Magazine*, November, 10–14.

Sampson, R.N., Wright, L.L., Winjum, J.K., Kinsman, J.D., Benneman, J., Kursten, E. and Scurlock, J.M.O. (1993) Biomass management and energy. *Water, Air and Soil Pollution* 70, 139–59.

Sandbach, F. (1980) *Environment ideology and policy* (Oxford: Blackwell).

Sanjour, W. (1994) The EPA fails to protect the environment. In C.P. Cozic (ed.) *Pollution* (San Diego: Greenhaven Press), 138–48.

Santee, M.L., Read, W.G., Waters, J.W., Froidevaux, L., Manney, G.L., Flower, D.A., Jarnot, R.F., Harwood, R.S. and Peckham, G.E. (1995) Interhemispheric differences in polar stratospheric HNO_3, H_2O, ClO and O_3. *Science* 267, 849–52.

Sattaur, O. (1989) The downward spiral of diversity. *Geographical Magazine* 61, 34–5.

Sauer, L. (1994) Making a habit of restoration: saving the eastern deciduous forest. In Ke Chung Kim and R.D. Weaver (eds) *Biodiversity and landscapes: a paradox of humanity* (Cambridge: Cambridge University Press), 209–27.

Savage, J.C., Lisowski, M. and Prescott, W.H. (1986) Strain accumulation in the Sumagin and Yakataga seismic gaps, Alaska. *Science* 231, 585–7.

Schimel, D., Enting, I.G., Heimann, M., Wigley, T.M.L., Raynaud, D., Alves, D. and Siegenthaler, U. (1995) CO_2 and the carbon cycle. In J.T. Houghton, L.G. Meira Filho, J. Bruce, Hoesung Lee, B.A. Callander, E. Haites, N. Harris and K. Maskell (eds) *Climate change (1994) Radiative forcing of climate and an evaluation of the IPCC IS92 emission scenarios* (Cambridge: Cambridge University Press), 35–71.

Schmidt, K. (1997) A drop in the ocean. *New Scientist* 154 No 2080, 40–4.

Schneider, D. (1996) Good wood. Can timber certification save the rainforest? *Scientific American* 274(6), 25–6.

Scholz, C.H., Sykes, L.R. and Aggarwal, Y.P. (1973) Earthquake prediction: a physical basis. *Science* 181, 803–9.

Schoon, N. (1996) Sea Empress was worst ever disaster for protected species. *Independent on Sunday*, 3 March, 4.

Schulze, E.-D. (1989) Air pollution and forest decline in a spruce (*Picea abies*) forest. *Science* 244, 776–83.

Schulze, R.E. and Kunz, R.P. (1995) Potential shifts in optimum growth areas of selected commercial trees and subtropical crops in southern Africa due to global warming. *Journal of Biogeography* 22, 679–88.

Schumacher, E.F. (1973) *Small is beautiful: a study of economics as if people mattered* (London: Abacus).

Schwartz, S.E. and Andreae, M.O. (1996) Uncertainty in climate change caused by aerosols. *Science* 272, 1121–2.

Scoging, H. (1991) Desertification and its management. In R. Bennett and R. Estall (eds) *Global change and challenge* (London: Routledge), 57–79.

Scott, A. (1990) *Ideology and the new social movements* (London: Unwin Hyman).

Scott, J. (1976) *The moral economy of the peasant* (New Haven: Yale University Press).

Sedjo, R.A. and Clawson, M. (1995) Global forests revisited. In J.L. Simon (ed.) *The state of humanity* (Oxford: Blackwell), 328–45.

Segerstahl, B. (1996) Chernobyl, 10 years later. *Environment* 38(3), 35–6.

Seleshi, Y. and Demaree, G.R. (1995) Rainfall variability in the Ethiopian and Eritrean highlands and its links with the southern oscillation index. *Journal of Biogeography* 22, 945–52.

Sen, A. (1981) *Poverty and famines: an essay on entitlement and deprivation* (Oxford: Clarendon Press).

Shankland, A. (1993) Brazil's BR-364 Highway: a road to nowhere? *The Ecologist* 23(4), 141–7.

Shaw, D.G. and Bader, H.R. (1996) Environmental science in a legal context: the Exxon Valdez experience. *Ambio* 25(7), 430–4.

Shcherbak, Y.M. (1996) Ten years of the Chernobyl era. *Scientific American* 274(4), 32–7.

Shea, C. (1988) *Renewable Energy* (Washington DC: Worldwatch Institute).

Sherratt, A. (1997) Climatic cycles and behavourial revolutions. *Antiquity* 71, 271–87.

Shields, R. (1991) *Places on the margins: alternative geographies of modernity* (London: Routledge).

Shindell, D.T., Rind, D. and Lonergan, P. (1998) Increased polar stratospheric ozone losses and delayed eventual recovery owing to increase greenhouse-gas concentrations. *Nature* 392, 589–92.

Shine, K.P. and Henderson-Sellars, A. (1983) Modelling climate and the nature of climate models: a review. *Journal of Climatology* 3, 81–94.

Shiva, V. (1987a) Chipko: rekindling India's forest culture. *The Ecologist* 17(1), 26–34.

Shiva, V. (1987b) Forestry myths and the World Bank. *The Ecologist* 17(4/5), 142–9.

Shiva, V. (1989) *Staying alive: women, ecology and development* (London: Zed Books).

Shiva, V. (1993) *The greening of global reach* (London: Zed Books).

Shiva, V. and Bandyopadhyay, J. (1986) *Chipko: India's civilisation response to the forest response to the forest crisis* (New Delhi: Intach).

Short, J. (1982) *An introduction to political geography* (London: Routledge & Kegan Paul).

Short, J.R. (1991) *Imagined country: society, culture and environment* (London: Routledge).

Shurmer-Smith, P. and Hannam, K. (1994) *Worlds of desire, realms of power: a cultural geography* (London: Edward Arnold).

Shyamsundar, P. and Kramer, R. (1997) Biodiversity conservation – at what cost? A study of households in the vicinity of Madagascar's Mantadia National Park. *Ambio* 26(3), 180–4.

Simmons, I.G. (1981) *The ecology of natural resources* (London: Edward Arnold).

Simmons, I.G. (1989) *Changing the face of the earth: culture, environment, history* (Oxford: Basil Blackwell).

Simmons, I.G. (1990a) The impact of human societies on their environments. In J. Silvertown and P. Sarre (eds) *Environment and society* (London: Hodder & Stoughton), 157–95.

Simmons, I.G. (1990b) The mid-Holocene ecological history of England and Wales and its relevance for conservation. *Environmental Conservation* 17, 61–9.

Simmons, I.G. (1991) *Earth, air and water: resources and environment in the late 20th century* (London: Edward Arnold).

Simmons, I.G. (1993a) *Interpreting nature. Cultural constructions of the environment* (London: Routledge).

Simmons, I.G. (1993b) Vegetation change during the Mesolithic in the British Isles: some amplifications. In F.M. Chambers (ed.) *Climate change and human impact on the landscape* (London: Chapman & Hall), 109–18.

Simmons, I.G. (1996) *The environmental impact of later Mesolithic cultures* (Edinburgh: Edinburgh University Press).

Simmons, I.G. (1997) *Humanity and Environment. A cultural ecology* (Harlow: Longman).

Simmons, I.G. and Innes, J.B. (1996a) An episode of prehistoric canopy manipulation at North Gill, North Yorkshire Moors, England. *Journal of Archaeological Science* 23, 337–41.

Simmons, I.G. and Innes, J.B. (1996b) Prehistoric charcoal in peat profiles at North Gill, North Yorkshire Moors, England. *Journal of Archaeological Science* 23, 193–7.

Simmons, I.G. and Innes, J.B. (1996c) Disturbance phases in the mid-Holocene vegetation at North Gill, North York Moors: form and process. *Journal of Archaeological Science* 23, 183–91.

Simon, J.L. (1980) Resources, population, environment: an oversupply of false bad news. *Science* 208, 1431–7.

Simon, J.L. (1981) *The ultimate resource* (Oxford: Martin Robertson).

Simon, J.L. (1991a) Global resource scarcity is not a serious problem. In M. Polesetsky (ed.) *Global resources: opposing viewpoints* (San Diego: Greenhaven Press), 24–8.

Simon, J.L. (1991b) Energy policies are unnecessary. In M. Polesetsky (ed.) *Global resources: opposing viewpoints* (San Diego: Greenhaven Press), 254–61.

Simon, J.L. (1994) More people, greater wealth, more resources, healthier environment. *Economic Affairs* 14(3), 22–9.

Simon, J.L. (1995) *The state of humanity* (Oxford: Blackwell).

Simon, J.L. and Boggs, R. (1995) Trends in the quantities of education – USA and elsewhere. In J.L. Simon (ed.) *The state of humanity* (Oxford: Blackwell), 208–23.

Simon, J.L. and Kahn, H. (eds) (1984) *The resourceful earth* (Oxford: Basil Blackwell).

Simon, J.L. and Wildavsky, A. (1995) Species loss revisited. In J.L. Simon (ed.) *The state of humanity* (Oxford: Blackwell), 346–61.

Sims, G.P. (1991) Hydroelectric energy. *Energy Policy* 19(8), 776–86.

Singer, F. (1984) World demand for oil. In J.L. Simon and H. Kahn (eds) *The resourceful earth* (Oxford: Basil Blackwell), 339–60.

Singer, S.F. (1995) Stratospheric ozone: science and policy. In J.L. Simon (ed.) *The state of humanity* (Oxford: Blackwell), 536–43.

Sioli, H. (1985) The effects of deforestation in Amazonia. *The Geographical Journal* 151, 197–203.

Sjöåsen, T., Ozolins, J., Greyerz, E. and Olsson, M. (1997) The otter (*Lutra lutra* L.) situation in Latvia and Sweden related to PCB and DDT levels. *Ambio* 26(4), 196–201.

Skåre, M. (1994) Whaling: a sustainable use of national resources or a violation of human rights? *Environment* 36(7), 12–20 and 30–6.

Skinner, J.A., Lewis, K.A., Bardon, K.S., Tucker, P., Catt, J.A. and Chambers, B.J. (1997) An overview of the environmental impact of agriculture in the UK. *Journal of Environmental Management* 50, 111–28.

Sloep, P. and van Dam-Mieras, M. (1995) Science on environmental problems. In P. Glasbergen and A. Blowers (eds) *Environmental policy in an international context: perspectives* (London: Arnold), 31–58.

Smil, V. (1997) Global population growth and the nitrogen cycle. *Scientific American* 277(1), 58–63.

Smith, A. (1994) Uneven development and the restructuring of the armaments industry in Slovakia. *Transactions, Institute of British Geographers* 19(4), 404–24.

Smith, A.G. (1970) The influence of Mesolithic and Neolithic man on British vegetation. In D. Walker and R.G. West (eds) *Studies in the vegetational history of the British Isles* (Cambridge: Cambridge University Press), 81–96.

Smith, A.G. (1975) Neolithic and Bronze Age landscape changes in Northern Ireland. In J.G. Evans, S. Limbrey and H. Cleere (eds) *The effect of man on the landscape: the highland zone*. Research Report 11, Council for British Archaeology, 66–74.

Smith, A.G. (1984) Newferry and the Boreal Atlantic Transition. *New Phytologist* 98, 35–55.

Smith, A.G. and Cloutman, E.W. (1988) Reconstruction of Holocene vegetation history in three dimensions at Waun Fignen Felen, an upland site in south Wales. *Philosophical Transactions of the Royal Society* (B) 322, 159–219.

Smith, B.D. (1995) *The emergence of agriculture* (New York: Scientific American Library).

Smith, B.D. (1997) The initial domestication of *Cucurbita pepo* in the Americas 10,000 years ago. *Science* 276, 932–4.

Smith, B.D. (1998) Between foraging and farming. *Science* 279, 1651–2.

Smith, D.R. and Flegal, A.R. (1995) Recent trends of lead in the biosphere. *Ambio* 24(1), 41–6.

Smith, G. (1989) Privilege and place in Soviet society. In D. Gregory and R. Welford (eds) *Horizons in human geography* (London: Macmillan), 320–40.

Smith, G. (1993) Ends, geopolitics and transitions. In R. Johnston (ed.) *The challenge for geography: changing world, changing discipline* (Oxford: Basil Blackwell).

Smith, K. (1993) *Environmental hazards: assessing risk and reducing disaster* (London: Routledge).

Smith, M. (1997) A short shell life. *Independent on Sunday*, 27 July, 37.

Smith, N. (1984) *Uneven development: nature, capital and the production of space* (Oxford: Basil Blackwell).

Smith, N. (1986) Imperialism. In R.J. Johnson, D. Gregory and D. Smith (eds) *The dictionary of human geography* (Oxford: Blackwell, 2nd edn), 216–17.

Smith, N. (1998) Nature at the millennium: production and re-enchantment. In B. Braun and N. Castree (eds) *Remaking reality: nature at the millennium* (London: Routledge), 271–85.

Smith, S. (1997) Environmental tax design. In T. O'Riordan (ed.) (1997) *Ecotaxation* (London: Earthscan), 21–36.

Smith, S. and Greene, O. (1991) The oceans. In P. Smith and K. Warr (eds) *Global environmental issues* (London: Hodder & Stoughton).

Smyth, C. and Jennings, S. (1990) Late Bronze Age-Iron Age valley sedimentation in East Sussex, southern England. In J. Boardman, I.D.L. Foster and J.A. Dearing (eds) *Soil erosion on agricultural land* (Chichester: John Wiley & Sons), 273–84.

Soja, E. (1989) *Postmodern geographies: the reassertion of space in critical social theory* (London: Verso).

Solem, T. (1989) Blanket mire formation at Haramsöy, Möre og Romsdal, Western Norway. *Boreas* 18, 221–35.

Soper, K. (1995) *What is nature?* (Oxford: Blackwell).

SORG (1996) *Stratospheric Ozone 1996. United Kingdom Stratospheric Ozone Review Group Sixth Report* (London: HMSO).

Soussan, J.G. (1992) Sustainable development. In A.M. Mannion and S.R. Bowlby (eds) *Environmental issues in the 1990s* (Chichester: John Wiley & Sons), 21–36.

Soussan, J.G. and Millington, A.C. (1992) Forests, woodlands and deforestation. In A.M. Mannion and S.R. Bowlby (eds) *Environmental issues in the 1990s* (Chichester: John Wiley & Sons), 79–96.

Sperling, D. (1996) The case for electric vehicles. *Scientific American* 275(5), 36–41.

Spizzirri, J. (1996) Warming shifts growing season. *Earth* 5(6), 11–12.

Sponsel, L.E., Headland, T.N. and Bailey, R.C. (eds) (1996) *Tropical deforestation: the human dimension* (New York: Columbia University Press).

Stanley, G. (1980) Wilderness carrying capacity: management and research in the United States. *Landscape Research* 5(3), 6–11.

Stannard, D. (1992) *American holocaust: Columbus and the conquest of the New World* (New York: Oxford University Press).

Stevens, P. (1996) Oil prices: the start of an era? *Energy Policy* 24(5), 391–402.

Stiling, P. (1992) *Introductory ecology* (London: Prentice Hall).

Stillman, P. (1983) The tradegy of the commons: a re-analysis. In T. O'Riordan and R.K. Turner, (eds) *An annotated reader in environmental management* (Oxford: Pergamon Press), 299–303.

Stolarski, R.S. (1988) The Antarctica ozone hole. *Scientific American* 258, 20–6.

Stolarski, R.S. and Cicerone, R.J. (1974) Stratospheric chlorine: possible sink for ozone. *Canadian Journal of Chemistry* 52, 1610–15.

Stone, P.H. (1992) Forecast cloudy. The limits of global warming models. *Technology Review* 95, 32–40.

Stonich, S. and DeWalt, B.R. (1996) The political ecology of deforestation in Honduras. In L.E. Sponsel, T.N. Headland and R.C. Bailey (eds) *Tropical deforestation: the human dimension* (New York: Columbia University Press), 187–215.

Street, P. and Miles, I. (1996) Transition to alternative energy supply technologies: the case of wind power. *Energy Policy* 24(5), 413–25.

Street-Perrott, F.A. and Roberts, N. (1994) Past climates and future greenhouse warming. In N. Roberts (ed.) *The changing global environment* (Oxford: Blackwell), 47–68.

Suau, A. (1985) Eritrea region in rebellion. *National Geographic* 168(3), 384–405.

Suplee, C. (1998) Unlocking the climate puzzle. *National Geographic* 193(5), 44–71.

Swain, A. (1997) Ethiopia, the Sudan and Egypt: the Nile River dispute. *Journal of Modern African Studies* 35(4), 675–94.

Swansea City Council (1978) *Lower Swansea Valley Facts Sheet* (Swansea: Swansea City Council).

Swinbanks, D. (1995) Waste shipment stirs debate over lack of nuclear store. *Nature* 374, 7.

Syers, J.K., Lingard, J., Pieri, C., Ezcurra, E. and Fauré, G. (1996) Sustainable land management for the semi-arid and sub-humid tropics. *Ambio* 25(8), 484–91.

Symington, F. (1996) Environmental regulations should be curtailed. In A. Sadler (ed.) *The environment: opposing viewpoints* (San Diego: Greenhaven Press), 185–92.

Szabo, M. (1993) New Zealand's poisoned paradise. *New Scientist,* 31 July, 29–33.

Szabolcs, I. (1986) Agronomical and ecological impacts of irrigation on soil and water salinity. *Advances in Soil Science* 4, 189–218.

t'Sas-Rolfes, M. (1994) Trade in endangered species: is it an option? *Economic Affairs* 14(3), 22–9.

Tabazadeh, A. and Turco, R.P. (1993) Stratospheric chlorine injection by volcanic eruptions: HCl scavenging and implications for ozone. *Science* 260, 1082–5.

Tang, X. and Madronich, S. (1995) Effects of increased solar ultraviolet radiation on tropospheric composition and air quality. *Ambio* 24(3), 188–90.

Tarr, J.A. and Ayres, R.U. (1990) The Hudson–Raritan Basin. In B.L. Turner II, W.C. Clark, R.W. Kates, J.F. Richards, J.T. Matthews and W.B. Meyer (eds) *The earth as transformed by human action* (Cambridge: Cambridge University Press), 623–40.

Taubes, G. (1995) Is a warmer climate wilting the forests of the north? *Science* 267, 1595.

Taylor, P. (1985) The world systems project. In R.J. Johnson and P. Taylor (eds) *A world in crisis: geographical perspectives* (Oxford: Basil Blackwell), 269–88.

Taylor, P. (1989) The error of developmentalism in human geography. In D. Gregory and R. Walford (eds) *Horizons in human geography* (London: Macmillan), 303–19.

Taylor, P. (1992a) Understanding global inequalities: a world systems approach. *Geography* 77, 1–11.

Taylor, P. (1992b) Science and values in the impending resumption of commercial whaling. *Ecos* 13(4), 40–6.

Taylor, P.J. and Buttel, F.H. (1992) How do we know we have global environmental problems? Science and the globalisation of environmental discourse. *Geoforum* 23(3), 405–16.

Ten Hove, H.A. (1968) The *Ulmus* fall at the transition Atlanticum–Subboreal in pollen diagrams. *Palaeogeography, Palaeoclimatology, Palaeoecology* 5, 359–69.

Teramura, A.H., Sullivan, J.H. and Ziska, L.H. (1990) Interaction of elevated ultraviolet-B radiation and CO_2 on productivity and photosynthetic characteristics in wheat, rice and soybean. *Plant Physiology* 94, 470–5.

Terhal, P. (1992) Sustainable development and cultural change. In J.H.B. Opschoor (ed.) *Environment, economy and sustainable development* (Groningen: Wolters-Noordhoff Publishers), 142.

Tetlow-Smith, A. (1995) Environmental factors affecting global atmospheric methane concentrations. *Progress in Physical Geography* 19(3), 322–35.

Thayer, jun., R.L. and Hansen, H. (1988) Wind on the land. *Landscape Architecture* 78, 69–73.

Thomas, A. and Potter, D. (1992) Development, capitalism and the nation state. In T. Allen and A. Thomas (eds) *Poverty and development in the 1990s* (Oxford: Clarendon Press), 116–41.

Thomas, K. (1984) *Man and the natural world: changing attitudes in England 1500–1800* (Harmondsworth: Penguin).

Thompson, E.P. (1967) Time, work discipline and industrial capitalism. *Past and Present* 38, 56–97.

Thompson, F.M.L. (1985) Towns, industry and the Victorian landscape. In S.R.J. Woodell (ed.) *The English landscape; past, present and future* (Oxford: Oxford University Press), 168–87.

Thompson, J. and Hinchcliffe, F. (1997) Sustaining the harvest. *People and the Planet* 7(1), 10–11.

Thompson, R.D. (1995) The impact of atmospheric aerosols on global climate: a review. *Progress in Physical Geography* 19(3), 336–50.

Thomson, K. and Dudley, N. (1989) Transnationals and oil in Amazonia. *The Ecologist* 19(6), 219–24.

Thornley, A. (1981) Pollen analytical evidence relating to the vegetation history of chalk. *Journal of Biogeography* 8, 93–106.

Thrift, N. (1987) Manufacturing rural geography. *Journal of Rural Studies* 3, 77–81.

Thrift, N. (1989) Images of social change. In C. Hamnett, L. McDowell and P. Sarre (eds) *The changing social structure* (London: Sage), 12–42.

Thrift, N. (1994) Inhuman geographies: landscapes of speed, light and power. In P. Cloke, M. Doel, D. Matless, M. Phillips and N. Thrift, *Writing the rural: five cultural geographies* (London: Paul Chapman), 191–248.

Tiffen, M., Mortimore, F. and Gichuki, F. (1994) *More people, less erosion: environmental recovery in Kenya* (Chichester: John Wiley & Sons).

Tisdell, C. (1985) World conservation strategy, economic policies and sustainable resource use in developing countries. *The Environmental Professional* 7, 102–7.

Tisdell, C. (1988) Sustainable development: differing perspectives of ecologists and economists, and relevance to LDCS. *World Development* 16(3), 373–84.

Tokar, B. (1991) Debt-for-nature swaps harm conservation. In M. Polesetsky (ed.) *Global resources: opposing viewpoints* (San Diego: Greenhaven Press), 267–71.

Toon, O.B. and Turco, R.P. (1991) Polar stratospheric clouds and ozone depletion. *Scientific American* 264(6), 40–7.

Toulmin, C. (1997) Sustaining Africa's soil. *People and the Planet* 7(1), 20–1.

Tourraine, A. (1974) *The post-industrial society: tomorrow's social history – classes, conflict and culture in the programmed society* (London: Wildwood House).

Townsend, J. (1993) Housewifisation and colonisation in the Colombian rainforest. In J. Momsen and V. Kinnaird (eds) *Different places, different voices: gender and development in Africa, Asia and Latin America* (London: Routledge), 270–87.

Townsend, J. and Wilson de Acosta, S. (1987) Gender roles in the colonisation. In J. Momsen and J. Townsend (eds) *Geography of gender in the Third World* (London: Hutchinson).

Trainer, F.E. (1995) Can renewable energy sources sustain affluent society? *Energy Policy* 23(12), 1009–26.

Treece, D. (1987) Brazil's Greater Carajas programme. *The Ecologist* 17(2), 75.

Treece, D. (1989) The militarisation and industrialisation of Amazonia: the Calha Norte and Grande Carajas programmes. *The Ecologist* 19(6), 225–8.

Tributsch, H. (1982) *When the snakes awake. Animals and earthquake prediction* (Cambridge, Massachusetts: MIT Press).

Troels-Smith, J. (1960) Ivy, mistletoe and elm. Climate indicators/fodder plants. *Danmarks Geologiske Undersøgelse IV* 4, 1–32.

Truett-Anderson, W. (1990) *Reality isn't what it used to be: theatrical politics, ready to wear religion, global myths, primitive chic and other wonders of the post-modern world* (San Francisco: Harper & Row).

Tsunogai, U. and Wakita, H. (1995) Precursory chemical changes in groundwater: Kobe earthquake, Japan. *Science* 269, 61–3.

Turco, R.P., Ackerman, T.P., Pollard, J.B. and Sagan, C. (1983) Nuclear winter: global consequences of multiple nuclear explosions. *Science* 222, 1283–92.

Turco, R.P., Toon, O.B., Ackerman, T.P., Pollack, J.B. and Sagan, C. (1984) The climatic effects of nuclear war. *Scientific American* 251(2), 23–34.

Turner, J. (1962) The *Tilia* decline: an anthropogenic interpretation. *New Phytologist* 61, 328–41.

Turner, J., Innes, J.B. and Simmons, I.G. (1993) Spatial investigation in the vegetation history of North Gill, North Yorkshire. *New Phytologist* 123, 599–647.

Turner, R.K., Pearce, D.W. and Bateman, I.J. (1993) *Environmental economics: an elementary introduction* (Hemel Hempstead: Harvester Wheatsheaf).

Turner, R.K. (1995) Environmental economics and management. In T. O'Riordan (ed.) *Ecotaxation* (London: Earthscan), 30–44.

Turner, R.K. (1995) Environmental economics and management. In T. O'Riordan (ed.) *Environmental Science for Environmental Management* (Harlow: Longman), 30–44.

Tyler, C. (1990) The sense of sustainability. *Geographical Magazine,* February, 8–13.

Uhl, C., Verissimo, A., Barreto, P., Mattos, M.M. and Tarifa, R. (1994) Lessons from the aging Amazon frontier: opportunities for genuine development. In Ke Chung Kim and R.D. Weaver (eds) *Biodiversity and landscapes: a paradox of humanity* (Cambridge: Cambridge University Press), 287–303.

United Nations (1972) *Development and environment. Reports and working papers of a panel of experts convened by the Secretary General of the United Nations* (Paris: Mouton).

United Nations Development Programme (1996) *Human development report* (New York: Oxford University Press).

Unruh, J. (1995) Post-conflict recovery of African agriculture: the role of critical resource tenure. *Ambio* 24(6), 343–8.

Unwin, T. (1983) Perspectives on 'development' – an introduction. *Geoforum* 14(3), 235–41.

Unwin, T. (1992) *The place of geography* (Harlow: Longman).

Unwin, T. (forthcoming) Changing societies: eastern Europe and the demise of the socialist bloc. In M. Phillips and H. Matthews (eds) *Contested worlds: an introduction to human geography* (London: Routledge).

Urry, J. (1990) *The tourist gaze: leisure and travel in contemporary societies* (London: Sage).

Urry, J. (1992) The tourist gaze and the environment. *Theory, culture and society* 9(3), 1–26.

Urry, J. (1995a) Is Britain the first post-industrial society? In J. Urry (ed.) *Consuming places* (London: Routledge), 112–25.

Urry, J. (1995b) The making of the Lake District. In J. Urry (ed.) *Consuming places* (London: Routledge), 193–210.

Urry, J. (1995c) A middle-class countryside? In T. Butler and M. Savage (eds) *Social change and the middle classes* (London: Routledge), 205–19.

Urry, J. (1995d) Tourism, travel and the modern subject. In J. Urry (ed.) *Consuming places* (London: Routledge), 141–51.

Usbourne, D. (1995) Canada enjoys moral glow over illegal trawler arrest. *Independent on Sunday*, 19 March, 12.

US Bureau of the Census (1997) *Statistical abstract of the United States* (Washington: US Bureau of the Census).

Van Andel, T.V. and Runnels, C.F. (1995) The earliest farmers in Europe. *Antiquity* 69, 481–500.

Van Blarcum, S.C., Miller, J.R. and Russell, G.L. (1995) High latitude river runoff in a doubled CO_2 climate. *Climate Change* 30, 7–26.

Van der Straaten, J. (1996) The distribution of environmental costs and benefits: acid rain. In P.B. Sloep and A. Blowers (eds) *Environmental problems as conflicts of interest 2: environmental policy in an international context* (London: Edward Arnold), 127–50.

Van Vliet-Lanoë, B., Helluin, M., Pellerin, J. and Valadas, B. (1992) Soil erosion in western Europe: from the last interglacial to the present. In M. Bell and J. Boardman (eds) *Past and present soil erosion* (Oxford: Oxbow Books), 101–14.

Van Voorst, B. (1994) Recycling: an overview. In C.P. Cozic (ed.) *Pollution* (San Diego: Greenhaven Press), 167–71.

Varela, F., Maturana, H. and Uribe, R. (1974) Autopoiesis: the organization of living systems, its characterisation and a model. *Biosystems* 5, 187–96.

Vavilov, N.I. (1992) *Origin and geography of cultivated plants* (Cambridge: Cambridge University Press).

Veiga, M.M. and Meech, J.A. (1995) Gold mining activities in the Amazon: clean-up techniques and remedial procedures for mercury pollution. *Ambio* 24(6), 371–5.

Veiga, M.M., Meech, J.A. and Hypolito, R. (1995) Educational measures to address mercury pollution from gold mining activities in the Amazon. *Ambio* 24(4), 216–20.

Velander, W.H., Lubon, H. and Drohan, W.N. (1997) Transgenic livestock as drug factories. *Scientific American* 276(1), 54–8.

Vesely, W.E. (1984) Engineering risk analysis. In P.F. Ricci, L.A. Sagan and C.G. Whipple (eds) *Technological risk assessment. NATO Advanced Science Institutes Series* (The Hague: Martinus Nijhoff Publishers).

Victor, D.G. and Salt, J.E. (1995) Keeping the climate treaty relevant. *Nature* 373, 280–2.

Vidal, J. (1992) The new waste colonialists. *The Guardian*, 14 February, 29.

Vidal, J. (1993a) Explode a condom, save the world. *The Guardian*, 10 July.

Vidal, J. (1993b) Judge grants reprieve to tree in path of bulldozers by changing it into a house. *The Guardian*, 12 November, 9.

Vidal, J. (1993c) Tory revolt looms over roads policy. *The Guardian*, 13 November, 8.

Vidal, J. (1993d) Twyford showdown on motorway trail. *The Guardian*, 5 July, 3.

Vidal, J. (1994) The real earth movers. *The Guardian*, Society Section, 7 December, 6–7.

Vidal, J. (1996) First man of the woods moved on in search of peace as protest grew. *The Guardian*, 9 March.

Vidal, J. and Bellos, A. (1996) Protest lobbies unite to guard rights. *The Guardian* 27 August, 5.

Vietmeyer, N.D. (1986) Lesser-known plants of potential use in agriculture and forestry. *Science* 232, 1379–84.

Vittori, E., Labini, S.S. and Serva, L. (1991) Palaeo-seismology: review of the state-of-the art. *Tectonophysics* 193, 9–32.

Vörösmarty, C.J., Sharma, K.P., Fekete, B.M., Copeland, A.H., Holden, J., Marble, J. and Lough, J.A. (1997) The storage and aging of continental runoff in large reservoir systems of the world. *Ambio* 26(4), 210–19.

Wace, N. (1990) Antarctica: a new tourist destination. *Applied Geography* 10, 327–41.

Walker, B. (1993) Rangeland ecology: understanding and managing change. *Ambio* 22(2–3), 80–7.

Walker, G. (1995a) Energy, land use and renewables: a changing agenda. *Land Use Policy* 12(1), 3–6.

Walker, G. (1995b) Renewable energy and the public. *Land Use Policy* 12(1), 49–59.

Walker, G. (1996) Kinder cuts. *New Scientist*, 21 September, 40–2.

Wallace, A. (1994) Sense with sustainable agriculture. *Communications in Soil Science and Plant Analysis* 25, 5–13.

Wallace, I. (1990) *The global economic system* (London: Unwin Hyman).

Wallerstein, I. (1974) *The modern world-system I: capitalist agriculture and the origins of the European world economy in the sixteenth century* (London: Academic Press).

Wallerstein, I. (1979) *The capitalist world-economy* (Cambridge: Cambridge University Press).

Wallerstein, I. (1983) *Historical capitalism* (London: Verso).

Walton, D. and Morris, E. (1990) Science, environment and resources in Antarctica. *Applied Geography* 10(4), 265–86.

Wang Futang and Zhao Zong-Ci (1995) Impact of climate change on natural vegetation in China and its implication for agriculture. *Journal of Biogeography* 22, 657–64.

Wang, B.L. and Allard, M. (1995) Recent climatic trend and thermal response of permafrost at Salluit, Northern Quebec, Canada. *Permafrost and Periglacial Processes* 6, 221–33.

Ward, B. and Dubos, R. (1972) *Only one earth: the care and maintenance of a small planet* (Harmondsworth: Penguin Books, with André Deutsch).

Warner, E. (1991) Tourists can save rainforests. In M. Polesetsky (ed.) *Global resources: opposing viewpoints* (San Diego: Greenhaven Press), 176–82.

Warner, F. (1996) Chernobyl, 10 years later. *Environment* 38(3), 5 and 35.

Warrick, R.A., Gifford, R.M. and Parry, M.L. (1986) CO_2, climatic change and agriculture. In B. Bolin, B.R. Doos, J. Jager and R.A. Warrick (eds) *The greenhouse effect, climatic change and ecosystems* (Chichester: John Wiley & Sons), 395–473.

Waton, P.V. (1982) Man's impact on the chalklands: some new pollen evidence. In M. Bell and S. Limbrey (eds) *Archaeological aspects of woodland ecology* (Oxford: British Archaeological Report, International Series 146), 75–91.

Watson, M. (1997) Where fish may safely graze. *New Scientist* 153 No 2069, 46.

Watson, R.T., Zinyowera, M.C., Moss, R.H. and Dokken, D.J. (eds) (1996) *Intergovernmental Panel for Climate Change: Climate change 1995. Impacts, adaptations and mitigation of climate change: scientific–technical analyses* (Cambridge: Cambridge University Press).

Watts, M. (1983) Hazards and crises: a political economy of drought and famine in Northern Nigeria. *Antipode* 15, 24–34.

Watts, M. (1984) State, oil and accumulation: from boom to crisis. *Environment and Planning D: Society and Space* 2, 403–28.

Watts, M. (1998) Nature as artifice and artifact. In B. Braun and N. Castree (eds) *Remaking reality: nature at the millennium* (London: Routledge), 243–68.

Webster, P.J. (1994) The role of hydrological processes in ocean–atmosphere interactions. *Reviews of Geophysics* 32(4), 427–76.

Weisman, J. (1996) Study inflames Ward Valley controversy. *Science* 271, 448–9.

Wellens, J. and Millington, A.C. (1992) Desertification. In *Environmental Issues in the 1990s* (Chichester: Wiley).

Wennan, D. (1996) Housing construction and environmental policies for sustainable development: Changzhou, China. *Ambio* 25(2) 78–81.

West, S., Charman, D.J., Grattan, J.P. and Cherburkin, A.K. (1997) Heavy metals in Holocene peats from south-west England: detecting mining impacts and atmospheric pollution. *Water, Air, Soil Pollution* 100, 343–53.

Western, S. (1988) Carrying capacity, population growth and sustainable development: a case study from the Philippines. *Journal of Environmental Management* 27, 347–67.

Whalen, S.C. and Reeburgh, W. (1992) Interannual variations in tundra methane emission: a four-year time-series at fixed sites. *Global Biogeochemical Cycles* 6, 139–59.

Whatmore, S. and Thorne, L. (1998) Wild(er)ness: reconfiguring the geographies of wildlife. *Transactions, Institute of British Geographers*, 23(4), 435–54.

Wheelwright, J. (1996) *Degrees of disaster: Prince William Sound – how nature reels and rebounds* (London: Yale University Press).

Whelan, E.M. (1995) The carcinogen or toxin of the week phenomenon. In J.L. Simon (ed.) *The state of humanity* (Oxford: Blackwell), 595–608.

Whipple, C.G. (1996) Can nuclear waste be stored safely at Yucca Mountain? *Scientific American* 274(6), 56–65.

White, G.F. (1996) Emerging isses in global environmental policy. *Ambio* 25(1), 58–60.

White, L. (1967) The historical roots of our ecological crisis. *Science* 155, 1203–7.

Whitmarsh, D.J. and Young, J.A. (1985) Management of the UK mackerel fisheries: an economic perspective. *Marine Policy* 9(3), 220–35.

Wigley, T.M.L. (1991) Could reducing fossil fuel emissions cause global warming? *Nature* 349, 503–6.

Wigley, T.M.L. and Raper, S.C.B. (1992) Implications for climate and sea level of revised IPCC emission scenarios. *Nature* 357, 293–300.

Wilcove, D.S. (1994) Turning conservation goals into tangible results: the case of the spotted owl and old-growth forests. In P.J. Edwards, R.M. May and N.R.Webb (eds) *Large-scale ecology and conservation biology* (Oxford: Blackwell), 313–29.

Wilkie, D.S. (1996) Logging in the Congo: implications for indigenous foragers and farmers. In L.E. Sponsel, T.N. Headland and R.C. Bailey (eds) *Tropical deforestation: the human dimension* (New York: Columbia University Press), 230–48.

Williams, C.T. (1985) *Mesolithic exploitation patterns in the Central Pennines. A palynological study of Soyland Moor* (Oxford: British Archaeological Report, British Series 139).

Williams, D.W. and Liebhold, A.M. (1995) Herbivorous insects and global change: potential changes in the spatial distribution of forest defoliator outbreaks. *Journal of Biogeography* 22, 665–71.

Williams, E. (1989) Dating the introduction of food production into Britain and Ireland. *Antiquity* 63, 510–21.

Williams, P. and Woessner, P.N. (1996) The real threat of nuclear smuggling. *Scientific American* 274(1), 26–31.

Williams, R. (1980) Ideas of nature. In R. Williams, *Problems in materialism and culture* (London: New Left Books).

Willis, K.J. and Bennett, K.D. (1994) The Neolithic transition – fact or fiction? Palaeoecological evidence from the Balkans. *The Holocene* 4(3), 326–30.

Wilson, G.A. and Bryant, R.L. (1997) *Environmental management: new directions for the twenty-first century* (London: UCL Press).

Wiltshire, P.E.J. and Moore, P.D. (1983) Palaeovegetation and palaeohydrology in upland Britain. In K.J. Gregory, *Background to palaeohydrology* (Chichester: John Wiley & Sons), 433–51.

Wise, J.P. (1995) Trends in food from the sea. In J.L. Simon (ed.) *The state of humanity* (Oxford: Blackwell), 411–15.

Witter, S.G. and Carrasco, D.A. (1995) Water quality: a development bomb waiting to explode. A Dominican example and possible solution. *Ambio* 25(3), 199–204.

Witze, A. (1995) The fall of the Maya: did climate change do them in? *Earth* 4(5), 63–4.

Woodin, S. and Skiba, U. (1990) Liming fails the acid test. *New Scientist* No 1707, 50–1.

Woodward, F.I. (1992) A review of the effects of climate on vegetation: ranges, competition and composition. In R.L. Peters and T.E. Lovejoy (eds) *Global warming and biological diversity* (New Haven: Yale University Press), 105–23.

Woodward, F.I., Smith, T.M. and Emmanuel, W.R. (1995) A global land primary productivity and phytogeography model. *Global Biogeochemical Cycles* 9, 471–90.

Wordsworth, W. (1798) Lyrical ballads, with a few other poems (London: J. and A. Arch); reprinted in Butler, J. and Green, K. (eds) *Lyrical ballads and other poems, 1797–1800 by William Wordsworth* (London: Cornell Unversity Press).

Wordsworth, W. (1952) *A guide through the district of the lakes* (Bloomington: Indiana University Press).

World Bank (1997) *World Bank Report, 1997* (New York: World Bank/IBRD).

World Commission on Environment and Development (1987) *Our common future* (Oxford: Oxford University Press).

World Conservation Monitoring Centre (1994) *Biodiversity Data Sourcebook* (Cambridge, UK: World Conservation Monitoring Centre).

World Development Report (1991) *Development and the environment* (Oxford: Oxford University Press).

World Development Report (1997) *The state in a changing world* (Oxford: Oxford University Press).

Worsley, A.T. and Oldfield, F. (1998) Palaeoecological studies of three lakes in the highlands of Papua New Guinea. II vegetational history over the last 1600 years. *Journal of Ecology* 76, 1–18.

Wright, K.I. (1994) Ground stone tools and hunter–gatherer subsistence in southwest Asia: implications for the transition of farming. *American Antiquity* 59, 238–63.

Wright, P. (1985) *On living in an old country* (London: Verso).

Wright, R.F. and Hauhs, M. (1991) Reversibility of acidification: soils and surface waters. *Proceedings of the Royal Society of Edinburgh* 97B, 169–91.

Wright, R.F. and Snekvik, E. (1978) Acid precipitation: chemistry and fish populations in 700 lakes in southernmost Norway. *Verh. Internat. Verein. Limnol.* 20, 765–75.

Wrigley, E.A. (1962) The supply of raw materals in the industrial revolution. *Economic History Revew* 15, 1–16.

Wrigley, E.A. (1988) *Continuity, chance and change: the character of the industrial revolution in England* (Cambridge: Cambridge University Press).

Wuethrich, B. (1995) Cascadia countdown. *Earth,* October, 24–31.

Wynne, B. (1989) Sheepfarming after Chernobyl: a case study in communicating scientific information. *Environment* 31, 11–39.

Xu, Z., Bradley, P. and Jakes, J. (1995) Measuring forest ecosystem sustainability: a resource accounting approach. *Environmental Management* 19(5), 685–92.

Yaliang, Y. (1996) Changzhou, China: water supply, sewage treatment and waste disposal strategies for sustainable development. *Ambio* 25(2), 86–9.

Yearley, S. (1991) *The green case: a sociology of environmental issues, arguments and politics* (London: HarperCollins).

Yong-Kwang Shin, Seong-Ho Yun, Moo-Eon Park and Byong-Lyol Lee (1996) Mitigation options for methane emissions from rice fields in Korea. *Ambio* 25(4), 289–91.

Youjing, S. and Jiadong, H. (1996) Sustainable development research for Changzhou city, China. *Ambio* 25(2), 82–5.

Younger, M. (1990) Will the sea always win? *Geography Review,* May, 2–6.

Zepp, R.G., Callaghan, T.V. and Erikson, D.J. (1995) Effects of increased solar ultraviolet radiation on biogeochemical cycles. *Ambio* 24(3), 181–7.

Zich, A. (1997) China: Three Gorges. *National Geographic* 192(3), 2–33.

Zorpette, G. (1996a) Hanford's nuclear wasteland. *Scientific American* 274(5), 72–81.

Zorpette, G. (1996b) Keeping the 'Tiger' at bay. *Scientific American* 275(1), 24–5.

Zvelebil, M. (1994) Plant use in the Mesolithic and its role in the transition to farming. *Proceedings of the Prehistoric Society* 60, 35–74.

Zvelebil, M. and Dulukhanov, P. (1991) The transition to farming in eastern and northern Europe. *Journal of World Prehistory* 5(3), 233–78.

Zvelebil, M. and Rowley-Conwy, P. (1984) Transition to farming in northern Europe: a hunter–gatherer perspective. *Norwegian Archaeological Review* 17(2), 104–28.

Index

Introduction.

The index covers Chapters 1 to 7. Index entries are to page numbers. Alphabetical arrangement is word-by-word, where a group of letters followed by a space is filed before the same group of letters followed by a letter, eg 'acid rain' will appear before 'acidification'. Initial articles, conjunctions and prepositions are ignored in determining filing order.